HEAVY CRUDE OILS

FROM GEOLOGY TO UPGRADING
AN OVERVIEW

Alain-Yves HUC
Expert Director, IFP Energies nouvelles

HEAVY CRUDE OILS

From Geology to Upgrading An Overview

Translated from the French
by Trevor Jones (Lionbridge)

2011

Editions TECHNIP 25 rue Ginoux, 75015 PARIS, FRANCE

FROM THE SAME PUBLISHER

- CO_2 Capture
  Technologies to Reduce Greenhouse Gas Emissions
  J. LECOMTE, P. BROUTIN, E. LEBAS
- Multiphase Production
  Pipeline Transport, Pumping and Metering
  J. FALCIMAIGNE, S. DECARRE
- Corrosion and Degradation of Metallic Materials
  Understanding of the Phenomena and Applications in Petrolem and Process Industries
  F. ROPITAL
- A Geoscientist's Guide to Petrophysics
  B. ZINSZNER, F.M. PERRIN
- Acido-Basic Catalysis (2 vols.)
  Application to Refining and Petrochemistry
  C. MARCILLY
- Petroleum Microbiology (2 vols.)
  J.P. VANDECASTEELE
- Physico-Chemical Analysis of Industrial Catalysts
  A Practical Guide to Characterisation
  J. LYNCH
- Chemical Reactors
  From Design to Operation
  P. TRAMBOUZE, J.P. EUZEN
- Petrochemical Processes (2 vols.)
  Technical and Economic Characteristics
  A. CHAUVEL, G. LEFEBVRE
- Marine Oil Spills and Soils Contaminated by Hydrocarbons
  Environmental Stakes and Treatment of Pollutions
  C. BOCARD

Printed in France

ISBN 978-2-7108-0890-9

Preface

Heavy oils and tar sands present in Canada and Venezuela are considered as the largest reserve of petroleum-like products in the world. Their in-place amount is estimated to be comparable to the total world reserves of conventional oil. Furthermore, there are occurrences of analogous deposits in several other countries, even if their inventory is generally not known. The major question is which part of these resources we will be able to make use of, through exploitation, transportation, and conversion to regular products requested by customers. These problems are mainly a consequence of viscosity and chemical composition of heavy oils.

Although heavy oils have been exploited in various ways for more than a century, their origin and structure were still in doubt thirty years ago. One can note that a comparable lack of knowledge on origin of traditional oil fields (exploited for more than a century) lasted until the early 1960's. During the three last decades, advances were made in elucidation of chemical compounds, present in heavy oils, and responsible for high carbon and aromatic content, and also of physical structures responsible for aggregates and high viscosity. The major advantage of the group of authors associated around A.Y. Huc is the close cooperation of geologists, geochemists, physical and chemical analysts, production and reservoir engineers, refiners, economists working in close cooperation, and frequently in the same research institution. That situation is typical of major advances on complex material, such as heavy oils, in many fields of science.

In particular, the physical structure of aggregates may explain how these macromolecular structures can be broken down and also the precise function of oil-water-rock interfaces during exploitation and separation of heavy oils. Based on this knowledge, the best techniques of production engineering could be adapted to the local characteristics of each deposit, in order to extract, mobilize and transport by pipeline the heavy oils. Technical conditions, such as heat and nature of surfactants to break the physical structure of aggregates, will then base an evaluation of what we could expect, with regard to the cost of the process as compared to the fraction of recovery. It is likely that some rather diversified techniques will be proposed, and applied to different conditions of depth, temperature, viscosity, etc. That is already the case in various areas of production (Canada, Venezuela), and the scientific advances on heavy oils may allow a larger choice of engineering techniques.

The last requirement in valorization of heavy oils is to manufacture in refineries the classical products, mainly heating and motor fuels, which are requested by consumers. Deasphalting may be easier, oncespecific surfactants have already been used for production. Chemical treatments, such as coking, visbreaking, hydrovisbreaking and possibly others, could be improved and more specific when based on a more precise knowledge of the heavy constituents. In general, the various processes in refineries could make important advances, based on a better knowledge of the chemical and physical structure of the heavy aggregates present in that particular petroleum.

Intensive exploitation of heavy oils on a larger scale might increase environmental problems already reported, and particularly the fate of mineral residues (sand, solid mineral byproducts), the optimal gestion of water, including residual water from process, and, in certain case, an increased emission of carbon dioxide. The group of authors discussed also that ultimate aspect regarding the future of heavy oils.

Bernard P. Tissot
Académie des Sciences, France

Table of Contents

PART 1

HEAVY CRUDE OILS

Editor: A.Y. Huc

Chapter 1

Heavy Crude Oils in the Perspective of World Oil Demand

N. Alazard-Toux

Chapter 2

Definitions and Specificities

A. Saniere

Chapter 3

Geological Origin of Heavy Crude Oils

A.Y. Huc

Chapter 4

Properties and Composition

I. Merdrignac, D. Espinat, I. Hénaut, J-F. Argillier

PART 2

RESERVOIR ENGINEERING AND PRODUCTION

Editor: G. Renard

Chapter 5

Reservoir Geology

R. Eschard

Chapter 6

Oil Sands: Mining and Processing

E. Delamaide

Chapter 7

Cold Production

G. Renard, J.F. Nauroy, E. Bemer

Chapter 8

Enhanced Recovery

G. Renard

Chapter 9
Heavy Oil Production: Pumping Systems
C. Wittrisch

Chapter 10
Examples of Large Heavy Oil Projects
E. Delamaide, A. Kamp, G. Renard, T. Rouaud, R. Kasprik

PART 3
SURFACE TRANSPORT

Editor: J-F. Argillier

Chapter 11
Heavy Oil Dilution

P. Gateau, I. Hénaut, J-F. Argillier

Chapter 12
Aqueous Emulsions

J-F. Argillier, I. Hénaut, D. Langevin

Chapter 13

Core Annular Flow

Y. Peysson

Chapter 14

Surface Pumps for Transport: Selection and Limitations

J. Falcimaigne

PART 4

UPGRADING

Editor: A. Quignard

Chapter 15

De-asphalting with Heavy Paraffinic Solvents

A. Quignard

Chapter 16

Visbreaking

A. Quignard, S. Kressmann

Chapter 17

Coking

A. Quignard, S. Kressmann

Chapter 18

Catalytic Hydrotreatment and Hydroconversion: Fixed Bed, Moving Bed, Ebullated Bed and Entrained Bed

J. Verstraete, D. Guillaume, M. Roy Auberger

PART 5
ENVIRONMENTAL ISSUES

Editor: A.Y. Huc

Chapter 19
Reservoir and Production

E. Delamaide, J.F. Nauroy, G. Renard, C. Dalmazzone, C. Noïk

Chapter 20

Upgrading

A. Quignard

Chapter 21

Greenhouse Gas Emissions

G. Renard, A. Saniere, E. Delamaide

Chapter 22

CO_2 Mass Balance: an Integrated Approach

J.B. Sigaud

PART 6
ONGOING TECHNOLOGICAL CHALLENGES

Editor: N. Alazard-Toux

Chapter 23
In Situ Upgrading of Heavy Oil and Bitumen

J.P. Héraud, A. Kamp, J.F. Argillier

Chapter 24
Process Workflows

J.P. Héraud

Acknowledgments

The authors gratefully acknowledge the support of the *IFP Energies nouvelles* management during the preparation of this book, which allowed us to mobilise a large team of contributors, most of them members or former members of the *IFP Energies nouvelles* group, including R&D teams, IFP School and IFP Technologies (Canada) Inc.

We are very honoured that B. Tissot, now a member of the French *Académie des Sciences*, agreed to introduce this book by underlining the importance of the heavy oil industry in the near-future energy supply and its contribution to the ongoing "energy transition" precluding the future full advent of the new energy technologies. The book's theme is therefore in line with one of B. Tissot's major concerns when he was Scientific Director at *IFP Energies nouvelles*.

Compiling a book on *heavy crude oil* with such a wide scope dealing with domains as varied as economics, geology, environment, production, transport technologies and refining would have been impossible without the coordinating and editing endeavour of senior specialists. In this respect, the scientific editor is highly indebted to the dedicated individuals responsible for coordinating each specific part of this book: N. Alazard-Toux, G. Renard, J.F. Argillier and A. Quignard. Working together as a team has been a fruitful exercise in terms of the mutual technical benefit as well as the human experience.

The coordinators and I recognise that such a book is obviously the result of integrating individual chapters written by specialists. To all the authors and co-authors responsible for these contributions we extend our warm thanks: E. Bemer, E. Delamaide, R. Eschard, D. Espinat, J. Falcimaigne, P. Gateau, D. Guillaume, I. Hénaut, J.-P. Héraud, A. Kamp, R. Kasprik, S. Kressmann, D. Langevin, I. Merdrignac, J.-F. Nauroy, C. Noïk, Y. Peysson, T. Rouaud, M. Roy Auberger, A. Saniere, J.-B. Sigaud, J. Verstraete and C. Wittrisch.

We acknowledge the support and the involvement of P. Boisserpe and M. Darthenay who did much more than just supervise the progress of the manuscript. Their contribution was invaluable for achieving our goal: produce a book.

Finally, we would like to thank the reviewers, I. Guibard, F. Morel, J.-P. Péries and especially C. Gadelle, with his comprehensive view of the field of heavy oil; his input has been decisive with major and fruitful remarks which were taken into account and which helped us improve the manuscript during the preparation of the book.

A.Y. Huc, Editor

List of authors

Nathalie Alazard-Toux
Engineer Degree from the IFP School (EGE cycle)
Engineer Degree from the Nancy *École Nationale Supérieure de Géologie*
Director, Economics and Information Watch and Management Division
IFP Energies nouvelles, 92852 Rueil-Malmaison Cedex
nathalie.alazard@ifpenergiesnouvelles.fr

Jean-François Argillier
PhD in Chemical Engineering from Paris University (1989)
Project Leader in Heavy Oil Production & EOR Technologies, Applied Chemistry and Physical Chemistry Division
IFP Energies nouvelles, 92852 Rueil-Malmaison Cedex
j-francois.argillier@ifpenergiesnouvelles.fr

Elisabeth Bemer
Engineer Degree from the *École Nationale des Ponts et Chaussées* (*École des Ponts ParisTech*)
PhD in Structures and Materials from the *École Nationale des Ponts et Chaussées* (*École des Ponts ParisTech*)
Research Engineer, Petrophysics Department, Reservoir Engineering Division
IFP Energies nouvelles, 92852 Rueil-Malmaison Cedex
elisabeth.bemer@ifpenergiesnouvelles.fr

Christine Dalmazzone
Engineer Degree and PhD in Process Engineering from the *Université de Technologie Compiègne* (UTC)
Holds a French national accreditation to supervise research
Research Engineer, Physical Chemistry of Complex Fluids Department, Applied Chemistry and Physical Chemistry Division
IFP Energies nouvelles, 92852 Rueil-Malmaison Cedex
christine.dalmazzone@ifpenergiesnouvelles.fr

Eric Delamaide
Engineer Degree from the *École Nationale Supérieure des Mines de Saint-Etienne*
Engineer Degree from the IFP School
Professional Engineer – Association of Professional Engineers, Geologists and Geophysicists of Alberta
General Manager, IFP Technologies (Canada) Inc. (www.ifp-canada.com)
810, 744 - 4th Avenue SW, Calgary, Alberta, T2P 3T4 - Canada
eric.delamaide@ifp-canada.com

Rémi Eschard
Geologist Engineer Degree from the IFP School
PhD in Geology from the Strasbourg University
Director, Geology-Geochemistry-Geophysics Division
IFP Energies nouvelles, 92852 Rueil-Malmaison Cedex
remi.eschard@ifpenergiesnouvelles.fr

Didier Espinat
Engineer Degree from the *École Centrale de Lille*
Engineer Degree and PhD from Paris VI University and the IFP School
Expert Director, Physics and Analysis Division
IFP Energies nouvelles, BP 3, 69360 Solaize, didier.espinat@ifpenergiesnouvelles.fr

Jean Falcimaigne
Engineer Degree from the *École Centrale de Paris*
Former Project Leader at *IFP Energies nouvelles* in the field of production equipment and systems for more than 30 years

Patrick Gateau
Research Engineer, Physical Chemistry of Complex Fluids Department, Applied Chemistry and Physical Chemistry Division
IFP Energies nouvelles, 92852 Rueil-Malmaison Cedex
patrick.gateau@ifpenergiesnouvelles.fr

Denis Guillaume
PhD in Chemistry – Petroleum Science from Pierre and Marie Curie University (Paris VI)
Engineer Degree from the IFP School
Director of the Catalysis and Separation Division
IFP Energies nouvelles - BP 3 - 69360 Solaize
denisjm.guillaume@ifpenergiesnouvelles.fr

Isabelle Hénaut
Engineer Degree from the *École Supérieure de Chimie Organique et Minérale* (ESCOM)
PhD in Materials Science and Engineering from the *École des Mines de Paris* (*Mines ParisTech*)
Master in Polymer Chemistry and Physical Chemistry from Paris VI University
Laboratory Manager in the field of Rheology, Applied Mechanics Division
IFP Energies nouvelles, 92852 Rueil-Malmaison Cedex
isabelle.henaut@ifpenergiesnouvelles.fr

Jean-Philippe Héraud
Engineer Degree from the *École Supérieure de Chimie Organique et Minérale* (ESCOM)
Engineer Degree from the IFP School
Master in Process Engineering at the *Université de Technologie Compiègne* (UTC)
Research Engineer, Process Design and Modeling Division
IFP Energies nouvelles, BP 3, 69360 Solaize
j-philippe.heraud@ifpenergiesnouvelles.fr

Alain-Yves Huc
PhD from Strasbourg Louis Pasteur University
Engineer Degree and PhD from *Institut National Polytechnique de Lorraine* (INPL) Nancy University
Expert Director
Deputy Director of the Geology-Geochemistry-Geophysics Division
IFP Energies nouvelles, 92852 Rueil-Malmaison Cedex
a-yves.huc@ifpenergiesnouvelles.fr

Arjan Kamp
Engineer Degree in Applied Physics from Eindhoven University of Technology, The Netherlands
PhD in Fluid Mechanics from the *Institut National Polytechnique de Toulouse*, France
Director, Open and Experimental Centre for Heavy Oil (CHLOE), affiliated to Pau University and sponsored by Total
arjan.kamp@univ-pau.fr

Rémi Kasprik
Postgraduate Degree in Scientific and Technical Information – Nancy I University
Information Watch Manager, Economics and Information Watch and Management Division
IFP Energies nouvelles, 92852 Rueil-Malmaison Cedex
remi.kasprik@ifpenergiesnouvelles.fr

Stéphane Kressmann
Engineer Degree from the *Ecole Nationale Supérieure de Chimie de Toulouse* (ENSCT), now *Ecole Nationale Supérieure des Ingénieurs en Arts Chimiques et Technologiques* (ENSIACET)
PhD in *Sciences Pétrolières* from Paris VI University – *Pierre et Marie Curie* (1991)
Project Manager
IFP Energies nouvelles, BP 3, 69360 Solaize
stephane.kressmann@ifpenergiesnouvelles.fr

Dominique Langevin
Docteur es Sciences, Paris VI University
Research Director, CNRS
Team Leader at the Orsay *Laboratoire de Physique des Solides*
langevin@lps.u-psud.fr

Isabelle Merdrignac
Engineer Degree from the *École Supérieure de Chimie Industrielle de Lyon* (ESCIL), now named *École Supérieure de Chimie Physique Electronique de Lyon* (CPE Lyon)
PhD in Analytical Chemistry and Organic Geochemistry from Strasbourg University (ULP)
Research Engineer in Process Experimentation Division
IFP Energies nouvelles, BP 3, 69360 Solaize
isabelle.merdrignac@ifpenergiesnouvelles.fr

Jean-François Nauroy
Geophysicist Engineer Degree from the IFP School
PhD from Paris VI University
Research Engineer, Petrophysics Department, Reservoir Engineering Division
IFP Energies nouvelles, 92852 Rueil-Malmaison Cedex
j-francois.nauroy@ifpenergiesnouvelles.fr

Christine Noïk
Engineer Degree from the *École Nationale Supérieure de Chimie de Toulouse* (ENSCT), now *École Nationale Supérieure des Ingénieurs en Arts Chimiques et Technologiques* (ENSIACET)
PhD in Polymer Physical Chemistry from Grenoble University
Research Engineer, Physical Chemistry of Complex Fluids Department, Applied Chemistry and Physical Chemistry Division
IFP Energies nouvelles, 92852 Rueil-Malmaison Cedex
christine.noik@ifpenergiesnouvelles.fr

Yannick Peysson
Engineer Degree from the *École Supérieure de Physique et de Chimie Industrielles de la Ville de Paris* (*ESPCI-ParisTech*)
PhD in Physics of Liquids from Pierre and Marie Curie University (Paris VI)
Project Manager in the field of complex flows in oil pipelines (2001-2007)
Project Manager in the field of CO_2 geological storage (since 2008), Reservoir Engineering Division
IFP Energies nouvelles, 92852 Rueil-Malmaison Cedex
yannick.peysson@ifpenergiesnouvelles.fr

Alain Quignard
Engineer Degree from the *École Supérieure de Chimie Industrielle de Lyon* (ESCIL), now named *École Supérieure de Chimie Physique Electronique de Lyon* (CPE Lyon)
Engineer Degree from the IFP School
Project Manager in the field of direct conversion of unconventional resources (extra heavy crudes, biomass, coal) into fuel, Process Design and Modeling Division
Expert "Conversion of Heavy Products"
IFP Energies nouvelles, BP 3, 69390 Solaize
alain.quignard@ifpenergiesnouvelles.fr

Gérard Renard
Engineer Degree from the Toulouse *École Nationale Supérieure d'Électrotechnique, d'Électronique, d'Informatique, d'Hydraulique et des Télécommunications* (ENSEEIHT)
Engineer Degree from the IFP School
Research Engineer in Reservoir Engineering Division
IFP Energies nouvelles, 92852 Rueil-Malmaison Cedex
gerard.renard@ifpenergiesnouvelles.fr

Thierry Rouaud
Engineer Degree from the *Conservatoire National des Arts et Métiers* (Paris)
General Secretary of Cedigaz
Cedigaz, 1 & 4 avenue de Bois-Préau, 92852 Rueil-Malmaison Cedex
thierry.rouaud@ cedigaz.org

Magalie Roy-Auberger
Engineer Degree from the *École Nationale Supérieure de Chimie de Paris* (*Chimie Paris-Tech*)
PhD in Catalysis from Claude Bernard University – Lyon 1
Research Engineer in Catalysis and Separation Division
IFP Energies nouvelles, BP 3, 69360 Solaize
magalie.roy@ifpenergiesnouvelles.fr

Armelle Saniere
Engineer Degree from the Nancy *École Nationale Supérieure de Géologie*
Master in Economics
Oil Supply Analyst, Economics and Information Watch and Management Division
IFP Energies nouvelles, 92852 Rueil-Malmaison Cedex
armelle.saniere@ifpenergiesnouvelles.fr

Jean-Bernard Sigaud
Former student at the *École Polytechnique*
Engineer Degree from the IFP School
Former Research Director at Total
Former Director in Refining, Petrochemistry & Gas at IFP School
TPA Professor
4, rue Rémi Belleau, 78590 Noisy le Roi
jean-bernard.sigaud@neuf.fr

Jan Verstraete
Engineer Degree in Chemical Engineering from Ghent University (Belgium)
PhD in Chemical Engineering from Ghent University (Belgium)
Project Manager from 1999 to 2004 in the field of catalytic cracking
Project Manager since 2004 in the field of ebullated bed hydroconversion
Project Manager from 2004 to 2006 in the field of fixed bed hydroprocessing
Project Manager since 2007 in the field of direct crude oil desulphurisation, Process Design and Modeling Division
IFP Energies nouvelles, BP 3, 69360 Solaize
jan.verstraete@ifpenergiesnouvelles.fr

Christian Wittrisch
PhD in Electronics from Montpellier Sciences University (1972)
Research Engineer, Mechanical Engineering Department, Applied Mechanics Division
IFP Energies nouvelles, 92852 Rueil-Malmaison Cedex
christian.wittrisch@ifpenergiesnouvelles.fr

Introduction and Synopsis

A.Y. Huc

Petroleum-related fluids have been intensively exploited since the end of the 19th century. They are found in the Earth's crust associated with sedimentary basins. They include a large range of substances – ranging from gases to solid asphaltites, with all the intermediates such as condensates, light oils, medium oils, heavy oils, extra-heavy oils and bitumens (also called "tar sands" when mixed with sands) – whose differences reflect a continuum of chemical compositions and physical properties. All of these "petroleums" actually share a common origin. During the geological history of the basin, they result from the natural thermal transformation of organic material derived from the residues of previously living organisms fossilized in specific layers of the sedimentary pile: the source rocks. Subsequently, the petroleum fluids generated in the source rock migrate throughout the sedimentary basin and accumulate in a trap, from which they can eventually be produced when discovered, assuming appropriate technical and economical feasibility.

In these regard, the more viscous crude oils (i.e. heavy oils, extra-heavy oils and bitumens) present a considerable challenge.

These viscous crude oils are the focus of the present volume and are referred to *via* the generic term of *heavy crude oils* [1]. This volume is designed to provide comprehensive technical information on this topic by placing it in the wider perspective of the multidisciplinary activities related to the *heavy crude oils* industry. This includes a first part addressing the economic relevance of the *heavy crude oils*, their definition and geological origin. Part 2 is devoted to the specificity of reservoir engineering when dealing with the production of such material, while Part 3 describes the challenges related to the surface transportation of such viscous fluids. Part 4 reviews the technologies involved in upgrading this feedstock, while

1. For the sake of clarity, it should be noted that oil shales refer to completely different geological objects and require different industrial treatments. Oil shales can be defined as source rocks which have never undergone an adequate thermal history (mainly due to lack of burial) sufficient for initiating the natural formation of oil and gas.
The exploitation of oil shales consequently relies on a high temperature (more than 350°C) pyrolysis treatment (retorting) of the mined rock in order to artificially mimic the processes controlling the natural generation of oil and gas in the sub-surface. More recently, pilots designed for *in situ* retorting are emerging. The product of this treatment is called shale oil.
To be qualified as an oil shale, the immature source rock should be easy to mine (i.e. at or near the surface) and sufficiently rich in organic matter of the right quality in order to produce a sufficient amount of shale oil on an energetic and economic basis. The topic of oil shales is beyond the scope of this volume.

Part 5 summarizes the environmental issues. Finally, Part 6 envisions the technological future of this industry.

Definition of the various *heavy crude oils* is based on two important physical properties: specific gravity and viscosity. In the petroleum industry, for the purposes of providing a commercial value for crude oils, specific gravity is expressed as a scale defined by the American Petroleum Institute: the "API gravity" or (API degree). The API gravity is a direct calculation from the specific gravity:

$$\text{API gravity} = (141.5/\text{Specific gravity}) - 131.5$$

This equation – which implies that the API° increases with the commercial value of the oil, which corresponds to a decrease in its specific gravity – is equal to 10 when the oil exhibits the specific gravity of pure water, which is given as 1.

According to the definition of the World Petroleum Congress (1980) and the United Nations through UNITAR (1982), the following definitions have been set forth:

- Conventional oils are characterized by an API gravity higher than 20°.
- Heavy oils are defined as oils having an API gravity in the range of 10°-20°.
- Extra-heavy oils and natural bitumens display an API gravity lower than 10°.

Note that the Canadian Centre for Energy defines *heavy oil* as having an API gravity between 22.3° and 10°.

- The difference between extra-heavy oils and bitumens is based on the **viscosity at reservoir conditions:** the viscosity of extra-heavy oils is lower than 10,000 mPa.s (cP).
- The viscosity of natural bitumens is higher than 10,000 mPa.s (cP).

As a result of their properties and composition, *heavy crude oils* are usually not directly suitable for transport by pipeline or as feedstock for conventional refineries. In this regard, *heavy crude oils* must be upgraded into a synthetic crude oil in order to increase their value (i.e. increasing API, lowering viscosity, reducing the amount of heavy ends, etc.). According to the University of Alberta (www.ualberta.ca/), the minimum objective of upgrading *heavy crude oils* is to allow expedition by pipeline without adding solvent, while the maximum objective is to process the *heavy crude oils* to generate a crude oil substitute of higher value which meets specific refinery input streams requirements.

Today, *heavy crude oils* appear to be major players in the outlook for the future of world energy (**Chapter 1**). As far as resources are concerned, their volume represents as much as all resources of conventional oils. However, due to detrimental properties (mainly viscosity), their exploitation – including production, transport and upgrading – requires adaptation of the current means of the petroleum industry and, more importantly, active development of innovative technologies (**Chapter 2**). Among the challenges faced by the *heavy crude oils* industry, controlling CO_2 emissions – which are potentially higher than for conventional oils – must be addressed.

Geologically speaking, *heavy crude oils* have the same origin as conventional oils. If we exclude immature oils generated by specific source rocks, a significant portion of them are actually derived from conventional oils *via* subsequent alteration mediated by bacteria (**Chapter 3**). This phenomenon – called biodegradation – occurs in specific geological settings. This can be exemplified by the largest deposits of heavy crude oils in the world: the

Orinoco Belt in Venezuela and the Alberta basin in Canada. These deposits are located at the outer rim of foreland basins at shallow depth where the oil has accumulated in a low-temperature environment allowing for the growth of bacterial communities feeding on the crude oil. This biodegradation is responsible for the properties of most of the heavy oils, extra-heavy oils and bitumens.

In fact, the alteration predominantly affects the light end of the oil, and more specifically the hydrocarbon molecules, resulting in a fluid that exhibits increased content of heavy molecules, concentrated in sulfur, nitrogen, oxygen and metals. The viscosity and specific gravity of the oils are controlled by the richness in non-hydrocarbon (e.g. containing heteroelements: nitrogen, sulfur and oxygen), high molecular weight moieties called resins and asphaltenes (also called NSO compounds). Definition of the latter compounds is not chemically sound, but is instead derived from an empirical preparative procedure [1].

Due to their crucial role in the chemical and physical properties of *heavy crude oils*, with implications for industrial treatment, the resins and asphaltenes have been extensively studied using a wide spectrum of techniques (**Chapter 4**). They are still ill-defined, but a conceptual model has begun to take shape involving a polydispersed distribution of molecules in terms of size and composition. Basic units rich in aromatics and containing functional groups tend to be prone to self-association and aggregation. The latter properties imply that physical and chemical conditions will determine the aggregation state. This situation is reflected in the industrial means used for the exploitation of *heavy crude oils* (dilution, thermal treatment, etc.). An important aspect of the structure of *heavy crude oils* is the instrumental role of resins in stabilizing the asphaltene aggregates within the hydrocarbon medium.

The high viscosity of the fluid hinders the displacement of *heavy crude oils* throughout the porous network of the reservoir hosting the accumulation, and consequently presents challenges in production. In this regard – and regardless of the production technology used – identifying and predicting the occurrence and architecture of very high porosity, very high permeability bodies (compared with reservoirs containing conventional bodies) within a produced reservoir is of paramount importance in optimizing the development of the field (**Chapter 5**). In the context of *heavy crude oil* production, the reservoir characterization protocol is thus a clear added-value, and often underused by operators.

As previously noted, many *heavy crude oil* accumulations are typically associated with the distal part of foreland basins. This is specifically the case for the two major deposits: the Orinoco Belt and Alberta. In these basins, the favored carrier beds allowing the long-distance migration of fluids, from the "oil kitchen" to the accumulating sites, are the first fluvial and deltaic sediments filling the evolving foreland basin. This situation explains why the fields in these two provinces are made of fluvial and deltaic sands with their specific characters (geometry, permeability barriers, petrography attributes, distribution, etc.), which must be considered when defining a development plan. Moreover, due to the shallow

1. Asphaltenes are precipitated out of the crude oil by adding a low molecular hydrocarbon (i.e. pentane or heptane). Resins and hydrocarbon molecules remain in the non-precipitated phase (maltenes or De-Asphalted Oil: "DAO"), and can be separated from each other in the laboratory by preparative chromatographic means.

context of these fields (see **Chapter 3**), which often have never been buried, the sands have not been transformed into sandstones and are still unconsolidated. This creates drilling and completion problems, as well as sand influxes during production.

The viscous character of the *heavy crude oils* is a strong limiting factor for their production, and the associated technologies need to be adapted from those routinely used for the exploitation of conventional oils.

In a situation where the *heavy crude oil* is a bitumen (too viscous to flow at reservoir conditions) and where the accumulations are located at shallow depth (less than 75 m), the overburden can be scratched away in order to access the bitumen deposit. The exposed bitumen-impregnated rock can then be mined using giant shovels and trucks (**Chapter 6**). This is typical of the mining operations that are currently performed in Alberta. The bitumen – which is mixed with sand (oil sand) – is extracted by subjecting a slurry (a mixture of oil sand and water) to air bubbling under specific temperature conditions to promote the formation of a froth (foam made of air and bitumen) separated from the settling sand. In Canada, the use of this mining procedure is forecast to triple by 2015.

At greater depth, similar to conventional oils, the exploitation of *heavy crude oils* must rely on well-based technologies.

The first approach, which does not involve specific enhancing processes, is referred to as cold production (**Chapter 7**). Due to the viscosity of the fluid, and even in favorable situations, the corresponding methodologies do not recover more than a few percent of the original oil in place. These methodologies include pumping in vertical wells, use of horizontal well technology and co-production of oil and sand (CHOPS: Cold Heavy Oil Production with Sand).

Several factors can contribute to the efficiency of cold production:

- Gas coming out of solution when the pressure drops, eventually promoting the formation of a lower viscosity foam. This gas is likely to be a byproduct of the bacterial activity involved in the biodegradation process itself.
- Compaction of the reservoir resulting from the subsidence of the overburden in shallow depth situations.
- Optimization of the fetch volume accounting for the reservoir's geological characteristics.
- Reservoir characterization workflow, contributing to better control of water coning when using complex well architectures allowed by horizontal well technology.
- Generation of high permeability zones – such as decompacted zones or wormholes – in the surroundings of the producing well when using CHOPS technology.

Cold production (primary production) can be economically attractive, but is far from sufficient for tackling the *heavy crude oils* production challenge. More demanding methodologies that enhance the recovery of these fluids must be implemented. They are mainly based on the use of heat, which strongly impacts the state of asphaltene/resin aggregates, and subsequently the oil viscosity (**Chapter 4**).

The processes applied or developed by operators involve the injection of steam into the reservoir (**Chapter 8**). This can be done on a cyclic basis according to the following sequence: injection of steam, waiting time for heat propagation in the reservoir, and production from the same well ("Cyclic Steam Stimulation" or "Huff and Puff technology"). This can also be done by steam flooding, in which the steam is introduced into the reservoir through injection wells in order to sweep the oil towards production wells. The success of such a method depends on good reservoir characterization associated with an optimized well location pattern (injectors and producers). A recently developed and promising concept is Steam Assisted Gravity Drainage (SAGD), which combines the use of steam with horizontal well technology. A producing horizontal well is completed at the base of the reservoir, parallel to a steam injector well located just above it. The geometry is designed to promote the creation of a heat chamber in which the warmed, less viscous oil flows downward to be collected by the producer well. *In Situ* Combustion (ISC) is another process, based on burning the oil itself within the reservoir in the presence of an air-containing gas in order to create a fire front. The generated heat allows the unburned crude to flow towards producer wells.

During the production of the highly viscous *heavy crude oils*, it is a common situation to deal with an insufficient natural pressure of the reservoir to achieve an economical recovery by natural pressure drive. Subsequently, production must rely on artificial lift (**Chapter 9**). The use of downhole pumps either driven by a motor at the surface (the popular sucker-rod pump), or by hydraulic pressure generated at the surface, is a practice of long date. The use of Progressive Cavity Pumps – which are based on the use of helical gears driven by a downhole electric motor – is a fast-growing technology due to its potential application in any type of well architecture (i.e. vertical, deviated and horizontal).

The number of *in situ* projects for the production of *heavy crude oils* is expected to quickly grow in the future. In addition to the two major countries which are aiming for a significant increase in the production of their considerable reserves – Canada and Venezuela – numerous other regions are involved in the development of *heavy crude oil* fields. Several of these projects are well-documented and can be used to evaluate the industrial dimensions of the various recovery processes implemented by the operating companies (**Chapter 10**).

With regard to cold production, management of the environmental impact is not very different from conventional oil operations (**Chapter 20**). However, the other technologies involve more specific concerns. Mining generates disturbed surface areas which need to be properly reclaimed. In this regard, Canadian legislation is enforced and lands are often in better condition and more carefully monitored after reclamation than before they were disturbed. The produced solid mineral byproducts – referred to as tailings – from oil sands mining are stored in terraced piles which, after reclaiming, are planted with vegetation. Yet the disposal of fine tailings is a more complex operation which still needs improvement. Water is a major concern due to the large quantities involved in operations, specifically for steam-based protocols. Typically, the production water is recycled after purification in order to generate the necessary steam (**Chapter 20**). CO_2 emissions are obviously an important issue when addressing the production of *heavy crude oil*, specifically when the technologies are based on steam generation. It is estimated that heat-based methods release an additional 40% of CO_2 as compared to the production of conventional oil or cold production of *heavy crude oils*. A tempting solution is to couple the production projects with the implementation of geological storage capabilities (CSC: Capture and Sequestration of CO_2) for accommodating the produced CO_2 (**Chapter 19**). Another solution could be to avoid CO_2 emissions by using a nuclear reactor to supply steam.

The viscous rheology of *heavy crude oils* interferes with their transport as such throughout surface pipeline facilities, as is done with conventional oils. Transport requires the use of appropriate technologies designed to either reduce viscosity or take advantage of an additional phase, thereby avoiding or reducing the detrimental expression of the rheology of *heavy crude oils* (**Part 3**).

Heat is known to reduce the viscosity of *heavy crude oils* (**Chapter 4**). The classical example of a warmed transport facility is the Alyeska pipeline in Alaska. However, energy consumption and technical issues currently limit the use of this technology.

Dilution of the crudes with relevant solvents including condensates, light crude oils or refinery products such as naphtha is the methodology most commonly used to reduce viscosity **(Chapter 11**). After transport, the diluent is usually recycled. The choice of solvent is a function of its regional availability and the stability of the blend. In this respect, blending tests must be performed before any operation to avoid mixtures which induce the precipitation of asphaltenes, causing partial plugging of the lines.

Partial upgrading processes (i.e. Solvent De-asphalting, Visbreaking or Coking – see **Chapters 15, 16** and **17**) near the production site – designed to thermally crack down the asphaltenes and resins – are an alternative solution for the viscosity reduction of transported crudes.

The use of an oil-in-water emulsion, possibly improved by chemical additives, is an alternative technology to bring the viscosity of the flowing polyphasic fluid within specifications for pipeline transport (**Chapter 12**). The emulsion can be directly used as feedstock for power plants or, if progress in technology allows, can be broken in order to recover the initial *heavy crude oil*.

For pipeline transport, other solutions are studied: e.g. the Core Annular Flow regime (**Chapter 13**). This technology involves water-lubricated transport of the heavy viscous oils. The less viscous fluid – water – migrates towards the area of high shear stress at the wall of the liner, allowing the viscous *heavy crude oil* to move as a continuous phase in the core of the pipe. Regardless of which technology is used for the transport of *heavy crude oils*, the progressive pressure drop along the line must be compensated for *via* the implementation of pumps along the line (**Chapter 14**). The choice and design of pumping facilities are a function of the required capacities and needed pressure rise. It should account for the type of transport, viscosity, occurring phases, exposure to corrosion and eventual presence of solid particles.

The bulk composition of *heavy crude oils* is far-removed from the specifications required by the market for petroleum-derived products. The detrimental properties include viscosity, low hydrogen content, high content in high molecular weight compounds rich in heteroelements (resins and asphaltenes), as well as high metal content (mainly nickel and vanadium). Subsequently, these types of crude oils must be substantially upgraded through refinery processes (**Part 4**). These processes rely on two main principles:

1) Separation of the valuable part of the crude by physical processes such as De-asphalting.

2) Transformation of the crude composition by deep thermal and catalytic processes with the eventual contribution of added hydrogen (Visbreaking, Hydrocracking, etc.).

The De-asphalting process involves precipitation of the heaviest part of the oil (asphaltenes and some of the resins) by the use of light paraffinic solvents which destabilize the asphaltene/resin aggregates (**Chapter 15**). Since the heavy resins and asphaltenes – with a high aromaticity (heavy polyaromatic structures), and highly enriched with heteroatoms such as sulfur, nitrogen and metals – are concentrated in the residual tar fraction, the resulting supernatant (de-asphalted oil) is strongly enriched in hydrogen, with saturate and naphthene hydrocarbons, very low metal content and – to some extent – lower sulfur and nitrogen content. As a result, this fraction is far more valuable. With regard to hydrocarbon purification, this process, which is based on a chemical structure fractionation, is far more specific than classical distillation processes based on the boiling point of compounds. By modifying operational parameters such as type of solvent, solvent-to-feed ratio and temperature, the quality and quantity of de-asphalted oil and residual tar can be adjusted by the operator. Thus, this process is very flexible.

Visbreaking is a mature non-catalytic thermal process using steam injection, in which *heavy crude oils* are cracked under rather mild temperature conditions to increase distillate yield and reduce the viscosity of distillation residues (**Chapter 16**). The thermal cracking involved in visbreaking modifies the very structure of the oil compounds. These changes involve a decrease in the size of the molecules and a subsequent decrease in viscosity. At the same time, the process results in aromatization of the asphaltenes and a correlative increase in the light saturates fraction of the maltenes, which may decrease their solubility power towards the asphaltenes. Such a situation potentially leads to destabilization and flocculation of the asphaltenes in the residual fraction. The engineering challenge of visbreaking is to adjust the thermal conditions in order to maximize the viscosity reduction and avoid the onset of asphaltene precipitation, as well as limiting the formation of gas and coke. Other thermal processes have been developed to increase the distillate yield without destabilizing the residual fraction. They may make use of hydrogen, hydrogen donor solvent or conditions close to or above the supercritical condition of water. None of them are currently used at an industrial scale.

Coking is a more severe non-catalytic thermal cracking process (**Chapter 17**). The objective is to use a disproportion reaction to recover the maximum quantity of hydrogen-rich distillates, along with a solid residue that has far lower hydrogen content: the coke. Moreover, the coke contains the unwanted metals and ashes, together with a substantial portion of the sulfur and nitrogen compounds. For commercial use, the distillates require further severe treatment or refinement in order to meet specifications regarding their olefin, nitrogen, sulfur and aromatic content.

The specifications and market demand for highly refined light and middle distillates resulted in the development of catalytic hydroconversion (**Chapter 18**). This involves the use of catalysts in the presence of high hydrogen pressure with the objective of removing the unwanted moieties – sulfur, metals and nitrogen – as well as promoting hydrogenation of the product, leading to higher transportation fuel and heating oil yields without coke rejection. For this purpose, various processes have been designed to function separately or in combination: removal of sulfur (hydrodesulfurization), removal of metals (hydrodemetallization), removal of nitrogen (hydrodenitrogenation), and increasing the hydrogen content of products (hydrogenation of aromatics and heterocycles, secondary cracking with formation of light fraction and coke residue). These processes are operated using various technologies, mainly differing in the design for the feed and catalyst bed contact: fixed bed, moving bed, ebullated bed and slurry (entrained) bed. The latter – probably the best adapted for the heaviest feedstocks – is still in the pilot stage, but will probably be industrially implemented in the near future.

At a time when energy is an issue of tremendous importance, the potential of the huge resources of *heavy crude oils* must be seriously examined. However, the substantial increase in the exploitation of this fossil fuel is challenged by major technical, economical and environmental difficulties. The environment issues are mainly related on the one hand to water management (production water, water supply for steam generation…) and solid byproducts management (more specifically acute for mining operations) (**Chapter 19**), and on the other hand to the control of the emission of greenhouse gases (**Chapter 21**) and of pollutants during refining (**Chapter 20**), the latter being constrained by strict regulations.

In addition to the currently growing technologies designed to improve reservoir characterization, production, transport and upgrading, other promising approaches are emerging. These include for instance:

- The development of prospective new technologies to achieve *heavy crude oil* upgrading within the reservoir itself (**Chapter 23**). In this regard, active R&D is currently in progress, exploring the use of solvent injection, implementation of hydrocarbon vapor in SAGD technology, *in situ* thermal conversion in the presence of hydrogen, association of combustion and gravity drainage, use of microbial communities generating methane and even catalytic conversion.
- The integration of the workflow – production with pre-treatment operations, transport and final upgrading – adjusted to optimize projects in terms of cost, energy consumption, markets needs and environmental protection (**Chapter 24**),

Amid concern for the anthropogenic increase in atmospheric CO_2 levels and global climate change, the impact of production and processing of carbon-rich fossil fuels such as *heavy crude oils* must be carefully considered (**Chapter 20**). However the impact of each individual project must be assessed *via* a pertinent approach in order to estimate the actual added CO_2 emissions as compared to a baseline situation without the project, or the situation if an alternative project would have been implemented **(Chapter 22**). It should be noted that each type of project will have a different mass balance in terms of carbon. Technologies with the greatest emissions are related to the production of steam (i.e. steamflooding and SAGD). These situations would require the deployment of mitigation means, such as capture and geological sequestration of produced CO_2 (Carbon Capture and Sequestration), or the supply of steam using a CO_2-free means of production (such as a nuclear reactor).

PART 1
Heavy Crude Oils

A.Y. Huc

Heavy crude oils – including heavy oil, extra-heavy oil and bitumen (often occurring as oil sands, which are a mixture of sand, water, clay and crude bitumen) – represent facets of what is usually known by the generic term "petroleum". As explained below, most of them are actually derived from conventional oils which have been altered by bacterial activity during their geological history.

Oil and gas are the result of the natural thermal evolution and transformation of sedimentary organic matter embedded in specific rocks: the so-called source rocks. Once generated, petroleum and gas are subsequently expulsed from this source rock and migrate through drainage pathways within porous/fractured permeable rocks (reservoir rocks) and along pervasive faults. On their way up towards the surface, the migrating hydrocarbons are eventually trapped by "defaults" of the draining system associated with structural or sedimentological features. These local accumulations result in the formation of fields, from which they can eventually be extracted *via* industrial means. During this upward migration from the generating source rock towards the surface, the hydrocarbons experience a progressive decrease in temperature. If the oil reaches a sufficiently shallow depth with a temperature enabling life to be sustained, bacterial activity can develop and feed on the petroleum, resulting in alteration of the migrating and accumulating oil, and promoting an increase in its viscosity. This is the main origin of "*heavy crude oils*": heavy oils, extra-heavy oils and bitumen.

It is important to notice that, in term of resources, the heavy oils, extra-heavy oils and tar sands which are the subject of this book account for an amount that is comparable to the so-called conventional oils. Unfortunately, this does not hold true for the reserves due to the difficulties in producing these highly viscous oils.

Increasing the reserves *via* the technical improvement of production procedures is clearly a major challenge for the future world energy supply.

The very nature of the produced oil – including high viscosity, low hydrogen content, metal content and acidity – requires specific transport technologies and sophisticated upgrading techniques in order to obtain products which meet market needs.

Understanding the composition of these "unconventional oils", as well as the relationship of this composition to the chemical and physical properties of heavy oils, is of prime importance in order to design and adjust the current and future technologies needed for production, transport and upgrading operations.

1 Heavy Crude Oils in the Perspective of World Oil Demand

N. Alazard-Toux

1.1 WORLD ENERGY BALANCE

Since the start of industrialization, world energy consumption has continued to increase. The demand for energy goes hand in hand with economic and social development. Started by Western countries, this model has spread to the entire planet since the end of the 20th century. Today, the countries of Latin America, India, Southeast Asia and above all China are part of this dynamic. Since energy is the driver of their development, they are increasing the demand for fossil energies. Thus, in 2004, China represented 42% of the world increase in energy demand, while European Union countries were only responsible for 10%.

Many reference energy scenarios which are regularly drawn up by national or international organizations such as the International Energy Agency (IEA), US Department of Energy (DOE), the European Commission, World Energy Council (WEC), or by private companies, show that in the coming years this trend should be confirmed. All agree on the fact that over the period 2020-2030, world primary energy demand will remain sustained. The trend scenarios drawn up by these organizations forecast an average growth rate of more than 1.5% per year. World consumption should therefore increase by 50%. Another significant fact: fossil energies (gas, oil and coal) will remain preponderant in the world energy balance, continuing to satisfy 80% of demand. Fears regarding climate change caused by greenhouse gas emissions and their impact on the environment, as well as the probable continuation of high oil prices, will undoubtedly result in the implementation of policies aimed at reducing or containing the consumption of fossil energies and oil. But considering the inertia of energy systems, it is unlikely that over the next 30 years fossil energies and oil will represent a significantly smaller portion of world consumption.

The 550 Policy Scenario published by the IEA illustrates this fact (Figure 1.1). In this scenario, which postulates an upper limit for greenhouse gas emissions by 2020, world primary energy consumption by 2030 is of course less than that of the Business as Usual scenario, but only by 10%, which still represents growth in primary energy demand of nearly 38% over the next 25 years. Emphasis is placed on improving energy efficiency in the industrial sector as well as the residential, services and transport sectors. The share of nuclear and renewable energies is larger in this scenario. Yet fossil energies retain a significant share in

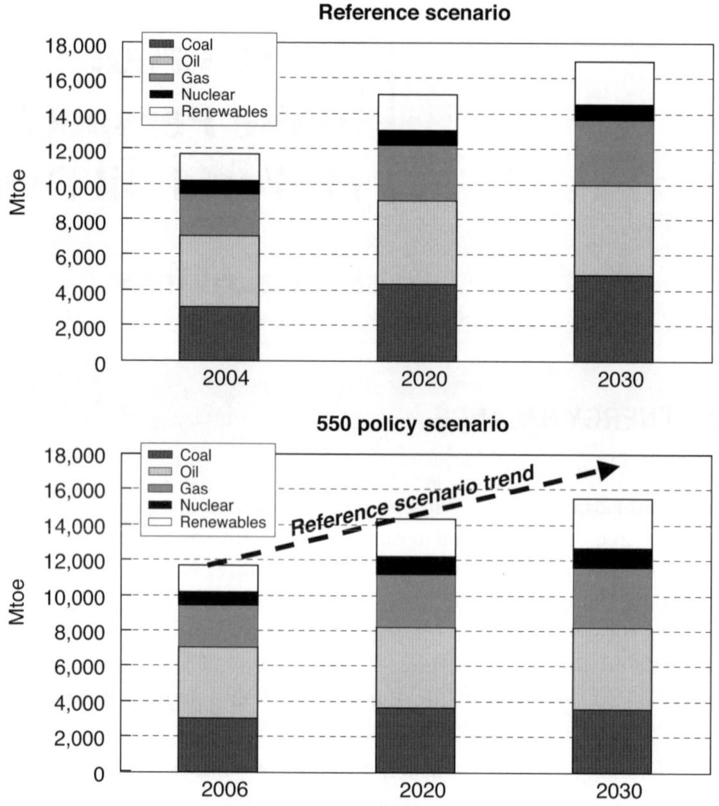

Figure 1.1

Evolution of World Primary Energy Demand Reference Scenario and Alternative Policy Scenario of the International Energy Agency.
Source: International Energy Agency, WEO (2008).

the world balance (on the order of 75%). Oil demand continues to grow by 1.2% per year on average throughout the period, or an increase of 30% by 2030.

Oil should therefore continue to be a dominant energy, even if it is likely that the future will bring a continuation of what has already been observed for several years: oil's fallback to its specific uses (petrochemicals and above all transport) which are still difficult to replace. Moreover, the transport sector should in the future be responsible for two-thirds or even three-quarters of the increase in world demand for petroleum products.

1.2 OIL FOR TRANSPORT

Over the past few decades in OECD countries, higher incomes, technological progress, improved infrastructure, greater leisure time and free circulation of goods have all resulted in a demand for greater mobility: more travel over larger distances, leading to a strong increase in road transport.

Over the last 20 years, the increase in transport has matched and even exceeded the increase in Gross Domestic Product (GDP) for industrialized countries. Although GDP has increased by more than 50%, distance traveled has grown by 70% and the automobile fleet by 60%. Three-quarters of goods and more than 90% of inhabitants of the OECD area move about today *via* road transport, and industrialized countries accommodate most of the current world automobile fleet, with four of every five vehicles.

This development of transport demand has resulted in increased demand for petroleum fuels, on which the sector's energy needs are currently highly-dependent.

On a global scale, fuels produced from crude oil currently constitute 95% of the energy used in road, sea and air transport, with biofuels, Liquefied Petroleum Gas (LPG) and – to a lesser extent – natural gas accounting for the remaining 5%. Solutions relying on alternative energies exist – some even for a very long time now – but their contribution to transport sector supply has remained marginal, mainly due to issues of cost, as well as energy density.

In the years to come, global demand for mobility should continue to grow, mainly due to developing countries.

As in OECD countries, the need for mobility in developing countries is tightly linked to income levels, distance separating workplace from home, mass tourism and other lifestyle changes that accompany access to consumer goods. This need for mobility should therefore increase, and road transport should continue its growth in the rest of the world.

In its report "Mobility 2030" (Table 1.1), the World Business Council for Sustainable Development (WBCSD) estimates that personal transport (expressed in passenger-kilometers per year) should increase on average worldwide by 1.6% per year between 2000 and 2030. This growth rate, which should be less than 1% per year in OECD countries, will overall be greater than 2% per year for developing countries, even reaching 3% for China. The transport of goods should increase even more significantly, with forecasts from the WBCSD's Reference Scenario showing an average annual growth rate of 2.5% in the number of ton-kilometers per year for the world. Here again, developing countries will be responsible for most of the increase, with average annual growth in the transport of goods remaining at less than 2% over the period for industrialized countries.

Currently, China has 3.2 vehicles per 1,000 inhabitants, while India has 4.5 and Europe nearly 500 vehicles per 1,000 inhabitants. It is estimated that by 2030, the total vehicle fleet in developing countries will triple.

In its Reference Scenario, the IEA estimates that oil consumption for transport use will then be smaller in the industrialized countries than in the rest of the world.

From a technical standpoint, solutions seeking to reduce oil consumption for transport are currently being studied or already exist operationally. Some of these solutions aim to replace petroleum with other fuels: natural gas or biofuels are examples in this area. Other solutions aim to reduce unit consumption by vehicles: this is the case for studies on engine downsizing or hybrid engines (combining an internal combustion engine with an electric motor). Finally, other solutions look further out into the future for their industrial and commercial development, and implement totally different technologies: this is the case for the

Table 1.1 Growth in Global Demand for Mobility over the Period 2000-2030.

	Average Annual Growth Rates 2000-2030	
	Personal Transport Activity (passenger-kilometers/year)	**Freight Transport Activity (ton-kilometers/year)**
Africa	1.9%	3.4%
Latin America	2.8%	3.1%
Middle East	1.9%	2.8%
India	2.1%	4.2%
Other Asia	1.7%	4.1%
China	3.0%	3.7%
Eastern Europe	1.6%	2.7%
Former Soviet Union	2.2%	2.3%
OECD Pacific	0.7%	1.8%
OECD Europe	1.0%	1.9%
OECD North America	1.2%	1.9%

Source: Word Business Council for Sustainable Development.

fuel cell/hydrogen association or the electric vehicle. None of these alternatives offers the same kind of answer to the problem of consuming less oil for transport, nor do they have the same degree of technical maturity. The production of biofuels has been developed actively during the last years. However, it is well known today that a massive development of present biofuels pathways at a world level can have a real negative impact on environment or food prices. At the other extremity of the range of technological solutions, the hydrogen/fuel cell combination allows for more radical gains in terms of oil consumption, but will not be widely distributable – industrially and commercially – at an acceptable cost for several years or even decades.

Thus, in the future, large quantities of oil and petroleum products will still be required for the needs of personal and goods mobility over the surface of the planet. These volumes should even continue to increase for many years to come.

1.3 PETROLEUM RESERVES AND RESOURCES

Within this outlook, the issue is how to meet needs, since the planet currently has proven reserves evaluated (depending on the source) at between 900 and 1,200 Gb (giga-barrels or a billion barrels), or the equivalent of 40 to 45 years at current production rates – and thus far less if we forecast an increase in demand.

To meet the demand for mobility in the absence of alternatives to traditional fuel/engine technologies that are rapidly and widely distributable on a worldwide scale, we will be required – even in a period of energy transition – not only to exploit current oil reserves, but to also mobilize other resources.

Regarding the mobilization of new resources, various pathways can be envisaged: make new discoveries, better recover the volumes of oil trapped in the subsurface, and finally, exploit atypical resources consisting of unconventional hydrocarbons which are more difficult and costly to produce.

New deposits of conventional oil remain to be discovered. They are undoubtedly smaller or located in prospects that are more difficult to detect than those in the past, but to identify them, the exploration sector benefits today from techniques which are highly improved. The more predictive approach, *via* basin modeling, or even the development of high-definition seismic surveys, will allow us to discover accumulations that were impossible to discover before, in order to explore more complex or deeper zones.

Increasing the recovery of oil in place in the subsurface is also a means of renewing reserves. Today, for every barrel brought to the surface, two are left in the reservoir. This is an average figure at a worldwide scale, since this ratio can vary highly from one deposit to another. Enhancing the recovery rate can significantly increase the reserves of a field. To do this, it is important to know more about the often very heterogeneous structure of the reservoir, in order to better position the wells and determine the preferred production technique and strategy. Increasing this rate can also mean wider use of enhanced recovery production techniques (drainage of fluids present in the deposit by injection of other fluids).

Finally, we cannot ignore the importance that so-called unconventional hydrocarbons may represent over the longer term as reserves. This term includes, among others, a group of resources: high density and viscous crude oils like tar sands and extra-heavy crude oils. These resources are more difficult and costly to produce than conventional crude oils, but they represent significant volumes. Resources in place of tar sands and extra-heavy crude oils identified on the planet have been estimated at 2,200 to 3,700 billion barrels.

The development of tar sands and extra-heavy crude oils is a reality. Currently, the production of very dense oils represents only a small share of world oil production, roughly 2 Mb/d, or barely 2 to 3% of world supply. But within an energy environment of continually increasing consumption, shortage of light crude and sustained high per-barrel prices for crude, we should see increased exploitation of unconventional crude oils. Most of the major oil companies – both national and international – have understood this. They are increasing investment in the development of these resources, for which exploitation is being carried out today at costs compatible with a barrel above US $50.

In addition to the large quantity of volumes in place, these oils have the advantage that they are not concentrated in the Middle East, but are instead abundant in another region of the world: the American continent, and more specifically in two countries, Canada and Venezuela. The Orinoco Belt in Venezuela and the forests and tundras located in northern Alberta, Canada contain the large majority of tar sands and extra-heavy crude resources identified on the planet. In addition to the economic appeal it may have, their development is often seen by consuming countries as a means of helping to maintain diversified supply sources. In the current context of geopolitical confrontation, where diversification of sources is paramount, the geographical location of these oils becomes a major advantage.

However, heavy or extra-heavy crude oils have as many disadvantages as advantages and enormous challenges remain to be met: technical challenges involved in the extraction, transport and refining of these viscous, high density liquids; ecological challenges as well, since current production techniques have a negative impact on the environment due among others to significant CO_2 release. Overall it is estimated that for a given volume of fuels, "well-to-wheel" CO_2 emissions (except for cold production, e.g. current Venezuelan projects) are twice as high for tar sands and extra-heavy crude oils as for conventional oil.

1.4 CARBON CONSTRAINT

Carbon dioxide is a greenhouse gas. Produced in large quantities by human activities (transport, residential and industry), this gas escapes mainly during the combustion of fossil energies (coal, oil or gas). In one century, concentrations of greenhouse gases in the atmosphere have increased by 50%, and those of CO_2 by 31%. Of course, the quantities of carbon dioxide emitted by humans (30 billion tons of CO_2 per year, or 8.1 billion tons of carbon) make up a small fraction of the total annual carbon cycle, but the biosphere and the oceans – natural carbon sinks – only absorb half of this. The surplus accumulates year after year in the earth's atmosphere while disturbing delicate climate mechanisms. Thus, since the start of the industrial era, Earth's temperature has increased by 0.6°C on average and sea levels have risen by 10 to 20 cm. Already, we are experiencing the initial negative effects: erratic weather, floods, hurricanes, heat waves, droughts, thawing glaciers, etc. Yet until recent years, there was no consensus in the international scientific community with regard to climate change – and particularly its anthropogenic causes – since models were not robust enough to verify the intuition of Arrhenius (the Swedish scholar who in 1895 was the first to understand the role of CO_2 in the greenhouse effect). We would have to wait for technological advances, as well as the acquisition of much data regarding past climate and the carbon cycle, to prove him right. In 1988, formation of the Intergovernmental Panel on Climate Change (IPCC) by the World Meteorological Organization (WMO) and the United Nations Environment Program (UNEP) marks a real turning point. Its 2,500 scientists are duly mandated by the United Nations with the mission of gathering and summarizing the relevant scientific data.

In March 1989 in The Hague, at the initiative of France, the Netherlands and Norway, 24 Heads of State symbolically commit to combat further strengthening of the greenhouse effect. On the same theme, in 1992 Rio de Janeiro (Brazil) hosts the Earth Summit (second United Nations conference on environment and development). It results in the United Nations Framework Convention on Climate Change (UNFCCC), signed by 166 countries. Its ultimate objective is to stabilize CO_2 concentrations in the atmosphere at a level that prevents any dangerous anthropogenic interference with the climate system. But this good intention was followed by no precise measure, other than the decision to regularly hold the Conference of Parties, the supreme body instituted by the Convention, to more precisely define the objectives and conditions of its application. For example, it will be decided that only developed countries – the main parties responsible for the greenhouse effect along with China – will be obligated to reduce their emissions.

In 1997 in Kyoto (Japan), the Parties to the Convention meet once again to negotiate quantified objectives for the reduction of greenhouse gases. The end of this conference witnessed the agreement and the signature of the historic accord which would become known as the Kyoto Protocol. It provides for an average reduction of 5.2% in greenhouse gas emissions of developed countries over the period 2008-2012 as compared to 1990 levels. To enter into force, this accord had to be ratified by at least 55 countries, representing 55% of the emissions of developed countries. The defection of the United States in 2001 will delay the process. The signature of Russia in 2004 will give it substance again. On February 16, 2005, the Kyoto Protocol finally entered into force.

But we would need to go even further still. The third report from the IPCC shows that we must limit global warming to 2°C and stabilize CO_2 concentrations at 450 ppmv (parts per million by volume) to prevent major catastrophes. This objective requires a decrease in world emissions by more than half, which for industrialized countries corresponds to division by a factor of 4 or 5. In this context, growth in the world production of carbon-rich heavy and extra-heavy crude oils must take place within a technological framework that limits releasing CO_2 into the atmosphere.

REFERENCES

Energy Information Administration/US Department of Energy (2006) International Energy Outlook 2006, DOE/EIA, Washington.

Intergovernmental Panel on Climate Change (2001) Climate Change 2001: Synthesis Report, www.ipcc.ch

International Energy Agency (2008) World Energy Outlook 2008, OECD/IEA, Paris.

Mathieu Y, Alazard-Toux N (2007) Un point sur les énergie fossiles. Responsabilité et Environnement n° 47, juillet 2007, pp 27-32.

World Business Council for Sustainable Development (2004) Mobility 2030: Meeting the challenges to sustainability, www.wbcsd.org

2 Definitions and Specificities

A. Saniere

2.1 DEFINITIONS

Heavy crude oil, extra-heavy crude oil and bitumen are naturally-occurring petroleum substances characterized by their high specific gravity.

In 1921, the American Petroleum Institute created the API Gravity Scale, initially to measure the specific gravity of liquids less dense than water, especially petroleum. Use of the API Gravity Scale is now extended to the whole range of specific gravity. The API gravity scale is recognized by the petroleum industry and today widely used. API gravity is graduated in degrees (API°) and the formula used to obtain the API gravity of petroleum liquids is:

$$\text{API gravity} = (141.5/\text{Specific gravity}) - 131.5$$

Specific gravity and API gravity evolve in opposite directions; thus, the smaller the API gravity, the heavier the fluid.

According to the API scale, different types of oil are defined:

- **Light crude**, which has an API gravity greater than 31.1° API, i.e. specific gravity less than 0.87.
- **Medium crude**, defined as having an API degree between 22.3 and 31.1° API, i.e. specific gravity between 0.87 and 0.92.
- **Heavy crude**, defined as having an API gravity less than 22.3° API, i.e. specific gravity greater than 0.92. In addition to its high specific gravity (equivalent to a low API°), heavy crude is defined as having high viscosity, generally above 10 centipoises (cP).

Note that there are various definitions of heavy crude, depending on the source used. The two main definitions used are that of the American Petroleum Institute described above, and that used by the World Petroleum Congress or the US Geological Survey, which states that heavy crude has an API° below 20. In the absolute sense, every company has its own definition of what it refers to as "heavy crude". Furthermore, industry professionals also have different definitions: typically, refining and exploration engineers do not have the same understanding of what is meant by "heavy crude". Thus, it is very important to specifically define "heavy crude" before beginning to work on it.

According to the Canadian Center for Energy, *heavy crude oil* is itself classified into different categories according to specific gravity and viscosity at reservoir conditions (Figure 2.1):

- **Heavy oil**, the API° degree of which is greater than 10; its viscosity is less than 10,000 cP (10 Pa.s) and it flows at reservoir conditions.
- **Extra-heavy oil**, the API° degree of which is less than 10 and the *in situ* level of viscosity is less than 10,000 cP (10 Pa.s), which means that it has some mobility at reservoir conditions.
- **Natural bitumen**, often associated with sands, and also referred to as **tar sands or oil sands**, the API degree of which is less than 10 and the *in situ* viscosity greater than 10,000 cP (10 Pa.s); it does not flow at reservoir conditions.

Note that the extra-heavy oil and bitumen have an API° less than 10, which means a specific gravity greater than 1: they are heavier than pure water.

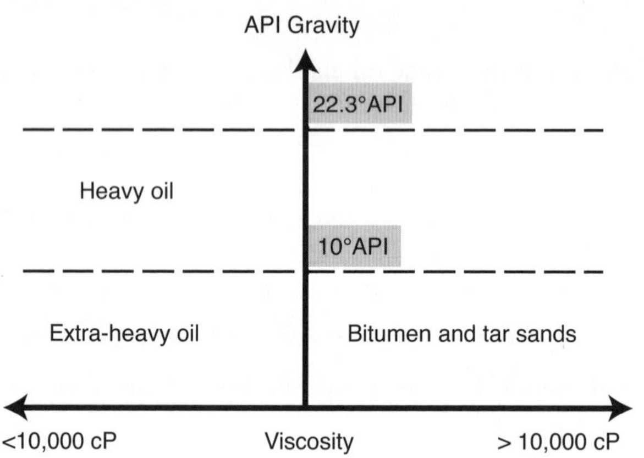

Figure 2.1

Definition of Conventional Heavy Oil, Extra-Heavy Oil and Bitumen.
Source: American Petroleum Institute & Canadian Center for Energy.

Most *heavy crude oil* is the result of bacterial alteration of conventional oil within the reservoir rock. It has different physical and chemical properties, and is generally degraded as compared to conventional light crude: high viscosity and high content of asphaltenes, heavy metals, sulfur and nitrogen (Table 2.1)).

These special properties require specifically-adapted technical solutions to be used throughout the development chain for these crudes, including extraction, upgrading, transport and refining. *Heavy crude oil* is difficult to exploit, but its volume is so significant that this fact alone justifies its interest. Due to its geographical location and the size of its resources, *heavy crude oil* development constitutes **a major economic and energy challenge**. In the current environment where the specter of peak oil is omnipresent, heavy crude exploitation will help to offset the decline in worldwide production of conventional oil. Its development on a large scale requires that **several technical challenges be met**.

Table 2.1 Comparison of the Composition of Extra-Heavy Oil and Light Oil.

	Zuata	Brent
Gravity (API°)	8.5	38.5
Viscosity (cSt at 60°C)	4,000	4
Sulfur (% by weight)	4.1	0.4
Ni (ppm)	94	1
V (ppm)	450	5
Acidity (mg KOH/g)	4.3	0.05

2.2 ECONOMIC AND ENERGY STAKES

2.2.1 Recoverable Volumes and Resources

Definitions: many definitions of hydrocarbon reserves and resources exist. Here, we will use the following:

- Resources refer to volumes of hydrocarbons present underground or volumes in place. They are of finite dimension.
  - Initial resources refer to volumes present before any exploitation.
  - Remaining resources refer to initial volumes from which cumulative production for the year in question is deducted.
- Recoverable volumes refer to volumes of hydrocarbons which can be extracted profitably with current techniques and economic conditions.
- Reserves refer to volumes of hydrocarbons contained only in deposits which are exploited or about to be exploited, and which can be extracted profitably with current techniques and economic conditions [Mc Keley, Brobstet Patt, 1973].

The identified in-place volumes of conventional heavy oil, extra-heavy oil and bitumen are estimated at about 5,000 Bb, i.e. the equivalent of the remaining in-place resources of conventional oils discovered up to now. Only about 1% of these heavy crude resources have already been produced. With current production technologies, recoverable resources are estimated at 700 Bb. About 56% of these resources are represented by bitumen in Canada and extra-heavy oils in Venezuela.

2.2.1.1 Bitumen

The identified volumes of bitumen in place are estimated at between 2,200 and 3,700 Bb, with the majority being located in Canada, where resources are estimated at between 1,600 and 2,500 Bb. Smaller volumes have been identified worldwide, mainly in Asia (270 Bb), Russia (260 Bb), Venezuela (230 Bb) and the United States (60 Bb) (in the states of Utah, Texas and California). Bitumen deposits are also present in Africa, where estimates of the resources in place range from 50 to 430 Bb, depending on the information source. In Russia, very large resources are present in the Lena-Tunguska basin of Eastern Siberia. Most of the other Russian deposits are located in the Tatar Republic, in the Timen-Pechora and Volga-Ural basins.

Canada's bitumen resources are located almost entirely within the province of Alberta with small extension within the province of Saskatchewan, with only minor oil sands deposits found on Melville Island in Canada's Arctic Island region. Alberta's oil sands deposits are grouped on the basis of geology, geography and bitumen content and are defined as the Peace River, Fort McMurray and Cold Lake Oil Sands Areas.

The Alberta Energy & Utilities Board (AEUB) estimates the **initial volumes-in-place** to be 1,600 Bb. The AEUB further estimates the ultimate volume in place, a value representing the volumes expected to be found by the time all exploratory and development activity has ceased, to be 2,500 Bb. They include:

- 140 Bb amenable to surface mining; this is located in the Fort McMurray Oil Sands Area.
- 2,400 Bb amenable to *in situ* recovery or underground mining methods.

According to the AEUB, **current technologies allow the recovery of roughly 178 Bb of bitumen**. With anticipated technologies, ultimate recoverable volume could be 300 Bb. About 20% (35 Bb) of the recoverable bitumen resources is located at a shallow depth and can be exploited using mining technologies. The other 80% (140 Bb) requires adapted petroleum technologies for its exploitation.

Recoverable volumes outside Canada are estimated at between 110 and 150 Bb.

2.2.1.2 Extra-heavy Oil

Worldwide extra-heavy oil resources are estimated at roughly 1,350 Bb. About 90% of those resources are located in the Orinoco Belt in Venezuela, where they are estimated at 1,200 Bb. Extra-heavy oil has also been identified in other countries, particularly in Ecuador (5 Bb), Iran (8 Bb) and Italy (1.5 Bb). In Russia, small amounts have been identified in the Volga-Ural and North Caucasus-Mangyshlak basins; accurate and timely data are insufficient to make sound estimates of the occurring quantities.

It is estimated that 20% of the resources in place in Venezuela is ultimately recoverable, which translates to roughly 240 Bb. With current technology and prices, the recoverable volumes are estimated at approximately 3% (36 Mb), according to the US Department of Energy. Recoverable volumes outside Venezuela are estimated at about 4 Bb.

2.2.1.3 Heavy Oil

Worldwide resources of oil at API degrees of between 10 and 22.3 can be estimated at roughly 1,000 Gb. Their geographical distribution is more diffuse than that of bitumens or extra-heavy oils. They have been found on all continents, but most are located in Venezuela, Mexico and Iran. Recoverable volumes are currently estimated at 200 to 250 Bb (Table 2.2).

Table 2.2 Remaining Volumes in Place in 2005 and Recoverable Volumes for Various Types of Crude Oil.

Type of Crude	Volumes in Place (Bb) in 2005	Recoverable Volumes (Bb) in 2005
Conventional heavy	1,000	200 to 250
Extra-heavy	1,350	40 to 244
Bitumen	2,650	290 to 450
Total heavy oils	***5,000***	***530 to 944***
Conventional crude	**4,800**	**1,032**

2.2.1.4 Unconventional Geographical Distribution

With 75% of resources located on the American continent, *heavy crude oils* have an unconventional geographical distribution, very different from that of conventional crude or even natural gas, which are centralized in the Middle East and the Former Soviet Union (FSU) (Figure 2.2). With time, the importance of these two zones in terms of conventional resources is increasing. Thus, *heavy crude oils* allow for geographical diversification of supplies for importing countries.

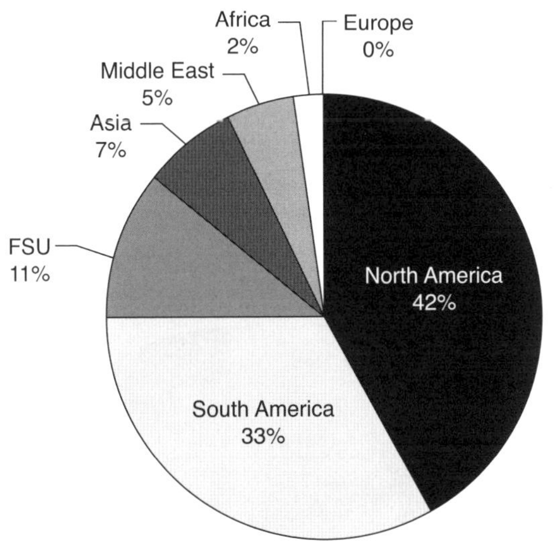

Figure 2.2

Geographical Distribution of *Heavy Crude Oils*.

2.2.2 Many Exploitation Projects

Worldwide production of heavy crude, extra-heavy crude and bitumen can be estimated for 2005 at nearly 8 Mb/d, representing 10% of total crude production which was 81.1 Mb/d in 2005. This production is distributed among the three types of crude as follows (Table 2.3).

Table 2.3

Type of Crude	Production in 2005 Mb/d
Conventional heavy	6.5
Extra-heavy	0.6
Bitumen	1
Total heavy oils	***8.1***
Total world production	**81.1**

Bitumen production is situated exclusively in Canada, just as that of extra-heavy crude is situated exclusively in Venezuela.

Total production of *heavy crude oils* is 8.1 Mb/d (Figure 2.3). 80% of this is located on the American continent: the leading producer is Mexico with nearly 2.5 Mb/d of conventional heavy oils. Canada is in second place with 1.5 Mb/d, 2/3 of which comes from bitumen and 1/3 from heavy oils. Venezuela, with production of 1.2 Mb/d, 50% of which comes from extra-heavy and 50% from heavy oils, comes in third place. Brazil is the fourth-leading producer of heavy oils, with roughly 0.7 Mb/d produced in 2005, most of which comes from deep offshore fields. Next is the United States with production on the order of 0.5 Mb/d, also from heavy oils. In other regions of the world, the main producers are China, Indonesia, Great Britain and Iran, all of whom produce heavy oils.

Production of bitumen and extra-heavy oils in 2005 was roughly 1.6 Mb/d, or 1.8% of worldwide crude oil production, although they represent 40% of the cumulative resources of conventional and *heavy crude oils*. Their portion of worldwide production should increase in the future if all planned projects are completed.

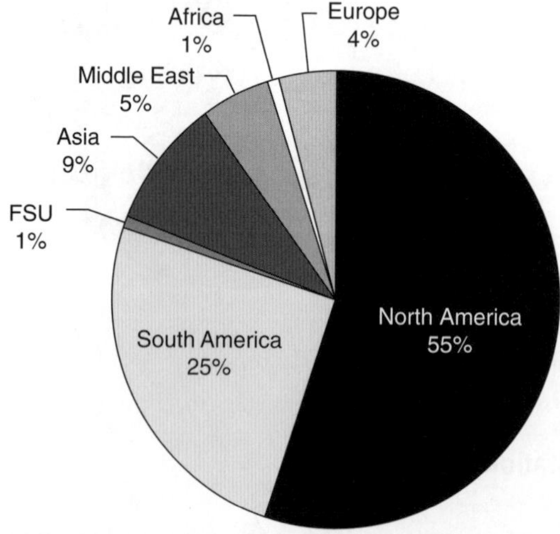

Figure 2.3

Worldwide Distribution of *Heavy Crude Oil* Production

2.2.2.1 Bitumen

Projects concerning bitumen exploitation are mainly located in Canada. In fact, the Alberta deposits are so concentrated that they are the only bitumen deposits which are economically recoverable. Only a minor amount of bitumen is still produced elsewhere for road materials and mastic, such as the Trinidad Pitch Lake deposit.

In the United States, no accumulations are being produced commercially. The geological conditions of the Utah deposits have made recovery difficult and expensive. Likewise, the Texan deposits, which are mostly deep and relatively thin, have also proved to be difficult to recover.

Tar sands resources in Canada are developed in very small proportions. In fact, according to the AEUB, 80% of possible oil sands areas are still available for exploration and leasing. This means that only 36 Bb of reserves are involved in ongoing or future development projects.

20% of recoverable bitumen resources is located at a shallow depth and can be exploited using mining technologies, and 8% of this volume has already been produced. The remaining 80% of recoverable resources can be produced with *in situ* technologies, and only 1% of this has already been produced.

2.2.2.2 Mining Production Projects

20% of recoverable bitumen resources is located at a shallow depth (less than 100 m) in the Fort McMurray Oil Sands Area and can be exploited using mining technologies. This production method currently provides 62% of Canadian bitumen production, i.e. 652,000 b/d in 2005.

For these very large projects, the installation of upgrading units dedicated to the exploitation is profitable. In all of the projects, the bitumen is upgraded on the production site and sold in the form of **Synthetic Crude Oil (SCO), characterized by an API degree between 29 and 36 and sulfur content between 0.1 and 0.2%.**

This activity is currently dominated by two companies: Syncrude and Suncor. Both companies have undertaken major projects to increase their bitumen output. Syncrude, which produced 262,000 b/d in 2005, forecasts a doubling of production by 2015 as part of its "Syncrude 21" project. The company will then be the mining industry leader, far ahead of its competitors. In 2005, Suncor produced about 200,000 b/d using mining methods. With its ongoing Project Millennium, the company's production should reach 325,000 b/d by 2010.

Shell Canada, under the Albian Sands Energy Company, has also been producing oil sands *via* mining methods since 2003 at Muskeg River, and is the third largest mining producer.

Five others projects are under development. By 2015, they should all be operating and the total production of synthetic crude oil should reach approximately 1.9 Mb/d (Table 2.4).

2.2.2.3 *In situ* Production Projects

80% of recoverable bitumen resources is located at a greater depth and must be exploited using *in situ* production technologies (i.e. recovery by adapted petroleum methods).

Table 2.4 Canadian Bitumen Mining: Ongoing Projects.

Project	Operator	2005 Production Kb/d	2010 Production Kb/d	2015 Production Kb/d
Syncrude 21	Syncrude [a]	262	382	507
Steepbank, Millenium, Voyager	Suncor	195	325	325
Athabasca Oil Sands Project (Muskeg River & Jackpine)	Albian Sands Energy Inc [b]	170	170	200
Horizon	CNRL		155	270
Fort Hills	Petro-Canada		30	170
Northern Lights	Synenco		3	100
Joslyn North Mine	Total		3	100
Kearl	Imperial Oil		0	130
TOTAL		**627**	**1,068**	**1,962**

[a] ***Syncrude** ownership: Canadian Oil Sands Trust (36.74%), Imperial Oil (25%), Petro-Canada (12%), ConocoPhillips (9.03%), Nexen (7.23%), Mocal (5%), Murphy Oil (5%).*

[b] ***Albian Sands Energy Inc**. has been created to operate Muskeg River on behalf of its joint venture owners: Shell (60%), Chevron (20%) and Western Oil Sands (20%).*

Roughly twenty projects, currently underway or being studied, are expected in the coming years. In 2005, *in situ* production of bitumen in Canada was about 400,000 b/d. It could reach 1,320 Mb/d in 2015. The largest exploitation underway is Cold Lake, led by Imperial Oil, which produced 150,000 b/d of bitumen in 2005. It should remain the largest exploitation by 2015 (Table 2.5).

In situ production projects are generally on a smaller scale than mining projects, and cannot support the cost of a dedicated upgrader. In most of these projects, the extra-heavy oil is blended with a lighter, less viscous hydrocarbon (diluent) and sold as **Bitumen Blend (BB), with an API degree of 21 and sulfur content between 2 and 4%**. Diluent typically constitutes 24-50% of the bitumen blend. Only two projects include **on-site upgrading** and produce SCO (Synthetic Crude Oil) instead of bitumen blend: Firebag (Suncor) and Long Lake (Nexen/OPTI).

All together, more than 25 Canadian projects dedicated to the exploitation of tar sands and bitumen have been developed or are about to be developed. **If completed, they will represent production of 2.05 Mb/d of synthetic crude and 1.06 Mb/d of bitumen to be blended by 2015. In other words, the 2005 Canadian heavy oil and bitumen production will be multiplied by a factor of 3 by 2015 and will represent about 3% of the world oil supply that year (Figure 2.4).**

Those volumes are mainly exported to the United States, but plans are underway to build pipelines from Alberta to deep water ports on the British Columbia coast (Prince Rupert or Kitimat) for tanker shipment to Chinese refineries. Both Endbridge and Terasen, Canada's dominant crude pipeline companies, are targeting Asian markets.

Table 2.5 Canadian *in situ* Bitumen: Ongoing Projects.

Project	Operator	2005 Production Kb/d	2010 Production Kb/d	2015 Production Kb/d
Fort MacMurray	***In situ* projects**	**91**	**570**	**685**
Kirby	CNRL		30	30
Surmont	ConocoPhillips		50	75
Joslyn	Total	2	60	60
Jackfish	Devon		35	35
Christina Lake	EnCana	13	70	70
Hangingstone	JACOS [a]	8	50	50
Long Lake	Nexen/OPTI	20	70	70
MacKay River	Petro-Canada	28	30	30
Meadow Creek	Petro-Canada		25	40
Lewis	Petro-Canada		30	60
Firebag	Suncor	20	120	165
Cold Lake oil sands	***In situ* projects**	**273**	**500**	**605**
Orion	Shell		10	20
Primerose/Wolf Lake	CNRL	85	135	135
Foster Creek	EnCana	38	100	100
Sunrise	Husky Energy		45	140
Tucker Lake	Husky Energy		30	30
Cold Lake	Imperial Oil	150	180	180
Peace River oil sands	***In situ* projects**	**28**	**32**	**32**
Seal	Shell	16	16	16
Peace River	Shell	12	16	16
TOTAL		**392**	**1,102**	**1,322**

[a] *JACOS: Japan Canadian Oil Sands.*

2.2.2.4 Extra-heavy Oils

Extra-heavy oil exploitation projects are concentrated in the Orinoco Belt of Venezuela, the largest deposit in the world with 1,350 Bb in place.

Four projects associating PDVSA with international companies exploit part of these volumes. The development patterns are similar in all four cases, i.e. cold production, transport *via* dilution over roughly 200 km to San José where the extra-heavy oil is upgraded to a greater or lesser extent depending on the project (see table below), then exported. In the Petrocedeno and Petropiar projects, extra-heavy oil is upgraded to reach 26 or 32°API, a lighter crude suitable for export to numerous refineries throughout the world. In contrast, in

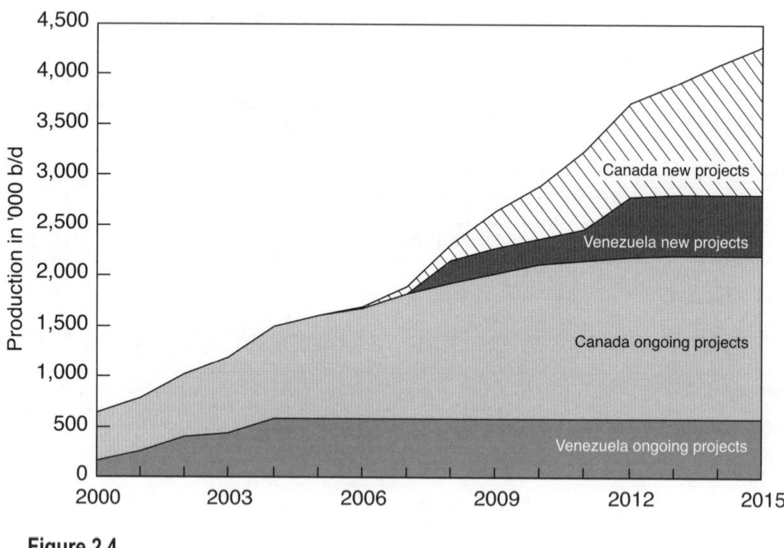

Figure 2.4

Bitumen and Extra-heavy Oil Production Forecasts.

the Petromonagas and Petroanzoategui projects, the crude oil is only partially upgraded and exported to the United States in refineries dedicated to heavy crude processing. The diluent added for transport is separated and returned to the production site for reuse.

The four projects were initially developed *via* strategic partnerships operated by international companies: Total, ConocoPhillips and ExxonMobil. The projects started in 1998 for Petrozuata, 2000 for Sincor and Cerro Negro, and 2001 for Hamaca. In May 2007, the Venezuela government nationalized operations linked to these four projects, which are now managed by joint ventures in which PDVSA holds a majority stake. International companies which fail to reach an agreement pull out of the projects, as is the case for Exxon and ConocoPhillips. Moreover, in August 2007, PDVSA changed the names of these projects (see table below).

At the end of 2005, the government of Venezuela unveiled a strategic plan requiring $15 billion in investment over the period 2006-2012 in order to increase production of extra-heavy oils in the Orinoco Belt. The goal is to double current production on the order of 600,000 b/d by 2012 (Figure 2.4). The first part of the plan consists in evaluating and certifying reserves in 27 blocks located in the four zones of the Orinoco Belt, i.e. Carabobo (formerly Cerro Negro), Ayacucho (formerly Hamaca), Junin (formerly Zuata) and Boyaca (formerly Machete). Several companies have already been awarded contracts to study 7 of these blocks, in the framework of the project "Magna Reserva": Oil & Natural Gas Corp (ONGC) of India, Lukoil and Gazprom of Russia, CNPC of China, Repsol-YPF of Spain, Petropars of Iran and finally Petrobras, the Brazilian state-owned oil company. After quantification and certification of the reserves, negotiations are expected between PDVSA and these companies for the development of blocks. An enhanced recovery process, probably steam injection, will be required to extract the quantity PDVSA hopes to recover, 20% of the volumes in place.

Table 2.6 Venezuela: Integrated Extra-Heavy Oil Projects, Ongoing as of September 2007.

Old Project Name and Partners	New Project Name and Partners	Reserves (in Gb)	Synthetic Crude Production (b/d)	Synthetic Crude °API	Investment (in US $ billions)
Sincor PDVSA – 38% Total – 47% Statoil – 15%	**Petrocedeno** PDVSA – 60% Total – 30.33% Statoil – 9.67%	2.5	180,000	32	4.2
Petrozuata PDVSA – 49.9% ConocoPhillips – 50.1%	**Petroanzoategui** PDVSA – 100%	1.6	104,000	19 to 25	4.8
Cerro Negro PDVSA – 41.6% ExxonMobil – 41.6% BP – 16.67%	**Petromonagas** PDVSA – 83.3% BP – 16.67%	1.8	108,000	16	2.5
Hamaca PDVSA – 30% ConocoPhillips – 40% Chevron – 30%	**Petropiar** PDVSA – 70% Chevron – 30%	2.2	180,000	26	4.4
	TOTAL	**8.1**	**572,000**		**15.9**

In Venezuela, one exploitation project was set up to produce Orimulsion, an emulsion composed of 70% extra-heavy oil and 30% water, and used as fuel in power stations. Bitúmenes Orinoco SA (Bitor), a wholly owned subsidiary of PDVSA, is the affiliate responsible for the production and international marketing of Orimulsion. At the end of 2005, production was 40,000 b/d, while it was roughly 100,000 b/d at the start of the year. Today, the future of Orimulsion production is uncertain. Given current petroleum prices, PDVSA announced that it wanted to stop production. Extra-heavy oils are in fact worth more when processed into crude, although still heavy (17°API), than sold in the form of Orimulsion. This decision was met with outcry from the companies – particularly Canadian and Italian – with whom PDVSA has long-term contracts for the supply of Orimulsion. Moreover, it seems that PDVSA still has a joint venture agreement with China National Petroleum Corp. to build an Orimulsion production unit of 6.5 million tons/yr.

2.2.3 Project Financial Specificities

2.2.3.1 Massive Investments

Exploitation projects for bitumen and extra-heavy crude oils require far greater investment than the most technical developments of conventional crude oils, e.g. those located offshore in extremely deep water. Investment in Canada for exploitation *via* mining techniques is roughly US $13 to 29 billion per project. In Venezuela, it is US $2 to 6 billion, and for petroleum exploitation projects in Canada, it is US $0.6 to 5 billion. By comparison, each development in deep offshore in the Gulf of Mexico or West Africa requires investment on the order of US $0.2 billion to 3 billion.

The ratio of investment to maximum production rate illustrates the capital-intensive nature of exploitation projects for very high density crude oils. Thus, in Venezuela, capital needs for the four exploitation projects of the Orinoco Belt vary from 17,000 to 32,000 US $/b/d of maximum capacity. This range is mainly due to the fact that not all operators have chosen the same option for bitumen upgrading: the more complex it is, the more capital-intensive the project.

In Canada, development projects based on petroleum techniques, which make use of steam injection to recover the crude but have no upgrader on-site, have capital needs of 18,000 to 29,000 US $/b/d. Mining projects, which combine complex recovery techniques with the construction of dedicated upgraders, have the highest capital needs, on the order of 40,000 to 50,000 US $/b/d.

At the same time, for crude oil exploitation projects in the deep offshore waters of West Africa or the Gulf of Mexico, capital needs are from US $8,000 to 14,000 US $/b/d, depending on the zone's maturity.

Given the high level of investment required for resource development, host countries have instituted specific financial terms designed to attract investment from petroleum companies. Thus, in Canada, the Royalty Regime is set at 1% as long as the return on investment has not been met, i.e. as long as the total cash position is negative. In general, for mining projects, the time needed for return on the investment is 12 to 15 years. At the start in Venezuela, companies operating the extra-heavy crude oils enjoyed a royalty rate of 1% and corporate taxes of 34%. In 2001, Caracas instituted a new Hydrocarbon law. Thus, the royalty rate increased to 16.6% in 2004, then to 33.33% in 2006. Corporate income tax was increased to 50% at the start of 2007.

Table 2.7 Capital Needs for Various Types of Petroleum Projects.

Type of Project	**Capital Needs in 2005 US $/b/d of Maximum Production Capacity**
Extra-heavy crudes in Venezuela	17,000-32,000
Mined canadian bitumen	40,000-50,000
In situ petroleum canadian bitumen	18,000-29,000
Offshore deeper than 1,000 m	8,000-14,000

2.2.3.2 Evolution of Production Costs

With the improved knowledge of operators and the development of new techniques, the production costs of bitumen and extra-heavy crudes are constantly decreasing over time. Thus, in the middle of the 1980s, production costs *via* mining techniques with upgrading for Canadian bitumen were roughly 32 US $/b (in 2004 dollars). Today, they are between 16 and 22 US $/b. In 2004, for Syncrude, operating costs were broken down as follows: 62% for extraction, 32% for upgrading and 6% for administrative costs and research and development.

For *in situ* production with cyclic steam injection, the costs, which were roughly 17 US $/b (in 2004 dollars) during the 1980s, are currently 9 to 13 US $/b. By using the technique of Steam Assisted Gravity Drainage (SAGD), they were between 10 and 16 US $/b in 2004. In these projects, steam is usually produced by burning natural gas. Natural gas consumption is on the order of 1 Mcf/b of bitumen produced, or 1.063 MM Btu/b. In 2004, the average price of Henry Hub natural gas was 6.17 US $/MMBtu, which means that the cost of purchasing the natural gas alone represented a cost per barrel of $6.5, or roughly 45% of the total operating expenses (OPEX) for projects. In 2005, the price of natural gas increased to an average of 9 US $/MMBtu, or 9.5 US $/b of bitumen produced and 55% of the OPEX for projects. In an environment of high natural gas prices, this strong dependency poses a problem for petroleum operators, who are currently seeking solutions to do away with it. The combustion of residue from coke or coal is being examined. Although these solutions have the advantage of doing away with natural gas, they pose other problems, particularly environmental (see specific section).

Regarding the Venezuelan projects of cold production combined with varying degrees of upgrading, the production costs are roughly 7 to 11 US $/b depending on the projects.

3 | Geological Origin of Heavy Crude Oils

A.Y. Huc

The vast majority of naturally-occurring viscous to very viscous petroleum – referred to as heavy oils and extra-heavy oils, or semi-solid petroleum material associated with sands (tar sands) – is the product of biological alteration of conventional fluid crude oils mediated by in-reservoir bacterial communities. This is specifically true for the enormous heavy oil resources of Western Canada and Eastern Venezuela. This phenomenon has been known and described since the last mid-century [Winters and Williams, 1969; Evans *et al.,* 1971; Deroo *et al.,* 1977; Blanc and Connan, 1993; Head *et al.,* 2003; Larter *et al.,* 2003; Larter *et al.,* 2006; Eschard and Huc, 2008].

Such biological alteration is potentially active as soon as the temperature of the geological environment hosting the oil is sufficiently low to allow the living processes of these microbial consortia to take place. Empirically, but based on a large set of observations, the microbial degradation of oil is assumed to be very active up to 65°C, but remains significant up to 80°C. If we consider that the geothermal gradient can vary on a regional basis from 20°C/km to 40°C/km, this corresponds to a maximum depth ranging from 1,500 m to 3,000 m.

Although quantitatively subordinate, it should be noted that factors other than biodegradation can be responsible for the formation of heavy oils. For instance, the heavy oils of California are explained by the very nature of the sedimentary organic matter from which the oil is sourced. This organic matter is thermally labile and releases petroleum at an early stage of the burial history of the source rock. This results in a viscous, sulfur-rich, "thermally immature" oil [Isaacs and Rullkötter, 2000].

3.1 IMPACT OF BIODEGRADATION ON OIL COMPOSITION

The alteration of a "conventional oil" *via* biodegradation results in a compositional change, producing a progressively denser heavy oil. On a bulk molecular level the effects are well documented and expressed as a quasi-sequential disappearance of specific classes of compounds, including successively the light n-alkanes, heavier n-alkanes, branched alkanes and cyclic alkanes. The term "quasi-sequential" refers to the fact that more resistant compound classes can start to be altered prior to the complete destruction of a less-resistant class, and limited overlapping is possible [Peters *et al.,* 2005]. This progressive removal of light hydrocarbons and saturates results in an increased relative concentration of remaining high molecular weight aromatics and NSO compounds (so-called resins and asphaltenes).

The industrial and economic implications of the biodegradation of oil are far-reaching and include an increase in viscosity, decrease in API gravity (Figure 3.1), relative increase in metal content (i.e. Nickel and Vanadium, since they are mainly present within the NSO fraction being concentrated), and increase in sulfur content. Moreover, the degradation of hydrocarbons by bacteria results in the formation of intermediate chemical species – including acid compounds – which results in increased oil acidity. This is reflected by the typically high TAN values measured in heavy oils (TAN: Total Acid Number, as determined in the laboratory, is the quantity of KOH (in mg) needed to neutralize 1 g of oil).

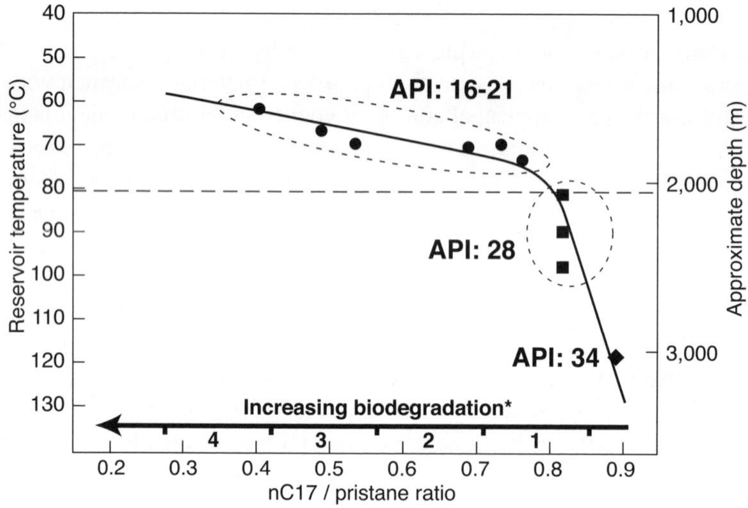

Figure 3.1

Change in nC_{17}/Pristane ratio as a marker for the extent of biodegradation versus reservoir temperature. Derived and modified from the Alba Trend fields in the Outer Witch Ground Graben, UK [Masson *et al.*, 1995]. Relationships with API gravity values. Relationships with the Peters and Moldowan ranking scale* [Peters *et al.*, 2005].

It should be noted that the increase in high molecular weight fractions is not likely due to relative concentration alone, following the selective removal of alkanes and aromatics (to a lesser extent). Numerous studies point to the fact that newly formed high molecular weight NSO compounds result from the biodegradation process (through oxidative reactions involving oxygen and sulfur species). In this respect and based on consideration of sulfur isotopes, the observed increase in sulfur content is in part due to the incorporation of additional sulfur. The latter is probably associated with microbial sulfate-reduction reactions, sustained by sulfates when present in the formation water.

3.2 ASSESSING THE LEVEL OF BIODEGRADATION

The selective and quasi-sequential removal of hydrocarbon classes during the process of microbial alteration is routinely used for evaluating the extent of biodegradation of crude oils. On this basis, Peters and Moldowan [Peters *et al.,* 2005] have proposed a scale ranging from 1 to 10, ranking the biodegradation level from none (0) to light (1-3), moderate (4-5), heavy (6-7), very heavy (8-9) and severe (10). Similarly, the relative resistance of the branched alkanes as compared to the n-alkanes is often accounted for by a convenient biodegradation parameter expressed as the following ratio: n-C_{17}/Pristane (Figure 3.1). In this ratio, nC_{17} is a representative molecule of the n-alkanes, the most labile series as far as biodegradation alteration is concerned, and pristane (2,6,10,14 tetramethylpentadecane), a C_{19} branched alkane belonging to the isoprenoids family, molecules which include a methyl group on every fourth carbon atom and exhibit increased resistance to biodegradation.

3.3 OIL BIODEGRADATION: QUANTITATIVE ASPECTS

In addition to a substantial change in the properties of the petroleum, the biodegradation alteration is actually responsible for the destruction of a sizable amount of the initial oils. For example, it should be noted that mass balance calculations suggest that biodegraded oils ranking "2" on the "Peters and Moldowan" scale have experienced a mass loss of 30-40% of the non-degraded oil, while those ranking "5" have experienced a loss of up to 70% [Peters *et al.,* 2005; Kowalewski *et al.,* in press]. With respect to these figures, it is significant to consider that the current resources of heavy oils, extra-heavy oils and tar sands represent half of the current total resources of petroleum on Earth. Since this altered petroleum represents the portion remaining after the biodegradation process, we can speculate that the quantity of initial resources of oil which have been altered amounted to twice the current resources of conventional oils.

Since the ultimate product of biodegradation is CO_2, the quantitative input of this gas into the natural system must be considered in any realistic carbon cycle assessment, and at a time scale which still must be accurately defined.

3.4 AGENTS OF DEEP SUBSURFACE BIODEGRADATION

In recent years, a major advance has been achieved by identifying anaerobic organisms as the main (and probably the only) agents of the biodegradation process in the deep subsurface. For years, the paradigm was that only aerobes were involved in the biodegradation of oils, or at least that anaerobes required aerobes to initiate degradation of petroleum [Jobson *et al.,* 1979]. This was based on a lack of knowledge in microbiology (the only known bacteria able to degrade hydrocarbons were aerobes), as well as observations reporting that biodegraded oil fields were often associated with active meteoric waters. The meteoric waters were supposed to carry the necessary dissolved oxygen and accompanying aerobes – charged at the outcrops – into the subsurface. The modern view, emphasizing the role of anaerobes, relies on:

- Mass balance calculations showing that the quantity of dissolved oxygen brought in by aquifers – even in the case of a very active aquifer – is largely insufficient to account for the degradation of a normal-sized oil pool [Horstad *et al.*, 1992].
- Actual identification of anaerobic microbial communities in formation waters associated with biodegraded fields, and sampled under careful microbiological conditions [Magot *et al.*, 2000].
- Occurrence of biodegraded oil pool settings in geological situations in which meteoritic water influx is unlikely (i.e. in certain deep offshore accumulations).
- Discovery of strains of anaerobic microorganisms capable of degrading saturated and aromatic hydrocarbons [Widdel and Rabus, 2001].
- Isolation – from biodegraded oils – of metabolites indicative of anaerobic hydrocarbon degradation [Aitken *et al.*, 2004].

This new insight has considerable implications regarding the understanding of the biodegradation process, and more importantly regarding the geological settings in which biodegraded oil fields can be discovered. The occurrence of meteoric circulation is no longer mandatory and consequently, biodegradation can potentially proceed in any hydrodynamic situation.

3.5 LIMITING FACTORS

It has been recognized for decades [Winters and Williams, 1969] that the main control for biodegradation is temperature (3.1). In fact, in order to proceed, this phenomenon requires that the environmental conditions are compatible with physiological processes. Although the theoretical limit of life is estimated to range between 120°C and 150°C (a strain living at 121°C has recently been identified), it seems that the maximum temperature for biodegradation in the deep subsurface is roughly 80°C, but that actually effective up to 65-70°C, and beyond 90°C the oil fields are "sterilized".

The discrepancy between the high theoretical values and observations in the subsurface has been discussed by [Head *et al.*, 2003], and tentatively explained by the low level of metabolic activity of the microorganisms living in the deep biosphere. This low metabolic rate is suspected to prevent the bacteria from renewing their heat-labile cell components at a sufficient rate, and consequently to strongly lower the temperature limits allowing for subterranean life.

A second important limiting factor is the availability of inorganic nutrients (i.e. phosphorus and potassium) in an environment that is strongly depleted of such elements. The main source for these nutrients likely involves the dissolution of minerals by the formation water, probably enhanced by the microbial activity itself [Bennett *et al.*, 2000]. This is a slow process, which is likely to explain the low rate of biodegradation observed in petroleum reservoirs. A figure of 10^{-4} kg of degraded hydrocarbons m^{-2} yr^{-1} at 40-70°C has been proposed [Head *et al.*, 2003]. In this respect, the observation that biodegradation is often associated with the invasion of surface-derived meteoric formation waters can be explained by the fact that these low-salinity waters are undersaturated in numerous inorganic chemical species and consequently more apt to release nutrients *via* the dissolution of minerals. This increased availability of nutrients is likely to increase the metabolic rate of the *in situ* microbial communities. To some extent, this would reconcile the new paradigm – according to which the occurrence of active meteoric water is not mandatory for biodegradation – with the empirically observed relationships between biodegradation and meteoric water influx.

3.6 GEOLOGICAL CONTROLS

Bacteria live in water due to the external hydrophilic properties of their membranes. Moreover, in order for them to have access to a sufficient supply of nutrients, this water should exhibit a continuous fluid phase. In this respect, the most favorable setting for biodegradation to proceed is at the base of the oil column, at the lower part of the oil-water transition zone (generically called the oil-water contact). Consequently, a crucial control for biodegradation efficiency is likely the residence time of the hydrocarbon molecules at this oil-water contact. In this regard, since the newly arriving oil molecules reaching the reservoir – after migration from the source rock – are incorporated into the field accumulation at the oil-water contact, biodegradation is likely to be concomitant with the filling of the reservoir. If this assumption held true, the filling conditions of a given field would be of paramount importance as far as biodegradation is concerned. This degree of importance is supported by the fact that the rate of biodegradation is estimated to be on the same order of magnitude as the rate of oil charging of a reservoir [Head *et al.,* 2003]. In such a dynamic system, the size, shape, sedimentary architecture of the reservoir and in-filling rate will directly control the surface area, pattern and rate of displacement of the oil-water contact surface, and consequently the biodegradation exposure time of fresh oil components arriving at the oil-water interface. The larger the oil-water contact surface and the lower the charging rate, the more favorable the conditions for effective biodegradation of an oil accumulation. On the other hand, smaller oil-water contact surfaces and higher charging rates will prevent the biodegradation from proceeding efficiently. This difference in the dynamics of oil charging might explain the variable level of biodegradation for fields with the same apparent temperature conditions, and for which a similar composition could have been expected.

In a given reservoir, the bacterial communities are suspected to be autochthonous, which means that as soon as the microbial population is killed by a sufficiently high temperature, replenishment by fresh bacteria is unlikely. In this respect, if during its geological history a reservoir experiences such a temperature before oil in-filling takes place, there is little chance that biodegradation can proceed, even if the current reservoir temperature would allow this phenomenon to occur. This is the case for reservoirs buried at temperatures higher than 90°C before being uplifted and filled by oil. This sterilization process prior to oil charging is termed "paleo-pasteurization", leading to unaltered petroleum in currently low-temperature reservoirs [Wilhelms *et al.,* 2000]. In the specific situation of multi-charged reservoirs, the oil accumulation can eventually be successively fed before and after the sterilization event. This is suggested as the origin of "mixed oils", exhibiting the characters of both biodegraded and unaltered oil.

Using basin modeling to reconstruct the temperature history of a target reservoir and the timing of oil charging is consequently of prime importance for operators to predict oil quality.

From a regional geology perspective, the heavy oils and tar sands are often located in the distal part of foreland basins (Figure 3.2).

This is specifically the situation of the two major provinces of heavy oils, i.e. Western Canada and Eastern Venezuela [Tissot and Welte, 1984]. These wedge-shaped basins are the result of regional loading of the sedimentary pile by compressive mountain building which, through erosion, concomitantly acts as a provider for the sediments deposited at the front of the growing range. The source rocks, deposited in the passive margin sequences preceding the compressive episode or, later, in the depression formed at the front of the orogenic belt, are deeply buried in the foredeep during the process of mountain building. Consequently, they reached adequate temperature conditions for oil generation, resulting in the formation of conventional oil fields after short distance oil migration. However, a common feature of such sedimentary basins is to exhibit a gentle dipping monoclinal pattern along the distal (i.e. from the mountain range) flank. This pattern is often accompanied by a lateral continuity of the sedimentary bodies, lacking major physical barriers, and by the occurrence of major unconformity surfaces. Continuous sedimentary layers (fluvio-deltaic) overlying the unconformities act as conductors for the migrating fluids. Subsequently, the oil which is not trapped in intermediate accumulations, on its way up towards the far end of the basin, can proceed over long distances (up to a hundred kilometers).

Eventually, the oil will reach depths which become progressively shallower [Demaison, 1977] and enter a temperature regime where life can be sustained and which becomes increasingly favorable to bacterial activity. Subsequently, along the migration pathway hydrocarbons will experience step-by-step biodegradation, resulting in oils becoming heavier when approaching the border of the basin. Both Western Canada and the Eastern Venezuela display such a geological situation.

The normal crude oils produced from the Early Cretaceous reservoirs of the Western Canadian basin are characterized by medium-high API gravity in the vicinity of the foredeep (i.e. 20-30°API in Bellshil lake area). Closer to the outcrops, the Lloydminster and Peace River area contains heavy oil (15-20°API), then the Cold Lake and Wabasca deposits reach 11°API. Finally, the outcropping Athabasca sands (tar sands) contain extra-heavy oil (6-8°API).

In Eastern Venezuela, the oils located at depths greater than 1,500 m are medium to light (25-40°API), low sulfur crudes (less than 1% S content) (i.e. in the Officina area). When approaching the southern rim of the basin, the oils become progressively shallower, heavier and biodegraded (10-25°API), as well as richer in sulfur (1-3% S content) (i.e. in the Tremblador/Pilon areas). Finally, they grade into the Orinoco Heavy Oil belt, a gigantic accumulation of sulfur-rich (4-5% S content) extra-heavy oil (less than 10°API).

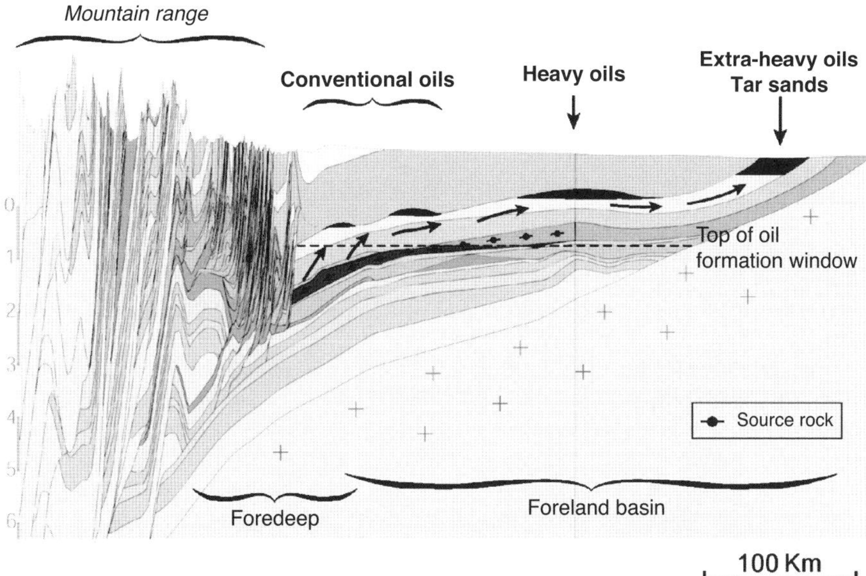

Figure 3.2

Schematic cross-section of a foreland basin providing a favorable geological situation for the formation of heavy oil accumulation. The world's major heavy oil provinces such as the Alberta province in Canada and in Eastern Venezuela are located in similar settings [modified after Deroo *et al.*, 1977, Demaison 1977 and Head *et al.*, 2003].

3.7 CONCLUSION

The vast majority of heavy oil accumulations are the result of biological activity altering formerly conventional oils. The main factors controlling the process are the physiological conditions required by the *in situ* bacterial communities, including adequate temperature scenario of the reservoir and availability of inorganic nutrients. On a case-by-case basis, understanding and predicting the occurrence and extent of biodegradation rely on deep insight into the geological circumstances surrounding the field accumulation formation, including static and dynamic information such as architecture and burial history of the reservoir, timing and rate of oil charging, hydrodynamism, occurrence and timing of meteoric water influx etc. It is significant that foreland basins are highly favorable geological settings for the accumulation of major heavy oil and tar sands deposits.

Current reservoir characterization and modeling workflows, as well as basin modeling packages, already provide useful tools for helping to understand and predict biodegradation in oil fields and prospects. Dedicated improvements to these approaches – e.g. local grid refining and adapted time-step resolution in basin simulators, will provide the opportunity to model the oil invasion at the scale of the field during charging (instead of the modelling of the oil displacement within producing fields as currently proposed by reservoir simulators) – would likely result in a significant increase in our ability to address the biodegradation aspect for the sake of exploration and production.

REFERENCES

Aitken CM, Jones, DM, Larter SR (2004) Anaerobic Hydrocarbon Biodegradation in Deep Subsurface Oil Reservoirs, *Nature*, **431**, pp 291-294.

Bennett PC, Hiebert FK, Rogers JR (2000) Microbial Control of Mineral-Groundwater Equilibria: Macroscale to Microsacle, *Hydrogeol. J.*, **8**, pp 47-62.

Blanc Ph, Connan J (1993) Crude Oils in Reservoirs: the Factors influencing their Composition in: *Applied Petroleum Geochemistry*, ed M.L. Bordenave, Editions Technip, Paris, pp 149-174.

Demaison GT (1977) Tar Sands and Supergiant Oil Fields, Am. Assoc. *Pet Geol. Bull.*, **61**, pp 1950-1961.

Deroo G, Powell TG, Tissot B, Mc Crossan RG (1977) The Origin and Migration of Petroleum in the Western Canadian Sedimentary Basin, Alberta, *Bull. Geol. Surv. Canada.*

Eschard R, Huc AY (2008) Habitat of Biodegraded Heavy Oils: Industrial Implications. *Oil and Gas Science and Technology*, **63**, 5, pp 587-607.

Evans CR, Rogers MA, Bailey NJL (1971) Evolution and Alteration of Petroleum in Western Canada, *Chem. Geol.*, **8**, pp 147-170.

Head IM, Jones DM, Larter SR (2003) Biological Activity in the Deep Subsurface and the Origin of Heavy Oil, *Nature*, **426**, pp 344-352.

Horstad I, Larter SR, Mills N (1992) A Quantitative Model of Biological Petroleum Degradation within the Brent Group Reservoir in the Gullfaks Field, Norwegian North Sea. *Org. Geochem.*, **19**, pp 107-117.

Isaacs CM, Rullkötter J (2000) The Monterey Formation, from Rocks to Molecules Columbia University Press, New York. 553 p.

Jobson AM, Cook FD, Westlake DWS (1979) Interaction of Aerobic and Anaerobic Bacteria in Petroleum Biodegradation, *Chemical Geology*, **24**, pp 355-365.

Magot M, Ollivier B, Patel BKC (2000) Microbiology of Petroleum Reservoirs, Antonie van Leeuwenhoek' *Int. J. Gen. Mol. Microbiol.*, **77**, pp 103-116.

Kowalewski I, Carpentier B, Magnier C, Huc AY (in press). Use of Heavy Metals Content for Quantitative Estimation of Biodegradation, In special publication *JPGE.*

Larter SR, Wilhelms A, Head IM, Koopmans M, Aplin AC, di Primio R, Zwach C, Erdmann C, Telnaes N (2003) The Controls on the Composition of Biodegraded Oils in the Deep Subsurface. Part I – Biodegradation Rates in Petroleum Reservoirs, *Org. Geochem.*, **34**, 2, pp 299-310.

Larter SR, Huang H, Adams J, Bennett B, Jokanola O, Oldenburg TBP, Jones DM, Head IH, Riediger C, Fowler MG (2006) The Controls on the Composition of Biodegraded Oils in the Deep Subsurface. Part II – Geological Controls on Subfurface Biodegradation Fluxes and Constraints on Reservoir Fluid Property Prediction, *AAPG Bull.*, **90**, 6, pp 921-938.

Mason PC, Burwood R, Mycke B (1995) The Reservoir Geochemistry and Petroleum Charging Histories of Paleogene-Reservoired Fields in the Outer Witch Ground Graben. In: JM Cubitt and WA England (eds), The Geochemistry of Reservoirs, *Geological Society* Special Publication, **86**, pp 281-301.

Peters KE, Walters CC, Moldowan MM (2005) The Biomarker Guide, 2nd edition, 2 volumes, Cambridge University Press.

Tissot BP, Welte DH (1984) Petroleum Formation and Occurrence, Springer-Verlag, 699 p.

Widdel F, Rabus R (2001) Anaerobic Biodegradation of Satured and Aromatic Hydrocarbons, *Curr. Opin. Biotechnol.*, **12**, pp 259-276.

Wilhelms A, Larter SR, Head I, Farimond P, di-Primio R, Zwach C (2000) Biodegradation of Oil in uplifted Basins prevented by Deep-Burial Sterilisation, *Nature*, **411**, pp 1034-1037.

Winters JC, Williams JA (1969) Microbial Alteration of Oil in Reservoirs, *Am. Chem. Soc. Div. Petr. Chem.* Preprints, **14**, pp E22-E31.

4 Properties and Composition

I. Merdrignac, D. Espinat, I. Hénaut, J-F. Argillier

The physico-chemical properties of *heavy crude oils* are still an active domain of research and the very nature of their constitutive moieties belongs to the frontier of science. Due to their crucial role in the chemical and physical properties of *heavy crude oils* – with implications for industrial issues including production, transport and upgrading – the resins and asphaltenes have been extensively studied using a wide spectrum of techniques. Although these substances remain ill-defined, a conceptual model begins to emerge consisting of a polydispersed distribution of molecules with regard to size and composition. Basic units rich in aromatics and containing functional groups tend to be prone to self-association and aggregation. Consequently, physical and chemical conditions control the aggregation state – a situation which is accounted for in the exploitation of the *heavy crude oils* (dilution, thermal treatment, etc.). An important aspect of the structure of *heavy crude oils* is the instrumental role played by resins in stabilizing asphaltene aggregates within the hydrocarbon medium.

The purpose of this chapter is to review the physical and chemical means used for characterization of the moieties which control the properties of *heavy crude oils*.

It should be noted that this chapter – which is rather specialized – can be read over without affecting the understanding of the other chapters.

Characterization of the chemical species contained in the heavy fraction of crudes remains a challenging task, despite the substantial progress achieved in recent decades. The complexity of the oil matrix increases as a function of its boiling point. Within the heavy ends, a wide variety of moieties are present in terms of structure and number of molecules [Strausz and Lown, 2003].

Many studies of heavy product analyses have primarily focused on asphaltenes, and to a lesser extent resins. Asphaltenes are defined by their insolubility in a given paraffinic solvent (i.e. nC5 or nC7), whereas resins constitute the most polar fraction soluble in the same given paraffinic solvent. Two types of characterization can be distinguished.

First, chemical characterization provides information on the elemental composition, chemical structure and functional groups of the macromolecules. The molecular characteristics of these species may vary from one crude to another. Within the same crude, results show that asphaltene molecules can be highly polydispersed. The main analytical tools used for chemical characterization of crude oils are liquid chromatography (saturates/aromatics/

resins/asphaltenes or SARA fractionation), elementary analyses, Nuclear Magnetic Resonance (NMR), InfraRed spectroscopy (IR), etc. From these analytical data, the representation of an average asphaltene molecule can be tentatively drawn using advanced molecular modeling approaches.

Second, colloidal characterization can provide insight to the state of dispersion of the asphaltenic entities viewed as macromolecules in a good solvent (usually aromatic) or in their own medium (crude). The required parameters are molecular mass, polydispersity in mass or in size and state of aggregation with aggregates of varying sizes. For this purpose, various techniques are deployed such as X-ray and neutron scattering (SAXS and SANS), Nuclear Magnetic Resonance (NMR), rheology and various fractionation techniques (ultracentrifugation, fractionation by solvents and Steric Exclusion Chromatography (SEC)).

4.1 HEAVY OILS: STRUCTURAL COMPOSITION

Heavy ends of crude oils can be defined as having a boiling point above 350°C and exhibiting an API density lower than 20. Associated molecules contain more than 25 carbon atoms (C_{25}), and present a structural complexity which increases with boiling point and molecular weight, density, viscosity and refractive index (aromaticity). These fractions are enriched in highly polar compounds such as resins and asphaltenes. They are composed of various chemical species of varying aromaticity, functional heteroatoms and metal content [Murgich *et al.*, 1996; Strausz *et al.*, 2002] as compared to the initial crude oils or their lighter fractions.

The saturated and aromatic hydrocarbons of heavy fractions usually include naphthenic and aromatic species with more than six alkylated cycles. The aromatic content increases with the boiling point, as well as the number of aromatic cycles in the structures. The aromatic distribution of Vacuum Gas Oil (VGO, 350-550°C) fractions is fairly centered on structures which include from one to three aromatic rings, whereas structures of residues (550°C$^+$) mainly contain five to six rings. The higher the boiling point of a fraction, the more enriched it is in heteropolyaromatic structures. Sulfur compounds contained in residues can be divided into five chemical classes: thiols, sulfides, disulfides, sulfoxides and thiophenes. The first four classes can be subdivided into cyclic and acyclic structures as well as alkyl-, aryl- and alkylaryl-derived species. Furthermore, thiophenes are condensed polyaromatic structures with benzo-, dibenzo-, naphthobenzo-thiophenes and other derived structures. In heavy fractions, the major sulfur species are thiophenic structures, followed by sulfide derivatives (cyclic and acyclic). Only few sulfoxide types are reported. Regarding nitrogen compounds, two classes are distinguished: basic and the neutral, with a ratio of basic to neutral compounds generally lower than one. Although nitrogen content in heavy crudes appears to be much lower than other heteroelements, it has a large impact on hydrotreatment processes due to catalyst poisoning. On the one hand, basic nitrogen compounds identified in the 350°C$^+$ fraction are mainly quinoline structures containing 2 to 4 aromatic rings with different configurations (peri- or catacondensed with various alkylation degrees) [Igniatiadis *et al.*, 1985]. On the other hand, carbazoles, benzo- and dibenzo-carbazole families exhibiting different alkylation degrees occur as neutral moieties [Dorbon *et al.*,

1982]. The distribution of neutral and basic families is strongly related to the geochemical type of oils. Differences correspond to relative abundance, alkylation degrees and isomer distributions [Merdrignac *et al.*, 1998]. Porphyrin structures are also present in heavy oil fractions, depending on the organic matter type from which they derive. Of extremely varied nature, they are generally complexed by nickel or vanadyl ions [Goulon *et al.*, 1984]. Finally, oxygenated structures are also present in oil fractions, but in small quantities. In the 350°C$^+$ fractions, they are represented by phenolic compounds and carboxyl functions (carboxylic acids, esters, ketones, amides and sulfoxides) [Moschopedis and Speight, 1976a].

4.2 HEAVY FRACTIONS: ADVANCED CHARACTERIZATION

Detailed structural characterization of heavy products is generally difficult to achieve, mainly due to analytical technique limitations.

Molecular analytical techniques generally used for lighter fractions (i.e. gas chromatography) cannot be directly applied. Consequently, preliminary simplification of matrices is necessary. Alternatively, structural information on isolated subfractions may be obtained by either chemical or colloidal characterization.

4.2.1 Fractionation Methods

Various fractionation schemes may be considered, as discussed below.

4.2.1.1 Distillation

Separation as a function of boiling point can be achieved. Thanks to this technique, a direct relationship between boiling point and certain properties such as viscosity or density can be obtained. However, boiling points are not directly proportional to average molecular weight, but rather depend on the chemical structure of compounds present in the fractions. It has been shown that polar compounds able to form aggregates are less volatile than the non-polar species of a similar given molecular weight. This may explain why polycondensed aromatic compounds enriched in heteroatoms (resins and asphaltenes) are mainly concentrated in high boiling point fractions. Several normalized methods of distillation are available.

The True Boiling Point method (TBP, ASTM D2892) and Potstill distillation (ASTM D5236) are used at preparative scale to obtain distillation cuts from initial point to 400°C with 5°C – 10°C cuts, and vacuum distillation cuts (400-550°C) with increments of 50°C, respectively. In addition, to control the quality of products or for routine analyses, simulated distillation is well suited (ASTM D2887, D6352) and can replace traditional preparative methods. Used at analytical scale, testing time can be reduced and smaller solvent volumes are used. This technique is based on the aptitude of hydrocarbons to be eluted according to their boiling point from a chromatographic column under specific conditions. Retention time of the species is converted into boiling point temperature *via* the use of a calibration curve, standardized from normal paraffin retention times. A profile of cumulative percentages as a function of the boiling point is obtained [Durand and Petroff, 1984; Bacaud *et al.*, 1998; Dahan *et al.*, 2004].

4.2.1.2 Chemical Type Fractionation

A. De-asphalting

De-asphalting is based on sample flocculation *via* contact with a paraffinic solvent (antisolvent) under stirring conditions for a given time and temperature. Parameters such as contact time, solvent/oil ratio, solvent type, temperature and washing steps can strongly influence precipitation [Speight, 1999; Ancheyta *et al.*, 2002]. The precipitation of asphaltenes can be achieved by using several standard analytical procedures (ASTM D893, D2006, D2007 and D3279). However, differences in the experimental conditions have a noticeable impact on the yield and chemical structure of asphaltenes.

B. SAR Separation

From a de-asphalted cut, SAR separation is used to obtain saturates, aromatics and resins (or polar) fractions eluted as a function of polarity. The techniques generally used are preparative liquid chromatography (flash chromatography), analytical chromatography (HPLC) or Thin Layer Chromatography (TLC). This type of separation is usually applied on the 300°C$^+$ cuts, as lighter products might be lost during purification steps.

Regarding flash chromatography, many methods have been developed and routinely applied in refineries and laboratories. The main advantage of this technique is the recovery of large fractions, allowing further characterization. However, round robin tests have shown that results may vary widely according to methodologies (solvents and active phases). Only few normalized ASTM methods have been developed in this field, such as ASTM D2007 [Peramanu and Pruden, 1999], or ASTM D4124, which was derived from the Corbett method.

As an alternative method, TLC is faster and more economical than flash chromatography. Since it is non-preparative, this technique is used to separate complex mixtures by silica support adsorption (plates or rods) [Sharma *et al.*, 1998; Cebolla *et al.*, 2003; Matt *et al.*, 2003].

SARA fractions may also be predicted through modelization [Akbarzadeh *et al.*, 2002] or from InfraRed (IR) or Near InfraRed (NIR) spectra [Aske *et al.*, 2001]. As compared to chromatography, the main advantages of IR/NIR are their simplicity and time savings. However, developing robust models adapted to a broad range of samples can be very time-consuming [Satya, 2004].

C. ABAN Fractionation

This approach has been proposed by various authors [Green *et al.*, 1984; Kim and Branthaver, 1996; Al-Zaid *et al.*, 1998; Mac Kay *et al.*, 1981] in order to separate samples into Acid, Basic, Amphoteric and Neutral fractions (ABAN) (Figure 4.1). This fractionation depends on the reactivities of compounds and their ability to form hydrogen bonds (acid and basic fractions) or complexes *via* charge transfer (-donor) (amphoteric fraction). Heteroatomic polar compounds are found in these various fractions. The acid fraction contains phenols, pyrroles and molecules having one carbonyl group. Nitrogen compounds like azaarenes and amides are collected in basic fractions. Polycondensed aromatic compounds containing heteroatoms constitute the majority of the amphoteric fraction.

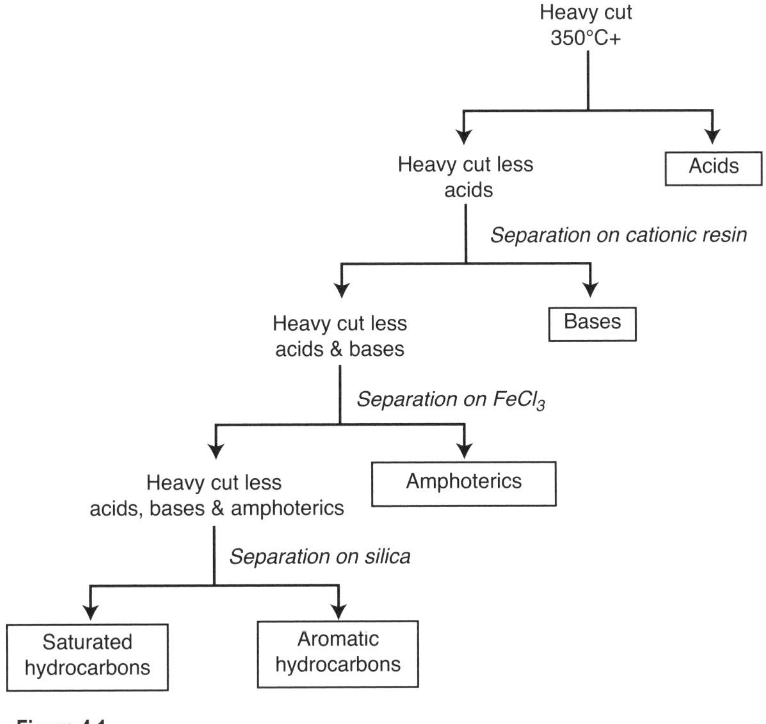

Figure 4.1

ABAN (Acid, Basic, Amphoteric, Neutral) Separation Scheme of a High Boiling Point Fraction (350°C$^+$).

D. Fractionation by Solvents of Different Polarities

This fractionation is based on asphaltene (already precipitated by *n*-C7) reprecipitation in toluene by various antisolvents of different polarities (normal paraffins, acetone and methanol). Szewczyk *et al.* (1996) have shown that subfractions obtained by various toluene and *n*-heptane mixtures exhibit high chemical polydispersity. It has been emphasized that the solubility of asphaltenes depends mainly on their aromaticity and aliphaticity properties, as well as asphaltene polarity [Andersen, 1997; Andersen *et al.*, 1997; Sharma *et al.*, 2000; Andersen *et al.*, 2001a; Neves *et al.*, 2001; Wattana *et al.*, 2002].

There are other types of fractionation, such as chemical derivations with acetylation of oxygenated compounds [Green and Reynolds, 1989], acid-base fractionations [Moschopedis *et al.*, 1976; Mc Kay *et al.*, 1976; Merdrignac, 1997], and sulfide oxidations [Payant *et al.*, 1989]. Selective complexations [Nishioka and Tomich, 1993; Altgelt and Boduszynski, 1994; Strausz and Lown, 2003], selective degradations (RICO – Ruthenium Ion Catalyzed Oxidation) [Boukir *et al.*, 1998; Su *et al.*, 1998; Strausz *et al.*, 1999a and b], and supercritical fluid extraction [Singh *et al.*, 1992; Shi *et al.*, 1997] have also been proposed in the literature.

4.2.2 Chemical Characterization

Following fractionation, the detailed description of functional groups contained in heavy oil fractions is the main challenge of chemical characterization. Numerous techniques may be used and we will focus on some of these below, while keeping in mind that the chemical composition of heavy petroleum molecules may exhibit large variations from one molecule to another.

4.2.2.1 High Performance Liquid Chromatography (HPLC)

The main application of HPLC is dedicated to the analysis of PolyAromatic Hydrocarbons (PAH) present in heavy petroleum compounds. Several types of detectors can be coupled (UltraViolet (UV), fluorescence, infrared or mass spectrometer). The detector generally used for petroleum applications is a UV detector. In this regard, aromatic compounds containing up to eight polycondensed aromatic structures can be detected, e.g. when monitoring catalytic hydrotreatments [Hatrik and Lehotay, 1994; Yan *et al.*, 1996; Fetzer and Kershaw, 1995; Simoneit and Fetzer, 1996].

4.2.2.2 ^{13}C and ^{1}H Nuclear Magnetic Resonance (NMR)

NMR is generally used to obtain structural information [Brown *et al.*, 1960; Clutter *et al.*, 1972; Bouquet and Bailleul, 1986]. Proton and carbon NMR spectra are illustrated in Figures 4.2a and 4.2b. Three types of protons can be identified: (i) aromatic protons (Haro), (ii) protons located in α of an aromatic cycle (Hα), and (iii) protons located in β and γ of an aromatic cycle (Hβ and Hγ). Using ^{13}C NMR, two major carbon types like aromatic and aliphatic may be distinguished (Figure 4.2b). With specific NMR data acquisition, other structural parameters may be determined:

- Quantification of aliphatic carbons: C_{quat} (quaternary carbons), CH, CH_2 and CH_3.
- Quantification of aromatic carbons: $C_{aro\text{-}total}$, C_{quat} ($C_{quat} = C_{quat\text{-}cond} + C_{quat\text{-}sub}$, $C_{quat\text{-}cond}$ quaternary condensed corresponds to carbons located between two aromatic nuclei next to each other, and $C_{quat\text{-}sub}$ corresponds to substituted carbons: aromatic carbons bonded with an alkyl chain), and C_{aro}-H.
- C/H aliphatic and C/H aromatic ratios.
- Substitution index (SI) = $\dfrac{C_{quat-sub}}{C_{aro-total} - C_{quat-cond}}$.
- Condensation index (CI) = $\dfrac{C_{aro-total} - C_{quat-cond}}{C_{aro-total}}$.

Such parameters are used to determine the structure of the studied fractions and structural models can be surmised, taking into account the aliphatic, aromatic, alkylated or polycondensed aromatic characters, number of aromatic cycles and length of alkyl chains.

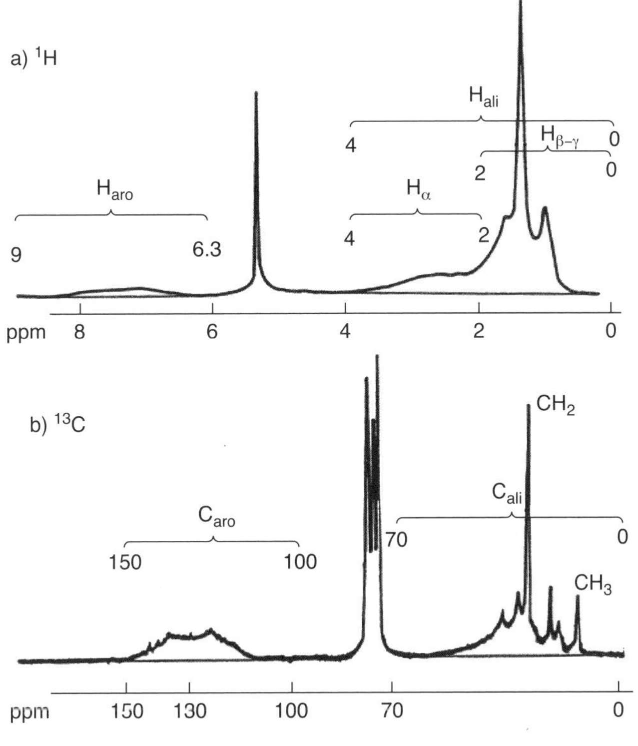

Figure 4.2

(a) ^{1}H and (b) ^{13}C NMR Spectra Obtained for an Asphaltenic Fraction.

4.2.2.3 Fourier Transform Infrared Spectroscopy (FTIR)

Infrared spectroscopy has provided important information concerning chemical functions present in asphaltenes or other petroleum compounds. Carboxylic, phenolic, ketone and ester functions have been highlighted [Moschopedis and Speight, 1976; Ritchie *et al.*, 1979; Speight and Moschopedis, 1981; Algelt and Boduszynski, 1994]. For heavy products, aromaticity, aliphaticity, length and ramification of alkyl chains can be determined [Boukir *et al.*, 1998; Nalwaya *et al.*, 1999; Buenrostros-Gonzales, 2001]. NMR and FTIR spectroscopies are complementary techniques for the chemical description of petroleum fractions.

4.2.2.4 XPS (X-ray Photo-Electron Spectroscopy) – EXAFS (Extended X-ray Fine Structure) – XANES (X-ray Absorption Near Edge Spectroscopy)

A significant issue of chemical characterization is the environmental determination of heteroelements such as sulfur, nitrogen, nickel and vanadium. XANES is a X-ray spectroscopy which provides data on the electronic structure of probed chemical elements (S, N, Ni or V). EXAFS spectroscopy is dedicated to the investigation of local atomic structure in close vicinity to the probed elements (interatomic distances of neighbors, and number of these neighbors). XPS is a classical surface analysis method but can also provide information on the state of oxidation of each element detected in the sample. The application of XANES to asphaltenes has shown different chemical forms for sulfur, such as thiophenes, sulfides (alkyls and di-sulfides) and oxidized species (sulfoxides and sulfones) [Sarret *et al.,* 1999; Gorbaty *et al.*, 2001]. For nitrogen compounds, pyridinic and pyrrolic structures have been detected [Mitra-Kirtley *et al.,* 1993]. Nickel and vanadium environments have also been investigated *via* EXAFS and XANES techniques [Goulon *et al.*, 1984]; they were successful in demonstrating the presence of porphyrin and non-porphyrin compounds [Biggs *et al.,* 1985a and b; Fish *et al.,* 1984]. Nevertheless, the structure of non-porphyrinic fractions still remains rather unclear.

As a conclusion to this brief review devoted to chemical characterization, we must refer to studies which have been carried out to propose structural models of average asphaltene or resin molecules [Bunger and Li, 1981; Strausz *et al.*, 1992]. Figure 4.3 shows one example of this kind of molecular construction. Complex software has been developed for this purpose [Kiet *et al.*, 1978; Oka *et al.*, 1977; Kowalewski *et al.*, 1996; Murgich, 2003]. The limitation of this approach is our ability to build a realistic average molecule with several simple blocks, e.g. including polynuclear aromatics or naphthenic rings. The proposed molecular structures involve large uncertainties because only average values are used in their construction. It has been suggested that a minimum of six model structures are necessary for a reasonable molecular representation [Mc. Caffrey *et al.,* 2003]. Nevertheless, a relatively good consensus exists for an average description of asphaltene entities *via* two models named after their overall pattern: (i) the "continental" type, in which asphaltenes are represented as large central aromatic regions (Figure 4.4a) and (ii) the "archipelago" type, in which asphaltenes contain smaller aromatic regions linked by alkyl chains (Figure 4.4b) [Yen *et al.*, 1961; Murgich *et al.*, 1996; Yen, 1998; Andersen *et al.*, 2001b; Yarranton *et al.*, 2002; Murgich, 2003].

4.2.3 Colloidal Characterization

Complementary to chemical characterization, the colloidal description of asphaltenes has been extensively investigated. Asphaltenes in solution with a good solvent, e.g. toluene or benzene, or in their natural medium, and surrounded by maltene molecules, can be considered as a system exhibiting heterogeneities. Sizes of these heterogeneities can vary from a few nanometers to a few microns. The measurement of asphaltene molecular weight has been extensively investigated. Significant variations in molecular weight measurements have been reported in the literature [Tissot, 1981; Speight *et al.*, 1985]. Several reasons may explain these data:

- Asphaltene polydispersity: asphaltene solutions form a heterogeneous mixture of highly polydispersed molecules in terms of size and chemical composition [Szewczyk *et al.*, 1996; Groenzin and Mullins, 2000; Sheu, 2002]. For such polydispersed systems, several average molecular weights can be defined [Heimenz, 1986], i.e. the number average molecular weight (M_n), and the weight average molecular weight (M_w). The techniques used for average molecular weights are in fact sensitive to the solute.
- Molecular associations: asphaltene molecules may associate to form aggregates of varying sizes. The aggregation state of the asphaltenes is strongly dependent on the operating conditions used for molecular weight determination, such as asphaltene

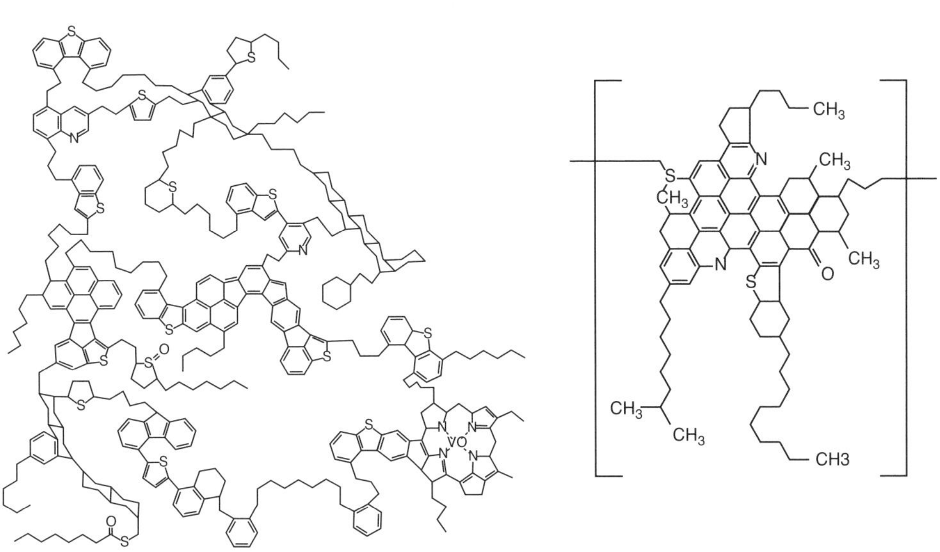

Figure 4.3

Different Asphaltene Models from (a) Venezuelan Crude Oil [Bunger and Li, 1981] and (b) Athabasca Asphaltene [Strausz *et al.*, 1992].

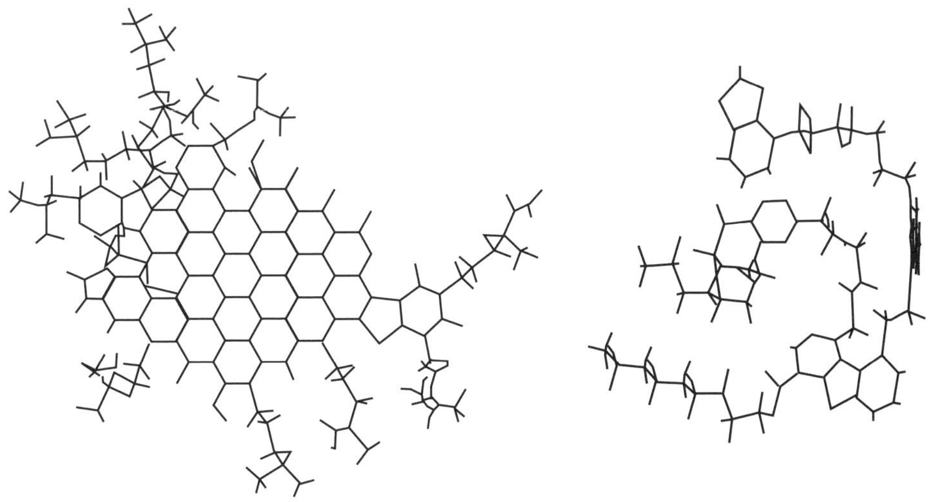

Figure 4.4

Different Asphaltene Models: (a) Continental Type and (b) Archipelago Type [from Murgish, 2003].

concentration, and temperature and nature of the solvent. The term *apparent* average molecular weight is usually more appropriate. Similar to micelle formation behavior, which is well-known for surfactant molecules in solution, for very low concentrations, it has been suggested that nearly complete dissociation of the aggregates occurs. The critical micelle concentration below which asphaltenes are fully-dissociated is likely to be obtained for a 0.1% concentration (w/w) [Andersen and Birdi, 1991; Sheu *et al.*, 1991; Sheu *et al.*, 1992a; Evans and Wennerström, 1998]. The formation of micelles, and in particular the forces involved at the start of the aggregation process, is still not well understood. However, some recent work suggests that self-association of asphaltenes follows a step-wise mechanism, independent of a critical micellar concentration [Murgich *et al.*, 2002; Merino-Garcia and Andersen, 2003, 2004a, 2004b]. This process corresponds to the formation of molecule aggregates of colloidal size, but not to the formation of actual micelles. In the "continental" configuration (Figure 4.4a), the individual aromatic layers can be stacked and form elementary particles. This was successfully illustrated in the famous Yen's model for asphaltene structure [Yen *et al.*, 1961; Yen, 1988] (Figure 4.5). Interaction forces include van der Waals and Coulomb interactions [Brandt *et al.*, 1995; Maruska *et al.*, 1987; Sheu *et al.*, 1994; Ignasiak *et al.*, 1977]. Repulsive interactions also exist and are considered to be responsible for the three-dimensional structure of asphaltene molecules. In the "archipelago" type, interactions involved in the association mechanisms are similar to those described for the "continental" type. However, several rotations of bridging chains are needed to reach the optimal conformation allowing for the aromatic region interactions. With this type of structure, intramolecular associations are then possible. The aggregate formation is consequently more complex with this "archipelago" model, since there is a combination between stacking, bridged interactions and hydrogen bonds [Murgich, 2005].

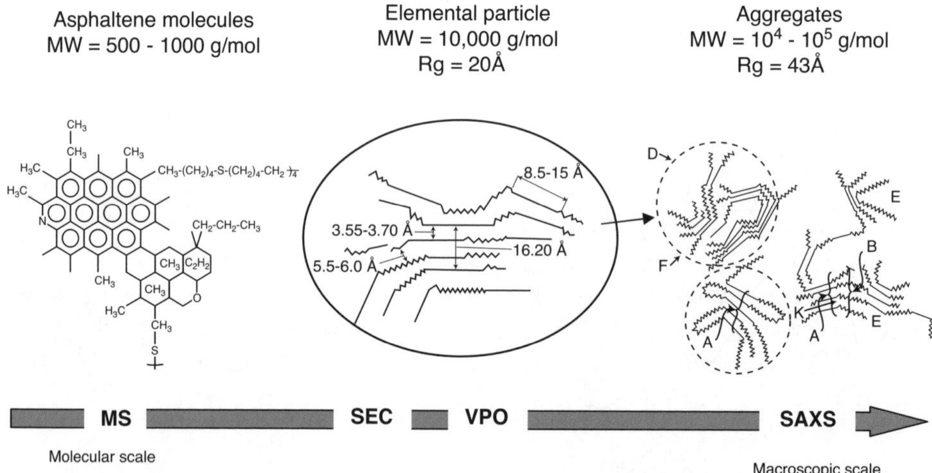

Figure 4.5

MW Measurement Domains of Various Analytical Techniques as a Function of the Aggregation State of Asphaltenes.

Many techniques have been deployed for macrostructural characterization of asphaltene solutions. Experimental approaches that allow measurement of colligative properties (boiling point modification, freezing point depression, vapor pressure and osmotic pressure) are sensitive to the number average molecular weight of the solute. For example, it has been shown that film balance, ultracentrifugation, viscosity or scattering measurements are more sensitive to high molecular weights as far as weight average molecular weight is concerned [Pfeiffer and Saals, 1940; Winniford, 1963; Ray *et al.*, 1957; Reerink *et al.*, 1973; Swanson, 1942].

We focus our review on several techniques which have been widely used for the colloidal characterization of heavy fractions, such as Vapor Pressure Osmometry (VPO), Size Exclusion Chromatography (SEC), Mass Spectrometry, Small Angle X-ray and Neutron Scattering (SAXS and SANS), and finally Nuclear Magnetic Resonance (NMR). We will also consider results regarding the rheological behavior of heavy petroleum fractions and asphaltene suspensions. However, it should be noted that the asphaltenic structures and colloidal aggregates formed by asphaltenes are different in each case, since the techniques are based on different theories and are performed according to different experimental conditions. [Speight *et al.*, 1985] (Figure 4.6, Table 4.1).

4.2.3.1 Vapor Pressure Osmometry (VPO)

This technique measures the *thermoelectric effect* of the *vapor tension decrease* of a *solution* compared to a pure solvent. This approach allows determination of an average molecular mass in number (M_n). This M_n measurement is strongly influenced by the association state of asphaltenes, which in turn depends on the solvent polarity, asphaltenic concentration and temperature conditions.

4.2.3.2 Size Exclusion Chromatography (SEC)

Size exclusion chromatography is designed to separate molecules according to their hydrodynamic volume. SEC is routinely used to describe mass distributions of heavy products, although the provided masses are not absolute [Kodera *et al.*, 2000; Bartholdy *et al.*, 2001; Guibard *et al.*, 2004].

However, the main limitations of this technique include (i) lack of a reliable standard: polystyrene is used for calibration and exhibits a chemical structure that is highly-unrelated to asphaltene, (ii) the tendency of asphaltenes to adsorb onto the chromatography phase, leading to underestimation of the molecular weight and (iii) the aggregation of asphaltenes depends on operating conditions, leading to overestimation of the molecular weight [Domin *et al.*, 1999; Merdrignac *et al.*, 2004a].

4.2.3.3 Mass Spectrometry (MS)

When applied to heavy products – and specifically to asphaltenes – mass spectroscopy must be designed with adequate ionization sources and analyzers. However, this method is limited due to the polydispersity, aggregation and low volatility of such material. Important analytical issues such as the ionization yield, possibility of multiple charge formation, and

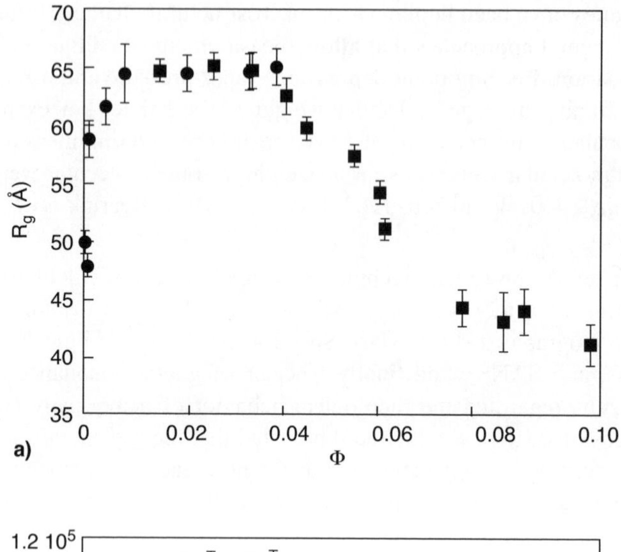

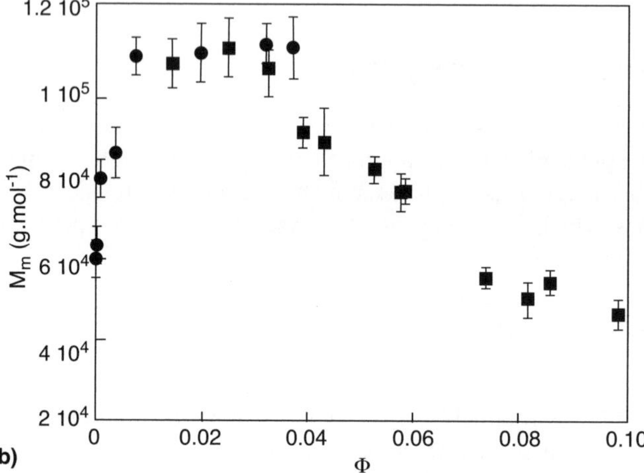

Figure 4.6

(a) Radius of Giration and (b) Average Molecular Weight Mw of Safanyia Vacuum Residue Asphaltenes in Toluene Solution as a Function of Asphaltene Concentration [from Fenistein, 1998].

Table 4.1 Asphaltene Molecular Weights Measured by Various Analytical Methods.

Analytical Method	Average Molecular Weight (g/mol)	
VPO	≈ 1,000-8,000	Mn
Ebulliometry	≈ 2,500-4,000	Mn
Cryometry	≈ 600-6,000	Mn
GPC	≈ 1,000-1,700	Mn
	≈ 6,000-13,000	Mw
MS	≈ 350-2,000	Mw
SAXS – SANS	≈ 50,000-1.10^6	Mw

presence or absence of fragmentation must be accounted for. The main challenge is the difficulty in selecting appropriate experimental conditions to release all the asphaltene molecules which are representative of the bulk sample composition. Three systems that seem to be best-suited for heavy petroleum fractions have been studied and described in the literature [Dutta and Harayama, 2001; Cunico *et al.*, 2002; Tanaka *et al.*, 2002; Strausz *et al.*, 2002; Robins and Limbach, 2003; Acevedo *et al.,* 2005]: (i) Matrix Laser Desorption Ionization/Time-Of-Flight (MALDI-TOF) (adapted for polar and aromatic compounds); (ii) Atmospheric Pressure Chemical Ionization/Quadripole/Time-Of-Flight mass spectrometry (APCI-Q-TOF) (adapted for weakly polar compounds); and (iii) Atmospheric Pressure Photoionization/Quadripole/Time-Of-Flight mass spectrometry (APPI-Q-TOF) (adapted for apolar compounds).

In order to decrease asphaltene polydispersity, these techniques can be coupled offline or online with SEC. Data obtained with the various ionization techniques has shown that they converge towards similar molecular weights in good agreement. Asphaltene mass profiles ranged from 350 Da and m/z = 2,000 Da with an average mass between 600 and 700 Da (maximum of SEC peak) depending on the technique [Merdrignac *et al.*, 2004b]. However, much larger mass distributions have been measured on the same types of samples by other authors [Herod *et al.*, 2000]. This may be the result of the formation of dimers, trimers or even polymer structures in the ionization system [Rodgers *et al.*, 2005].

Even when being fractionated, asphaltene subfractions are still a very complex mixture. Co-elutions and unresolved peaks are predominant in such distributions when analyzed with these types of instruments. To obtain a higher degree of information, authors have used Fourier Transform Ion Cyclotron Resonance MS (FTICR MS), in particular using an electrospray ionization source. With this very powerful technique of extremely high resolution, more than 10,000 compositionally-distinct compounds can be resolved enabling assignment of chemical formulas ($C_CH_HN_NO_OS_S$). In addition, various compound families like class ($N_NO_OS_S$), type (Z in $C_CH_{2c+Z}N_NO_OS_S$) and alkylation series may be identified. Information regarding the molecular weight and structural identification of heavy products can be thus achieved [Qian *et al.*, 2001a; Rodgers *et al.*, 2001; Qian *et al.*, 2001b; Hughey *et al.*, 2002; Marshall *et al.*, 2004; Klein *et al.*, 2006 and literature cited therein].

4.2.3.4 Pulsed Field Gradient Spin-Echo ^{1}H NMR (PFG-^{1}H NMR)

The self-diffusion coefficient (D) is an interesting characteristic from which information on the aggregation state and molecular weight of studied species in solution can be obtained [Stilbs, 1987]. PFG- ^{1}H NMR can be applied to complex systems like asphaltenic compounds. Comparatively fast and applicable to dark mixtures, this technique is used to gain information on the geometry of asphaltenic molecules by measuring the diffusion associated with either the asphaltenes or the solvent [Evdokimov *et al.*, 2003; Östlund *et al.*, 2002; Östlund *et al.*, 2003; Östlund *et al.*, 2004a; Östlund *et al.*, 2004b]. Results show that the average diffusion coefficient tends to decrease when the concentration of asphaltenes in solution increases [Johannesson and Halle, 1996; Norinaga *et al.*, 2004]. This may be due to obstruction effects which can be rationalized by a non-spherical structure. It has been proposed that the asphaltenic particles have a rather disc-like shape [Östlund *et al.*, 2001].

4.2.3.5 Small Angle X-ray and Neutron Scattering (SAXS, SANS)

Measurement of the scattering of radiation (X-ray or neutron) at small angles is an interesting technique for investigating the colloidal structure of heavy oil products. These techniques are sensitive to density fluctuations or the occurrence of heterogeneities. The scattered intensity after crossing the sample is measured as a function of the scattering vector Q ($Q = \frac{4\pi}{\lambda}\sin\theta$), where θ is the angle of observation (deviation from the incident direction) and λ is the wavelength of the incident radiation [Espinat *et al.*, 1998]. For asphaltene solutions, two parameters can be obtained from scattering experiments: the weight average molecular weight (M_w) and radii of gyration of the asphaltene macromolecule [Espinat, 1990]. Many studies have tried to define the most reasonable shape of this molecule in agreement with the scattering curves, but it seems difficult to discriminate against various simple models, like sphere, rod or disk-like particles. Additional parameters characterizing the polydispersity of the asphaltenes can be introduced in the model [Herzog *et al.*, 1988; Ravey *et al.*, 1988; Overfield *et al.*, 1989; Sheu and Storm, 1995; Thiyagaran *et al.*, 1995; C. Bardon *et al.*, 1996; Barre *et al.*, 1997; Espinat *et al.*, 1998; Kilpatrick and Gawrys, 2005]. Since asphaltenes can be considered as solvated aggregates, scattering data were fitted using scattering models of fractal objects of spherical particles [Xu *et al.*, 1995; Roux *et al.*, 2001; Gawrys *et al.*, 2002]. X-ray radiation is a unique technique allowing the investigation of natural products such as *heavy crude oils* and vacuum residues. Important results on the colloidal behavior of asphaltenes in solution in good solvents obtained from SAXS and SANS experiments are shown on Figure 4.6. Weight average molecular weight (M_w) and radius of gyration (R_g) are plotted as a function of asphaltene concentration [Espinat *et al.*, 1998; Fenistein, 1998]. Three domains can be highlighted:

- For very low concentrations, a decrease of M_w and R_g is observed, but it should be emphasized that for these asphaltene concentrations, scattering intensity remains very low and M_w and R_g determinations are difficult.
- Up to 4% v/v, M_w and R_g are constant as asphaltene concentration increases.
- Above 4% v/v, a progressive decrease of M_w and R_g takes place.

This scattering behavior has been explained according to polymer solutions theory [De Gennes, 1979]. For low concentrations, asphaltene aggregates are far from each other, corresponding to the so-called diluted regime (Figure 4.7). As the concentration increases, aggregates become close to each other, either by steric repulsions or overlapping of the asphaltene entities. In this last regime (semi-dilute regime), the scattering intensity measured does not depend on M_w and R_g, but is characteristic of a length much smaller than the average size. The threshold concentration between the two regimes, noted C* (see Figure 4.7), is strongly dependent on the asphaltene origin and the dispersion solvent. In Figure 4.6, concerning Safaniya vacuum residue asphaltenes in toluene solution, C* is close to 4% (w/w), but literature reports have mentioned values between 2% and 8%, according to the solvent used [Roux *et al.*, 2001; Sheu *et al.*, 1992b]. We can see, in Figure 4.6, that M_w is close to 110,000 $g.mol^{-1}$ and R_g close to 65 angstroms. According to the literature, taking into account the differences between experimental conditions, the measured asphaltene radii of gyration may vary from 30 to 200 angstroms, and M_w from 50,000 to 1.10^6 $g.mol^{-1}$.

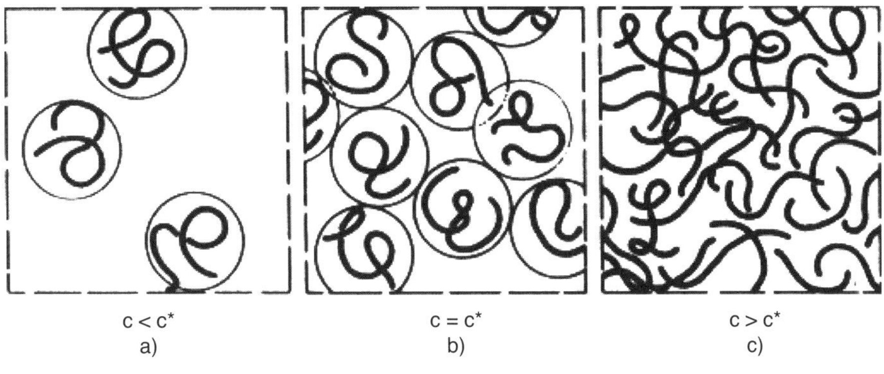

Figure 4.7

(a) Illustration of (a) Dilute, (b) Onset of Overlap and (c) Semi-Dilute Solutions [from De Gennes, 1979].

In addition to the nature of the dispersive medium and the temperature, other parameters have also been found to exert a strong influence on M_w and R_g. The aggregate sizes decrease with the polarity of the solvent used [Ravey and Espinat, 1990; Fenistein *et al.*, 2000; Fenistein and Barré, 2001], whereas progressive decreases in R_g and M_w of the asphaltene entities are involved when temperature is increased [Espinat, 1991; Espinat *et al.*, 2004; Tanaka *et al.*, 2003].

To progress in the understanding of such a system, complementary approaches have been achieved by reducing polydispersity using ultracentrifugation techniques. Fractions of different aggregate sizes have been obtained. Analyses have shown an increase in metal content as the molecular weight increases, whereas a rather simple relationship between M_w and R_g has been proposed ($M_w = 27\ R_g^2$) [Fenistein, 1998].

Small angle X-ray scattering has been also used to characterize asphaltenes in real systems. Pure Safanyia vacuum residue was studied (Figure 4.8) in comparison with its maltenes fraction (soluble part of the crude oil in *n*-heptane). Large heterogeneities (or density fluctuations) exist in the vacuum residue macrostructure [Espinat *et al.*, 1998]. Radii of gyration were measured on artificial vacuum residues, created by adding varying amounts of asphaltenes into the recovered maltenes. This was an elegant way to vary the asphaltene concentration in a natural system [Hénaut *et al.,* 2001]. Rheological measurements have also been carried out on these artificial materials. Figure 4.9 displays the results of the two techniques. We noted similar behavior of the gyration radius as compared to one of the asphaltenes in toluene solution (see Figure 4.6). It seems that the macrostructure of asphaltenes in their natural medium is fairly close to the one described in a good solvent (toluene). All these data can improve the precision of the aggregation model of asphaltenes in bitumen, as proposed by Pfeiffer and Saal (1940).

4.2.3.6 Heavy Oils: Rheological Properties

As mentioned above, the measurement of rheological properties can help to understand the behavior of asphaltenes in solution. Rheology is defined as the study of deformation of

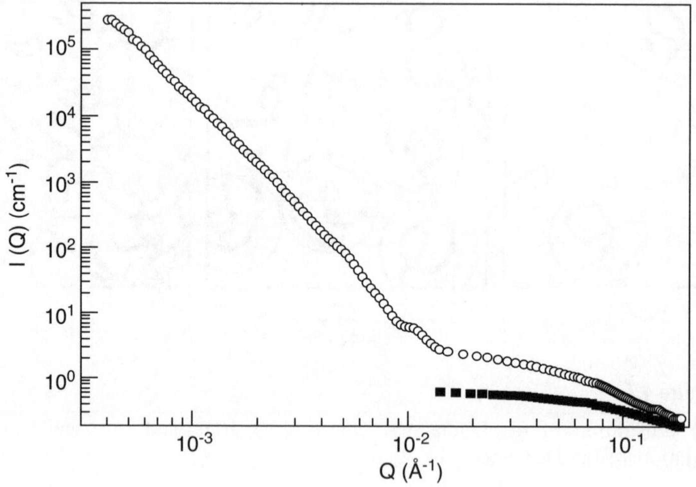

Figure 4.8

Small Angle X-ray Scattering of (o) Safanyia Vacuum Residue and (■) Its Maltenes [from Espinat *et al.*, 1998].

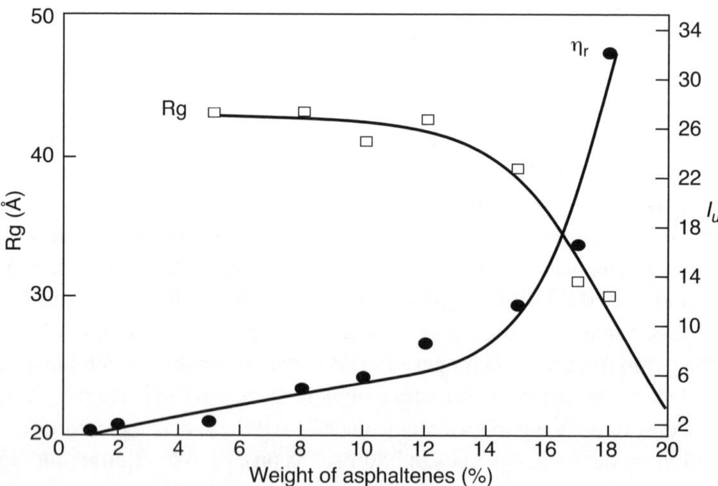

Figure 4.9

Rheological Behavior Compared to SAXS Measurements of Radius of Gyration for Pure Artificial Crude (Asphaltenes Added to Maltenes at Various Concentrations) [from Hénaut *et al.*, 2001].

matter resulting from the application of a force. The three most widely used parameters in rheology are defined below:

- Stress σ: applied force per unit area.
- Rate of deformation $\dot{\gamma}$: a stress applied to a material causes a relative deformation γ. The rate of deformation is defined by $\dot{\gamma} = \frac{d\gamma}{dt}$.
- Viscosity η: the resistance to flow of a material is stated by its viscosity which is defined as the stress divided by the rate of deformation.

The rheological response of a colloidal dispersion strongly depends on its physical and chemical characteristics: strength of particle interactions, volume fraction, size and shape of particles, etc. Rheology is therefore helpful to study formulation parameters. The common approach consists in following the variation of viscosity as a function of particle type and particle concentration. These measurements can be done at a large range of temperatures and pressures. This possibility also makes the rheological technique very advantageous and efficient in the characterization of a product under various external conditions.

Three different examples concerning heavy oil characterization are provided below. They respectively account for the type of solvent used in de-asphalting and influence of resins, thermal behavior of heavy oils, and pressure-viscosity response of heavy oils.

- Type of solvent used in de-asphalting and influence of resins:

As previously discussed (4.2.1), the asphaltene structure varies with the type of de-asphalting solvent. The precipitation of asphaltenes is commonly performed by using *n*-alkanes. It is well known that if a longer *n*-alkane is employed, fewer asphaltenes are obtained. This means that the asphaltenes precipitated with a long *n*-alkane can be considered as the purest ones [Coustet, 2003; Hénaut *et al.*, 2003]. Compared to a long *n*-alkane, a short *n*-alkane precipitates pure asphaltenes and co-precipitated resins. In order to evaluate the difference between these two kinds of asphaltenes and to illustrate the role of resins, rheological measurements have been performed on solutions prepared with pentane (short alkane) asphaltenes and heptane (long *n*-alkane) asphaltenes. Both asphaltenes are derived from the same Venezuelan crude oil. They were added to xylene in various concentrations. The viscosity of each sample was determined at a temperature of 20°C with the rheometer AR2000 using a double concentric cylinder. This was plotted as a function of asphaltene concentration in Figure 4.10. The experimental curve confirms that above a critical concentration C*, the viscosity of the asphaltene solutions increases strongly. This phenomenon has been attributed to the overlapping of asphaltene entities (Figure 4.7). The most noticeable result is that the "heptane asphaltenes" overlap at a lower concentration than the "pentane asphaltenes" in xylene. This suggests that co-precipitated resins prevent pure asphaltenes from overlapping. This observation agrees with the peptization ability of resins and its structural effect. The experiments using "heptane asphaltenes" and "pentane asphaltenes" were repeated with maltenes instead of xylene. The maltenes that were used come from the de-asphalting of the oil made with pentane. The viscosities that were measured are shown in Figure 4.11. As previously performed, they were plotted as a function of asphaltene concentrations. Due to the resins contained in the maltenes, the "heptane asphaltenes" can be peptisized like the "pentane asphaltenes" containing co-precipitated resins. When added in maltenes, these two kinds of asphaltenes therefore have the same critical concentration C*. This illustrates the existing interactions between asphaltenes and resins. It also confirms that the difference between the asphaltenes in xylene is due to a structural effect and not to a dilution effect. If dilution effect was a controlling factor, the curves obtained for maltene and xylene solutions would not have matched.

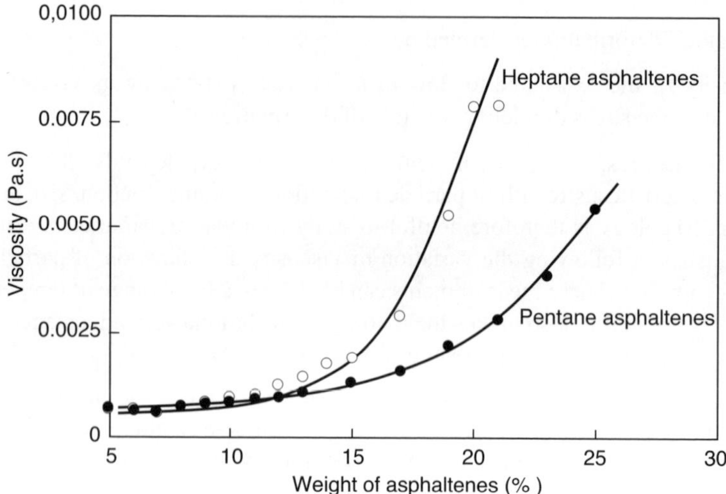

Figure 4.10

Viscosity of Asphaltene Solutions: Case of Xylene, T = 20°C.

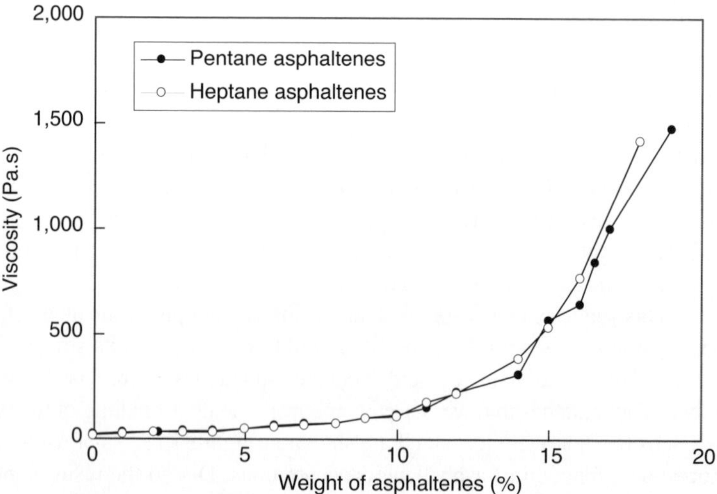

Figure 4.11

Viscosity of Asphaltene Solutions: Case of Maltenes, T = 20°C.

- Thermal behavior of heavy oils:

As previously mentioned, the forces involved in asphaltene interactions include van der Waals, Coulomb interactions and repulsive forces. An increase in temperature provides kinetic energy to the molecules; this in turn generates thermal expansion and consequently a reduction of the effect of the forces between molecules. Viscosity is particularly sensitive to temperature. For a conventional crude oil, this dependence is accurately described by the Arrhenius law [Civan, 2006].

$$\mathrm{Ln}\eta = \frac{\mathrm{Ea}}{\mathrm{RT}} + \mathrm{k}$$

Ea: activation energy
R: universal gas constant
T: absolute temperature
η: viscosity at T
k: constant

This correlation was applied to the viscosity of a Venezuelan crude oil between 3 and 80°C (Figure 4.12). Two main features can be extracted from the heavy oil response:

- Existence of two activation energies: Ea = 116 kJ/mol for T < 40°C and Ea = 83 kJ/mol for T > 40°C. This phenomenon can be explained by stronger interaction forces below 40°C.
- Particularly high value of Ea.

Both features have been observed for all the studied *heavy crude oils* [Argillier *et al.*, 2002]. The experiments that have been carried out confirm the strong impact of the thermal dependence of heavy oil viscosity. This crucial influence has to be taken into account when assessing the basic process of thermal operations commonly used to produce heavy oils.

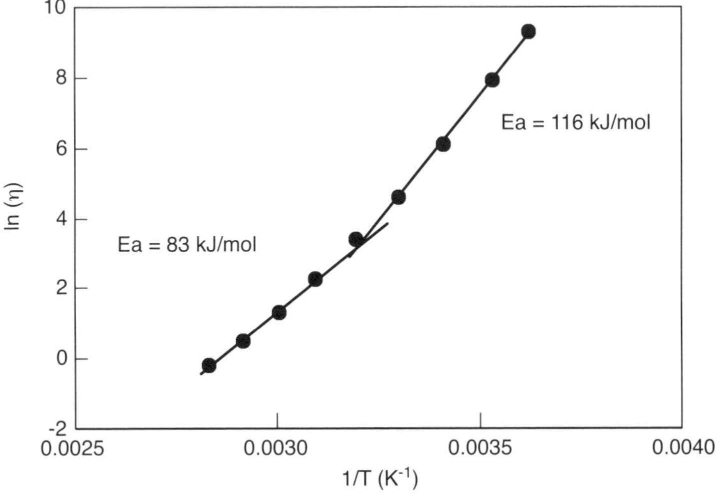

Figure 4.12

Arrhenius Plot of *Heavy Crude Oil*: Existence of Two Activation Energies.

- Pressure-viscosity response of heavy oils:

Based on free volume considerations, it can be shown that as a rule, the viscosity of a fluid increases with increasing pressure [Schmelzer *et al.*, 2005]. This dependence was assumed to be purely exponential by Dowson and Higginson (1966), who introduced a simple pressure-viscosity coefficient α:

$$\alpha = \frac{1}{\eta}\frac{d\eta}{dP}$$

Although the majority of heavy oils are located at shallow depth with depth limit pressure, it seems interesting to evaluate their piezoviscosity to better predict pressure loss in production wells. The pressure viscosity results shown in Figure 4.13 were obtained on a rheometer equipped with a pressure cell using magnetic coupling [Abivin *et al.,* 2008 & 2009]. They were carried out on a Venezuelan heavy oil at a temperature of 20°C with a pressure ranging from atmospheric pressure to 10 MPa. The experimental curve shows that the viscosity follows an exponential law with $\alpha = 5.3\ 10^{-8}\ Pa^{-1}$. These measurements were also carried out on the corresponding propane de-asphalted oil and led to a value of $\alpha = 3.6\ 10^{-8}\ Pa^{-1}$. These results again show how asphaltenes influence heavy oil properties.

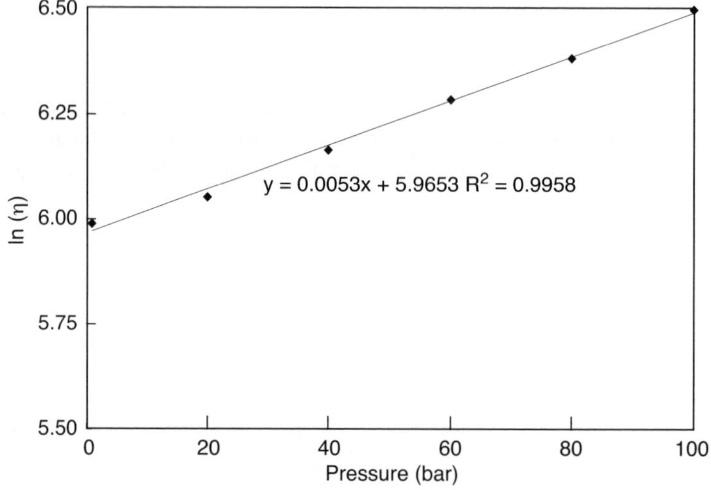

Figure 4.13

Pressure-Viscosity Response of a Venezuelan Heavy Oil at 20°C.

4.3 CONCLUSION

High boiling point petroleum fractions (350°C$^+$) are known to be enriched in the most polar compounds such as resins and asphaltenes. Asphaltenes contain molecules of variable aromaticity with varying contents of heteroatoms, metals and functional groups. An asphaltene fraction is a complex mixture with polydispersed molecules in terms of size and chemical composition. Such structures cannot be represented by a single model molecule. Several models have been proposed in the literature to describe these structures (e.g. "continental" and "archipelago" types). From these highly functionalized mixtures, specific properties emerge like self-association/aggregation of asphaltenic structures, requiring tedious characterization. Average structural information can be obtained. However, it cannot be representative of all the chemical and structural variety that such matrices may contain. Property measurements of asphaltenes (e.g. MW) strongly depend on the analytical technique and operating conditions used (concentration, temperature, solvent and pressure). The asphaltenic structures are not measured in the same aggregation state. This is why the aggregate size of asphaltenes may be different and may vary by several magnitude orders depending on the method used. The analytical scheme generally applied to characterize heavy petroleum fractions may be divided into two parts: (i) preparative fractionation steps to simplify the initial mixture (too complex for a direct detailed characterization), and (ii) chemical or colloidal characterization (depending on the desired information) applied on the fractionated sample. An illustration of the analytical procedure and the information obtained is summarized in Table 4.2.

Table 4.2 Summary of Analytical Procedures Used for Characterization of Heavy Oil Fractions.

Fractionation Methods	**Chemical Characterization**		**Colloidal Characterization**	
	Techniques	**Information**	**Techniques**	**Information**
• Boiling point • Chemical classes • Degradation • Solubility	• Pyrolysis – Gas Chromatography – Mass Spectrometry (Py-GC/MS)	General composition	• Vapor Pressure Osmometry (VPO)	MW
	• High Pressure Liquid Chromatography (HPLC)	Function groups	• Size Exclusion Chromatography (SEC)	Hydrodynamic volume distribution
	• ^{13}C Nuclear Magnetic Resonance (NMR)	Structural motives	• Mass Spectrometry (MS)	MW
	•InfraRed Spectroscopy (IR)	Function groups	• Small Angle X-ray and Neutron Scattering (SAXS – SANS)	$R_G \rightarrow$ MW
	• X-ray Photoelectron Spectroscopy – Extended X-ray Absorption Fine Structure – X-ray Absorption Near Edge Spectroscopy (XPS -EXAFS – XANES)	Environment and chemical functionalities	• Pulsed Field Gradient Spin-Echo NMR (PFG SE- ^{1}H NMR)	D $\rightarrow$ hydrodynamic radius
			• Rheology	C* thermal dependence Ea pressure viscosity coefficient α

REFERENCES

Abivin P, Argillier J-F, Hénaut I, Moan M (2008) Viscosity Behavior of Foamy Oil: Experimental Study and Modeling, *Petroleum Science and Technology*, **26**, 13, 1545-1558.

Abivin P, Hénaut I, Argillier J-F, Moan M (2009) Rheological Behavior of Foamy Oil, *Energy & Fuels*, **23**, 3, pp 1316-1322.

Acevedo S, Gutierrez Luis B, Negrin G, Pereira JC, Mendez B, Delolme F, Dessalces G, Broseta D (2005) Molecular Weight of Petroleum Asphaltenes: A Comparison Between Mass Spectrometry and Vapor Pressure Osmometry, *Energy & Fuels*, **19**, pp 1548-1560.

Akbarzadeh K, Ayatollahi Sh, Nasrifar Kh, Yarranton HW, Moshfeghian M (2002) Equations Lead to Asphaltene Deposition Predictions, *Oil and Gas Journal*, **51**.

Altgelt KH, Boduszynski M (1994) In *Composition and Analysis of Heavy Petroleum Fractions.* Ed Dekker, New York.

Al-Zaid K, Khan ZH, Hauser A, Al-Rabiah H (1998) Composition of High Boiling Petroleum Distillates of Kuwait Crude Oils, *Fuel*, **77**, 5, pp 453-458.

Ancheyta J, Centeno G, Trejo F, Marroquin G, Garcia JA, Tenerio E, Torres A (2002) Extraction and Characterization of Asphaltenes From Different Crude Oils and Solvents, *Energy & Fuels*, **16**, pp 1121-1127.

Andersen SI, Birdi KS (1991) Aggregation of Asphaltenes as Determined by Calorimetry, *J. of Colloid and Interface Science*, 1, **42**, 2, pp 497-502.

Andersen SI (1997) Separation of Asphaltenes by Polarity using Liquid-Liquid Extraction, *Pet. Sci. Technol.*, **15**, 1&2, pp 185-198.

Andersen SI, Keul A, Stenby EH (1997) Variation in Composition of Subfractions of Petroleum Asphaltenes, *Pet. Sci. Technol.*, **17**, 7&8, pp 611-645.

Andersen SI, Lira-Galeana C, Stenby EH (2001a) On the Mass Balance of Asphaltene Precipitation, *Pet. Sci. Technol.*, **19**, 3&4, pp 457-467.

Andersen SI, Del Rio-Garcia JM, Khvostitchenko D, Shakir S, Lira-Galeana C (2001b) Interaction and Solubilization of Water by Petroleum Asphaltenes in Organic Solution, *Langmuir*, **17**, pp 307.

Argillier J-F, Coustet C, Hénaut I (2002) *SPE International Thermal Operations and Heavy Oil Symposium*, Calgary, Canada, 4-7 November, 2002.

Aske N, Kallevik H, Sjöblom J (2001) Determination of Saturate, Aromatic, Resin, and Asphaltenic (SARA) Components in Crude Oils by Means of Infrared and Near-Infrared Spectroscopy, *Energy & Fuels*, **15**, pp 1304-1312.

ASTM D893, *Annual Book of ASTM Standards*, American Society for Testing and Materials.

ASTM D2006, *Annual Book of ASTM Standards*, American Society for Testing and Materials.

ASTM D2007, *Annual Book of ASTM Standards*, American Society for Testing and Materials.

ASTM D2887, *Annual Book of ASTM Standards*, American Society for Testing and Materials.

ASTM D2892, *Annual Book of ASTM Standards*, American Society for Testing and Materials.

ASTM D3279, *Annual Book of ASTM Standards*, American Society for Testing and Materials.

ASTM D4124, *Annual Book of ASTM Standards*, American Society for Testing and Materials.

ASTM D5236, *Annual Book of ASTM Standards*, American Society for Testing and Materials.

ASTM D6352, *Annual Book of ASTM Standards*, American Society for Testing and Materials.

Bacaud R, Rouleau L, Cebolla VL, Membrado L, Vela J (1998) Evaluation of Hydroconverted Residues. Rationalization of Analytical Data Through Hydrogen Transfer Balance, *Catal. Today*, **43**, 3&4, pp 171-186.

Bardon C, Barré L, Espinat D, Guille V, Li MH (1996) The Colloidal Structure of Crude Oils and Suspensions of Asphaltenes and Resins, *Fuel Science and Technology Int.*, **14**, pp 203-242.

Barré L, Espinat D, Rosenberg E, Scarsella M (1997) Colloidal Structure of Heavy Crudes and Asphaltene Solutions, *Rev. Inst. Fr. Pet.*, **52**, pp 161-175.

Bartholdy J, Lauridsen R, Mejlholm M, Andersen SI (2001) Effect of Hydrotreatment on Product Sludge Stability, *Energy & Fuels*, **15**, pp 1059-1062.

Biggs WR, Fetzer JC, Brown RJ, Reynolds JG (1985a) Characterization of Vanadium Compounds in Selected Crudes I. Porphyrin and Non-Porphyrin Separation, *Liquid Fuel Technology,* **3**(4), pp 397.

Biggs WR, Fetzer JC, Brown RJ, Reynolds JG (1985b) Characterization of Vanadium Compounds in Selected Crudes II. Electron Paramagnetic Resonance Studies of the First Coordination Spheres in the Porphyrin and Non-Porphyrin Fractions, *Liquid Fuel Technology,* **3**(4), pp 423.

Boukir A, Guiliano M, Doumenq P, El Hallaoui A, Mille G (1998) Structural Characterization of Crude oil Asphaltenes by Infrared Spectroscopy (FTIR). Application to Photo-Oxidation, *Comptes rendus de l'Académie des Sciences*, série II, fascicule C-chimie 1, **10**, pp 597-602.

Brandt HCA, Hendriks EM, Michels MAJ, Visser F (1995) Thermodynamic Modeling of Asphaltene Stacking, *J. Phys. Chem.*, **99**, pp 10430.

Buenrostro-Gonzales E, Andersen SI, Garcia-Martinez JA, Lira-Galeana C (2002) Solubility/Molecular Structure Relationships of Asphaltenes in Polar and Nonpolar Media, *Energy & Fuels*, **16**, pp 732-741.

Bunger JW, Li NC (1981) In *Chemistry of Asphaltenes*, Advances in Chemistry Series 195, American Chemical Society.

Cebolla VL, Membrado L, Matt M, Galvez M, Domingo MP (2003) Thin-Layer Chromatography for Hydrocarbon Characterization in Petroleum Middle Distillates, In *Analytical Advances for Hydrocarbon Research*, Ed. CS Hsu, Plenum Publishers, New York, Ch 5, pp 95-112.

Civan F (2006) Viscosity-Temperature Correlation for Crude Oils using an Arrhenius-Type Asymptotic Exponential Function, *Petroleum Science and Technology*, **24**, 5-6, pp 699-706.

Coustet-Pierre C (2003) PhD thesis, Université de Bretagne Occidentale, France.

Cunico RL, Sheu EY, Mullins OC (2002) *Proc. of Int. Conf. on Heavy Org. Dep.*, Puerto Vallarta, Mexico.

Dahan L, Thiebaut D, Bertoncini F, Espinat D, Quignard A (2004) *Preprints, ACS,* Div. Fuel Chem., **49**, 1, pp 18.

De Gennes PG (1979) In *Scaling Concepts in Polymer Physics*, Cornell University Press.

Domin M, Herrod A, Kandiyoti R, Larsen JW, Lazaro MJ, Li S, Rahimi P (1999) A Comparative Study of Bitumen Molecular-Weight Distributions, *Energy & Fuels*, **13**, pp 552-557.

Dorbon M, Schmitter JM, Arpino P, Guiochon G (1982) Extraction and Characterization of Carbazoles and Lactams from Petroleum, *J. Chromatogr.*, **246**, pp 255-269.

Dowson D, Higginson GR (1966) Fundamentals of Roller and Gear Lubrication, *Elastohydrodynamic Lubrication*, Pergamon Press Oxford.

Durand JP, Petroff N (1984) *Collect. Colloq. Semin. (Inst. Fr. Pet.)*, **40**, pp 200.

Dutta TK, Harayama S (2001) Time-of-Flight Mass Spectrometric Analysis of High-Molecular-Weight Alkanes in Crude Oil by Silver Nitrate Chemical Ionization after Laser Desorption, *Anal. Chem.*, **73**, 5, pp 864-869.

Espinat D (1990) Application of Light, X-ray and Neutron Diffusion Techniques to the Study of Colloidal Systems. 1. Theoretical Description of 3 Techniques, *Rev. Inst. Fr. Pet.*, **45**, 6, pp 775-820.

Espinat D (1991a) Application of Light, X-ray and Neutron Diffusion Techniques to the Study of Colloidal Systems. 2. Research on Different Systems – Polymers in Solution in the Solid-State, Micellar Solutions, Fractal Systems, *Rev. Inst. Fr. Pet*, **46**, 5, pp 595-635.

Espinat D (1991b) Application of Light, X-ray and Neutron Diffusion Techniques to the Study of Colloidal Systems. 3. Microemulsions, Solid Materials, Liquids and Miscellaneous Systems with Industrial Uses, *Rev. Inst. Fr. Pet*, **46**, 6, pp 759-801.

Espinat D, Fenistein D, Barré L, Frot D, Briolant Y (2004) Effects of Temperature and Pressure on Asphaltenes Agglomeration in Toluene. A light, X-ray, and Neutron Scattering Investigation, *Energy & Fuels*, **18**, pp 1243-1249.

Evans DF (1998) Wennerström. In *The Colloidal Domain*, 2nd, Ed. Wiley, New York.

Evdokimov IN, Eliseev NY, Akhmetov BR (2003) Initial Stages of Asphaltene Aggregation in Dilute Crude Oil Solutions: Studies of Viscosity and NMR Relaxation, *Fuel*, **82**, pp 817-823.

Fenistein D (1998) *PhD thesis,* Université Paris VI, France.

Fenistein D, Barré L, Frot D (2000) From Aggregation to Flocculation of Asphaltenes, a Structural Description by Radiation Scattering Techniques, *Oil & Gas Sci. and Technol*, **55**, pp 123-128.

Fenistein D, Barré L (2001) Experimental Measurement of the Mass Distribution of Petroleum Asphaltene Aggregates using Ultracentrifugation and Small-Angle X-ray Scattering, *Fuel*, **80**, pp 283-287.

Fish RH, Komlenic JJ, Brian K, Wines (1984) Characterization and Comparison of Vanadyl and Nickel Compounds in Heavy Crude Petroleums and Asphaltenes by Reverse-Phase and Size-Exclusion Liquid-Chromatography Graphite-Furnace Atomic-Absorption Spectrometry, *Anal. Chem.*, **56**, pp 2452-2460.

Gawrys KL, Spiecker PM, Kilpatrick PK (2002) *Preprints, ACS,* Div Pet. Chem., **47**, 4, 332.

Gorbaty ML, Kelemen SR (2001) Characterization and Reactivity of Organically Bound Sulfur and Nitrogen Fossil Fuels, *Fuel Process. Technol.,* **71**, pp 71-78.

Goulon J, Retournard A, Friant P, Goulon-Ginet C, Berthe C, Muller JF, Poncet JL, Guilard R, Escalier JC, Neff B (1984) Structural Characterization by X-ray Absorption Spectroscopy (EXAFS/XANES) of the Vanadium Chemical Environment in Boscan Asphaltenes, *J. Chem. Soc. Dalton Trans.*, pp 1095.

Green JB, Hoff RJ, Woodward PW, Stevens LL (1984) Separation of Liquid Fossil-Fuels into Acid, Base and Neutral Concentrates. 1. An improved Nonaqueous Ion-Exchange Method, *Fuel*, **63**, pp 1290-1301.

Groenzin H, Mullins OC (2000) Asphaltene Molecular Size and Structure from Many Sources*, Energy & Fuels*, **14**, pp 677.

Guibard I, Merdrignac I, Kressmann S (2004) In *Heavy Hydrocarbon Resources: Characterization, Upgrading and Utilization*, American Chemical Society, Symposium Series 895, Ed. M Nomura, Oxford Univ., Chapter 4, 51.

Heimenz PC (1986) In *Principles of Colloid and Surface Chemistry*, Second Edition, Marcel Dekker Inc., **46**.

Hénaut I, Argillier JF, Pierre C, Moan M (2003) *SPE Offshore Technology Conference*, Houston, Texas, U.S.A., 5-8 May, 2003.

Hénaut I, Barré L, Argillier JF, Brucy F, Bouchard R (2001) *SPE Int. Sym. on Oilfield Chemistry*, pap 65020, Houston, Texas, February 2001.

Herod AA, Lazaro MJ, Domin M, Islas CA, Kandiyoti R (2000) Molecular Mass Distributions and Structural Characterisation of Coal Derived Liquids, *Fuel*, **79**, pp 323-337.

Herzog P, Tchoubar D, Espinat D (1988) Macrostructure of Asphaltene Dispersions by Small-Angle X-ray Scaterring, *Fuel*, **67**, pp 245-250.

Hughey CA, Rodgers RP, Marshall AG (2002) Resolution of 11,000 Compositionally Distinct Components in a Single Electrospray Ionization Fourier Transform Ion Cyclotron Resonance Mass Spectrum of Crude Oil, *Anal. Chem.*, **74**, pp 4145-4149.

Ignasiak T, Strausz OP, Montgomery DS (1977) Oxygen Distribution and Hydrogen-Bonding in Athabasca Asphaltene, *Fuel*, **56**, pp 359-365.

Ignatiadis I, Schmitter JM, Arpino P (1985) Separation and Identification by Gas-Chromatography and Gas Chromatography Mass-Spectrometry of the Nirogen-Compounds from a De-Asphalted Heavy Oil – Evolution of their Distribution after a Catalytic Hydrotreatment *J. Chromatogr.*, **324**, pp 87-111.

Johannesson H, Halle B (1996) Solvent Diffusion in Ordered Macrofluids: A Stochastic Simulation Study of the Obstruction Effect, *J Chem. Phys.*, **104**, 17, pp 6807-6817.

Kiet HH, Malhotra SL, Blanchard LP (1978) Structure Parameter Analyses of Asphalt Fractions by a Modified Mathematical Approach, *Anal. Chem.,* **50**, pp 1212-1218.

Kilpatrick PK, Gawrys KL (June 2005) The Use of Small Angle Neutron Scattering in the Study Asphaltene Agregation, *Preprints The 6th International Conference on Petroleum Phase Behavior & Fouling*, Amsterdam, Netherland.

Kim S, Branthaver JK (1996) Polarities of SHRP Asphalts Calculated from Ion Exchange Chromatography Separation Data, *F. Sci. & Technol. Int'l.*, **14**, pp 365-393.

Klein GC, Rodgers RP, Marshall AG (2006) Identification of Hydrotreatment-Resistant Heteroatomic Species in a Crude Oil Distillation Cut by Electrospray Ionization FT-ICR Mass Spectrometry, *Fuel*, **85**, pp 2071-2080.

Kodera Y, Kondo T, Saito I, Okegawa K (2000) Continuous-Distribution Kinetic Analysis for Asphaltene Hydrocracking, *Energy & Fuels*, **14**, pp 291-296.

Kowalewski I, Vandenbroucke M, Huc AY, Taylor MJ, Faulon JL (1996) Preliminary Results on Molecular Modeling of Asphaltenes using Structure Elucidation Programs in Conjunction with Molecular Simulation Programs, *Energy & Fuels* **10**, pp 97-107.

Matt M, Galvez EM, Cebolla VL, Membrado L, Bacaud R, Pessayre S (2003) Improved Separation and Quantitative Determination of Hydrocarbon Types in Gas Oil by Normal Phase High-Performance TLC with UV and Fluorescence Scanning Densitometry, *J. Sep. Sci.*, **26**, pp 1665-1674.

Marshall AG, Rodgers RP (2004) Petroleomics: The Next Grand Challenge for Chemical Analysis, *Acc. Chem. Res.*, **37**, pp 53-59.

Maruska HP, Rao BML (1987) The Role of Polar Species in the Aggregation of Asphaltenes, *Fuel Sci. Tech. Int.*, **5**, 2, pp 119.

McCaffrey WC, Sheremata JM, Gray MR, Dettman H (2003) *Proc. The 4th International Conference on Petroleum Phase Behavior & Fouling*, Trondheim, Norway, June 2003.

Mc Kay JF, Amend PJ, Harsberger, Cogswell TE, Latham DR (1981) Composition of Petroleum Heavy Ends. 1. Separation of Petroleum Greater Than 675-Degrees-C Residues, *Fuel*, **60**, pp 14-16.

Merdrignac I (1997) Composition des structures azotées dans les pétroles. Implication pour leur réactivité au cours des procédés d'hydrodézatotation, *PhD,* Université Louis Pasteur, Strasbourg, France.

Merdrignac I, Béhar F, Briot P, Vandenbroucke M, Albrecht P (1998) Quantitative Extraction of Nitrogen Compounds in Oils: Atomic Balance and Molecular Composition, *Energy & Fuels*, **12** (6), pp 1342-1355.

Merdrignac I, Truchy C, Robert E, Guibard I, Kressmann S (2004a) Size Exclusion Chromatography: Characterization of Heavy Petroleum Residues. Application to Resid Desulfurization Process, *Pet. Sci. Technol.*, **22**, 7&8, pp 1003-1022.

Merdrignac I, Desmazières B, Terrier P, Delobel A, Laprevote O (2004b) Analysis of Raw and Hydrotreated Asphaltenes using Off-Line and On-Line SEC/MS Coupling, *Proc. of Int. Conf. on Heavy Org. Dep.*, Los Cabos, Baja California, Mexico, Nov. 2004.

Mitra-Kirtley S, Mullins OC, Van Elp J, George SJ, Chen J, Cramer SP (1993) Determination of the Chemical Structures in Petroleum Asphaltenes using XANES Spectroscopy, *J. Am. Chem. Soc.,* **115**, pp 252.

Merino-Garcia D, Andersen SI (2003) Isothermal Titration Calorimetry and Fluorescence Spectroscopy Study of Asphaltene Self-Association in Toluene and Interaction with a Model Resin, *Pet. Sci. Techn.*, **21**, pp 507-525.

Merino-Garcia D, Andersen SI (2004a) Interaction of Asphaltenes with Nonylphenol by Microcalorimetry, *Langmuir*, **20**, pp 1473-1480.

Merino-Garcia D, Andersen SI (2004b) Thermodynamic Characterization of Asphaltene-Resin Interaction by Microcalorimetry, *Langmuir*, **20**, pp 4559-4565.

Moschopedis SE, Fryer JF, Speight JG (1976a) Investigation of Carbonyl Functions in a Resin Fraction from Athabasca Bitumen, *Fuel*, **55**, pp 184-186.

Moschopedis SE, Speight JG (1976b) Investigation of Hydrogen-Bonding by Oxygen Functions in Athabasca Bitumen, *Fuel*, **55**, pp 187-192.

Mullins OC, Sheu EY (1998) Structure and Dynamics of Asphaltenes, Plenum Press.

Murgich J, Rodriguez JM, Aray Y (1996) Molecular Recognition and Molecular Mechanics of Micelles of some Model Asphaltenes and Resins, *Energy & Fuels*, **10**, pp 68-76.

Murgich J (2003) Molecular Simulation and the Aggregation of the Heavy Fractions in Crude Oils, *Molecular Simulations,* **29**, pp 451-461.

Murgich J (2005) Molecular Mechanics Study of the Selectivity in the Interaction Between some Typical Resins and Asphaltenes, *Proc. The* 6th *International Conference on Petroleum Phase Behavior & Fouling*, Amsterdam, Netherlands, June 2005.

Nalwaya V, Tangtayakom V, Piumsomboon P, Fogler S (1999) Studies on Asphaltenes Through Analysis of Polar Fractions, *Ind. & Eng. Chem. Res.*, **38**, 3, pp 964-972.

Neves GBM, De Sousa MD, Travalloni-Louvisse AM, Lucas EF, Gonzales G (2001) Characterization of Asphaltene Particles by Light Scattering and Electrophoresis, *Pet. Sci. Technol.*, **19**, 1&2, pp 35-43.

Norinaga K, Wargardalam VJ, Takasugi S, Lino M, Matsukawa S (2001) Measurementr of Self Diffusion Coefficient of Asphaltene in Pyridine by Pulsed Field Gradient Spin-Echo ^{1}H NMR. *Energy & Fuels*, **15**, pp 1317.

Oka M, Chang HC, Gaualas GR (1977) Computer-Assisted Molecular-Structure Construction for Coal-Derived Compounds, *Fuel,* **56**, pp 3-8.

Östlund JA, Andersson SI, Nyden M (2001) Studies of Asphaltenes by the Use of Pulsed-Field Gradient Spin Echo NMR, *Fuel*, **80**, 1529-1553.

Östlund JA, Löfroth JE, Holmberg K, Nyden M (2002) Flocculation Behavior of Asphaltenes in Solvent/Nonsolvent Systems, *J of Coll and Interface Sci*, **253**, pp 150.

Östlund JA, Nyden M, Auflem IH, Sjöblom J (2003) Interactions Between Asphaltenes and Naphthenic Acids, *Energy & Fuels*, **17**, pp 113-119.

Östlund JA, Wattana P, Nyden M, Fogler HS (2004a) Characterization of Fractionated Asphaltenes by UV-Vis and NMR Self-Diffusion Spectroscopy, *J of Coll. and Interface Sci.*, **271**, 2, pp 372-380.

Östlund JA, Nyden M, Stilbs P (2004b) Component-Resolved Diffusion in Multicomponent Mixtures. A Case Study of High-Field PGSE-NMR Self-Diffusion Measurements in Asphaltene/Naphthenic Acid/Solvent Systems, *Energy & Fuels*, **18**, pp 531-538.

Overfield RE, Sheu EY, Sinha SK, Liang KS (1989) Sans Study of Asphaltene Aggregation, *Fuel Sci. and Technol. Int.*, **7**, 5-6, pp 611.

Peramanu S, Pruden BB (1999) Molecular Weight and Specific Gravity Distributions for Athabasca and Cold Lake Bitumens and their Saturate, Aromatic, Resin, and Asphaltene Fractions, *Ind. Eng. Chem. Res.*, **38**, 8, pp 3121-3130.

Pfeiffer JP, Saal RN (1940) Asphaltic Bitumen as Colloid System, *Journal of Physical Chemistry*, **44**, pp 139.

Qian K, Rodgers RP, Hendrickson CL, Emmett MR, Marshall AG (2001a) Reading Chemical Fine Print: Resolution and Identification of 3000 Nitrogen-Containing Aromatic Compounds from a Single Electrospray Ionization Fourier Transform Ion Cyclotron Resonance Mass Spectrum of Heavy Petroleum Crude Oil, *Energy & Fuels*, **15**, pp 492-498.

Qian K, Robbins WK, Hughey CA, Cooper HJ, Rodgers RP, Marshall AG (2001b) Resolution and Identification of Elemental Compositions for more than 3000 Crude Acids in Heavy Petroleum by Negative-Ion Microelectrospray High-Field Fourier Transform Ion Cyclotron Resonance Mass Spectrometry, *Energy & Fuels*, **15**, pp 1505-1511.

Ravey J.C, Ducouret G, Espinat D (1988) Asphaltene Macrostructure by Small-Angle Neutron-Scattering, *Fuel*, **67**, 11, pp 1560.

Ravey JC, Espinat D (1990) Macrostructure of Petroleum Asphaltenes by Small Angle Neutron Scattering, *Progress in Colloid & Polymer Sci.*, **81**, pp 127.

Ray BR, Witherspoon PA, Gorin RE (1957) *J. Phys. Chem.*, **61**, 1296.

Ritchie RGS, Roche RS, Steedmann W (1979) Pyrolysis of Athabasaca Tar Sands – Analysis of the Condensable Products from Asphaltene, *Fuel*, **58**, pp 523-530.

Reerink H, Lijzenga (1973) *J. Inst. Pet.*, **59**, pp 211.

Robins C, Limbach PA (2003) The Use of Nonpolar Matrices for Matrix-Assisted Laser Desorption/ Ionization Mass Spectrometric Analysis of High Boiling Crude Oil Fractions, *Rapid Commun. Mass Spectrom.*, **17**, 24, pp 2839-2845.

Rodgers RP, Hendrickson CL, Emmett MR, Marshall AG, Greabey M, Qian K (2001) Molecular Characterization of Petroporphyrins in Crude Oil by Electrospray Ionization Fourier Transform Ion Cyclotron Resonance Mass Spectrometry, *Can. J. Chem.*, **79**, pp 546-551.

Rodgers RP, Purcell JM, Schaub TM, Marshall AG (2005) First Experimental Evidence for Buckybowls in Nature: Field Desorption and APPI FT ICR MS of the Aromatic Fraction of Crude Oil, *Preprints The* 6th *International Conference on Petroleum Phase Behavior & Fouling*, Amsterdam, Netherlands, June 2005.

Roux JN, Broseta D, Demé B (2001) SANS Study of Asphaltene Aggregation: Concentration and Solvent Quality Effects, *Langmuir*, **17**, pp 5085-5092.

Sarret G, Connan J, Kasrai M, Bancroft GM, Charie-Duhaut A, Lemoine S, Adam P, Albrecht P, Eybert-Berard L (1999) Chemical Forms of Sulfur in Geological and Archeological Asphaltenes from Middle East, France, and Spain Determined by sulfur K- and L-edge X-ray Absorption Near-Edge Structure Spectroscopy, *Geochim. Cosmochim. Acta.*, **63**, 22, pp 3767-3779.

Satya S (2004) *Proc. Int. Conf. on Heavy Org. Dep.*, Los cabos, Baja California, Mexico, Nov. 2004.

Sharma BK, Sarowha SLS, Bhagat SD, Tirawi RK, Gupta SK, Venkataramani PS (1998) Hydrocarbon Group Type Analysis of Petroleum Heavy Fractions using the TLC-FID Technique, *Fres. J. of Anal. Chem.,* **360**, 5, pp 539-544.

Sharma BK, Tyagi OS, Aloopwan MKS, Bhagat SD (2000) Spectroscopic Characterization of Solvent Soluble Fractions of Petroleum Vacuum Residues, *Pet. Sci. Technol.*, **18**, 3&4, pp 249-272.

Sheu EY, De Tar MM, Storm DA, DeCanio J (1992a) Aggregation and Kinetics of Asphaltenes in Organic-Solvents, *Fuel*, **71**, pp 299-302.

Sheu EY, Liang KS, Sinha SK, Overfield RE (1992b) Polydispersity Analysis of Asphaltenes Solutions in Toluene, *J. of Coll. and Interface Sci.*, **153**, 2, pp 399-410.

Sheu EY, De Tar MM, Storm DA (1994) Dielectric Properties of Asphaltene Solutions, *Fuel*, **73(1)**, pp 45-50.

Sheu EY, Storm DA (1995) In *Asphaltenes – Fundamentals and Applications*, Edited by EY Sheu and OC Mullins, Plenum Press, 1.

Sheu EY (2002) Petroleum Asphaltene-Properties, Characterization, and Issues, *Energy & Fuels*, **16**, pp 74-82.

Speight JG, Moschopedis SE (1981) *Preprints Am.Chem. Soc., Div. Petrol. Chem,* **26**, 4, pp 907.

Speight JG, Wernick DL, Gould KA, Overfield RE, Rao BML, Savage DW (1985) Molecular-Weight and Association of Asphaltenes – A Critical-Review, *Rev. Inst. Fra. Pet.*, **40**, 1, pp 51-61.

Speight JG (1999) In *The Chemistry and Technology of Petroleum*, Marcel Dekker Inc., New York.

Stilbs P (1987) Fourier Transform Pulsed-Gradient Spin-Echo Studies of Molecular Diffusion, *Prog. NMR Spectrosc.*, **19**, pp 1-45.

Strausz OP, Mojelsky TW, Lown EM (1992) The Molecular-Structure of Asphaltene – An Unfoldfing Story, *Fuel*, **71**, pp 1355-1363.

Strausz OP, Peng P, Murgich J (2002) About the Colloidal Nature of, *Energy & Fuels*, **16,** pp 809-822.

Strausz OP, Lown EM (2003) In *The Chemistry of Alberta Oil Sands, Bitumen and Heavy Oils.* AERI, Calgary.

Swanson JM (1942) A Contribution to the Physical Chemistry of the Asphalts, *Journal of Physical Chemistry*, **46**, pp 141-150.

Szewczyk V, Behar F, Behar E, Scarsella M (1996) Evid, *Rev. Inst. Fra. Petr*, **51**, 4, pp 575-590.

Tanaka R, Hunt JE, Winans RE, Thiyagarajan P (2003) Aggregates Structure Analysis of Petroleum Asphaltenes with Small-Angle Neutron Scattering, *Energy & Fuels*, **17**, pp 127-134.

Thiyagaran P, Hunt JE, Winans RE, Anderson KB, Miller JT (1995) Temperature Dependent Structural Changes of Asphaltenes in 1-Methylnaphthalene, *Energy & Fuels*, **9**, pp 829.

Tissot B (1981) Present Knowledge on Heavy Consituents of Crude Oils, *Rev. Inst. Fra. Pet.*, **36**, 4, pp 429-446.

Wattana P, Fogler HS, Yen A, Del Carmen Garcia M, Carbognani L (2002) Characterization of Polar based Asphaltene Subfraction. *Proc of Int. Conf. on Heavy Org. Dep.*, Puerto Vallarta, Mexico, Nov. 2002.

Winniford RS (1963) The Evidence for Association of Asphaltenes in Dilute Solutions, *J. Inst. Petrol.*, **49**, 475, pp 215.

Xu Y, Koga Y, Strausz OP (1995) Characterization of Athabasca Asphaltenes by Small-Angle X-ray-Scattering, *Fuel*, **74**, pp 960-964.

Yarranton HW, Beck L, Alboudwarej H, Svrrcek WY (2002) *ACS*, Div. Petrol. Chem., **47**, 4, pp 336.

Yen TF, Erdman JG, Pollack SS (1961) Investigation of Structure of Petroleum Asphaltenes by X-ray Diffraction. *Anal. Chem*, **33**, pp 1587.

Yen TF (1988) In *Encyclopedia of Polymer Science and Engineering*, M Grayson and JJ Krochwitz (Eds).

PART 2

Reservoir Engineering and Production

G. Renard

Heavy oils, extra-heavy oils and bitumens were defined in Chapter 2.1. These are mixtures of liquid hydrocarbons with low gas saturation, generally trapped in reservoirs at shallow depth located from several tens to several hundreds of meters deep, and rarely deeper than 1,500 m. Consequently, the temperature in these reservoirs is several tens of degrees Celsius. The higher the viscosity of a crude oil, the lower its API degree and the lower the temperature at which it is found. In reservoir conditions, a *heavy crude oil* will thus have a relatively limited – or zero – ability to flow naturally. When a well is put into production in a heavy oil reservoir, primary production with no additional means other than the energy linked to the fluid pressure will rapidly drop, and very quickly a means of pumping or providing energy will be needed to achieve and maintain an economically acceptable flow rate. For a very shallow reservoir depth, the crude oil may be so viscous that it cannot even flow. The rocks of shallow reservoirs are generally unconsolidated sands, i.e. the grains of sand that form the solid framework of the reservoir are not linked to each other. Consequently, grains of sand can be produced with the crude oil, unless means are implemented to prevent this production.

The objective of this chapter is to present the various methods currently used to produce reservoirs of *heavy crude oils*. These methods depend on several parameters:

- Reservoir characteristics (depth, thickness, temperature, type of rock, permeability – i.e. capacity for fluid flow, and degree of complexity linked to the sedimentology of the reservoir and its geological history).
- Whether an aquifer is present.
- Crude oil characteristics – including viscosity – linked to the geological history of fluids and thermodynamic conditions.

To correctly produce a reservoir, the first step to perform is its characterization in order to know the quantities of hydrocarbons in place and the complexity of its structure for the flow of fluids. This is true for all reservoirs, whether they hold a light or *heavy crude oil*; but it is even more critical with a heavy oil that, as discussed, flows with difficulty in the reservoir. Chapter 5 describes the specific production geology of *heavy crude oil* reservoirs.

Production techniques for *heavy crude oils* are then discussed in Chapters 6 to 8. They are classified into three major categories, based first on reservoir depth, and then on whether the crude oil can be produced by pumping alone.

When the reservoir depth is less than 70-80 m and there is no pressure in the reservoir, mining extraction of the oil-impregnated rock is generally used; this is followed by rock-bitumen separation. Described in Chapter 6, this method is used successfully in Canada to produce Athabasca bitumens in zones in which the thickness of overfill is limited. When the ratio of impregnated layer thickness to overburden thickness is no longer favorable, mining extraction is no longer economical and other extraction techniques must be planned.

Production by pumping is used when possible and economical, at least in the initial phase of heavy oil production. This is referred to as cold production. The techniques currently used in this framework are described in Chapter 7. When reservoirs are unconsolidated – which is generally the case resulting from their low compaction due to limited burial during the geological history – producers complete the wells with special equipment to avoid the production of sand coming from the reservoir itself. However, it has been observed – particularly in Canada – that concomitant production of sand in some cases significantly increased the heavy oil recovery rate. This process is called CHOPS (Cold Heavy Oil Production with Sand). Another phenomenon described in Chapter 7 concerns the foamy character of some crude oils, considered to be beneficial for cold production.

Even when it is possible, cold production only allows for the recovery of a very limited portion of the oil in place – rarely more than 10%. Due to the high viscosity of crude oil as compared to that of water or gas, the injection of water or gas is not technically efficient. Techniques have been developed which aim to reduce crude oil viscosity in order to increase its ability to flow and be produced. These include techniques for injecting steam, solvent or air, which are described in detail later. These techniques are constantly evolving. Thus, the development of horizontal wells has enabled new schemes for producing heavy oils and even bitumens.

The rising of effluents from the bottom to the surface in the case of heavy oils requires special techniques. These are explained in Chapter 9.

Field application of the main techniques for heavy oil production described in Chapters 6 and 7 is illustrated in Chapter 10 by the description of several major industrial projects for heavy oil exploitation implemented in various regions of the world.

5 | Reservoir Geology

R. Eschard

5.1 HEAVY OIL PETROLEUM HABITAT THROUGHOUT THE WORLD

5.1.1 Specificity of the Petroleum System

Reservoirs with heavy oil can be encountered in various basins and depositional settings. The occurrence of heavy oil is first linked to a specific evolution of the petroleum system, with biodegradation being the first factor controlling heavy oil formation (see Chapter 3). Bacterial degradation of the oil is only possible at low temperature, and the process is then only efficient at shallow depth or in surface conditions. Degradation of the oil is an irreversible mechanism, and secondary cracking of the degraded heavy oil remains very limited.

At shallow depth, specific early mature source rock also favors the formation of heavy oil products. Bitumen is the first product of kerogen cracking. An example is the rich source rock of the Monterrey Formation in California, which sources the surrounding reservoirs with low mature heavy oil products.

Active hydrodynamism also favors biodegradation and bacterial nutrients are also provided by the invasion of meteoric waters. Periods of uplift and erosion then favor degradation of the oil *via* the invasion of meteoric water.

Water washing also helps to increase viscosity, as the gas fraction can be dissolved and pushed further in the drainage system. Loss of the volatile part of the hydrocarbon can also occur during migration in the subsurface when the reservoir suffers a complex multiphase history with dismigration of the lightest products through the seals or drains. In surface conditions, meteoric fresh water can invade the reservoir from surface exposures and progressively degrade the oil in the reservoir. Meteoric water can also percolate from the surface to the reservoir through thin sedimentary covers, with the sealing capacity of the shales often being reduced at shallow depth. Finally, loss of the volatile hydrocarbons to the atmosphere can also enhance degradation of the oil when oil-bearing reservoirs are exhumed to the surface.

For all these reasons, most of the heavy oil reservoirs around the world are encountered at a shallow depth, and frequently in recent late Tertiary clastic sediments (with the noticeable exception of the Cretaceous Canadian heavy oil belt).

5.1.2 Foreland Basins

Heavy oil accumulations are frequently encountered in a foreland basin context. Foreland basins are formed ahead of a mountain belt. Sediments are deeply buried just ahead of the uplifted mountain belt in the foredeep part of the basin. Stratification then progressively reaches the surface away from the mountain belt (the forebulge). Sediments that were eroded from the mountain belt are deposited in the basin as the mountain belt is uplifted (synorogenic deformation). In foreland basins, heavy oil reservoirs can be found in the forebulge part of the basin, in the thrust belt itself at shallow depth, or in the subsurface at greater depth (Figure 5.1). Thus, many heavy oil plays found around the world are associated with a foreland basin context: the Persian foreland basin in Iran [Bashari A, 1988; Moshtaghian A *et al.*, 1988; etc.] and in Saudi Arabia and Kuwait [Wani MR and Al-Kabli SK, 2005], Florida-Cuba foreland basin in Cuba [Magnier C *et al.*, 2004], Junggar basin in China [Niu J and Hu J, 1999], Adriatic depression in Albania or Italy, etc.

The two main heavy oil provinces of Venezuela and Canada are found over the forebulge part of the Caribbean or Rocky Mountain forelands, respectively. In the forebulge setting, biodegradation occurred in the reservoirs which were deposited and remained at a shallow depth during the entire geological history. The oil kitchen (i.e. the area where specific pressure and temperature conditions enabled generation and expulsion of hydrocarbons from the source rock) was located in the deepest part of the basin in the Canadian or Caribbean foreland basins, and the hydrocarbons migrated laterally towards the forebulge where biodegradation occurred at shallow depth. The drainage system which enabled long-distance migration consisted of the first fluvial sediments filling the foreland basins. These basal synorogenic fluvial sands (i.e. the sediments which were deposited during the basin deformation) formed a drainage system with great lateral continuity, onlapping the basal foreland unconformity. These fluvial or deltaic sands marked the beginning of the deformation in the foreland basin. The increase of the sediment supply is due both to the uplift of the thrust belts and erosion of the forebulge. In Venezuela and Canada, a gradient in the heavy oil products is observed, with the bituminous and extra-heavy oil found in the outcrop passing progressively to heavy then light oil towards the foredeep. The seal above the reservoirs consists of marine shales, deposited during the increase of the subsidence rate as the foreland continued to develop. Bitumen also plugs porosity at the outcrops, making a regional seal laterally.

Heavy oil reservoirs can also be found in fluvial and deltaic synorogenic sediments in the thrust belt itself. For example, this is the case in the Miocene sediments of the Patos-Marinze field in Albania [Roure F *et al.*, 2004] or in the Junggar basin in China [Niu J and Hu J, 1999]. The oil was generated from a source rock in the deep part of the basin and vertically dismigrated through the faults or following the unconformity at the base of the synorogenic sediments. Biodegradation was then active in near-surface conditions. Bitumen may also form a plug at the exposure trapping lighter product downdip.

In foreland basins, some heavy oil plays can also be found in the deep part of the basins. As stated above, early migration and oil trapping followed by biodegradation occurred when the reservoir was at shallow depth over the forebulge. Then, the reservoir was later buried together with the degraded heavy oil products. As the foreland basin migrated while the deformation front progressed, these units were later buried and involved in the deformation. For example, this is the case for the carbonate reservoirs in Iran [Bashari A, 1988; Moshtaghian A *et al.*, 1988; etc.] or Cuba [Magnier C *et al.*, 2004], which were probably charged then biodegraded while they were at a shallow depth in the forebulge setting, just below the basal unconformity. As stated above, a late uplift may also bring the impregnated reservoirs close to the surface where biodegradation can re-occur (Figure 5.1).

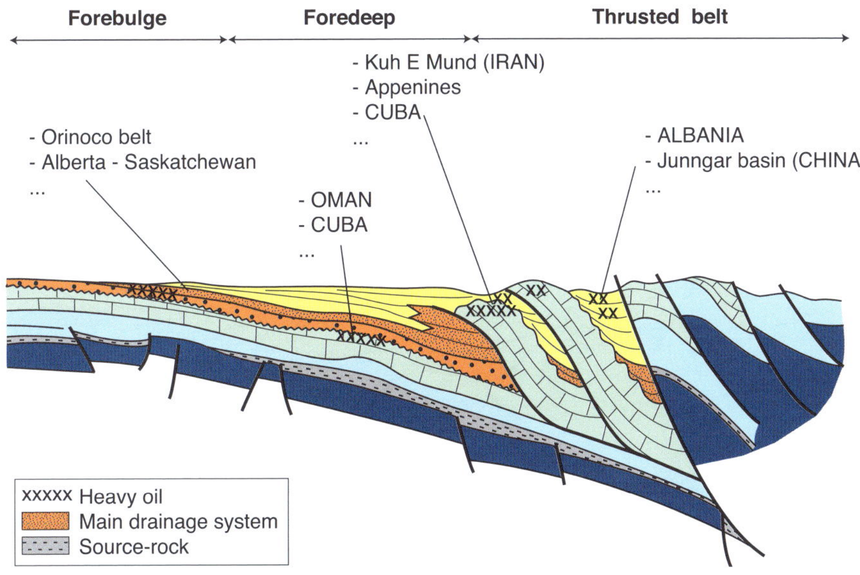

Figure 5.1

Types of habitats of heavy oil fields in foreland basins, such as those found in the Orinoco and Canadian heavy oil belts, and in various basins throughout the world. The oil migrated from the foredeep to the forebulge through fluvial sands draping the foreland basal unconformity. Biodegradation occurred here at shallow depth. Reservoirs may have been latterly buried in the foreland itself or uplifted in the thrust belt with their heavy oil products. Vertical dismigration and surface alteration may also occur in the thrusted units. Reservoirs consist either of fluvio-deltaic clastic sediments of the synorogenic succession, or of the carbonates which were deposited before the mountain belt formed.

5.1.3 Deep Offshore Passive Margin and Rift Systems

Heavy oil reservoirs are frequently encountered in the deep offshore of the passive margins of Brazil [Pinto AC *et al.,* 2001], West Africa or the Tertiary post-rift turbidites of the North Sea [Jayasekera AJ and Goodyear SG, 2000]. The thermal gradient is particularly low in a passive margin, enabling the biodegradation of oil in the reservoir even at significant depth. Furthermore, in the deep offshore setting, the sediment supply rate is low and burial remains relatively limited for recent sediment. The reservoirs then remained at a shallow depth after an early oil charge. This context then favors *in situ* biodegradation of the oil.

Heavy oil can also be found in the rift systems which were latterly uplifted and even exhumed during geological history. In the Madagascar western rift margin, major heavy oil accumulations are trapped in the Permo-Triassic Karoo reservoirs and exposed at the surface. The Permian and Mesozoic rift and post-rift successions were uplifted from the late Cretaceous, and the margin was tilted westward, then eroded in the hinterland. The heavy oil prospects probably result from updip migration of the oil following the main tilt, the oil being biodegraded when it approached the surface (Figure 5.2).

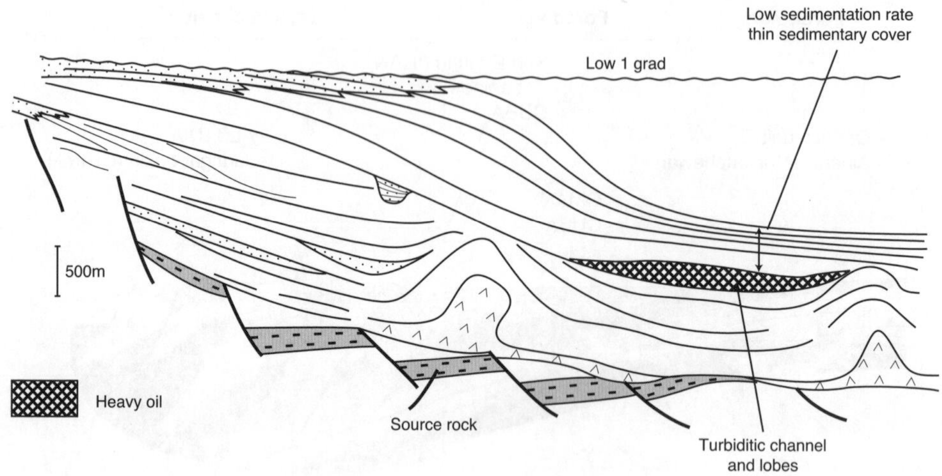

Figure 5.2

Types of habitats of the main heavy oil fields in passive margin context: an early migration occurred in the turbiditic reservoir in a context where sedimentation rate and thermal gradients are low, making biodegradation processes possible.

5.1.4 Highly Subsident Pull-Apart or Back-Arc Basins

Another geological setting where biodegraded oils can be found at shallow depth is highly subsident basins filled with clastic fluvial sediments. A source rock generated the hydrocarbons in the present day or in recent times in the deepest part of the basin. In the case of continental rift, pull-apart of back-arc basins, where sediment supply and subsidence can be very high, the basin-fill is mostly made of sand-rich fluvial sediments and vertical connectivity within the whole stratigraphic succession is high. The lack of continuous shale layers greatly reduces the seal efficiency. Faults also compartmentalize the reservoirs. The oil migrated vertically through the clastic reservoirs or through the faults from the source rock to the surface. The oil can finally be trapped at a shallow depth below a local seal in a compartment or even reach the surface, were the biodegradation process can occur. For example, this setting is found in the pull apart basins of California (although some of the heavy oil in California also results from moderately buried immature source rock) [Williams LL *et al.*, 2001], and in the back-arc basins of Indonesia [Waite MW *et al.*, 1997]. A final possibility for biodegradation to occur is when a reservoir with an initial light oil product was uplifted during geological history, with the oil being latterly biodegraded at shallow depth. This last setting is not very common, the exhumed trap rarely keeping its sealing capacity intact during the uplift, and a dismigration generally occurring during the uplift (Figure 5.3).

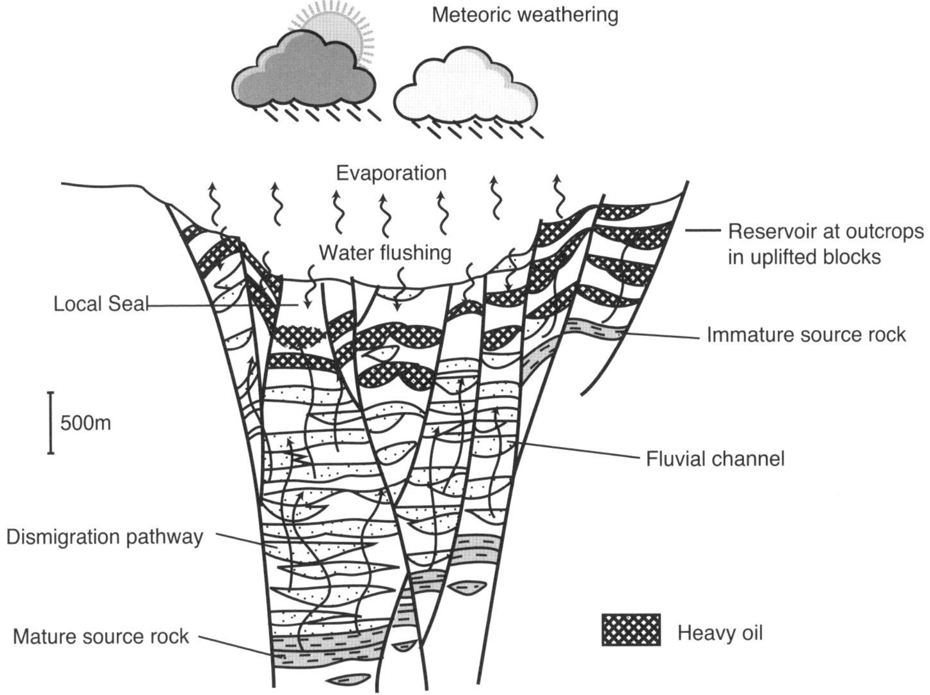

Figure 5.3

Types of habitats of the main heavy oil fields in pull apart or back-arc basin filled by clastic sediments, with high subsidence rates. Dismigration occurred from the deep basin towards shallow horizons where the oil was degraded. In areas with limited burial (on the right-hand side of the diagram), immature source rock may also charge the reservoirs with heavy oil products.

5.2 SPECIFICITY OF HEAVY OIL RESERVOIR GEOLOGY

5.2.1 Introduction and Reservoir Workflow

In heavy oil reservoirs, the high viscosity of the fluid makes its displacement difficult during cold or enhanced production. Generally speaking, the drainage efficiency of a well is fairly reduced and rarely exceeds a few hundred meters at maximum. In the Kern River field in California, the typical steam injection pattern is a 5-spot covering 1 ha (2.5 acres) and in the Duri field in Indonesia, the injection pattern covers an area of around 6.25 ha (15.5 acres) [Curtis C *et al.*, 2002].

The reservoir development workflow which is used for producing heavy oil fields is strongly dependent on the depth of the prospect and on economic factors. Shallow prospects are mainly developed with wells just logged and without using any seismic data because the cost of seismic acquisition is much higher than the cost of a few shallow wells. As the reservoir targets become deeper, the cost of the wells increases and well spacing will consequently be greater. Input of the seismic data now becomes critical to predict the reservoir extension and its heterogeneity, especially when horizontal wells are planned for the prospect development. In the case of small fields with limited reserves, numerical models are hardly used, the reservoir being managed with conventional reservoir engineering techniques. When significant reserves are expected, the reservoir development workflow of a heavy oil reservoir would follow the classic steps of a reservoir simulation. After description of the wells and interpretation of seismic data, a reservoir layering is proposed, taking into account specific porosity/permeability cut-offs (see section below). Gridded numerical reservoir models are then constructed to better manage long-term development of the reservoir: tests of various recovery processes, optimization of well spacing, production forecasts, etc. The reservoir model is then of great added value for long-term management of the field. In giant heavy oil accumulations where long-term management of the field is critical, a 4D seismic monitoring approach can also be implemented to image the evolution of the fluid saturation. This technique is particularly efficient when thermal recovery processes are used, since steam saturation modifies the petro-acoustic properties of the reservoirs and the seismic character. Such techniques can be used to image the reservoir parts swept by the steam and identify bypassed areas where significant oil reserves are still present and can be identified [Waite MW *et al.,* 1997; Zhang W *et al.*, 2005]. Such techniques will likely be developed in the near future in the giant heavy oil provinces.

Because of the low mobility of heavy oil, the reservoir heterogeneity distribution strongly influences well productivity and development of the steam zone. Siltstones and mudstones create conductivity barriers and induce bypass zones. Finally, the impact of the heterogeneity may also change over time, depending on the recovery process that is implemented. For instance, vertical permeability barriers will have a low impact during cold production of horizontal wells, but a stronger impact during gravity drainage if steam injection or the SAGD process is used to produce the heavy oil (SAGD is explained in Section 8.1.5).

5.2.2 High Porosity and Permeability Values

The oil viscosity is the main limiting factor for heavy oil production. In order to compensate the detrimental properties of these oils, the reservoirs must present very high porosity and permeability values as compared to conventional oil plays in order to make the recovery economical. Typically in producing a heavy oil field, the porosity range must be between 20 and 30%, or even more in the bitumen belts at outcrops, and permeability must always be sufficiently high, e.g. above 1 Darcy. Theses high ranges of porosity and permeability are preferentially found in shallow clastic reservoirs, where porosity was preserved from compaction. Only the most porous facies will efficiently contribute to production. Coarse-grained fluvial or tidal channel-fills then create most of the reservoirs of the heavy oil producing fields. Sands with a finer grain size or siltstone levels will not contribute significantly to production and even form a permeability barrier to steam injection, for example. Very coarse and porous sands can also be found in turbiditic channels and lobes in the deep offshore, where burial was limited. One of the technical challenges that the industry will have to face in the near future will be the development of recovery processes to produce heavy oil in facies with a lower porosity range, which will greatly increase the reserves of heavy oil belts.

Carbonates, which are lithified earlier than clastic sediments and rapidly lose porosity during early burial, rarely present sufficient ranges of poro-perm to produce through the matrix porosity, with a few exceptions of vuggy carbonates formed during early dissolution processes. Carbonates must be fractured to make potential heavy oil reservoirs, but even so, oil production in such setting is challenging. Until now, the experience of heavy oil production in fractured carbonate reservoirs has rarely been attempted.

5.2.3 Unconsolidated Sands

Clastic reservoirs found at shallow depth and which remained at shallow depth (not more than a few hundreds of meters) during geological history are generally unconsolidated. Specific drilling and completion problems are thus frequent due to the unconsolidated sands. On the one hand, sand-influxes may constitute a major problem for producing these reservoirs, decreasing production, plugging the bottomhole, and eroding the well and surface equipment. On the other hand, some specific recovery processes (e.g. CHOPS described in Section 7.3) may benefit from sand influxes, with the sands being produced jointly with the oil. Unconsolidated reservoirs must face mechanical problems during depletion. With the high ranges of porosity mentioned above, a significant compaction which is positive for oil recovery may occur in the reservoir during depletion. Ground surface subsidence is frequently observed in depleted shallow fields, e.g. as is the case in heavy oil fields located along or below the Maracaibo Lake where dykes had to be built to avoid the risk of land submersion due to subsidence. The wells are often also deformed, if not integrally cased. Another issue is the sealing capacity of shales above the reservoir which is often questionable in shallow settings, and must be carefully evaluated before increasing pressure in the reservoir during injection of a fluid for EOR.

5.3 EXAMPLE OF HEAVY OIL FLUVIAL RESERVOIRS OF THE MANNVILLE GROUP IN CANADA

5.3.1 Fluvial Reservoirs: Reservoir Architecture

Fluvial and fluvio-deltaic reservoirs contain most of the heavy oil in place in the two main provinces of Canada and Venezuela, but also in California, Indonesia, etc. In the following sections we will describe the type of heterogeneity which can be found in the fluvio-estuarine reservoirs of the Canadian heavy oil belt. The model presented in the following sections can also be easily applied to the Orinoco heavy oil belt, whose reservoirs present organization and architecture that are very similar to those found in Canada. Similarly, heavy oil fluvial reservoirs can also be found in Albania, California, China, Indonesia, etc.

As stated above, fluvial reservoirs are sand-rich, but very heterogeneous at every scale. Generally speaking, the best net to gross ratios are found in the channel-fills themselves in which the coarser-grained material was deposited. The overbank deposits, such as crevasse splays or levees, which contribute to production in conventional oil and gas reservoirs, are not considered as reservoir facies in a heavy oil context because of their finer grain-size and lower porosity than the channel-fills. Siltstone beds also constitute a permeability barrier to heavy oil production in the channel-fills themselves [William LL *et al.*, 2001], which is not the case with lighter hydrocarbon products.

The architecture of fluvial reservoirs is mostly controlled by the rate of incision and amalgamation of the channels. Fluvial systems are typically organized cycles corresponding to the variation of the base level through time, which itself results from the complex interaction of subsidence and climate (itself controlling the water influx and sediment supply in the river system).

In proximal settings, far from the sea or a permanent lake, the variation of the base level is registered by changes in the rate of amalgamation of the channels. During periods in which the space available for sediment was low, the channels were amalgamated to form a sand-rich channel belt. The width of the channel belt varies from a few kilometers to hundreds of kilometers. They form sand-sheets with a high sand-shale ratio, but also with a high degree of heterogeneity, each channel cross-cutting the underlying ones. Lateral migration of the channels eroded and washed out the floodplain fines basinwards. As the accommodation rate started to increase in response to a base-level rise, floodplain sediments vertically aggraded and the floodplain shales were preserved between the channels. As a result, the vertical connectivity between individual channels will decrease.

The architecture of the fluvial reservoirs varies in a cycle. Basal sand-sheets, which present a significant lateral continuity and high net to gross ratios, will be preferentially developed by horizontal wells to get the maximum pay in the wells (Figure 5.4). A fine-scale heterogeneity may locally affect well productivity and may also strongly impact development of a steam zone. Above, in the aggrading floodplain, individual channels are rarely connected and will be produced by putting wells parallel to the channel orientation, or by drilling deviated wells to statistically increase the chances of cross-cutting several channels in the same well.

5.3.2 Incised Valley Models

Generally speaking, during periods of base level fall, rivers incised the floodplain shales, creating a network of incised valleys (Figure 5.5). As the base level started to rise, high-energy braided channels filled the basal part of the valley with coarse-grained material. The continuous rise of the base level then induced a flattening of the depositional profile and a decrease of the transport energy, and then meandering channels filled the valley above the braided channels. When the river system was connected to the sea, the tidal influences progressively increased in the valley system as the relative sea-level rose, depositing heterolithic sandstones and mudstones. Finally, open marine sediments can be deposited in the valley itself, or overwhelm the interfluves. In the case of endoreic basins connected to a permanent lake system, a lacustrine transgression can also develop during the maximum rise of the base level.

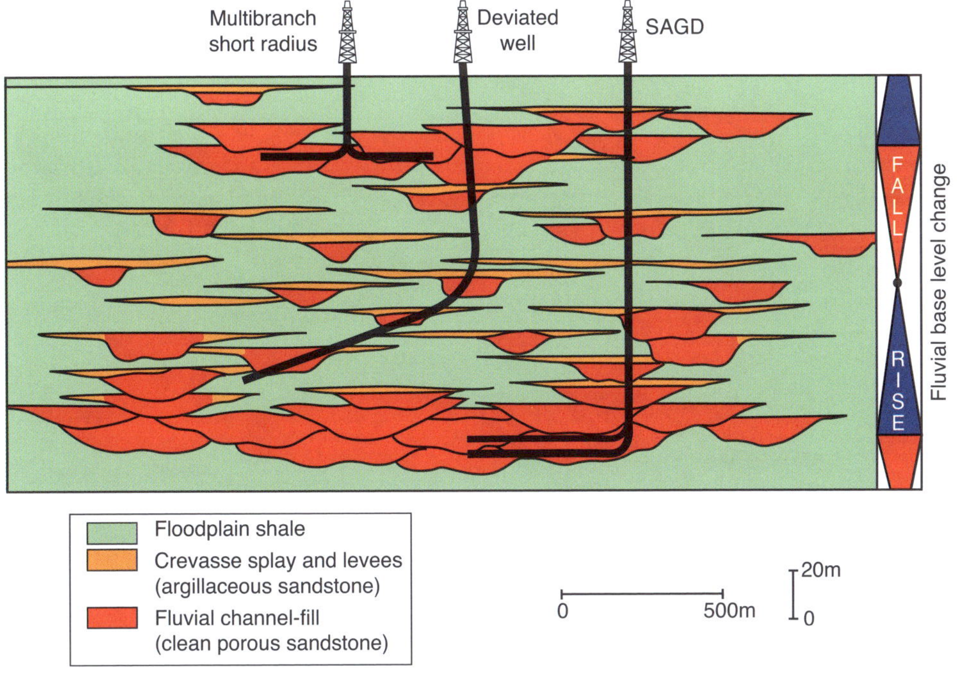

Figure 5.4

Typical organization of fluvial reservoirs and possible well configuration for producing heavy oil depending on channel connectivity. At the base of the fluvial cycle, the channels are amalgamated in channel belts forming a continuous reservoir unit. As the base-level rose, the floodplain shale was better preserved and reservoir vertical and lateral connectivity also decreased.

The regressive part of the cycle is marked by an overall progradation of a delta plain, facies being more and more proximal upward before a new incision. Such cycles are commonly found in the Cretaceous succession of Alberta and Saskatchewan (Canada), or in the Miocene succession of the Orinoco (Venezuela). Each step of the base level cycle is marked by a specific organization of the facies. Generally speaking, the level of heterogeneity increases upwards in the valley-fill. The basal part of it is sand-rich and relatively homogeneous, while the meandering then the tidal channels are less continuous and highly heterogeneous at a fine-scale, which makes the production of heavy oil far more challenging in the upper units.

5.3.3 Regional Setting of the Mannville Group in Canada

The Mannville Group in Western Canada has been chosen to illustrate the type of sedimentary heterogeneity which can be encountered in fluvial systems in heavy oil reservoirs. As previously stated, the model of heterogeneity distribution of the Canadian heavy oil reservoirs can also be applied in many other similar geological contexts, especially in the Orinoco heavy oil belt (Venezuela). The reservoirs of the Mannville Group in Saskatchewan

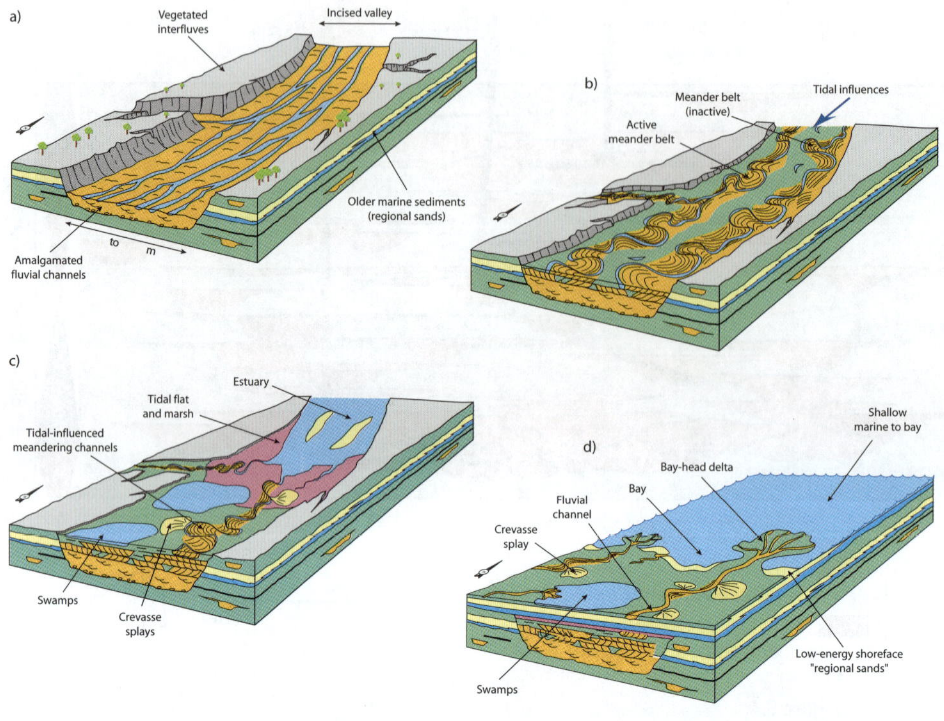

Figure 5.5

Reservoir architecture of a fluvial system during a base level cycle **a)** valley incision and initial infill phase by braided channels **b)** meandering channel belt **c)** estuarine systems **d)** offshore shales deposited during the Maximum Flooding Surface, then delta progradation during the regressive part of the cycle.

and Alberta constitute one of the most prolific heavy oil provinces of the world. The deposition of Mannville group was contemporaneous to the formation of a foreland basin during Middle Cretaceous times. The succession is 700 m (2,300 ft) thick in the foothill area westward, while it progressively thins out to a few tens of meters when approaching the outcrops belt eastward. Strata are gently dipping westward at a regional scale towards the foredeep. The sediments of the Mannville Group onlap a major unconformity eroding the underlying strata whose age ranges from Lower Cretaceous in the foothills to lower Paleozoic in the Eastern area [Hayes BJR *et al.*, 1994]. The sedimentation was also influenced by local depocenters (Peace River arch, Liard basin, etc.).

The Lower Mannville group mostly consists of fluvial sediments onlapping the unconformity. A network of large incised fluvial valleys was developed above the unconformity, the location of which was partly controlled by the differential erosion of the underlying Paleozoic strata and by local tectonic movements. Interfluves remained emerged during the infill of the valleys and were later onlapped by the Upper Mannville sediments. The valley river systems were flowing northward towards the boreal ocean, the fluvial systems progressively passing to estuarine and shoreface sediments. During the Albian time, a relative sea-level rise induced an overall backstep of the fluvial system, depositing estuarine sediments in the valleys, then marine sediments themselves overwhelmed the valley and blanketed the whole area, forming the seal of the Lower Mannville reservoirs.

The Upper Mannville Group in Saskatchewan consists of an alternation of shallow marine, brackish and fluvial sediments. At least five transgressive – regressive cycles can be identified and correlated regionally in the Upper Mannville Group. Each cycle began with a relative sea-level drop of variable amplitude, forming a network of incised valleys. The valley width is often less than a kilometer, for a depth which can reach 30 m (100 ft). They incised during short-lived relative sea-level drops, and filled during the subsequent sea-level rise. The correlative lowstand wedges can be found northward in the form of shoreface sequence [Hayes BJR *et al.*, 1994; Mc Phee D and Pemberton SG, 1994]. The valleys were mostly filled with fluvial sediments passing upward to tidal deposits. Above, the transgression deposited a layer of open marine to brackish sediments of regional extension above which prograded shoreface and mouthbar, forming "regional sands" a few meters thick. On top of the prograding marine sands, floodplain shales with coal beds and fluvial channels are sometimes preserved. The cycle ends with a new incision of fluvial valleys during the next sea-level drop. A North-South cross-section summarizing the stratigraphic architecture of the Mannville Group across Saskatchewan is shown in Figure 5.6, while a typical well section of the Mannville group in central Saskatchewan is displayed in Figure 5.7.

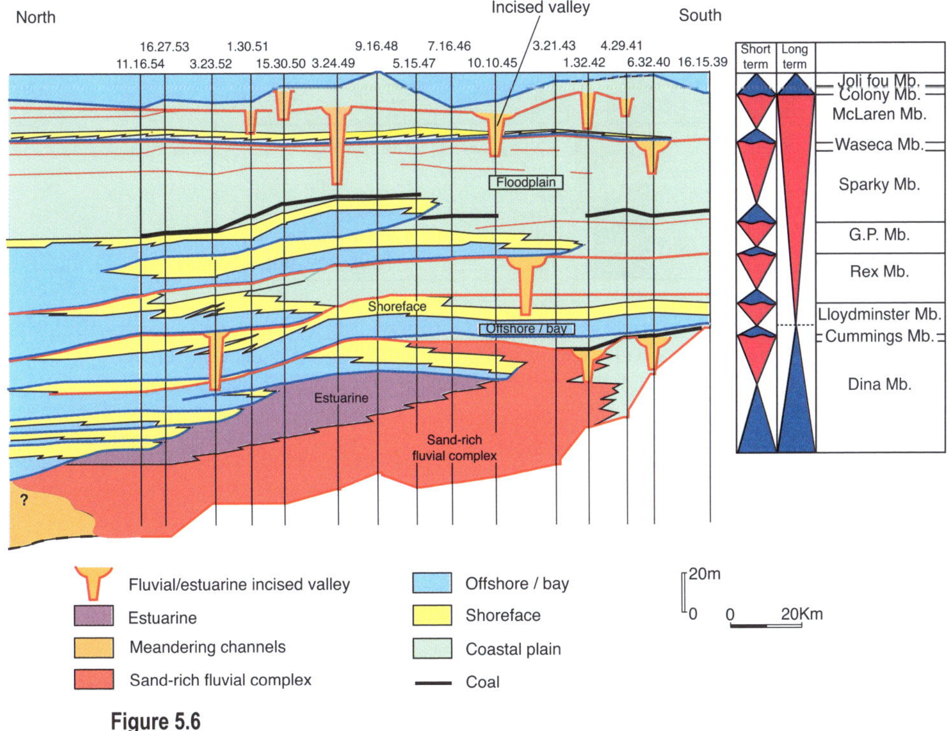

Figure 5.6

Regional North cross-section across Saskatchewan reconstructing the stratigraphic architecture of the Mannville group.

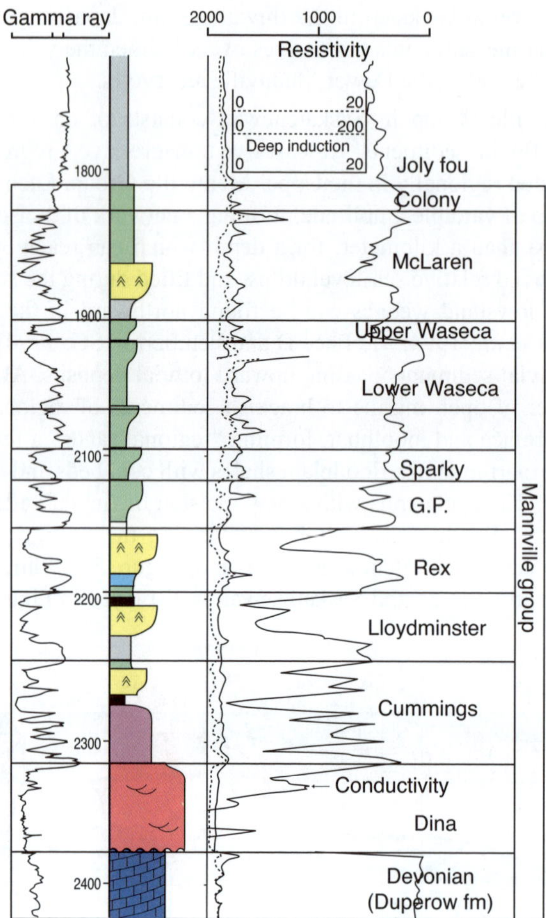

Figure 5.7

Logs signature of a well in the Lloydminster area in Saskatchewan. Fluvial valleys can be incised at different stratigraphic levels of the Upper Mannville group.

5.3.4 Reservoir Heterogeneity in the Fluvial Reservoirs of Saskatchewan

In Central Saskatchewan, the basal sands of the Dina Member consist of very coarse-grained unconsolidated sands. They were deposited in amalgamated braided channels (Figures 5.8 and 5.9). The main fluvial pathways may have been guided by the morphology of the basal unconformity. Several cycles of incisions may have occurred during the infill of the valley-fill, adding a complexity to the reservoir architecture of this unit. The braided channel-fills display two types of heterogeneity. First, mud-clasts are draping the channel basal incision

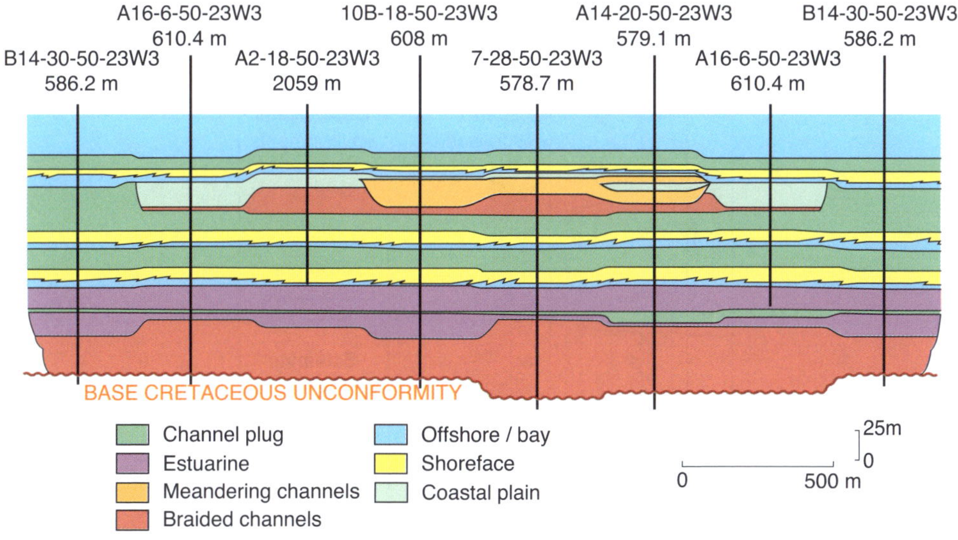

Figure 5.8

Cross-section across a producing field in Central Saskatchewan showing the reservoir geometry in the Mannville Group. Note the sand-rich continuous fluvial basal layer overlaid by the heterolithic estuarine facies, and the valley with a limited extension in the upper part of the section. Seismic time slices of the lower and upper reservoirs can be seen in Figures 5.10 and 5.12.

surface. These mud-clast layers are discontinuous, and do not create any significant barrier to the flow of in place or injected fluids, but help to reduce the vertical permeability of the layer. A second type of heterogeneity is the development of thin siltstone or mudstone layers corresponding to abandonment facies preserved on top of the channel sequence. They may reach a few hundreds of meters of lateral extension for a few meters or less of thickness. The porosity abruptly drops in these layers. Such heterogeneity significantly reduces the vertical connectivity between different channels constituting the basal sand-sheet and reduces the sweeping efficiency of fluids needed by EOR. Moreover, when such siltstone layers extend close to a steam injector, they may act as a thermal screen and affect development of a steam zone. For example, this can dramatically affect the efficiency of well pairs in the SAGD process. An illustration of the distribution of heterogeneity from geological information extracted from 3D seismic data is shown in the Figure 5.10.

Above the basal fluvial sand-sheet, meandering channels and tidal facies testify to the backstepping evolution of the fluvial system during a transgression. They form a very heterogeneous layer (the Cummings Formation) in which tidally-influenced point bars are developed at the base and overlain by estuarine channels and tidal flats. The layer is very heterogeneous at every scale: meandering channels interfinger with floodplain shales and coals, whereas coarse-grained but narrow tidal channels are encased in mud- and sand-flat facies in its upper part.

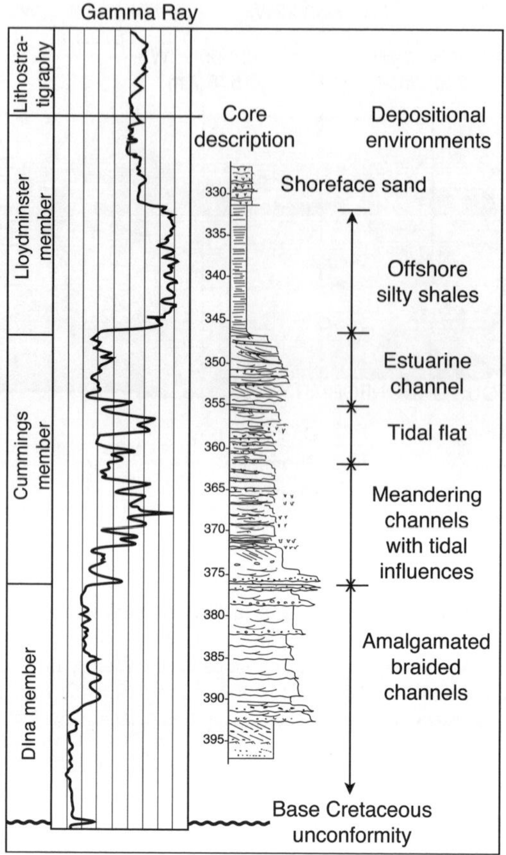

Figure 5.9

Log and core description of a heavy oil well of the Lower Mannville Group, Central Saskatchewan. The valley-fill typically displays a fining upward evolution. The coarse-grained basal deposited in braided channels (Dina Member) are overlaid by bioturbated point bars of meandering channels, then by tidal deposits (Cummings member) before the transgression of the Lloydminster offshore shales.

The lateral accretion surfaces of the meandering point bars were draped with thin mud-layers and also affected by bioturbation. The finer-grained material of the point bars and the mud drapes significantly reduce the porosity and permeability of the meandering channel-fill. The facies is named Inclined Heterolithic Surface (I.H.S.) by Canadian geologists, and producing such facies with a low porosity still remains a technical challenge. Mud-plugs, corresponding to the abandonment of a meander bend, are also present laterally to the point bars. They form another type of heterogeneity in this layer.

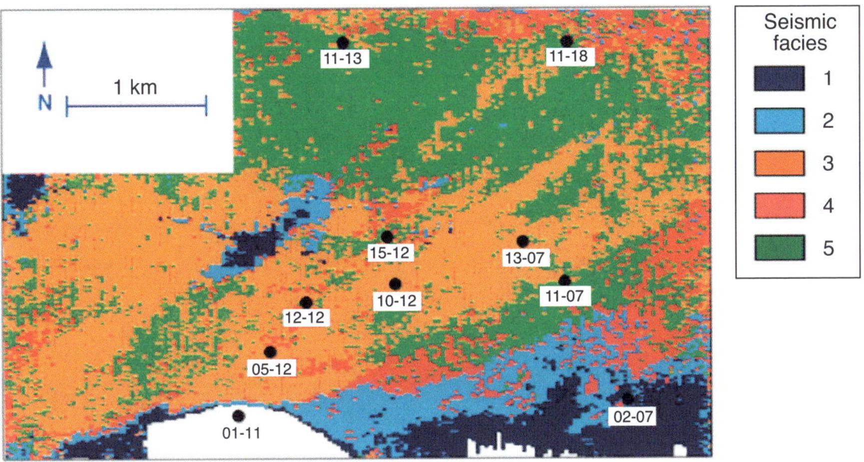

Figure 5.10

Seismic facies interpretation of the basal Dina sand in the Senlac field (Canada), showing the reservoir heterogeneity distribution along a time slice [modified from Dequirez PY *et al.*, 1995]. The seismic facies 1 and 2 (dark and light blue) are interpreted as thin reservoirs on the margin of the incised valley, facies 3 (orange) is correlated to very porous sands, facies 4 (red) presents poor reservoir quality, and facies 5 (green) is a mixture of porous sands and interbedded heterolithic facies.

The tidal and estuarine channels are generally formed by coarse- to medium-grained sands. Their porosity is however reduced by thin mud laminations, typical from the tidal processes, draping the sigmoidal bedding filling the channels. The channels are incised in tidal flats, which display wavy bedding, and thin rippled sandstone beds alternating with mud layers at a fine-scale. They are locally called the "interbedded facies". As a result, the porosity and permeability of these layers abruptly drop, making production of the heavy oil in this layer problematical because of the fine-scale heterogeneity.

A theoretical model of the distribution of heterogeneity in the reservoirs of the Lower Mannville group is shown in the Figure 5.11. Such a model is representative of most of the fluvio-estuarine reservoirs producing heavy oil around the world, especially in the Orinoco heavy oil belt in Venezuela.

In the Upper Mannville group, the incised valley systems are much narrower than in the Dina Formation. An incised fluvial channel of the Colony member (Figure 5.12) has maximum thickness of 33 m (100 ft) and is less than a kilometer wide (0.6 miles). The valley fill displays a fining upward sequence. Massive blocky basal sands are overlain by more heterolithic sands towards the top of the valley-fill. The mean porosity in the basal sand is 33% and mean permeability is around 4 Darcies.

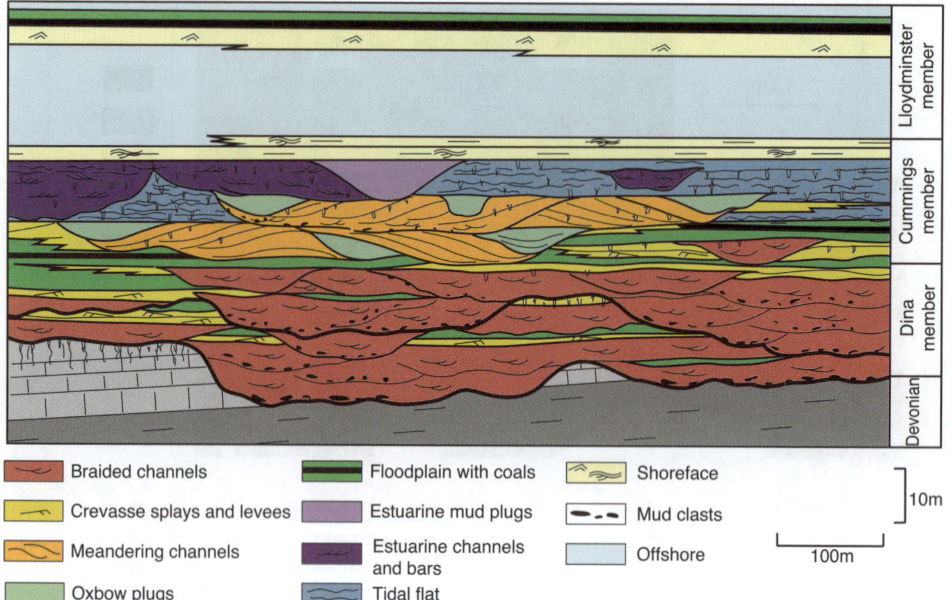

Figure 5.11

Conceptual model of the heterogeneity distribution in the Lower Mannville Group. The basal braided channels are overlaid by meandering channels which in turn pass to tidal flats and estuarine channels. The reservoir heterogeneity consists of locally preserved floodplain mudstones and siltstones, mud-clast layers, abandoned channel plugs and mud drapes developed on the point bars and in the tidal facies.

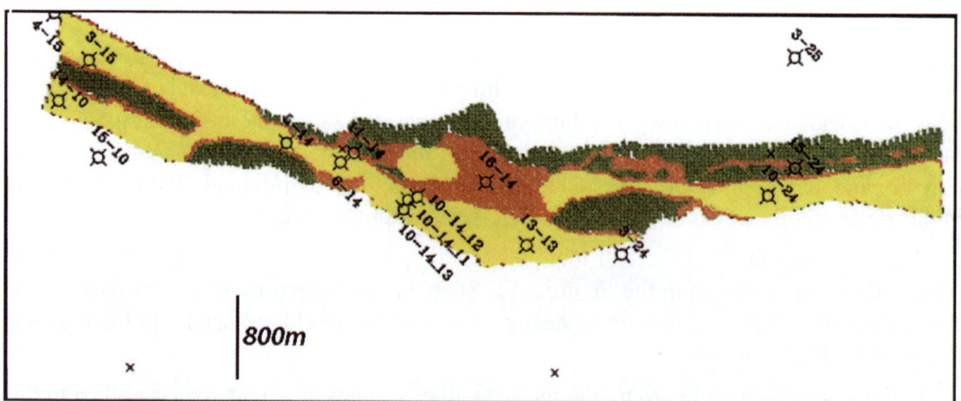

Figure 5.12

Seismic facies in an incised valley of the Upper Mannville Group, Saskatchewan, Canada (yellow: clean porous sands; orange: argillaceous sands; green: shale). The valley is 30 m thick and corresponds to the upper reservoir shown in the cross-section in Figure 5.8.

5.4 CONCLUSIONS AND PERSPECTIVES

High viscosity of oil makes its displacement difficult in the porous media of the reservoir. The drainage area of a single well and its productivity index are also particularly reduced in heavy oils. For vertical wells, close well spacing (hundreds of meters, down to tens of meters in some cases) is used to develop the fields. But more generally, horizontal wells or complex multidrains are used to increase the well production (see section 7.2). Predicting the net pay along the well trajectory is then critical.

The effect of heterogeneity on the sweep efficiency is critical when a fluid is injected to heat and displace heavy oil. The reservoir facies of heavy oil fields only correspond to the most porous and permeable sands. As a result, even facies with a medium range of porosity will act as barrier to fluid flow. The impact of reservoir heterogeneity on thermal recovery is still poorly understood, and some lithologies may act as barriers to injected fluids due to their properties. This point is particularly important in the case of SAGD steam chamber development [Ehlig-Economides CA *et al.*, 2001]. Phenomenological fluid-flow models can also help to understand reservoir behavior in heterolithic layers during production, increasing the potential oil volume to be produced from facies with a lesser range of porosity and permeability. The reservoir model must also be flexible enough to be quickly updated when a large number of wells are drilled in a short time, e.g. as is the case in the Orinoco heavy oil prospects in Venezuela [Uzcategui E, 2001].

Finally, the production of heavy oil carbonate reservoirs is virtually unknown as yet. As stated above, carbonate reservoirs must have a high matrix porosity to be produced with thermal processes, and the dynamic behavior of the fractures must be carefully evaluated with phenomenological models before starting production.

For these reasons, well-managed field development in heavy oil reservoirs should integrate a more detailed reservoir description than is usually used in conventional oil plays [Williams LL *et al.*, 2001]. This is not often the case since low-cost drilling does not incite to undertake detailed reservoir modeling studies or acquire expensive 3D seismic data sets. However, seismic data is increasingly used to map the channels pathways in fluvial reservoirs in order to optimize well trajectory and image small-scale heterogeneity which may affect development of a steam chamber [Kopper R *et al.*, 2001]. Also, cross-well seismic and 4D time-lapse seismic is still a challenging approach to better monitor the evolution of a steam zone *versus* time. The identification of potential bypassed areas can be visualized by comparing the evolution of the seismic attributes which are modified by pressure variations and changes in steam saturation [Waite MW *et al.*, 1997; Zhang W *et al.*, 2005].

REFERENCES

Bashari A (1988) Occurrence of Heavy Crude Oil in the Persian Gulf. Paper 44 (presented at the *4th UNITAR/UNDP International Conference on Heavy Crude and Tar Sands*, 7-12/08/88, Edmonton, Alberta, Canada).

Curtis C, Kopper R, Decoster E, Guzman-Garcia A, Huggins C, Knauer L, Minner M, Kupsch N, Linares LM, Rough H, Waite M (2002) Heavy-Oil Reservoirs. *Oilfield Review,* **14**, 3, pp 30-51.

Dequirez PY, Fournier F, Blanchet C, Feuchtwanger T, Torriero D (1995) Integrated Stratigraphic and Lithologic Interpretation of the East-Senlac Heavy Oil Pool. Expanded Abstracts (presented at the *65th Annual International Meeting of the Society of Exploration Geophysicists*, 8-13/10/95, Houston, USA).

Ehlig-Economides CA, Fernandez B, Economides MJ (2001), Multibranch Injector/Producer wells in Thick Heavy Oil-Crude Reservoirs. *SPE Reservoir Evaluation Journal,* **4**, 3, pp 195-200.

Hayes BJR, Christopher JE, Rosenthal L, Los G, McKercher B, Minken D, Tremblay YM, Fennell J (1994) Cretaceous Mannville Group of the Western Canada Sedimentary Basin. In: Geological Atlas of the Western Canada Sedimentary Basin, G.D. Mossop and I. Shetson (comp.), *Canadian Society of Petroleum Geologists and Alberta Research Council*, Calgary, Alberta.

Jayasekera AJ, Goodyear SG (2000) The Development of Heavy Oil Fields in the United Kingdom Continental Shelf: Past, Present and Future. *SPE Reservoir Evaluation,* **3**, 5, pp 371-379.

Kopper R, Kupecz J, Curtis C, Cole T, Dorn-Lopez D, Copley J, Munoz A, Caicedo V (2001) Reservoir Characterization of the Orinoco Heavy Oil Belt: Miocene Oficina Formation, Zuata field, Eastern Venezuela. Paper SPE 69697 (presented at the *2001 SPE International Thermal Operations and Heavy Oil Symposium*, 12-14/03/01, Porlamar, Margarita Island, Venezuela).

Mc Phee D, Pemberton SG (1994) Sequence Stratigraphy of the Lower Cretaceous Mannville Group of East Central Alberta. *Canadian Society of Petroleum Geologists*, Exploration Update 1994, Program with Abstracts, pp 27-28.

Magnier C, Moretti I, Lopez JO, Gaumet F, Lopez JG, Letouzey J (2004) Geochemical Characterization of Source Rocks, Crude Oils and Gases of Northwest Cuba. *Marine and Petroleum Geology,* **21**, 2, pp 195-214.

Moshtaghian A, Malekzadeh R, Azazpanah A (1988) Heavy Oil Discovery in Islamic Republic of Iran. Paper 99 (presented at the *4th UNITAR/UNDP International Conference on Heavy Crude and Tar Sands*, 7-12/08/88, Edmonton, Alberta, Canada).

Niu J, Hu J (1999) Formation and Distribution of Heavy Oil and Tar Sands in China. *Marine and Petroleum Geology,* **16**, 1, pp 85-95.

Pinto AC, Guedes SS, Bruhn CHL, Gomes JAT, Sá AN, Fagundes Neto JR (2001) Marlim Complex Development: A Reservoir Engineering Overview, Paper SPE 69438 (presented at the *SPE Latin American and Caribbean Petroleum Engineering Conference*, 25-28/03/01, Buenos Aires, Argentina).

Roure F, Naszaj S, Mushka K, Fili I, Cadet JP, Bonneau M (2004) Kinematics and Petroleum Systems – An Appraisal of the Outer Albanides. In: KR McKlay ed., Thrust Tectonics and Hydrocarbon systems, *AAPG Memoir,* **82**, pp 474-493.

Uzcategui E (2001) Reservoir Characterization and Exploitation Scheme in the Orinoco Belt, SPE 69698 (presented at the *2001 SPE International Thermal Operations and Heavy Oil Symposium*, 12-14/03/01, Porlamar, Margarita Island, Venezuela).

Waite MW, Sigit R, Rusdibiyo AV, Susanto T, Primaldi H, Satriana D (1997) Application of Seismic Monitoring to Manage an Early-Stage Steamflood. *SPE Reservoir Engineering,* **12**, 4, pp 277-283.

Wani MR and Al-Kabli SK (2005) Sequence Stratigraphy and Reservoir Characterization of the 2nd Eocene Dolomite Reservoir, Wafra Field, Divided Zone, Kuwait-Saudi Arabia. Paper SPE 92827 (presented at the *2005 SPE Middle East Oil and Gas Show and Conference*, 12-15/03/05, Kingdom of Bahrain).

Williams LL, William SF, Kumar M (2001) Effects of Discontinuous Shales on Multizone Steamflood Performance in the Kern River Field. *SPE Reservoir Evaluation and Engineering,* **4**, 5, pp 350-357.

Zhang W, Youn S, Doan Q (2005) Understanding Reservoir Architectures and Steam Chamber Growth at Christina Lake, Alberta, by using 4D Seismic and Crosswell Seismic Imaging. Paper SPE 97808 (presented at the *SPE/PS-CIM/CHOA International Operations and Heavy Oil Symposium*, 1-3/11/05, Calgary, Alberta, Canada).

6 | Oil Sands: Mining and Processing

E. Delamaide

Although oil mining was first observed and recorded in the 13th century by Marco Polo and has since been done in several regions of the world, it is not an intuitive concept and is far from the popular image of oil gushing from a derrick. Today, the technique is used at a large scale only in Canada, where it accounts for 22% of the country's oil production [CAPP (Canadian Association of Petroleum Producers), 2006]. However, it is a business that is growing fast: with capital investments of about US $20 to $25 billion until 2010, it is estimated [CAPP, 2006] that nearly 29% of Canada's oil production will come from oil sands mining by 2010, and 38% by 2015. With proved reserves of 5.6 billion m^3 (35.2 billion barrels) of oil [Alberta Energy and Utilities Board, 2006], mineable oil sands represent slightly less than total Nigerian crude oil reserves (ranked 10th in the world, [Oil and Gas Journal, 2005]) and 50% more than total US crude oil reserves (ranked 11th). These reserves are within an area of 3,400 km^2, along the Athabasca River north of the city of Fort McMurray in northern Alberta (Canada).

The surface that is currently considered as mineable by the Alberta government ("Surface Mineable Area") is limited to the area where the total overburden – i.e. the layer of soil composed of sand, clay and silt that needs to be removed in order to gain access to the oil sands themselves – does not exceed 75 m (246 ft); it corresponds to 37 contiguous townships (land parcels of 9.654 km × 9.654 km [6 miles × 6 miles]). There, the oil that is contained in the Wabiskaw-McMurray deposits of Lower Cretaceous age is so viscous that it does not flow under natural reservoir temperature conditions.

Today, three mines are in operation: those of Suncor, Syncrude and Shell Albian Sands [Albiansands website]. In total, these three operations accounted for approximately 93,000 m^3/d (584,000 bopd) of crude bitumen in 2005.

A number of new projects are expected to come online in the next few years: the Canadian Association of Petroleum Producers forecasts [CAPP, 2006] that oil sands mining production will grow to 162,000 m^3/d (1,019,000 bopd) by 2010 and to 278,000 m^3/d (1,750,000 bopd) by 2015, which means that it will nearly double by 2010 and triple by 2015. The potential and importance of oil sands mining cannot therefore be underestimated.

To reach this point, it took over 130 years from the first expedition dedicated to the oil sands, and more than 70 years since the first project started. It also took the vision and dedication of a few visionary individuals.

6.1 HISTORY

6.1.1 Early History

The potential of the Athabasca oil sands has been known for quite some time [Ferguson BG, 1985; Athabasca Regional Issues Working Group website]; following their discovery in the late 1700s, the oil sands were studied during successive expeditions by the Geological Survey of Canada starting in the early 1870s. The first expedition dedicated to the Athabasca region was led in 1882 by John Macoun, who made some preliminary observations on oil sands. In 1882, Robert Bell, a renowned geologist, went back to the Athabasca and made detailed observations on the geology of the oil sands on outcrops; he was probably the first to understand their potential commercial value. In his footsteps followed people such as Sidney Ellis and Karl Clark, scientists who pioneered the industrial exploitation of oil sands.

6.1.2 Bitumount and Abasand

Following a difficult period of pilot plant construction and operations, the first truly commercial bitumen extraction plant was constructed in 1937 by the International Bitumen Co. It successfully produced good quality bitumen, but was plagued by breakdowns and operational problems. In severe financial difficulties, the company finally ran out of money in 1938.

Another commercial-scale plant was built in 1935-36 by Abasand Oils Ltd. with a capacity of 250 tons of oil sand per day, but it was not until 1941 that commercial operations fully started for a few months, until the plant was completely destroyed by a fire in November 1941.

This partial technical success was encouraging at a time when Canada's wartime requirements for oil products were high. However, the economics of the process was still questionable.

6.1.3 Suncor, Syncrude and the Future

Several unsuccessful efforts followed, and the situation remained virtually unchanged until 1953, when the Great Canadian Oil Sands consortium led by the US company Sun Oil was formed. The consortium built a large plant with a capacity of 7,150 m^3/d (45,000 bopd); the plant was completed in September 1967 and started operations at that time. In 1979, the consortium gave birth to Suncor Inc. which continues operations to this day.

In the meantime, the Syncrude consortium had been formed in 1964 by a number of oil companies. After nine years of studies, design and engineering, construction of the Syncrude plant began in 1973 and operations started in 1978.

The next project – the Athabasca Oil Sands Project (AOSP) operated by Albian Sands Energy Inc. (now Shell Albian Sands) – did not start operations until 2002.

The various phases of a mining project are now described.

6.2 PHASES OF A PROJECT

6.2.1 Introduction

Oil sands projects are major undertakings that are only viable over the long-term. As will be seen below, ten years can elapse between the initial conception of the project and the first commercial production. The capital invested is on the magnitude of billions of dollars per project, and ensuring adequate return on capital in these conditions requires large production volumes.

An oil sands mining project can be split into several phases:

- Planning and construction.
- Operations.
- Reclamation.

6.2.2 Planning and Construction

6.2.2.1 Resource Delineation

Oil sands mining is done in open pits. As mentioned above, it is practical only for depths of 75 m (246 ft) and less of overburden; for oil sands located deeper than this, *in situ* recovery must be considered.

The thickness of the overburden is evaluated by core holes during the planning stages of mining operations. For instance, Shell Canada/Albian Sands drilled nearly 1,500 test wells at the Muskeg River Mine, while UTS Energy drilled over 1,000 core holes for their Fort Hills Project. In areas of high interest, is it not uncommon to drill one core hole every 100 m (328 ft).

Core holes are also used to appraise the thickness of the oil sand and its bitumen content, as well as identify the richest areas and establish the mining program accordingly.

The Alberta Energy Utilities Board (AEUB), which regulates oil sands mining in Alberta, considers that an oil sand thickness of 3 m (10 ft) is the minimum appropriate for mining. Average pay thickness is in the 30 m (100 ft) range [AEUB, 2001].

The bitumen content of the sand is of course of high interest. The AEUB considers that 7% weight is the minimum mineable bitumen content, and typical numbers for average bitumen content are in the 10-12% weight range.

Another important parameter is TV/BIP, which corresponds to the ratio of total volume (TV) of material to be mined in order to obtain a given volume of bitumen (Bitumen in Place: BIP). The lower this ratio, the better; the AEUB considers that maximum or "cut-off" TV/BIP is about 12 m^3/m^3, which means that in order to obtain one cu. m of bitumen, no more than 12 cu. m of material should be mined. For instance, the Syncrude Aurora North Mine has a TV/BIP ratio of 7.2 and is one of the best leases in the area [Canadian Oil Sands Trust, 2006].

6.2.2.2 Planning, Application and Regulatory Approval

Once the resources in place are sufficiently appraised, the planning can start. It needs to address the mine plan – where and when to mine – as well as the facilities for extraction, upgrading (when applicable) and reclamation. It must also include consultations with the local authorities and communities, as well as environmental impact assessments.

The whole process culminates with the Application to the Board – the Alberta Energy and Utilities Board – that regulates oil and gas operations in Alberta.

The regulatory process involves several provincial and federal agencies as well as the local communities, and can take from one to two years.

6.2.2.3 Engineering, Procurement and Construction

Detailed engineering can start even before the application is approved. Next come the procurement and construction phases which, with projects as large as oil sands mining, can take over two years.

The first phase of the construction is drainage of the surface water. In Northern Canada, low ground is often waterlogged and spongy, and covered by dead plants in various stages of decomposition – this is called muskeg. Draining this water is an operation which can take a few years in itself. Then, construction of the facilities can start.

Some of the components are built off-site and then transported by road in gigantic loads that require serious planning and coordination efforts.

As will be seen later, keeping a project on track, on time and on budget is a major undertaking, especially with so many different projects ongoing at the same time. Finding the specialized manpower – construction people, welders and electricians – is increasingly becoming an issue in Alberta.

6.2.3 Operations

The operations occur in three successive phases [Alberta Chamber of Resources, 2004]:

- Mining.
- Extraction.
- Upgrading.

6.2.3.1 Mining

The first step of mining operations is the removal of the muskeg and the overburden, which is a huge task. The volumes to be removed are very large: for instance, CNRL reported moving about 10.8 million m^3 (381 million cu. ft.) of overburden in 2006 – this represents a cube of 221 m × 221 m × 221 m (725 ft × 725 ft × 725 ft) [CNRL, 2006]. Syncrude reported that in 2004 its cost of overburden removal was CDN $184 million, which corresponds to about 20% of its operating costs [Syncrude, 2004].

The overburden is not disposed of, but rather partly used for road building and other construction work; the rest is stored since it will be used for land reclamation purposes once mining operations have been completed.

Once the oil sands themselves are exposed, mining can begin. In the "old days" mining was conducted with huge bucketwheels and draglines but the maintenance costs were prohibitive, and now the industry is mostly using giant shovels and trucks (Syncrude still uses some bucketwheels but they are being phased out). Shovels are less costly and also more flexible.

The trucks used are the largest in the world (Figure 6.1); they can carry between 350 to 400 tons of sand. As tall as a two-story house, with tires that are nearly 4 m tall and weigh as much as 2 adult elephants (5.3 tons), these trucks can nonetheless be filled to capacity in a little over two minutes. This is only because the shovels used are equally gigantic; the bucket can carry up to 100 tons in one "scoop".

Once the trucks are loaded, they deliver their load to crushers and rotary breakers, where the sand is broken into small pieces of less than 5 cm, and rocks are also removed. The oil sand is then transported to the next phase of operations – extraction – by a combination of conveyor belts and hydrotransport.

Hydrotransport was developed in the 1990s by Syncrude as a way to improve energy efficiency. It consists in mixing the oil sand with water; the resulting slurry is then transported by pipeline to the extraction plant. It has the added advantage of beginning the sand/bitumen separation which is also done using hot water.

Figure 6.1

A Giant Truck and its Operator (Courtesy of Suncor Energy Inc.).

6.2.3.2 Extraction

The aim of the extraction process is to separate the sand from the bitumen. As seen earlier, there is only about 10 to 12% by volume of bitumen, so there is a lot of sand to dispose of.

A. Primary Separation

The details of the process – and in particular the temperature – vary depending on the operator [Websites], but the basic principles remain the same. Air is added to the slurry, which is allowed to settle in tanks (the Primary Separation Vessels). The air enables the bitumen to separate from the sand grains, and the air/bitumen mixture forms a froth that floats to the top of the vessel where it is skimmed off. The sand falls to the bottom of the vessel (and becomes what is known as tailings) and in the middle remains a mixture of sand, clay, water and bitumen that is called middlings.

Sometimes caustic soda is added to improve the separation. Syncrude is using the so-called Low Energy Extraction Process in its Aurora operations; this process operates at low temperature (25 °C, 77°F) but uses kerosene and frother chemicals [application by Syncrude for the Aurora Mine] instead of hot water and soda.

B. Secondary Separation

It is of course important to recover as much bitumen as possible, thus the middlings are processed in the secondary separation stage to extract the bitumen they still contain. The principle is mostly the same as for the primary separation; more air is added to promote the formation of more froth, which contains most of the remaining bitumen. This froth joins the froth resulting from the primary separation for further processing. The remaining mixture joins the tailings. About 2 to 5% extra bitumen can be recovered from the middlings.

C. Froth Treatment

The froth is extracted and proceeds to the next stage of the process. At that point it is still a mixture of mostly bitumen (50%), but also water (40%) and air in addition to a small quantity of clay. The water, solids and air must be removed before the bitumen is sent to the upgrader or the refinery.

Froth treatment varies from operator to operator, but in general froth is de-aerated and heated to remove air and water, then the resulting bitumen is diluted with naphtha. It then either goes through centrifuges or inclined plate settlers (which allows for gravity segregation of particles) to separate the solids from the liquid; the resulting product goes through a set of centrifuges in order to minimize as much as possible the solid and water content in the bitumen. The naphtha is also recovered at the end of this process.

At this stage, the bitumen is as clean as it can be made and it is then transported to the upgrading/refining facilities.

D. Tailings

The sand, clay, water and residual bitumen that are extracted during the whole separation process constitute what is called the tailings. Tailings treatment is addressed in Chapter 19.

6.2.3.3 Upgrading

The bitumen is a very heavy, viscous crude and most operators choose to upgrade it into a "synthetic" crude oil before shipping it. Upgrading will be discussed in detail in Part 4.

The full cycle to produce the Synthetic Crude Oil (SCO), i.e. bitumen after upgrading, from the oil sands as planned for instance in the CNRL's Horizon Project [CNRL website] is shown in Figure 6.2. Typically, to produce 0.159 m^3 of SCO (1 bbl) requires removing approximately 2 tons of soil/rock above the deposit, 2 tons of oil sands, 0.32 to 0.8 m^3 (2 to 5 bbl) of water, and 24 m^3 (850 ft^3) of natural gas.

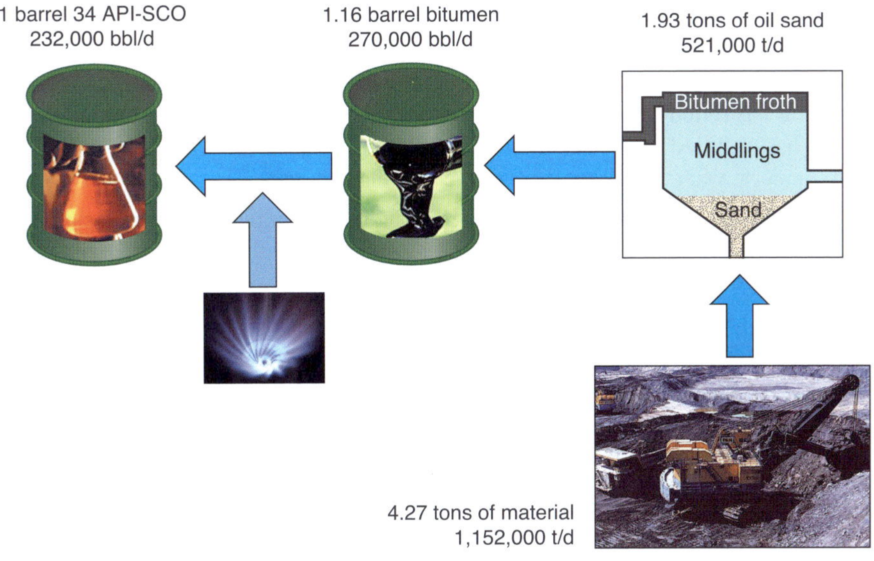

Figure 6.2

Typical Extraction Cycle of Bitumen.
Source : CNRL.

6.3 CHALLENGES AND OUTLOOK

6.3.1 Challenges

Although the industry has come a long way since the early days of Bitumount and Abasand, it is still facing numerous challenges. The impact of the Kyoto Protocol will not be as severe as initially feared, with an incremental cost of only CDN $0.20 – $0.30/bbl. But pressure is always strong to do more for the environment, in particular with regard to tailings management (see Chapter 19), but also water consumption.

Inflating capital costs and the lack of manpower – which are partly linked – are also some of the major issues.

Oil sands mining projects were plagued by delays and cost overruns in 2004-2006. Syncrude had to raise its budget for the construction of its Stage 3 expansion from an initial CDN $4.1 billion to CDN $8.1 billion – a 90% increase. The Athabasca Oil Sands Project experienced the same problem; initial estimates in September 2004 for the cost of the expansion were CDN $4.0 to CDN $4.5 billion, but Shell Canada Ltd. later announced a revised budget of CDN $7.3 billion in 2005 and CDN $11 billion in July 2006.

A major part of the cost overruns is due to severe shortages in manpower. With so many projects in construction at the same time, this is a very significant issue.

6.3.2 Research & Development: the Key to Reducing Costs

From the very start, with Karl Clark and his successors, and from Bitumount to Abasand, oil sands mining has relied heavily on research to make operations more economical and more efficient. That push led to the abandonment of the original bucketwheels and draglines for the more modern, versatile and cost-effective shovels and trucks.

This technological progress has succeeded in reducing mining operating costs from over US $30/bbl to less than US $10/bbl, thus making them competitive as compared to conventional oil production.

Some of the technological improvements have been small – redesigning pieces of equipment or processes so that they perform more reliably – but others have been real step-changes: for instance, in the 1990s Syncrude piloted the use of hydrotransport to replace conveyor belts and the technology has since become the industry standard; Syncrude also introduced the Low Energy Extraction Process [Alberta Government website] at its Aurora project that operates at approximately 25°C (77°F). At these temperatures, the new process consumes about one-third the energy of the traditional 80°C (176°F) process, bringing huge cost and environmental benefits. Today, UTS Energy is pilot-testing the BITMIN [UTS website] extraction process, which uses a counter-current desander for the bitumen-sand separation. The BITMIN uses a rotary drum to effect the separation of bitumen from sand and clays, while largely maintaining the physical integrity of the clays. It is used to significantly reduce the volume of fluid fine tailings, compared with previous oil sands extraction processes. The process has the potential to reduce energy and water needs as well as the tailings footprint, thereby providing further environmental and economic advantages.

Companies continue to work on improving the separation process, to recover more bitumen at lower cost, and also to make their operations more environmentally friendly.

The Alberta Oil Sands Technology and Research Authority (AOSTRA) was formed in 1974 at the initiative of the province of Alberta with the aim of helping the industry to do research and develop oil sands resources at low cost. Now, research is done both by the companies on their own or in partnership with others; the Canadian Oil Sands Network for Research and Development (CONRAD [Conrad website]) is an R&D network that promotes and encourages collaborative research on the oil sands.

A number of mining projects are currently in progress. They are described in Chapter 10.

REFERENCES

AEUB (2001) Interim Directive 2001-7.

Alberta Chamber of Resources (Jan. 30, 2004) Oil Sands Technology Roadmap – Unlocking the Potential.

Alberta Energy and Utilities Board (May 2006) Alberta's Energy Reserves 2005 and Supply/Demand Outlook 2006-2015: ST98-2006.

Alberta Government website: www.aet.alberta.ca/technology.aspx

Albiansands website: www.albiansands.com

Application by Syncrude for the Aurora Mine, Application # 960552.

Athabasca Regional Issues Working Group: www.oilsands.cc/publications/history.asp

Canadian Association of Petroleum Producers (May 2006) Canadian Crude Oil Production and Supply Forecast 2006 – 2020.

Canadian Environmental Law Association website: www.pollutionwatch.org/pressroom/releases/20061011.jsp

Canadian Oil Sands Trust (June 14, 2006) CAPP Investment Symposium.

Canadian Oil Sands Trust website: www.cos-trust.com

CNRL press release, April 26, 2006.

CNRL website: www.cnrl.com/horizon/

Conrad website: www.conrad.ab.ca

Ferguson BG (1985) Athabasca Oil Sands – Northern Resource Exploration, 1875 – 1951.

Oil Sands Discovery Center website: www.oilsandsdiscovery.com

Oil Sands Vegetation Reclamation Committee (October 1998) Guidelines for Reclamation to Forest Vegetation in the Alberta Oil Sands Region.

Oil and Gas Journal, Dec. 19, 2005.

Pembina Institute (May 2006) Troubled Waters, Troubling Trends. Pembina Institute website www.pembina.org

Suncor website: www.suncor.com

Suncor 2005 Annual Report.

Syncrude Canada website : www.syncrude.ca

Syncrude Canada Ltd. (2004) Sustainability Report.

Syncrude Canada Ltd. (2005) Sustainability Report.

Syncrude website: www.syncrude.com

UTS website: www.uts.ca/forthills/mining/bitmin.html

7 | Cold Production

G. Renard, J.F. Nauroy, E. Bemer

Heavy and extra-heavy oils are characterized by their high or extremely high viscosity at reservoir conditions. Mobility – *i.e.* the ability to flow through a porous medium – is expressed by the ratio of the effective permeability of the porous medium to the oil viscosity, and is therefore low or very low. Thus, primary recovery of heavy and extra-heavy oils is generally low – limited to a few percent of the Original Oil In Place (OOIP).

The viscosity which must be taken into account is the viscosity at reservoir conditions. For example, the viscosities at 25, 50 and 200°C (77, 122 and 392°F) for three different oils are shown in Table 7.1. The first two temperatures correspond to reservoir conditions at a depth of 600 m (~2,000 ft) in Canada for the lowest value (T_1) and in Venezuela for the highest value (T_2). Oil A in two identical reservoirs differing by temperature alone is considered to be an extra-heavy oil at a reservoir temperature of 25°C (77°F), and a heavy oil at a reservoir temperature of 50°C (122°F). Oil C is considered to be a heavy oil for both temperatures. According to the ratio of viscosity $\mu(T_1)/\mu(T_2)$ respectively at 25°C and 50°C, a well producing oil A in a reservoir at 50°C will produce with a rate roughly 20 times higher than an identical well in a reservoir at 25°C. This ratio is lower for oils B and C, but nonetheless remains high. As explained in Chapter 8, decreasing oil viscosity by heating the reservoir allows for faster production; viscosities at 200°C (392°F) are very low. However, this heating is not always technically feasible or economically viable. Thus, cold production is a widely-used production method. For such production, a diluent is often injected at the bottom of the production wells to reduce the friction losses in the wellbore (see Chapter 9).

This chapter describes the three most commonly used methods to recover heavy and extra-heavy oils at reservoir temperature. The first method is simply pumping in vertical wells. The second method is linked to the development of horizontal well technology, which emerged in the 1980s. It is now a proven and efficient technique to produce heavy and extra-heavy oils. It is also widely used, regardless of oil type. As its name indicates, a

Table 7.1 Viscosity μ *versus* Temperature for Different Oils.

Oil	T_1 = 25 °C Res. Conditions	T_2 = 50 °C Res. Conditions	T_3 = 200 °C Heating	$\mu(T_1)/\mu(T_2)$
A	100,000	5,624	10	17.8
B	10,000	1,000	6.3	10
C	1,000	178	4.0	5.6

third technique – called CHOPS for Cold Heavy Oil Production with Sand – favors sand production to increase the productivity of the well, followed by oil production.

7.1 COLD PRODUCTION USING CONVENTIONAL WELLS

Regardless of whether it is light or heavy, several physical phenomena occur when oil is withdrawn from a reservoir:

1) Pressure falls and oil expands.
2) Dissolved gas comes out of solution as soon as the bubble point is reached.
3) Dissolved gas flows to the wells but can also form a secondary gas cap.
4) If water is present below the oil pay, it can also flow quite rapidly to the wells.
5) Compaction of the reservoir rock is promoted, inducing a reduction of the pore volume.

The production mechanism associated with the first two phenomena is commonly referred to as solution-gas drive. The three others are described as gas-cap drive, water drive and compaction drive, respectively.

Oil recovery by solution-gas drive is generally much less efficient for heavy than for light oils, for at least two reasons: gas is less soluble in heavy oils, and heavy oil reservoirs are frequently less deeply buried. Thus, the pressure is lower – which means that there is less dissolved gas. The approximate effect of *in situ* oil viscosity on ultimate recovery by solution-gas drive is shown in Figure 7.1. For an oil viscosity greater than 100 cP, the ultimate primary recovery by this mechanism is at most on the order of 7 to 8 percent of OOIP.

In heavy oil reservoirs, water drive or gas drive are not favorable since the mobility of water or gas is too high as compared to the mobility of oil. Thus, the displacement efficiency is poor. On the contrary, compaction drive can have a huge impact on primary recovery of heavy oils. This is especially true when the reservoirs are large and not very deep. An example of this involves the reservoirs of the Bolivar Coast of Lake Maracaibo in Venezuela. Figure 7.2 shows the impact of compaction drive in the Tia Juana field, both in terms of maximum recovery and subsidence (also see Chapter 10).

If there were no compaction drive in the Bolivar reservoirs, the recovery would be limited to 5 percent of OOIP. However, as the pore pressure is decreased due to fluid withdrawal, the reservoir compacts, and additional oil is produced but subsidence of the ground occurs at the surface. In the Bolivar Coast, compaction has had a strong impact on the environment since continuous subsidence at ground level necessitated the construction of dykes to prevent flooding of large areas of land. Compaction drive is expected to occur in reservoirs with matrices which are sensitive to increased stress.

For many decades, most heavy oil reservoirs in California (Coalinga, Kern River, Midway Sunset, San Ardo and South Belridge), Venezuela (Tia Juana, Jobo, Bachaquero and Lagunillas), Russia, etc. were produced through primary production using vertical wells until an economic limit was reached. In the 1960s, steam stimulation or flooding was used to improve oil recovery in these reservoirs (see Chapter 10). Fortunately, in the 1990s the implementation of horizontal wells also allowed cold production to be used to produce most heavy oil reservoirs.

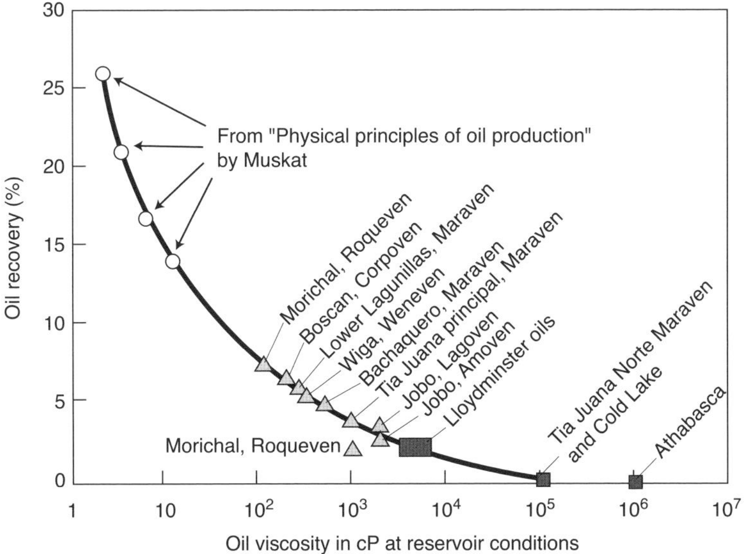

Figure 7.1

Approximate Effect of Oil Viscosity on Recovery for Solution-Gas Drive (after [Borregales CJ, 1977] and [Butler RM, 1994]).

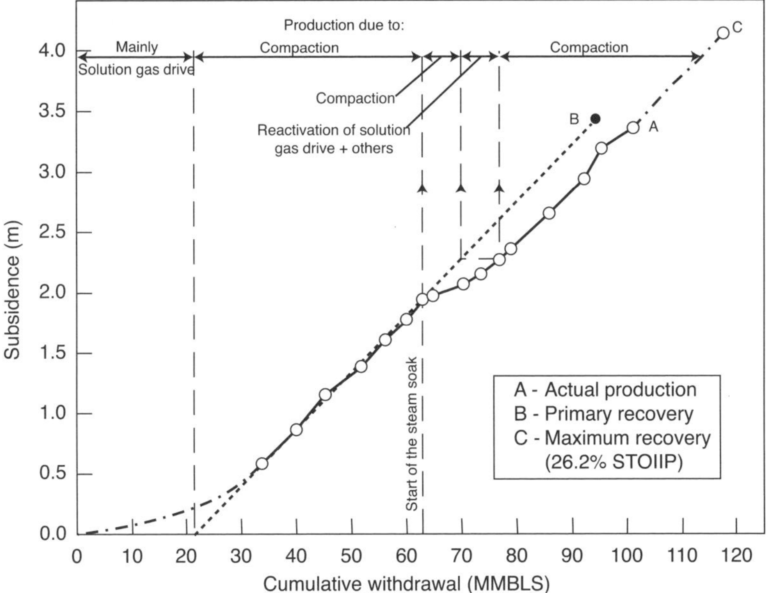

Figure 7.2

Subsidence at Surface due to Compaction during Primary Production and Steam Soak in the Tia Juana Field (after [Borregales CJ, 1977]).

7.2 COLD PRODUCTION USING HORIZONTAL WELLS

7.2.1 Horizontal Wells: a Brief History

At the beginning of the 1960s, general theoretical studies were conducted in the USSR to evaluate the productivity of complex drainage systems. They provided a basis to calculate the productivity indices of horizontal wells, either isolated or as arrays. Since the results of these studies were promising, several horizontal wells were drilled, generally with great difficulties and mitigated production success.

The main difficulties encountered involved the location of the horizontal section inside the reservoir and completion of the drainhole. In 1978, Elf (now Total) and IFP (now *IFP Energies nouvelles*) launched a large research project whose objectives were (1) to assess the technology as completely as possible and (2) to drill several pilot wells. The project was motivated by the development of the Rospo Mare field, a heavy oil karstic reservoir in the Adriatic Sea. At the end of 1984, four horizontal wells were completed and put on-stream. One of these wells was a considerable success, exceeding the predicted performance.

In 1985, horizontal drilling and production entered the industrial age. After a four-year period of acceptance, the boom occurred in 1989. This evolution is typical of a major technological breakthrough. The boom occurs after a period of observation during which the technical and economical validity of the technology needs to be proven. Once this step has been reached, the number of applications sharply rises.

The success of horizontal wells remained consistent since the first field-scale applications. The explosive increase in the number of horizontal wells drilled in many countries over the last two decades is remarkable. Improvements in drilling technology have led to lateral lengths of several kilometers with a toe placement accuracy to within a few meters.

Now, "advanced wells" are also available. "Advanced wells" refers to wells that have complex geometries and architectures. The most common (Figure 7.3) are cluster wells (slanted or curved branches drilled with different azimuths from the same vertical hole), stacked wells, multilateral wells (composed of several horizontal arms drilled from the same horizontal drains), re-entry wells and 3D wells.

Advanced wells may be considered as a new tool in the toolbox of reservoir engineers. Instead of developing new methods to move the oil to the wellbore, the wellbore can now be cost-effectively taken to the oil by drilling as many laterals as necessary to access trapped oil. In this respect, advanced wells may be considered as an IOR (Improved Oil Recovery) technique.

Horizontal wells are now the norm rather than the exception where they can provide advantages over conventional wells, e.g. increased productivity, accelerated recovery and reduced coning tendencies.

The advantages of horizontal wells as compared to conventional wells are threefold:

1) Increased production rates due to higher productivity.
2) Accelerated oil recovery due to these higher rates.
3) Increased oil recovery per well since very often the economic limit for producing a well is a minimum oil rate which is reached later with a horizontal well than with a vertical one.

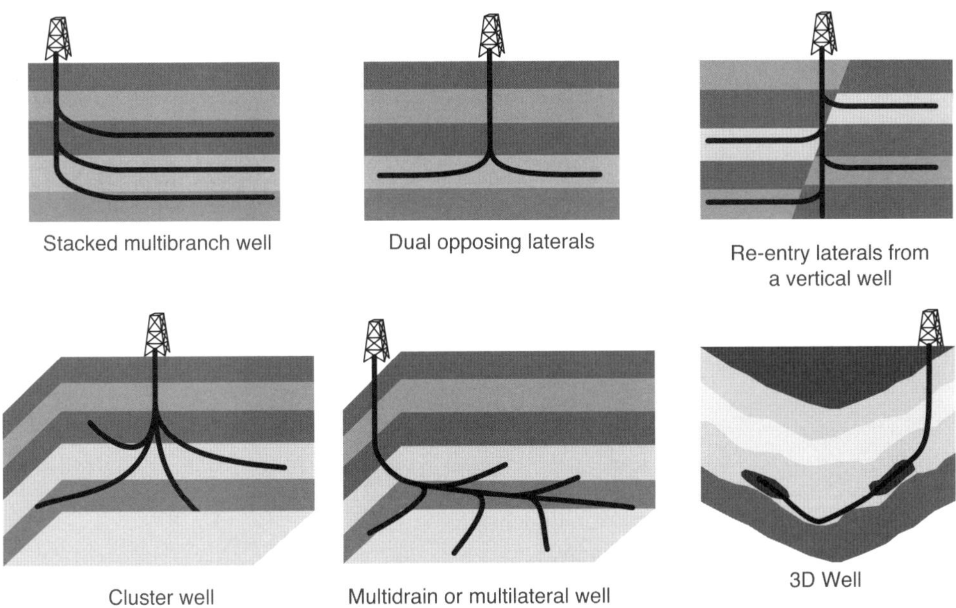

Figure 7.3

Multilateral Well Architectures Made Possible by the Advent of Horizontal Well Technology.

7.2.2 Productivity Evaluation

7.2.2.1 Theory – Principle

Horizontal wells (Figure 7.4) are differentiated from conventional wells by their larger area of contact within the reservoir. Therefore, using horizontal wells reduces the near-wellbore pressure drop that is characteristic of conventional vertical wells. This is particularly true for very thin reservoirs (a few meters thick) where a horizontal well can stay in the pay zone for the several hundred meters of its length L while a conventional vertical well will be completed only through the thickness h of the reservoir. This does not imply that the productivity of a horizontal well is greater than that of a conventional well by the ratio of L/h. For instance, the productivity of a horizontal well would be greatly impaired if the vertical permeability of the reservoir is very small, while the productivity of a vertical well would not be so affected by this permeability value.

To evaluate the potential of replacing conventional wells with horizontal wells, the easiest and most commonly used method is to compare their respective PIs (Productivity Index), Jv and Jh. Jv (vertical well productivity index) and Jh (horizontal well productivity index) are derived assuming steady-state flow conditions.

For a vertical well with radius R_w draining a cylindrical volume of radius R_{ev}, the productivity index of a vertical well is defined by [Muskat M, 1937]:

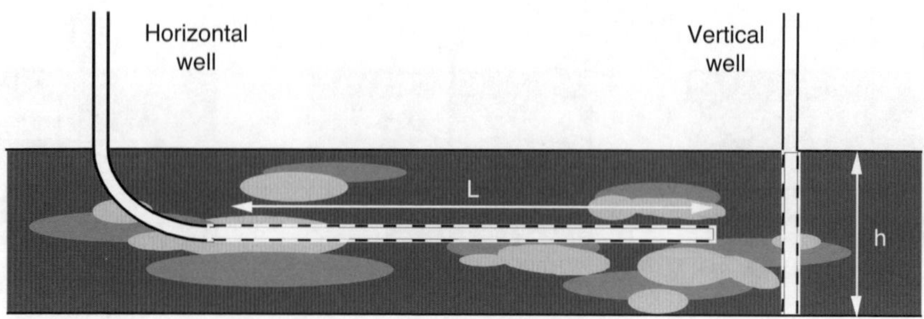

Figure 7.4

Schematic Representation of a Horizontal and a Vertical Well in a Reservoir.

$$J_v = \frac{Q}{P_e - P_w} = C\frac{2\pi K_h h}{B_o \mu_o} \frac{1}{\ln\left[R_{ev}/R_w\right] + S}$$

where B_o Oil volume factor, dimensionless
h Reservoir thickness, L
J_v Vertical well productivity index, $M^{-1}L^4T$
K_h Horizontal permeability, L^2
μ_o Oil viscosity, $ML^{-1}T^{-1}$
P_e Outer drainage pressure, $ML^{-1}T^{-2}$
P_w Wellbore bottom hole pressure, $ML^{-1}T^{-2}$
Q Well production rate, L^3T^{-1}
R_{ev} Reservoir outer radius, L
R_w Wellbore radius, L
S Skin factor (possible added resistance to flow, limited to the region around the wellbore)

with $C = 1$ if units are written in dimensionally-consistent form, or
$C = 0.08537$ for Q in m^3/d; K_h in mD; h in m; μ_o in cP or mPa.s; P_e and P_w in MPa.

Steady-state flow of fluid to a horizontal well leads to the following equation to evaluate its productivity index:

$$J_h = \frac{Q}{P_e - P_w} = C\frac{2\pi K_h h}{B_o \mu_o} \frac{1}{\cosh^{-1}(X) + \frac{\beta h}{L}\ln\left[\frac{h}{2\pi R'_w}\right] + S}$$

with $X = \frac{2a}{L}$ if the drainage area is ellipsoidal

or $X = \frac{\cosh\left[\pi a / 2b\right]}{\sin\left[\pi L / 2b\right]}$ if the drainage area is rectangular with a lateral feed at $x = a/2$

where J_h Horizontal well productivity index, $M^{-1}L^4T$
a, b Half the major and minor axes of drainage ellipse, or sides of drainage rectangle parallel and perpendicular to the horizontal well, L

K_v Vertical permeability, L^2
L Horizontal well length, L
β Permeability anisotropy ratio, $\sqrt{K_h/K_v}$
R'_w $R_w (1 + \beta)/(2\beta)$
$\cosh^{-1}$ Inverse hyperbolic cosine function,

$$\cosh^{-1}(X) = \left(X + \sqrt{X^2 - 1}\right) \quad (X \geq 1)$$

C Same value as in previous formula.

As explained in [Renard G and Dupuy JM, 1991]:

- The first term in the denominator of this equation expresses the contribution of flow far from the well such as flow to a vertical fracture of the same length as the horizontal well.
- The second term gives the flow contribution in the immediate vicinity of the well due to the convergence of streamlines towards the well.

This equation means that wellbore damage, well spacing, well length, and above all, reservoir permeability anisotropy are the most important factors determining the absolute effectiveness of a horizontal well. The computation of J_h/J_v for particular reservoirs will quickly indicate the potential of using horizontal wells instead of conventional wells or the replacement ratio to tap heavy oil reservoirs:

$$J_h/J_v = \frac{\ln\left[R_{ev}/R_w\right] + S}{\cosh^{-1}(X) + \frac{\beta h}{L}\ln\left[\frac{h}{2\pi R'_w}\right] + S}$$

The greater this ratio is, the greater the advantage of using horizontal wells. An economic assessment is also required since horizontal wells are more costly than vertical wells. An example of the calculation is provided in the next section.

7.2.2.2 Example: Pelican Lake Field

Pelican Lake field (Canada) – originally operated by CS Resources [Fontaine T *et al.*, 1991] – shows the natural initial evolution of horizontal well technology with anticipated results.

The Pelican Lake area, 300 km (188 miles) north of Edmonton in the Wabasca region of Alberta, Canada, covers a 230 km^2 area. The primary development focus is the exploitation of oil reserves in the Wabiskaw "A" which is a thin (4-6 m [13-20 ft]), shallow (409 m [1,340 ft] true vertical depth TVD), unconsolidated sand with 26% porosity and 3 darcies average horizontal permeability. The oil has a viscosity between 1,000 and 10,000 cP at reservoir conditions.

Over the period 1988-1996, CS Resources drilled 36 horizontal wells in the Wabiskaw formation (Figure 7.5), among which 3 open holes and 3 completed multilateral wells. Each of the wells was drilled using a long radius technology with horizontal sections for the main holes ranging from 448 m (1,470 ft) to more than 1,560 m (5,100 ft).

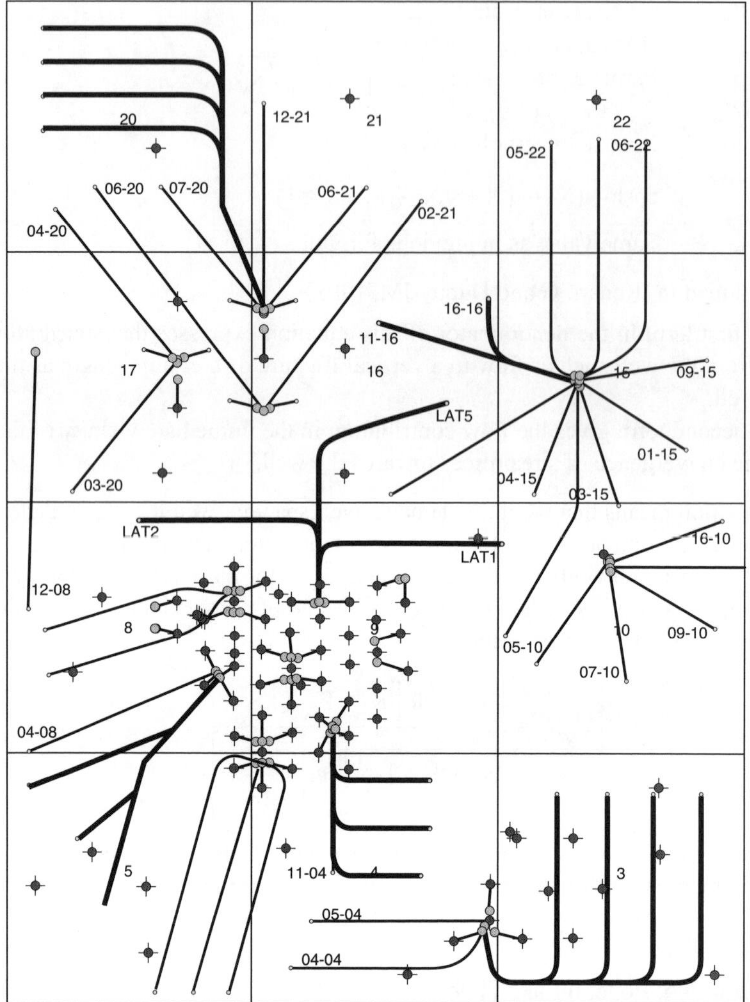

Figure 7.5

Part of Pelican Lake Field (as of 1997) showing First Horizontal (thin) and Multilateral Wells (thick).

Estimating that a typical vertical well at Pelican Lake drains a radial surface of 6 ha (14.8 acres), i.e. R_{ev} = 138 m (453 ft), and assuming that a 1,000 m (3,280 ft) long horizontal well drains a rectangular surface 300 m (984 ft) wide by 1,300 m (4,265 ft) long, the steady-state equations previously presented can be used to estimate a theoretical PI improvement with horizontal wells:

With $R_w = 0.1$ m (4 in.), $\beta = \sqrt{K_h/K_v} = \sqrt{2}$, h = 3 m (10 ft), S = 0, L = 1,000 m (3,280 ft), a = 300 m (984 ft), b = L + a = 1,300 m (4,265 ft), we get $J_h/J_v = 13.2$, value that is consistent with field values for 1,000 m (3,280 ft) long horizontal wells.

Another example with a = 300 m (984 ft), L = 1,300 m (4,265 ft), b = L + a = 1,600 m (5,250 ft), gives $J_h/J_v = 16.6$ (Figure 7.6).

The relative performance of horizontal wells *versus* vertical wells, and now of multilaterals *versus* horizontal wells, is clearly established at Pelican Lake. For this type of reservoir produced under solution-gas drive, the longer the horizontal borehole in a good quality reservoir, the higher the productivity and producible reserves with equivalent interwell spacing and drainage areas.

In 1991, CS Resources was the first in Canada to drill an open hole lateral arm off a horizontal well. Through the use of chemical tracers, oil production from the end of the lateral arm was confirmed. In 1993, CS Resources innovated the successful drilling of the first multilateral horizontal well utilizing the Lateral Tie-Back System (LTBS™), jointly developed by Sperry-Sun, CS Resources and IFP (now *IFP Energies nouvelles*) [Smith RC *et al.*, 1994]. Several multilateral wells were successfully drilled with total exposure to the reservoir of more than 8,000 m (26,250 ft) with several laterals of more than 1,400 m (4,593 ft). Figure 7.7 shows the acceleration in production brought about by horizontal and later by multilateral wells.

The placement of the wells in Pelican Lake has evolved since the start of the project. First horizontal wells were drilled in a spoke-wheel fashion. Now, horizontal wells are drilled parallel to each other to allow for better reservoir management with better possible drainage architecture to sustain production in the future. It must be recognized that even if economic, primary recovery of thin heavy oil reservoirs with horizontal wells will reach only a few percent of the oil in place. Thus, reservoir management using horizontal wells should take into account the possibility of using these wells to implement secondary or tertiary recovery methods.

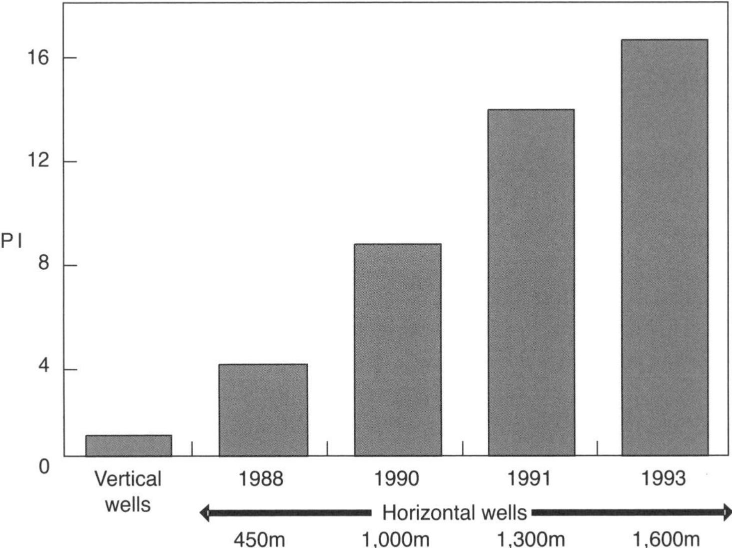

Figure 7.6

Comparison of PI of Horizontal & Vertical Wells in the Pelican Lake Field.

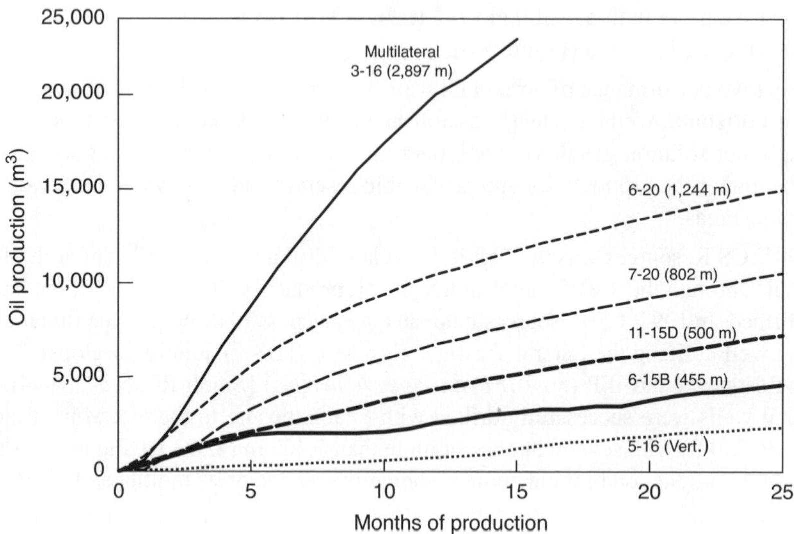

Figure 7.7

Acceleration in Oil Production with Horizontal and Multilateral Wells in the Pelican Lake Field.

Since the 1980s, horizontal wells and multilateral wells have found wide acceptance in the petroleum industry. As an example, this new technology has enabled the development of the huge heavy oil reserves of Venezuela (Orinoco Belt) and Canada (Alberta). Figure 7.8 shows the very complex well architecture used by Petrozuata in the Orinoco Belt to set and keep the well branches in the saturated zones of the very heterogeneous fluvio-deltaic reservoir to be produced.

7.2.3 Reservoirs in the Presence of a Bottom Aquifer

7.2.3.1 Theory

As previously discussed, horizontal wells generally have higher productivity indices than vertical wells. This results in horizontal wells being able to produce either at the same flow rate with lower pressure drawdown, or at a greater flow rate with the same pressure drawdown as compared to vertical wells. Moreover, the pressure gradients within the reservoir will be less pronounced when horizontal wells are utilized. Horizontal wells can be drilled and positioned near the top of the reservoir utilizing Measurements While Drilling (MWD) and Logging While Drilling (LWD), thereby maximizing the distance from the aquifer. This allows for an efficient method to economically produce heavy oil reservoirs which are subject to water coning problems.

Coning/cresting occurs when bottom water migrates upward through the oil column to the well (Figure 7.9).

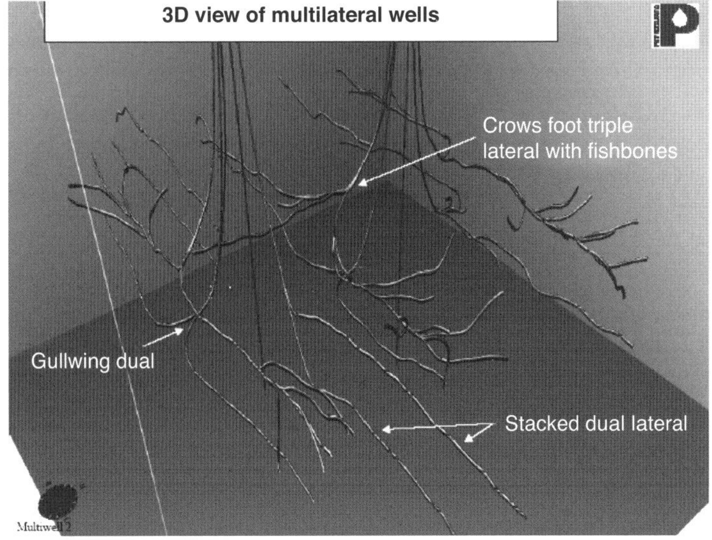

Figure 7.8

Example of Complex Well Architecture used by Petrozuata to Produce Heavy Oil in the Orinoco Belt. Origin: L. Summers – Petrozuata C.A.

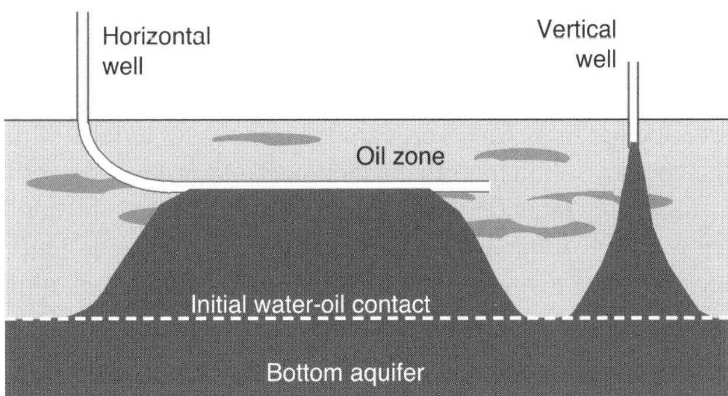

Figure 7.9

Less Coning/Cresting Effect Experienced with Horizontal Wells as Compared to Vertical Wells.

In the last twenty years, many authors have addressed the issue of water coning or cresting toward vertical and horizontal wells [Joshi SD, 1991]. The first analytical models aimed at determining the critical (or maximum) rate for which water would not migrate and break through to the well, and the breakthrough time if the well was produced at supercritical rates (higher than the critical rate). In heavy oil reservoirs, the critical rate and breakthrough time are irrelevant as the critical rates are very low for both the vertical and horizontal wells and consequently, water breakthrough is rapid (commonly a few weeks or months). In heavy oil reservoirs, an important aspect of water coning or cresting is to evaluate the evolution of the water-cut after breakthrough and determine the influence of main parameters on the ultimate recoverable reserves: horizontal well length, production flow rate, water/oil mobility ratio, well spacing, and reservoir thickness. Generally this evaluation is performed to an economic limit usually defined by a minimal oil rate.

A semi-analytical approach, which is well suited for homogeneous reservoirs, was proposed to forecast the production of vertical or horizontal wells subject to water coning [Pietraru V and Cosentino L, 1993]. The solutions are based on the hypothesis of gravity drainage and transient flow. In addition to the usual parameters – critical oil rate and breakthrough time – the proposed methodology can be used to determine the water-cut for a variable oil rate and the maximum oil rate variable with time for an imposed water-cut (production forecast).

For heterogeneous or layered reservoirs, analytical solutions do not apply and numerical modeling is required. Heterogeneities, such as impermeable horizontal shale streaks, can greatly affect the performance of horizontal wells in the presence of a bottom aquifer [Lara A and Renard G, 1993]. Numerical modeling is also required for simulating the production of horizontal wells which have complex trajectories with low points near the aquifer zone that create a higher degree of susceptibility to the coning/cresting phenomena.

7.2.3.2 Example: Winter Field

The Winter pool is about 75 km (47 miles) southeast of the City of Lloydminster in Saskatchewan, Canada [Sharp DA *et al.*, 1991]. The productive interval is the Cummings Formation, which is an unconsolidated sandstone occurring at a depth of approximately 750 m (2,460 ft). Its average porosity is 30% and its permeability is 6 darcies. The reservoir has a total thickness of 30 m (98 ft), half of which is impregnated by a 14°API oil (viscosity of 3,000 cP at reservoir conditions) over bottom water aquifer.

Historical oil production rates from vertical wells demonstrated that further utilization of conventional wells would result in marginal economics. Oil production per vertical well was very limited with initial oil rates between 2 and 7 m^3/d (13 and 44 bopd). After a short period of production at low water cut (3 to 9 months), the water production rate quickly increased in all vertical wells to an average of 70 percent. Cumulative oil production at this point was less than 0.3% of the original oil in place.

In mid-1988, the operator undertook the drilling of five horizontal wells. The existing 14 vertical wells which were previously drilled revealed rapid change in geological facies within some of the wells. In Winter-type reservoirs, geological control remains an important factor and considerable attention must be devoted to structural and stratigraphic interpretations to ensure that a majority of the horizontal section is placed within the most favorable portion of the reservoir. In the case of the first well, a portion of the horizontal section penetrated a previously undetected shale channel, resulting in approximately 50% of the well being non-productive. To improve the placement of the following horizontal wells, a 3D seismic survey was acquired (Figure 7.10). The interpretive results offered a substantially improved understanding of reservoir configuration and facilitated the process of determining appropriate horizontal well orientation and standoff within the reservoir (Figure 7.11). To enhance productivity, the horizontal sections of new wells were increased from 500 m to 750 m (1,640 to 2,460 ft). As a result, these wells were much more successful than the first one and reached the economic viability of production from this pool.

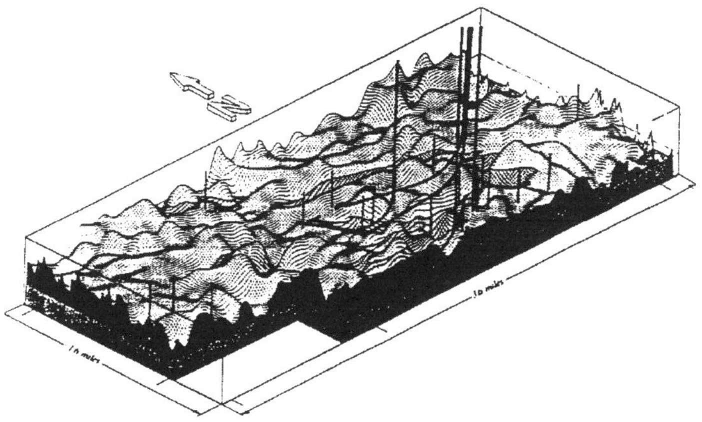

Figure 7.10

Winter field - Perspective View of the Reservoir from 3D Seismic.

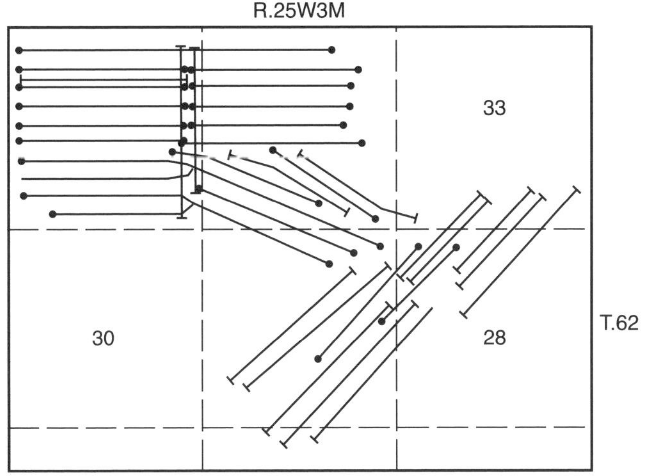

Figure 7.11

Winter Project Area (as of Dec. 1994).

Following the results of the initial horizontal well program, a history match using numerical reservoir simulations was carried out to quantify the estimated reserves per well and optimal well spacing. Results indicated that a 500 m long horizontal well (1,640 ft) producing at 60 m^3/d (380 bopd) would recover approximately 19,000 m^3 (119,000 bbl) before the water-cut reaches 90%. This is about eight times greater than the cumulative oil production obtained from a vertical well in the Winter field. Other important results from simulations were that the production would be accelerated by increasing the length of the horizontal wells, and the optimal spacing would be 75 m (246 ft) for parallel horizontal wells. The ultimate oil recovery was estimated to be 10.5% of original oil in place. Figure 7.12 presents the production performance of a typical horizontal well compared to a typical vertical well. The production data from the Winter field is plotted in Figure 7.13.

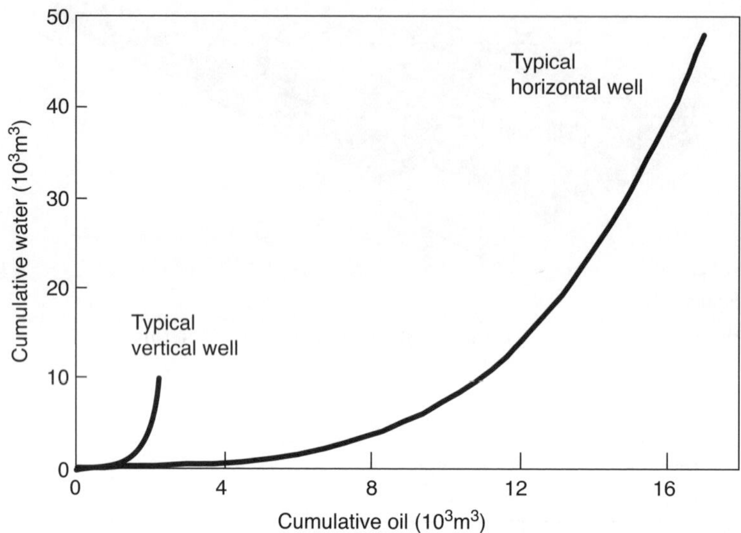

Figure 7.12

Comparison of Production Performance of Typical Horizontal and Vertical Wells.

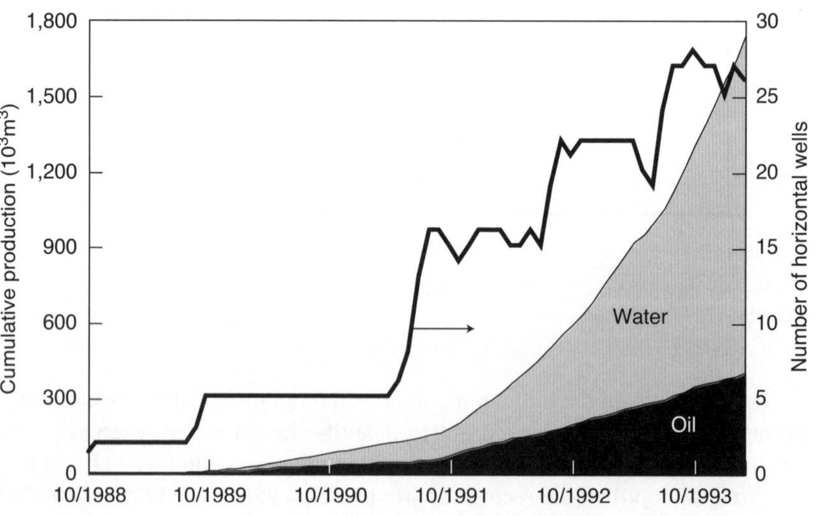

Figure 7.13

Winter Field (Canada). Performance of Horizontal Wells.

7.3 COLD HEAVY OIL PRODUCTION WITH SAND (CHOPS)

7.3.1 What is CHOPS?

In heavy oil reservoirs, promoting sand to enter the wellbore along with fluids has resulted in significant improvements in production by factors of 10 or more [Renard G *et al.*, 2000]. This technique is called CHOPS (**C**old **H**eavy **O**il **P**roduction with **S**and) and has found numerous applications, mainly in Canada.

Four mechanisms are thought to be responsible for the surprising oil rate enhancement in CHOPS wells [Dusseault M, 2002]:

1) Fluid flow rate is increased if the sand matrix is allowed to move because the Darcy velocity relative to the solid matrix increases with matrix movement.
2) As sand is produced from the reservoir, a zone of enhanced permeability is generated and grows outward, allowing a greater fluid flux to the wellbore.
3) In a highly viscous oil, a pressure drop below the bubble point pressure leads to gas trapping and the generation of a "foamy oil" zone (see 7.4), which is supposed to generate continuing sand destabilization and drive solids and fluids toward the wellbore.
4) Solids movement in the near-wellbore environment eliminates fines trapping, asphaltene deposition, or scale development on the formation matrix outside the casing. At the same time a large amount of sand is produced.

Each of these mechanisms has been investigated by past and ongoing theoretical work. Brief reviews of these mechanisms are presented in the following sections.

7.3.2 Field Observations

Voluntary sand production has been made possible by the use of Progressive Cavity Pumps (PCP). In fact, this type of pump allows for the production of effluents containing a significant fraction of solids. Sand production is significant at the start of well production, possibly reaching 30 to 50% of the initial flow rate (by volume). It then gradually decreases and stabilizes at a few percent after several months. Cumulative sand production reaches several hundred – and sometimes several thousand – cubic meters after several years of production. For example, Figure 7.14 shows the production of oil and sand recorded from a well in the Frog Lake field in Canada [Huang WS, 1997].

The cause-effect relationship between sand production and improved oil productivity and production is now understood in poorly consolidated heavy oil reservoirs in Canada [Renard G *et al.*, 2000]: sand production drives the formation of zones of increased permeability around the wells, but the geometry of these high-permeability zones has not been determined.

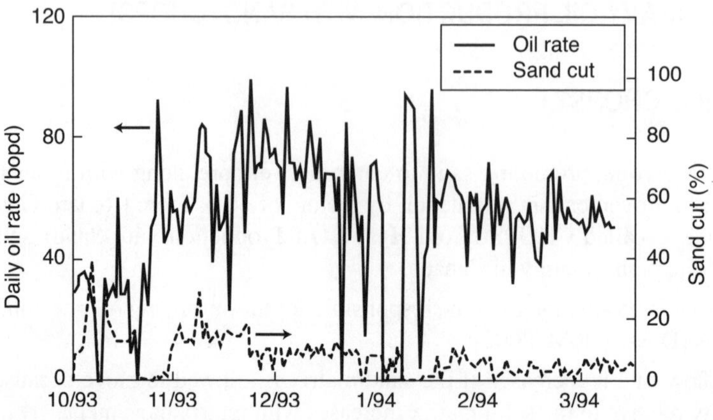

Figure 7.14

Production of Oil and Sand from a Well in the Frog Lake Field [Huang WS, 1997].

7.3.3 Theoretical Explanations

For continuous sand production, a "liquefied" zone around the wells must grow in order to supply the sand. Two scenarios are possible. In the first scenario, the liquefied zones around the perforations grow spherically and fuse, forming a liquefied cylindrical zone around the wells. This is the "compact growth" scenario.

In the second scenario, each of the zones – or at least some among them – starts to spread longitudinally in the producing layer, creating a network of high-permeability channels: wormholes. In this case, the boundary between the reservoir's intact and remolded zones cannot be rigorously defined.

However, the physical mechanisms are similar in both cases: the loss of cohesion and liquefaction of sand are caused by the hydraulic gradient.

Some authors even propose a combination of both schemes (Figure 7.15) [Tremblay B. *et al.*, 1998]. Field data results do not allow one hypothesis to be chosen over the other. Tests using tracers between wells have demonstrated the existence of rapid "communication" between the wells [Elkins L *et al.*, 1988]. Some authors believe that this "communication" is the result of wormholes, while others believe these are simply fractures. But logs perpendicular to some wells show very high porosities (greater than 50%) which may be the trace of remolded zones.

7.3.3.1 Compact Growth of the Remolded Zone

This hypothesis assumes simultaneous dual flows, with one consisting of viscous fluid and the other of granular matrix, and including interaction between the two [Geilikman *et al.*, 1994; Geilikman *et al.*, 1995]. When fluid pressure in the wells is decreased below a critical threshold, the conditions for shear rupture are met and a yielded and flowing zone starts to

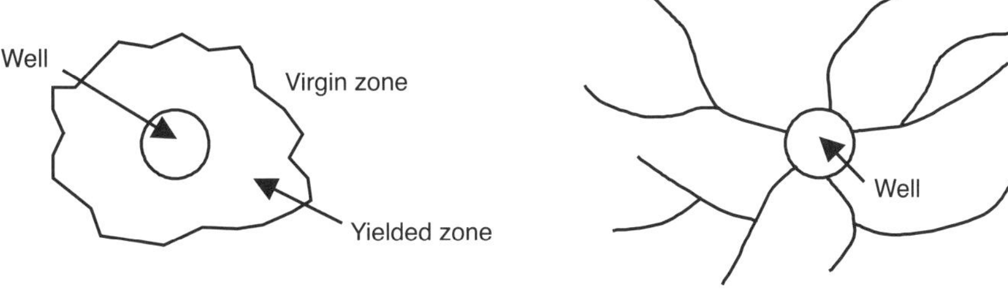

Figure 7.15

Geometry of Zones of High Permeability (Wormholes Assumed in the Figure on the Right).

grow around the wells. The drawdown (difference between pressure in the reservoir and pressure in the wells) is the key parameter involved. Production of solids is assimilated to a phenomenon of fraying and disentangling. The boundary zone between the intact zone and the yielded zone propagates in the opposite direction of the flow. As long as the drawdown is above a certain threshold, the front continues to propagate. Since reservoir pressure decreases with time, the drawdown decreases and falls below the critical threshold. The sand then undergoes compaction. At this stage, the front stops propagating.

The application of models based on this approach results in productivity gains on the order of 2, lower than those currently observed in the field, which are between 10 and 100.

7.3.3.2 Growth by Wormholes

This approach assumes that wormholes have a stable structure that is approximately cylindrical and of constant cross-sectional area along their length. The wormhole is filled with a slowly flowing suspension and the tips of wormholes continue to propagate. This approach is supported by much experimental work undertaken at ARC [Tremblay B *et al.*, 1998] and by the analogy with internal erosion schemes observed in earthfill dams. The diameter of the wormholes can be highly variable (from 5 cm to 1 m).

The contact surface area for the wormholes could be far larger than that for compact growth.

7.3.3.3 New Approaches

A new approach has been developed based on the theory of macroscopic damage [Shao JF and Marchina P, 2002]. The zone around wells which contains heterogeneities (wormholes, cavities, etc.) is replaced by a homogenous "damaged" zone. The change in porosity results in increased permeability. At the same time, mechanical properties gradually deteriorate. Numerical simulations with finite element computer code result in the same oil and sand production profiles observed in the field.

Another original poromechanical approach has also been proposed, and makes no prejudgment of an initial erosion schema [Yalamas T *et al.*, 2004]. This approach distinguishes two zones: an intact reservoir zone exhibiting poroelastic behavior, and a mixed sand – oil zone exhibiting viscous fluid behavior. No hypothesis is made regarding the geometry of the boundary between these two zones. An algorithm can be used to predict the evolution of the boundary according to coupled hydraulic and poromechanical conditions [Servant G *et al.*, 2006].

7.4 FOAMINESS OF HEAVY OILS

As a reservoir is produced, the pressure decreases. When the pressure reaches the bubble point pressure of the oil, the lightest components of the oil (mainly methane) are released and gas bubbles are liberated. As previously discussed, this process – known as solution-gas drive – helps in oil recovery by expansion of the gas released from the crude. When considering a heavy oil, the process can be described in three steps:

1) **Gas nucleation**: corresponding to the release of the dissolved light components into a free gas phase when pressure is decreased below the bubble point.
2) **Bubble growth**: corresponding to mass transfer by molecular diffusion of the dissolved light components to the free gas phase.
3) **Gas mobilization**: Above a given gas saturation S_{gc}, called critical gas saturation, the gas phase becomes connected and is produced preferentially because of its higher mobility in comparison to oil.

The oil viscosity has an important effect on the coalescence dynamics of gas bubbles during bubble nucleation, as well as during later stages when bubbles grow primarily by expansion. On the one hand, for a light oil, the gas bubbles rapidly coalesce to form a continuous gas phase. For a heavy oil, the gas bubbles remain dispersed in the oil phase for a very long time. On the other hand, the presence of gas trapped and dispersed in the oil has an influence on its viscosity. Understanding the role played by the foamy character of an oil on its rate of production and ultimate recovery by solution gas-drive has attracted much interest in the petroleum industry since, if it can be explained, it will open new ways to optimize production.

Studies performed on foamy oils have been conducted on many aspects:

- Influence of oil chemistry on oil viscosity and ability to trap the gas bubbles or let them form a continuous phase [Bauget F *et al.*, 2001].
- Influence of the presence of gas bubbles on the oil viscosity [Abivin P *et al.*, 2006].
- Influence of bubble formation mechanisms under pressure decrease. A critical review of these mechanisms [Bauget F, Lenormand R, 2002] has demonstrated that the pre-existence of bubbles in bulk or in crevices is the only theory that is physically justified and can entirely explain the available experimental observations.
- Influence of depletion rate on final oil recovery [Maini B, 1999].

The concept of pre-existing bubbles has been used to develop a dynamic model [Bauget F *et al.*, 2003] that is able to match the experimental results as shown in Figure 7.16. The parameters of the model are determined from the two extreme depletion experiments performed at 53 and 9 bar/d respectively, and the third intermediate set of data performed at 28 bar/d is simulated with these parameters.

By matching the experimental data, the dynamic model can be used to determine the specific set of relative permeabilities required for each depletion rate, and then to determine the set of relative permeabilities for any other depletion rate, especially the depletion rate used in the field.

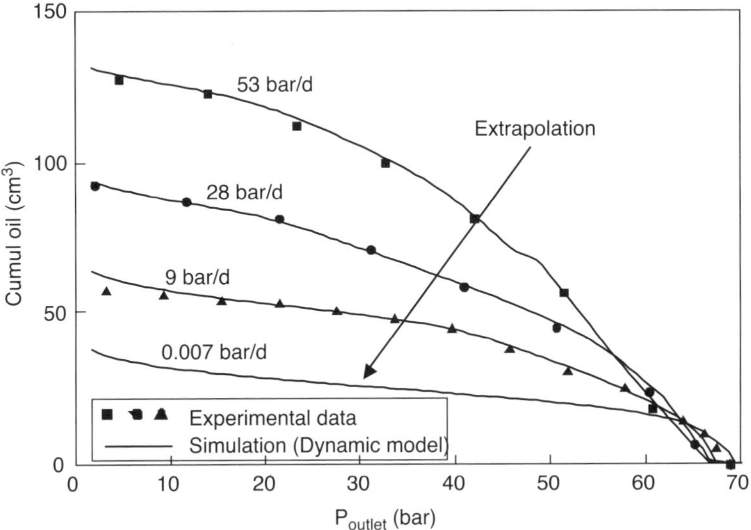

Figure 7.16

Simulation of the Experiments Performed Using a Venezuelan Heavy Oil (295 cP) with the Dynamic Model of Pre-Existing Bubbles [Bauget F *et al.*, 2003].

7.5 CONCLUSIONS

Cold production of heavy and extra-heavy oils is generally limited to a few percent of Original Oil In Place (OOIP). However, it is used primarily in most reservoirs producing through solution-gas drive before implementing any enhanced recovery method.

Horizontal well technology has been very efficient for the development of many assets where heavy oils are present, such as Canada and Venezuela. Horizontal wells can not only improve productivity and production, but also reduce coning tendencies in reservoirs where a bottom aquifer is present.

The CHOPS process is successful in some places. The role of foaminess and the technology of sand production, however, need to be more thoroughly investigated in order to optimize this process and extend the domain of its application.

However, since it leads to low recovery, cold production needs to be followed by enhanced recovery methods. The methods usually implemented are described in the next chapter.

REFERENCES

Abivin P, Hénaut I, Argillier J-F, Moan M (2006) Viscosity Behaviour of Foamy Oil: Experimental Study and Modeling (presented at the *Annual European Rheology Conference*, Hersonissos, Crete, 27-29 April, 2006).

Bauget F, Langevin D, Lenormand R (2001) Effects of Asphaltenes and Resins on Foamability of Heavy Oil. SPE 71504 (presented at the *SPE Annual Technical Conference*, New Orleans, 30 Sept. – 3 Oct., 2001).

Bauget F, Lenormand R (2002) Mechanisms of Bubble Formation by Pressure Decline in Porous Media: a Critical Review. SPE 77457 (presented at the *SPE Annual Technical Conference*, San Antonio, 29 Sept. – 2 Oct., 2002).

Bauget F, Egermann P, Lenormand R (2003) A New Model to Obtain Representative Field Relative Permeability for Reservoirs Produced Under Solution Gas Drive. SPE 84543 (presented at the *SPE Annual Technical Conference*, Denver, 5-8 Oct., 2003).

Borregales CJ (1977) Steam Soak on the Bolivar Coast (published in Canada-Venezuela Oil Sands Symposium, Edmonton, 30 May-4 June, 1977. Reprinted in *The Oil Sands of Canada-Venezuela*, 1977, CIM Special volume 17).

Butler RM (1991) Thermal Recovery of Oil and Bitumen, GravDrain Inc.

Dusseault M (2002) CHOPS – Cold Heavy Oil Production With Sand in the Canadian Heavy Oil Industry (Available on www.energy.gov.ab.ca/oilsands/1189.asp).

Elkins L, Morton D, Blackwell W (1972) Experimental Fireflood in a Very Viscous Oil-Consolidated Sand Reservoir. SPE4086 (presented at the *1972 ATCE Conference*, San Antonio, 8-11 October, 1972).

Fontaine T, Hayes L, Reese G (1993) Development of Pelican Lake Area Using Horizontal Well Technologies. *JCPT*, **32**, 9, pp 44-49.

Geilickman M, Dusseault M Dullien, F.A (1994) Sand Production as a Viscoplastic Granular Flow. SPE 27343 (presented at the *SPE Formation Damage Control Symposium*, Lafayette, 7-10 Feb., 1994).

Geilickman M, Dusseault M, Dullien FA (1995) Dynamic Effects of Foamy Fluid Flow in Sand Production Instability. SPE 30251 (presented at the *SPE International Heavy Oil Symposium*, Calgary, 19-21 June,1995).

Huang WS, Marcum BE, Chase MR, Yu CL (1997) Cold Production of Heavy Oil From Horizontal Wells in the Frog Lake Field. SPE 37545 (presented at the *SPE ITOHOS Symposium*, Bakersfield, 10-12 February, 1997).

Joshi SD (1991) Horizontal Well Technology. PennWell Publishing Company.

Lara A, Renard G (1993) Selective Tests in Horizontal Wells for a Better Reservoir Characterization. SPE 26447 (presented at the *68th SPE ATCE Conference*, Houston, 3-6 Oct., 1993).

Maini B (1999) Laboratory Evaluation of Solution Gas Drive Recovery Factors in Foamy Oil Reservoirs. Paper 99-44 (presented at the *CSPG and Petroleum Society Joint Convention*, Calgary, 14-18 June, 1999).

Muskat M (1937) The Flow of Homogeneous Fluids Through a Porous Media. Reedited by Intl. Human Resources Development Corp., Boston.

Pietraru V, Cosentino L (1993) A New Analytical Approach to Water and Gas Coning for Vertical and Horizontal Wells. *Revue de l'Institut Français du Pétrole*, **48**, 5, Sept. 1993, pp 501-513.

Renard G, Dupuy JM (1991) Formation Damage Effects On Horizontal-Well Flow Efficiency. *JPT*, **30**, 1, pp 786-789, 868-869.

Renard G, Nauroy J-F, Deruyter Ch, Moulu J-C, Sarda J-P, Le Romancer J-F (2000) Production froide des huiles visqueuses – Première partie: Observations sur champ. Oil & Gas Science and Technology – *Revue IFP*, **55**, 1, pp 35-66.

Servant G, Marchina P, Peysson Y, Bemer E, Nauroy J-F (2006) Sand Erosion In Weakly Consolidated Reservoirs: Experiments And Numerical Modelling. SPE 100023 (presented at the *SPE/DOE IOR Symposium*, Tulsa, 22-26 April, 2006).

Shao JF, Marchina P (2002) A Damage Mechanics Approach for the Modelling of Sand Production in Heavy Oil Reservoirs. SPE/ISRM 78167 (presented at the *SPE/IRSM Rock Mechanics Conference*, Irving, 20-23 Oct., 2002).

Sharp DA, Coffin Ph, Giannesini J-F, Lessi J (1991) The Application of Horizontal Production Systems. *13th World Petroleum Congress*, Buenos Aires, Oct, 1991.

Smith RC, Hayes LA, Wilkin JF (1994) The Lateral Tie-Back System: The Ability to Drill and Case Multiple Laterals. IADC/SPE27436 (presented at the *IADC/SPE Conference*, Dallas, Feb. 1994).

Tremblay B, Sedgwick G, Vu D (1998) CT Imaging of Sand Production in a Horizontal Sand Pack using Live Oil. CIM 98-78 (presented at the *49th CIM Annual Technical Meeting*, Calgary, 8-10 June, 1998).

Tremblay B, Sedgwick G, and Vu D (1998) CT imaging of Wormhole Growth Under Solution-Gas Drive. SPE 39638 (presented at the *SPE/DOE IOR Conference*, Tulsa, 19-22 April, 1998).

Yalamas T, Nauroy J-F, Bemer E, Dormieux L, Garnier D (2004) Sand Erosion in Cold Heavy Oil Production. Paper SPE 86949 (presented at the *SPE International Thermal Operations and Heavy Oil Symposium*, Bakersfield, 16-18 March, 2004).

8 | Enhanced Recovery

G. Renard

As indicated in the previous chapter, primary production of heavy and extra-heavy oils is generally limited to a few percent of Original Oil In Place (OOIP). Therefore, increasing their recovery efficiently and economically, *via* the implementation of specific enhanced processes, is required. The following sections describe the most salient characteristics of thermal and chemical "Enhanced Oil Recovery" processes applied by operators to increase recovery of heavy and extra-heavy oils, as well as new techniques such as SAGD (Steam Assisted Gravity Drainage) designed to recover bitumen.

8.1 HOT FLUID INJECTION

As indicated in Table 7.1, high to very high viscosity of heavy oil or bitumen sharply decreases with temperature. As a general rule, and for a given increase in temperature, the more viscous the oil, the more significant its viscosity decrease. Heat injection into the reservoir *via* the injection of hot fluids, hot water, steam or hydrocarbon solvent is therefore one of the most widely applied techniques to decrease oil viscosity *in situ* and allow for better oil flow towards production wells ([Prats M, 1982]; [Burger J *et al.*, 1985]; [Bavière M *et al.*, 1991]; [Sarathi PS and Olsen DK, 1992]). Reduction of oil viscosity is not the only effect of hot fluid injection. However, as indicated in Figure 8.1, viscosity reduction by heat injection is largely predominant for heavy oils.

Hot fluid injection can be performed by various methods, the details of which are described in the following sections.

8.1.1 Hot Water Injection

There has been some economic success for hot waterflooding, the most significant being at the Schoonebeek field in the Netherlands where oil viscosity at reservoir conditions is moderate. However, hot water has lower heat content than steam as the transformation of water to steam requires energy – known as latent heat – which can then be released into the reservoir as the steam condenses in order to more efficiently heat the reservoir and the fluids it contains. Furthermore, field applications of hot water flooding have been plagued by shale swelling, severe channeling and high water-to-oil ratios, generally implying low sweep efficiency. It has also been observed that the residual oil level that can be achieved with a steam

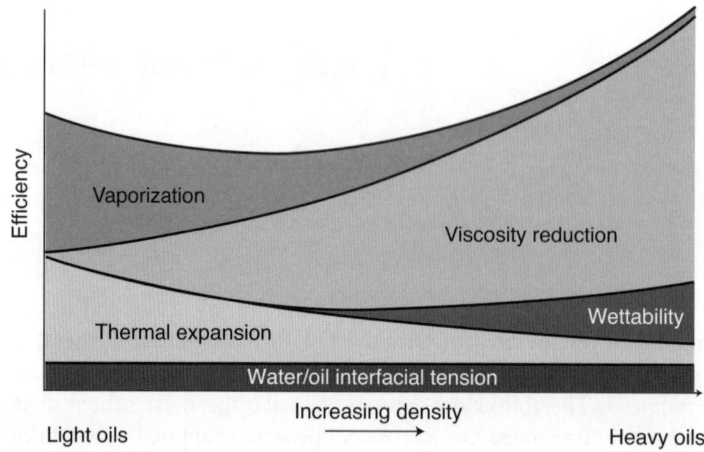

Figure 8.1

Oil Displacement by a Hot Fluid: Contribution of Various Mechanisms [Burger *et al.*, 1985].

flood is markedly lower than that found with hot water flood – even at the same temperature. Therefore, it seems that poor recovery efficiency makes waterflooding of very viscous crudes uneconomical even if hot water is used. For these reasons, steam injection is preferred to water injection and hot water flooding is more widely applied as a follow-up treatment to steamflooding.

8.1.2 Cyclic Steam Stimulation (CSS)

8.1.2.1 Description of Process and Mechanisms

Steam soak – another term that is frequently used for Cyclic Steam Stimulation (CSS) or "Huff & Puff" – was discovered as a promising production method rather accidentally in 1959 by Shell [Giusti LE, 1974] in Venezuela during early steam drive testing in the Mene Grande Tar Sands (Figure 8.2). When steam erupted at the surface due to breakdown of the overburden, the injection wells were backflowed to relieve the reservoir pressure. This resulted in high oil production rates, all the more impressive since the reservoir was unproducible by primary means.

It was concluded that injection of limited amounts of steam might be a very effective method for stimulating production of heavy-oil wells.

Following this finding, there was very rapid growth in the use of steam stimulation, particularly in several reservoirs in California: Kern River, Midway Sunset, South Belridge, San Ardo, etc, where steamflooding was used intensively afterwards.

In the CSS process, several cycles are performed in the same well, successively injector and producer. Figure 8.3 shows a schematic view of two successive steam stimulation cycles.

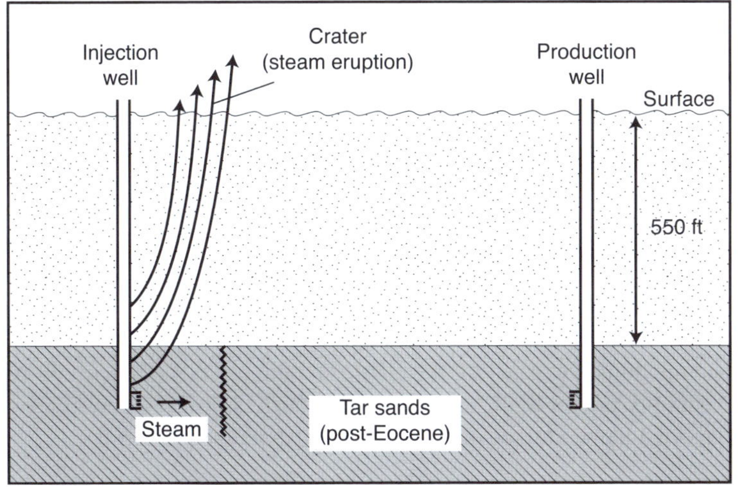

Figure 8.2

Birth of the Steam Soak Process at Mene Grande Tar Sands Steam Drive [Giusti LE, 1974].

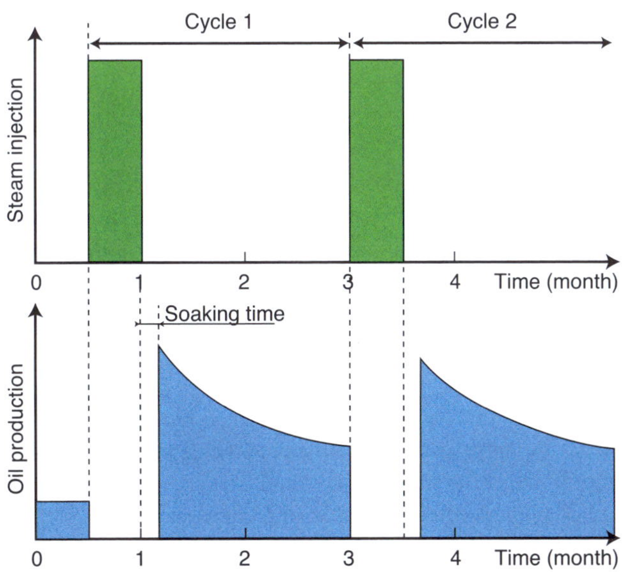

Figure 8.3

Schematic Representation of Two Successive Cycles of Steam Stimulation.

Each cycle consists of:

- An injection period – generally lasting one to three weeks – during which a slug of steam is injected into the reservoir layer. The slug size depends mainly on reservoir thickness and permeability, as well as oil viscosity (i.e. $h \times k/\mu$). It usually varies from a few thousand up to ten to fifteen thousand cubic meters of CWE (Cold Liquid Water Equivalent). The steam is injected into the reservoir at rates on the order of 40 to 100 m^3/d/m of thickness (85 to 210 bspd/ft) CWE.
- A soaking period of a few days during which the well is shut in to allow heat transfer from the condensing steam to the *in situ* fluids and rock.
- Finally, a production period allowing the well to flow back naturally and then be pumped. Production usually lasts from half a year to one year per cycle. It ends when an economic limit is reached for the oil-steam ratio or OSR (the term for its reciprocal – the steam-oil ratio or SOR – is also used), which is the ratio between the volume of oil produced and the volume of steam injected, expressed in CWE.

Usually, after two to three cycles, subsequent cycles become less effective. It is important to note that the economic limit is field-dependent.

When cyclic steam injection is used, the injection pressure is usually limited during the steam injection period to prevent formation fracturing. Imperial Oil has implemented the cyclic steam stimulation process to produce bitumen in its Cold Lake field in Canada. Injection pressure is intentionally above fracturing pressure to allow steam and hot water to penetrate deeper into the reservoir. This development started with small-scale pilots in the early 1960s and has continued to large field scale, with current bitumen production in excess of 13,000 m^3/d (81,800 bopd).

Cyclic steam has been tested in horizontal wells in many reservoirs, but results have not been conclusive.

8.1.2.2 Performance Prediction

A significant feature of steam stimulation is that the injected heat is concentrated near the well bore. Steam during flow-back also has a cleansing effect near the wellbore *via* removal of heavy residues. Injectivity and productivity of the stimulated wells are therefore increased. The radius invaded by cyclic steam is limited to 15 to 20 m (50 to 65 ft) around the wellbore. Cyclic steam stimulation is generally necessary to reduce the injection pressure when continuous steam injection between wells in a pattern-like fashion is used (see next section).

In Venezuela, the amount of steam injected per cycle – [roughly 8,000 m^3 (around 50,000 bbl)] – is generally larger than in Californian fields (from 700 to 2,200 m^3 [4,400 to 14,000 bbl]), mainly due to differences in pay thickness, reservoir permeability and oil viscosity. The quantity of oil recovered per cycle in Venezuela is also larger than that in California. The higher Oil Steam Ratios (OSR) common in the Bolivar Coast are considered to be significantly influenced by compaction of the oil-bearing sands during production. Figure 8.4 shows an example of typical cyclic steam stimulation in a Venezuelan field [Borregales CJ, 1977].

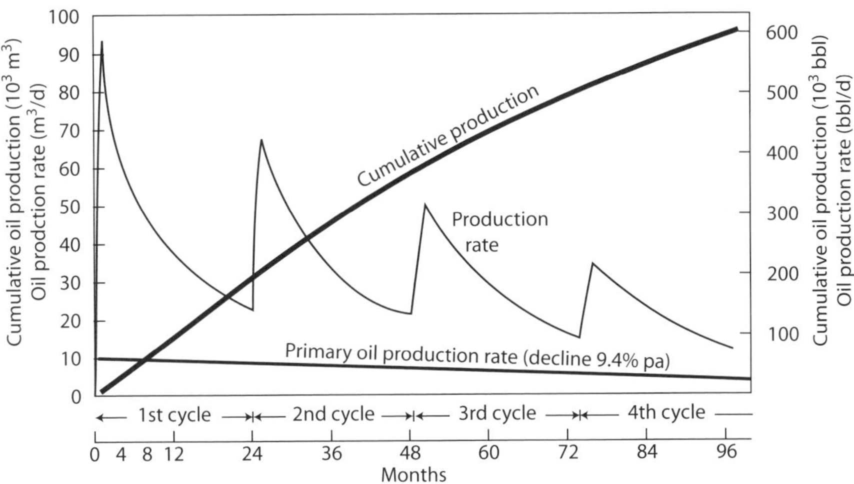

Figure 8.4

Typical Cyclic Steam Stimulation in a Bolivar Coast Field of Venezuela [Borregales CJ, 1977].

8.1.2.3 CSS: Favorable Factors

Reservoirs are generally considered good candidates for cyclic steam stimulation when their Cumulative Oil Stream Ratio (COSR) is at least 5 to 10 times the OSR of 0.075 m^3/m^3 (thus, between 0.37 and 0.75 m^3/m^3). This OSR corresponds to a thermal efficiency of 1. The thermal efficiency is the ratio of the heat in the produced oil to the heat required in the steam generators. A value of 1 means that all the oil produced is burned in the generators.

Favorable factors that lead to such high COSR for cyclic steam injection are:

- Shallow to limited reservoir depth (< 800-1,000 m [2,600-3,300 ft]) in order to limit heat losses in the injection wells and inject steam of sufficient quality into the reservoir. Available latent heat present in the injected steam will be higher in shallow reservoirs having lower pressures (in this respect, reservoir pressure and depth are interrelated screening criteria. For instance, the Boscan field is at a depth of about 2,000 m [6,560 ft], but it is depleted).
- Sufficiently thick reservoir (i.e. greater than 6 to 8 m [20 to 26 ft]) to prevent high heat losses to the surrounding formations, but not so thick as to create high steam segregation effects (i.e. less than 30 m [100 ft]). It is the combination of depth and thickness that must be considered, i.e. thin but very shallow reservoirs will be possible candidates – even if their COSR is not as high as good reservoir candidates – if their operating costs are sufficiently low (mainly steam generation).
- Sufficiently high permeability (i.e. greater than 250 mD), since injectivity must be high enough to prevent excessive heat losses in the injection wells. The use of horizontal wells can be positive in this respect. However, this option must be carefully studied to evaluate steam distribution along the wells and the consequences for further steam drive using such wells.

- Sufficient oil in place ($\phi \times S_{oi} > 0.16$; e.g. $S_{oi} > 0.65$ at the start of steam injection and porosity > 25%, where ϕ is reservoir porosity and S_{oi} is initial oil saturation).
- Available water to generate steam – inexpensive and as clean as possible to decrease treatment costs.
- Fuel or gas available at low cost for steam generation. In some cases, associated gas is available in regions where there is no gas market. In such cases, references to the calorific value of produced oil are not relevant.

8.1.2.4 CSS: Detrimental Factors

Factors that are detrimental for cyclic steam injection include:

- Poor injectivity, very viscous crude, too great reservoir depth (heat losses in the well) or too high pressure (low heat content).
- Presence of a strong aquifer water drive. However, in the Canadian Pikes Peak reservoir – where there is more than 3 m (10 ft) of bottom water – large steam slugs on the order of 24,000 m^3 CWE (151,000 bbl) on later cycles generated oil rates with an OSR > 0.33 m^3/m^3 (SOR < 3.0), while small steam slugs of less than 2,500 m^3 CWE (15,700 bbl) resulted in poor production responses. Success with the much larger steam slugs is credited to creation of an effective aquifer block, similar to a technique described in a patent by M. Prats [Prats M, 1977].
- Presence of a gas cap. While theoretically relevant, this issue must be carefully studied on a case-by-case basis to determine its real impact.
- Low net to gross pay fraction. The reason essentially involves the energy loss for heating interbedded layers over the full pay thickness. Here again, a case-by-case approach is needed to assess the real impact of thermal barriers on a steam stimulation project.
- When the reservoir has several layers, treating each layer independently is preferable since it is difficult to determine the distribution of steam between the layers when it is injected in commingled fashion.
- Thick layers do not respond as well as could be expected because the steam does not penetrate to the bottom of the reservoirs.
- Extensive fractures.

These beneficial and detrimental factors should be used as general guidelines. However, each case is unique and cyclic steam injection must be carefully studied according to the specific data for a given reservoir. To predict recovery and define the optimal parameters for each cycle, simple analytical models such as Boberg and Lantz's model [Boberg TC and Lantz RB, 1966] can be used. However, numerical models in which a better characterization of the reservoir is provided will likely result in better history matching, prediction and optimization.

Steam soak is also important as a starting process for steam drive projects where initial injectivity is low, and remains the major process used in reservoirs with poor continuity. Also, it is still widely used in reservoirs with good gravity drainage or reservoir compaction drive.

CSS is mainly successful in reservoirs where natural recovery mechanisms (gravity drainage, pressure depletion and solution gas drive) are ineffective because of low oil mobilities. Recovery by CSS is usually on the order of 10 to 15 percent of Original Oil In Place (OOIP).

8.1.3 Continuous Steam Injection

8.1.3.1 Description of Process and Mechanisms

Continuous steam injection – also known as steamflood or steam drive – is a process in which steam is continuously injected into one or preferentially several wells and oil is driven to separate production wells. Figure 8.5 shows the water saturation, temperature and steam saturation profiles during a one-dimensional displacement of oil by steam.

Due to the presence of a condensable gas phase, the behavior of steam drive is very different from the behavior of water injection. The presence of the gas phase causes distillation of the light components of the oil and their movement forward toward the cold part of the reservoir. When the steam condenses, these light components do likewise, thus generating a "solvent" bank at the condensation front. Of course, oil viscosity is reduced by the temperature increase, and oil mobility is improved. Very low values of residual oil saturations (< 5%) have been reported in some field cases.

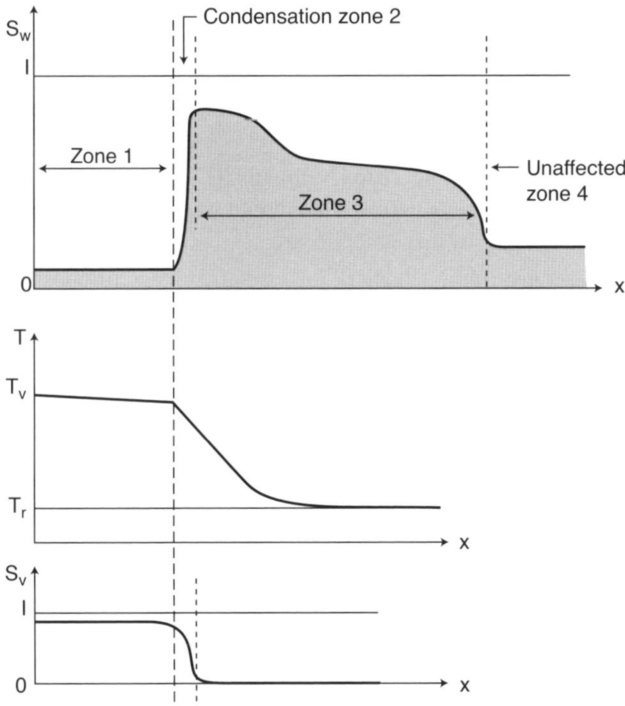

Figure 8.5

Water Saturation (Sw), Temperature (T) and Steam Saturation (Sv) during Steamflood.

8.1.3.2 Performance Prediction

When the practice of steamflooding was begun in the 1960s, this process was essentially considered to be a displacement process under differential pressure between injectors and producers; thus, the term "drive" was coined to qualify this process.

Early analytical tools available to steamflood designers for estimating heat losses and growth of the steam chest were based on the frontal advance concept (Marx and Langenheim [Marx JW and Langenheim RN, 1959], Myhill and Stegemeier [Myhill NA and Stegemeier GL, 1978]). In this concept, higher steam injection rates reduce heat losses to the underburden and overburden, and result in a more rapid recovery of oil.

Post-flood cores drilled in observation wells showed that gravity override was a significant factor. Thus, at times there were significant inconsistencies between the analytical models and the physics of the process. However, for many years numerous engineers still perceived steam injection in terms of a displacement process in which higher injection rates meant higher oil production rates and faster recoveries. The predominant philosophy was "if you want more oil, inject more steam".

In the mid-1980s, with more extensive use of numerical models, it became obvious that the main mechanisms for oil recovery by steamflood emphasize steam override and gravity drainage. In fact, the injected steam rises rapidly to the top of the injected interval and then travels within a thin layer towards a production well. After steam breakthrough occurs at the producer, only a very small differential pressure can be maintained between injector and producer. Thus, the oil is not being produced by a "drive", but rather by a "drag". With the very low pressure gradient between injection and production wells, the pressure in injection wells can therefore be limited.

At the time of these findings, Neuman [Neuman CH, 1985] developed a new analytical model based upon gravity override effects. This model was later simplified by Vogel [Vogel JV, 1982]. Some key concepts of these models are:

- The dominant production mechanism for viscous oil steamfloods is gravity drainage of oil.
- Oil rates are largely unaffected by steam rates higher than the minimum required to maintain the steam chest.
- Injection rates should be proportional to project area (not project volume).
- Decreasing injection rates with time minimizes heat losses by produced fluids and casing blow.

Figure 8.6 shows a schematic view of gravity drainage of hot oil and slight steam drag in sufficiently thick layers with good vertical permeability.

The concept of varying steam injection rates during the life of a steamflood was independently demonstrated by both reservoir simulation and field testing in the early 1970s. Conclusions from reservoir simulation studies [Chu C and Trimble AE, 1975] indicated that: 1) the optimal constant steam rate is proportional to pattern size rather than sand thickness, 2) economics of a project can be improved over the constant rate case by using higher steam rates in the initial stages and then decreasing the rate with time, and 3) decreasing rate in a hyperbolic fashion appeared superior to a linear decrease. All of these observations are consistent with current understanding of the steamflood process.

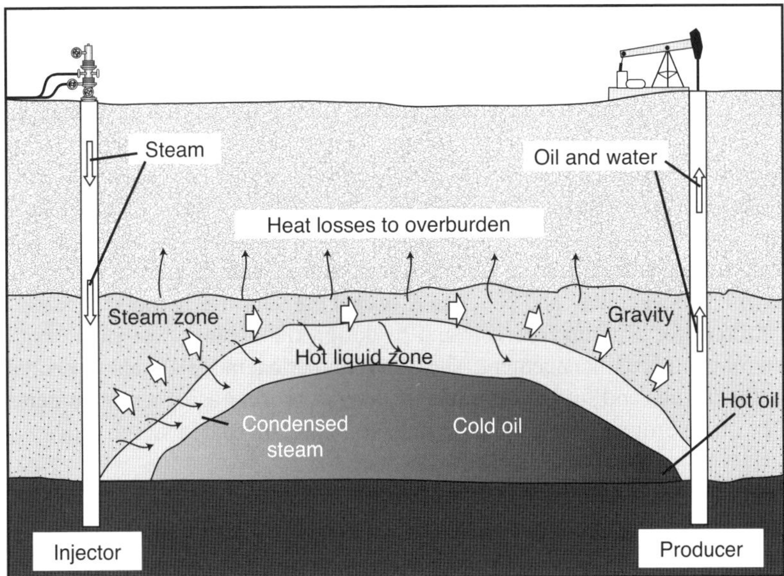

Figure 8.6

Schematic Cross-Section of Continuous Steam Injection [Vogel JV, 1982].

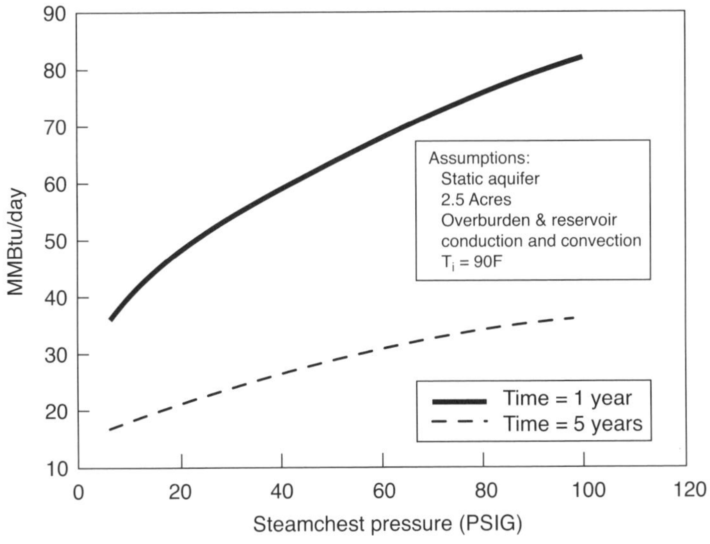

Figure 8.7

Heat Required to Maintain a Steam Chest in a Californian Reservoir [Hanzlik EJ and Mims DS, 2003].

Figure 8.7 shows an example of the difference in heat requirements to maintain a steam chamber at year one *versus* year five of a typical California steamflood pattern [Hanzlik EJ and Mims DS, 2003]. The heat requirement in the fifth year is less than half the requirement in year one. To confirm these results, Hanzlik and Mims presented several examples of heat management by California operators (Coalinga, Kern River, Midway Sunset and South Belridge). With current management practice, they showed that the rates used in the first steamflood projects were quite reasonable. However, in the 1970s the rates were increased without any improvement in oil production. Heat management is therefore very essential for the profitability of steamfloods and their design is of prime importance. The analytical models developed by Myhill & Stegemeier and Neuman & Vogel are very useful tools to assess the interest of steamflood for a given reservoir and to design the evolution of steam rate over the process life for good heat management. The use of numerical models is also essential to interpret production results and further optimize the overall process using a more precise description of reservoir heterogeneities which can play an important role in sweep efficiency.

8.1.3.3 Continuous Steam Injection: Favorable Factors

Making reference to thermal efficiency (as we did above for cyclic steam), steam drive projects are considered successful when the COSR is between 0.15 and 0.375 m^3/m^3, which corresponds to a thermal efficiency between 2 and 5. Unlike steam soaks, steam drives do not respond until built-up oil banks and heat reach the production wells. Since peak production rates may not be observed for several years after the start of injection, implementing pilots is expensive and expansion to full scale is somewhat hazardous.

Factors which are favorable for continuous steam injection include:

- Good response of cyclic steam injection. Injectivity is a real concern for steam drive since insufficiently low flow rates lead to high heat losses in the injection wells. Decreasing the oil saturation in a sufficiently wide radius around the future injection wells is thus necessary. At the same time, the oil displacement between injectors and producers must be fast enough to limit heat losses to the overburden.
- Lateral continuity of pay zones. Selected patterns are usually tested assuming a small impact of reservoir heterogeneities and therefore symmetry in oil displacement toward the production wells of the patterns. As noted in most field cases, steam front is usually not regular and some anisotropy in areal permeability or strong heterogeneities such as channels can strongly modify the areal sweep efficiency. Techniques that allow forecasting of steam breakthrough or bad sweeping, or detection of the steam front, are therefore of prime importance. One technique – extensively used in the Duri field for this purpose – is repeated 3D seismic *versus* time (Time lapse 3D seismic), also called 4D seismic. This 3D technique provides instantaneous planar views of steam invaded zones and applied *versus* time allows monitoring of steam injection in wells in order to correct non-uniform steam front advances.
- Adapted patterns and spacing between wells which need simulation studies and pilots.
- Monitoring of steam injection with a conversion to cold or hot water at a later time when too much steam is produced and oil production is reduced. This monitoring can be forecast using numerical models when enough data is available on the geological characteristics of the reservoir. Use of 4D seismic can be of great help, as previously mentioned.

8.1.3.4 Continuous Steam Injection: Detrimental Factors

Factors which are detrimental for continuous steam injection are quite similar to those presented for cyclic steam stimulation (see 8.1.2.4). Other detrimental factors include:

- Strong aquifer drive. However, it must be carefully studied.
- Presence of a gas cap. It must also be carefully studied, since it is not always a detrimental factor.
- Extensive fractures. Fractures can connect injectors to producers and lead to premature steam breakthrough with no means to obtain good sweep efficiency.

Continuous steam injection has been extensively practiced worldwide where heavy oils are present: Canada (Saskatchewan), China, Indonesia, Russia, Trinidad, USA (California) and Venezuela (see examples in Chapter 10). Reported recoveries have ranged from 5% to more than 65% of OOIP with COSR varying from 0.05 to 0.6 m^3/m^3.

8.1.4 Usual Strategies when using Cyclic and/or Continuous Steam Injection

Since the 1960s, the general field development strategy for cyclic and/or continuous steam injection has been the following:

- For big fields such as Duri in Indonesia, Karamay in China, Kern River in California, etc., the reservoir is divided into several areas or assets according to oil viscosity – which varies locally from one place to another, mainly due to reservoir depth – or other factors such as level of heterogeneity of the flow units (some assets were or are still operated by the same oil company (e.g. Duri field) or by several (e.g. Kern River field)). Each area is then developed through the following stages: the most promising area is produced first and so on until the least promising which has the worst characteristics.
  - Primary recovery is started and continued for as long as it is economical. As indicated in the previous chapter, the recovery mechanism is either due to solution gas drive and/or compaction drive. Regular patterns (line or staggered) are considered at this stage.
    When an aquifer is present, however, production wells are invaded quite rapidly by water and primary production is stopped. In recent decades, horizontal wells have been successfully used for such a case to improve primary recovery (as illustrated previously by the Winter field case).
  - After primary production is completed in an area, steam injection patterns with vertical wells are defined in a pilot zone of this area. Most often, several patterns with varying well spacing and/or numbers of wells (5, 7 or 9-spots) are implemented at the same time and their performance compared after steam injection to eventually select the best pattern to be used to fully develop the area.
    The most regular pattern (Figure 8.8) is the regular 7-spot, where each production well is surrounded by six injectors located at the same distance from the central well of the pattern (in an inverted pattern, the central well is an injector). The same can be said of the 5-spot pattern. However, with the 5-spot pattern, there are fewer wells

surrounding the central well in a pattern (4, as compared to 6 in the 7-spot pattern) and the areal and volumetric sweep efficiency by steam displacement is expected to be (and effectively is) less. The third pattern normally used is the 9-spot pattern. It has an advantage over the others since it involves more wells (8) around each central well. However, it suffers from great dissymmetry in sweep efficiency since the central well in this pattern is not at the same distance from the peripheral wells.

- As stated previously, all wells of the pilot patterns are then steam-stimulated several times. The cycles are operated in order to initiate some pressure effect between a well in which steam is injected and its surrounding neighbors.
- As soon as steam stimulation by cyclic injection does not give a sufficiently high response, continuous steam injection is initiated and conducted for several years until the operator generally decides to convert steam injection into hot or cold water injection to reduce energy consumption.
- When several pattern types are tested simultaneously, their results are then compared and the best pattern applied to the undeveloped areas.
- To reduce steam segregation and increase flow efficiency, several operators have tried in the past to inject foams, with some success [Delamaide E and Kalaydjian F, 1996].
- When steam breaks through to the producing well, it creates high-temperature problems that include wasted heat and reduced well productivity. To limit steam override and improve vertical sweep, Water-Alternating-Steam Process (WASP) – similar to WAG (Water Alternate Gas) – has been tried with success. One benefit of WASP is the reduction of steam breakthrough severity [Hong KC, 1999].

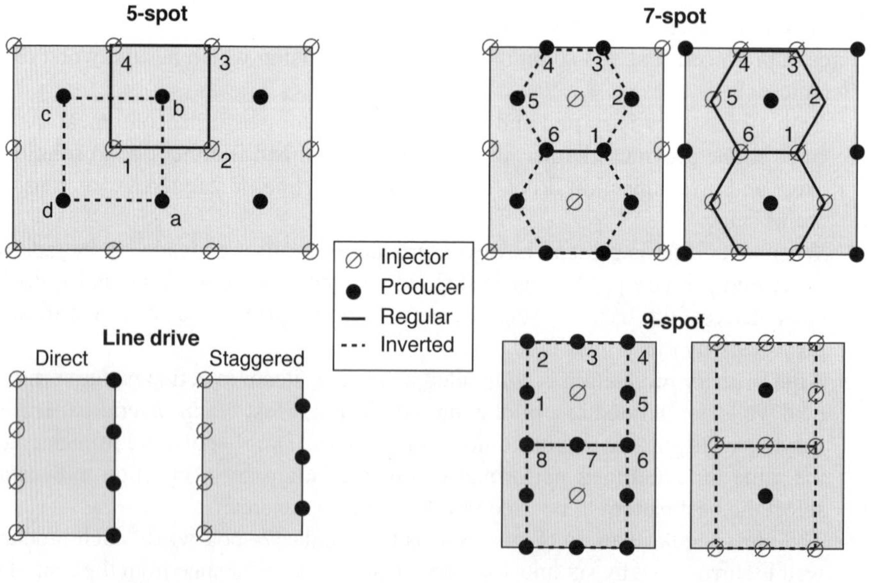

Figure 8.8

Various Patterns Considered for Continuous Steam Injection.

- This general field development strategy is usually applied when the reservoir is made up of one or two layers, and when there is no risk associated with the presence of an aquifer.
  - When several saturated layers are present in a reservoir (e.g. as is the case in the Kern River field which contains a tenth of oil layers), implementation of steam injection is performed first in the deepest layer, then in the oil layer just above and so on up to the shallowest layer, in order to benefit from heating of the surrounding formations when a layer is steamed and produced, and also to take advantage of vertical discontinuities which may exist in the separating layers and can allow limited steam migration from one oil layer to the upper layers, resulting in some preheating of the layers.
  - When an aquifer is present, special attention is required in order to avoid injecting a large amount of steam into the water leg.
  - When the reservoir is significantly dipping (greater than 10°), steamflood line drive (Figure 8.8) is preferred to usual patterns. In such reservoirs, steam tends to flow updip because of gravity, while condensed hot water flows downdip. As a result, the heat arrives first at the updip producers, causing oil production primarily from the updip section of the reservoir. Therefore, oil production from this type of reservoir is strongly affected by injector and producer location. Control can be maintained by shutting in nearby updip producers when steam breaks through, or by reducing steam injection after significant heating has occurred updip. If the reservoir contains a significant gas cap, either initially or as a result of steam injection and oil drainage, steam cycling can occur in the updip portion of the reservoir and can waste the injected heat. Updip injection of a non-condensable gas can reduce or eliminate this waste and improve the steamflood efficiency.
- In recent years, operators have tried to use horizontal wells to implement cyclic and continuous steam injection in order to improve recovery from their fields. Horizontal wells have theoretical advantages over vertical wells, mainly involving greater injectivity or productivity. However, they present several drawbacks when the reservoir is heterogeneous and steam injection or oil production must be monitored to improve areal or volumetric sweep efficiency.

  In the Tangleflags reservoir [Jespersen PJ and Fontaine T, 1993] in Saskatchewan, Canada, which is underlain by a strong aquifer, a steam drive was operated between vertical injection wells and horizontal producers. The success of steam injection in this case is due to the high quality of the reservoir with a thick oil layer and high horizontal and vertical permeabilities. Direct steam drive initiated in the presence of a gas cap which facilitated the injectivity and displacement of steam above the thick oil layer resulted in gravity-dominated displacement of the heated oil by steam from top to bottom with limited loss of heat in the aquifer.

Figure 8.9 shows the life cycle of a heavy oil reservoir under steam injection recovery. After steam chest expansion, heat management must be carefully optimized throughout the entire mature flood period, as previously outlined.

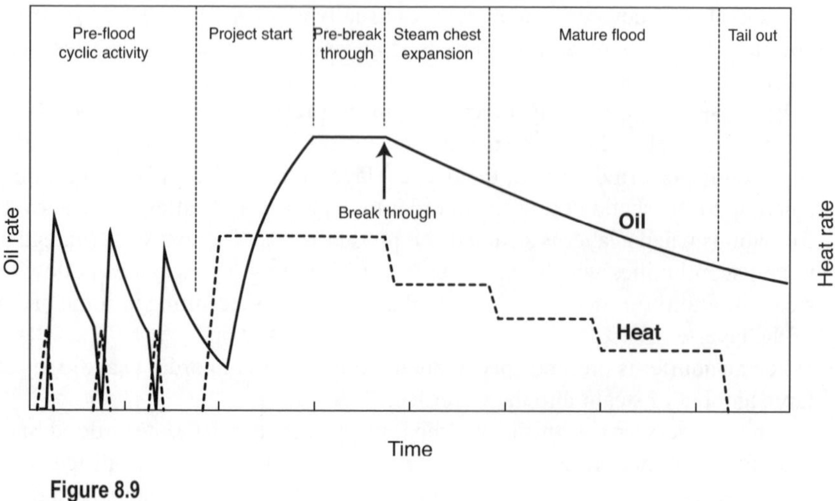

Figure 8.9

Steamflood Operation Lifecycle [Zahedi A *et al.*, 2004].

8.1.5 Steam Assisted Gravity Drainage

8.1.5.1 Description and Mechanisms

Steam Assisted Gravity Drainage (SAGD) combined with horizontal well technology is certainly one of the most important concepts developed in reservoir engineering in the last two decades. Gravity drainage in itself is not new. However, its use to unlock heavy oil and bitumen reserves to profitable recovery was not so obvious. The concept of SAGD was first suggested and studied by Dr. R. Butler [Butler RM, 1991], whose intention in developing this process was to devise a means whereby heavy oil or bitumen could be removed in a systematic manner in order to give a more complete recovery than is possible in conventional steamflooding processes. Butler proposed using gravitational forces assisted by steam to move oil to a production well. The geometry of SAGD, in its general form, is quite simple (Figure 8.10).

Butler reasoned that if steam were injected in a horizontal well above but close to a horizontal production well that was completed at the bottom of the reservoir, the steam would tend to rise and the condensate, together with warmed oil, would fall. The oil and condensate would be removed continuously from the production well. If these liquids were not removed too quickly, then the tendency of the steam to flow directly to the production well – and thus bypass the reservoir – could be reduced or possibly even eliminated. In conventional steamflooding, as seen in the previous section, the oil that is displaced by injected steam is cooled ahead by in place fluids and is hard to push towards the production well. In SAGD, the oil remains heated as it flows around the steam chamber. Butler developed the gravity drainage theory that predicts the rate at which the process will occur and confirmed the viability of the concept by lab experiments. The unique features of SAGD are:

- Use of gravity as the primary motive force for moving oil.
- Large production rates obtainable with gravity using horizontal wells.

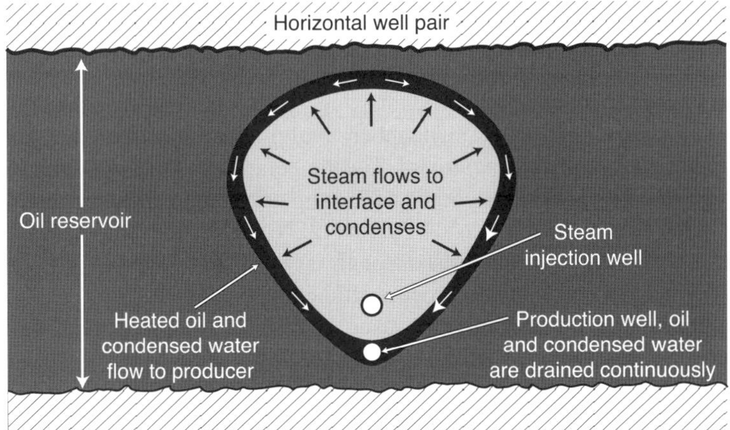

Figure 8.10

Steam Assisted Gravity Drainage Process [Butler RM, 1991].

- Flow of the heated oil directly to the production well without having to displace uncontacted oil.
- Faster oil production response as compared to steamflooding (especially when the heavy oil is initially mobile at reservoir conditions).
- High recovery efficiency (up to 70% of OOIP) – even for bitumen.
- Low achievable Cumulative Steam Oil Ratio (CSOR) due to the large potential injection/production rates limiting the heat losses (can be lower than 2.0 m^3/m^3 with steam at 80% quality).
- Low sensitivity to limited shale intervals and interbedding.

Following the initial trial done in the 1980s at the UTF site (near Fort McMurray in Canada) from underground mines [Edmunds N, 1999], SAGD was first used experimentally with wells drilled from the surface in the East Senlac field in 1995 [Chakrabarty C *et al.*, 1998]. Currently, several field-scale implementations of SAGD are currently underway in Canada (see chapter 10).

8.1.5.2 Implementation of the SAGD Process: Phases Involved

Four important phases during SAGD must be performed carefully in order to complete a successful operation:

- **Drilling and completion:** Drilling must be handled carefully in order to drill the production and injection wells at a constant distance from each other, insofar as possible. For long drains in such close proximity, a new technique to pilot the wells had to be developed in order to prevent the wells from intersecting each other. The injection well is thus drilled relative to the production well.
- **Start-up:** The start-up phase is critical for the future results of a well pair when oil viscosity is very high and does not allow initial flow of fluids (steam, water or oil) between the injector and the producer. Thus, the start-up consists in pre-heating the interval between the injection and production wells to allow the oil to flow towards the producer by gravity.

This is achieved by steam circulation both in the injector and producer for a period of several weeks to several months. In the East Senlac field, oil is mobile at reservoir conditions, so it is possible to inject steam directly into the reservoir after a short preheating period.

- **Steam chamber monitoring:** After start-up, steam injection is controlled to maintain a constant pressure in the steam chamber; total production – as well as pressure at the production well – are also monitored. The temperature in the production well is generally monitored with a thermocouple and care is taken to avoid steam breakthrough by keeping the temperature 30 to 50°C (86 to 122°F) below the temperature of the steam (steam trap control). Steam breakthrough is prevented to avoid loss of steam and energy which is detrimental to process efficiency.
- **Production:** Gas lift is sometimes used to activate the production well, involving technical issues associated with the unstable phenomenon of geysering. As hot fluids are flowing to the surface, production water flashes in the tubing under lower pressure and high temperature, decreasing the average fluid density and reducing the need for gas in the gas lift. The lift rate must therefore be monitored carefully since it has a direct impact on production and thus on the operations of the well pairs.

Numerical modeling is important to evaluate and optimize an SAGD operation [Egermann P *et al.*, 2001], as illustrated in Figure 8.11. Use of dynamic sub-gridding, with fine grid blocks to discretize the steam chamber interface, can help to decrease the high CPU time usually required to perform 2-D or 3-D SAGD computations [Lacroix S *et al.*, 2003].

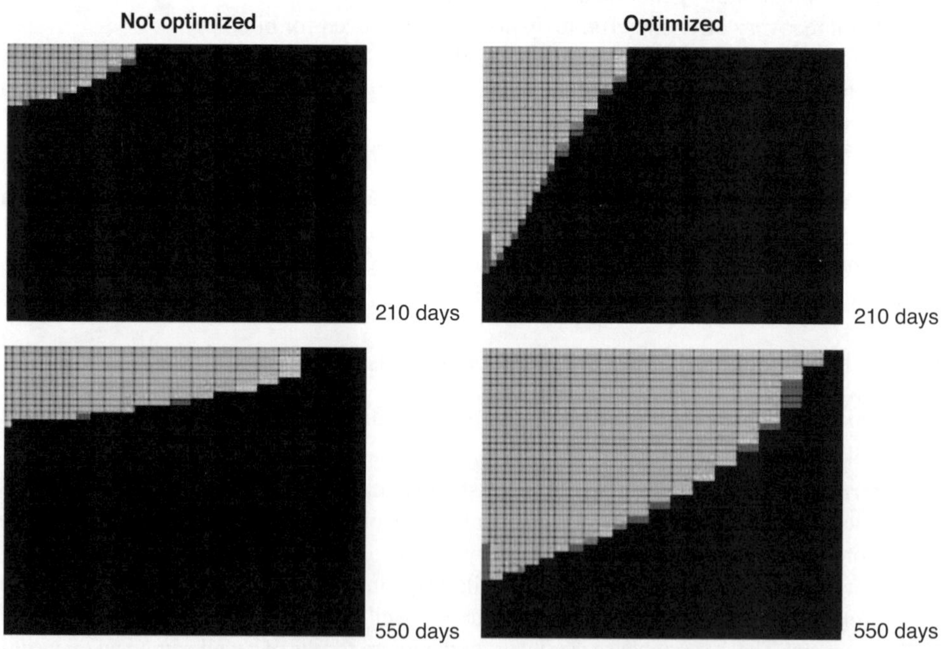

Figure 8.11

Modeling of Celtic SAGD
Steam Saturation *versus* Time (2D XZ view) [Egermann P *et al.*, 2001].

8.1.5.3 SAGD: Favorable Factors

Factors that are favorable for SAGD include [Palmgren C and Renard G, 1995]:

- Sufficiently thick reservoir (i.e. greater than 12 m [50 ft]) to allow the creation of the steam chamber. The well pair is located at the bottom of the reservoir, a few meters from each other.
- High permeability both in horizontal and vertical directions (i.e. greater than 1.5 darcies). Steam chamber development needs high permeability.

8.1.5.4 SAGD: Detrimental Factors

Factors that are detrimental for SAGD include:

- Reservoirs which are too thin hinder the development of an optimal steam chamber and induce highly detrimental heat losses towards the overburden. Steam chamber monitoring would also be very difficult to prevent steam breakthrough at the production well.
- Reduced vertical permeability which does not allow the flow of fluids (condensed steam and water) to the lower producer at a sufficient rate since recovery is controlled by gravity.
- Presence of significant heterogeneities. Barriers to the flow of steam or hot fluids along the steam chamber can limit process efficiency. There are no published studies on the minimum size of heterogeneities above which SAGD is no longer attractive. Numerical modeling on a case-by-case basis is required to investigate their real impact.
- High pressure associated with deep reservoirs (i.e. greater than 800 m [2,600 ft]). The higher the reservoir pressure, the higher the operating costs and the lower the latent heat of steam available to heat the rock and in place fluids.
- Presence of a bottom aquifer. However, SAGD can be successfully operated even with an aquifer present below the oil zone. Numerical modeling is required to design a project (placement of wells, flow rate for injection and production, etc.).
- Presence of a gas cap. This is qualified as a thief zone since the gas cap can act as a sink for injected steam. Here again, a project can be successfully implemented by optimizing the process *via* numerical modeling to determine the pressure at which it must be operated.

8.1.5.5 Suggested Improvements to Original SAGD

As for other processes involving steam injection, SAGD suffers from the need for large volumes of water to generate steam (typical production results indicate that 1 cu. m of oil requires at least 2.5 cu. m or more – currently up to 4 cu. m – of water). Steam generation, on another hand, is a source of a large amount of CO_2 production. It has been determined that production of 16,000 m^3/d (100,000 bopd) of bitumen with an SOR of 2.5 m^3/m^3 would produce 14,300 t/d of CO_2. To reduce the amount of steam – or more generally reduce operating costs – several improvements to the SAGD process have been proposed and tested [Shin H and Polikar M, 2004]:

- **Single Well SAGD (SW-SAGD)**: This technology uses a single horizontal well to inject steam and produce oil instead of two horizontal wells as in the conventional SAGD, also referred to as Dual Well SAGD. The single horizontal well is located at the base of the reservoir. Steam is injected into the insulated tubing and fluid production flows through the annulus. SW-SAGD has been tested in several Canadian fields but has not been extended to field scale and therefore its viability is still questionable.
- **Steam and Gas Push (SAGP)**: This recovery method was introduced to enhance SAGD efficiency by adding a small amount of Non-Condensable Gases (NCG) such as natural gas or nitrogen. The effect of the NCG is to reduce the amount of steam to be injected. However, the timing of the NCG injection is very important. When it is done during the steam chamber rising period, cumulative oil production and oil rate are decreased.
  When the gas is injected during the late period of the SAGD process, the process is also referred to as ***SAGD Wind-down***. At this time, the SOR can be reduced without severely reducing oil production since the injected NCG gases rise to the top of the reservoir and then help the steam chamber to propagate horizontally. The reservoir is still hot and the energy in place can be utilized.
- **Expanded Solvent SAGD (ES-SAGD)**: The aim of this process is to combine the benefits of steam and solvent in the recovery of heavy oil and bitumen. In this process, the solvent is injected together with steam in a vapor phase. It condenses around the interface of the steam chamber and dilutes the oil; solvent in conjunction with heat reduces oil viscosity. ES-SAGD – also called SAP (Solvent Aided Process) – has been tested in the Senlac field [Gupta S *et al.*, 2003] and has led to promising results to lower water and energy requirements while improving oil production and SOR.
- **Fast-SAGD**: This process combines the SAGD and CSS processes. It aims to help propagate the steam chamber laterally. It requires an SAGD well pair of horizontal wells and several additional horizontal producers without their respective injectors. The interest in this variation of SAGD is to drill in fewer wells.

8.1.6 Steam Injection Processes: Operational Issues

In steamflooding, a great amount of the energy consumed to generate steam (at least 20%) is lost without providing useful heat in the reservoir. There are losses in the steam generator, steam surface lines, well bore, and to under- and overburden. Within the reservoir, it is also necessary to heat the rock surrounding the fluids in place.

More common operational problems with steam injection are:

- Technical hitches with steam generators.
- Stand-out of well completions to high temperature which require special cements that are not always efficient, with possible fluid leakage behind the completion.
- Well sanding out.
- pH that can result in well plugging with silicate.
- Formation of sulfur gases.
- Clay swelling.
- Formation of stable emulsions.

8.2 *IN SITU* COMBUSTION

8.2.1 Principle of ISC

In situ is the literal Latin term for "in place". Therefore, *In Situ* Combustion (ISC) is simply the burning of fuel where it exists, i.e. in the reservoir. The term ISC is applied to recovery processes in which air, or more generally an oxygen-containing gas, is injected into a reservoir, where it reacts with organic hydrocarbons acting as fuels. The heat generated is then used to help recover unburned crude (the fuel actually burned is not the crude oil but rather the carbon-rich residue resulting from thermal cracking and distillation of the residual crude near the combustion front).

With ISC, heat losses are concentrated at the fire front. ISC features thus have the potential to be more efficient and economical than the use of steam, although energy must be provided to compress the air. This energy is much lower than that required for steam generation. Moreover, the process does not require water and involves lower CO_2 emissions.

8.2.2 Description and Mechanisms

The first step in a combustion operation is ignition. If the oil in place is sufficiently oxidizable under bottom hole conditions, spontaneous ignition may occur in the formation near the injection well after several days of oxidizing gas injection. This phenomenon occurs more often in formations in which the reservoir temperature is approximately 60°C or greater (pressure level is also important). Otherwise, ignition takes place by artificially heating the formation near the injection well bore *via* electric heater, burner, hot-fluid injection or injection of highly oxidizable chemicals.

The most widely used form of the combustion process is simple air injection. It is referred to as dry combustion (Figure 8.12) to distinguish it from wet combustion (Figure 8.13), in which water and air are injected into the reservoir.

With continuous air injection, the combustion zone propagates outward from injectors, thereby displacing reservoir fluids toward producers.

In dry forward combustion, the combustion front pushes the unburned fractions of crude oil along ahead of it. The heavier fractions are transformed into a carbonaceous deposit with low hydrogen content, which is often improperly called "coke". This deposit is burned with the oxygen from the injected gas.

It is convenient to describe the process by considering four main zones, as indicated in Figure 8.12:

- Zone 1. Combustion has already taken place here. The injected gas heats up upon contact with the matrix and recovers a small amount of the heat released by the combustion.
- Zone 2. Combustion zone: oxygen is used up in the combustion reactions involving hydrocarbons and the "coke" retained on the rock. The temperature attained in this zone essentially depends on the nature and quantity of various organic components remaining in the porous medium; the peak temperature is generally in the 400-600°C range.

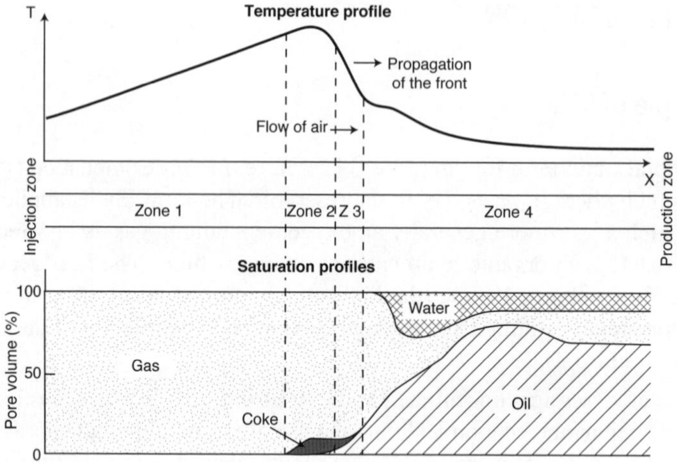

Figure 8.12

Dry Forward Combustion: Temperature and Saturation Profiles.
Source: Taken from [Bavière M *et al.*, 1991].

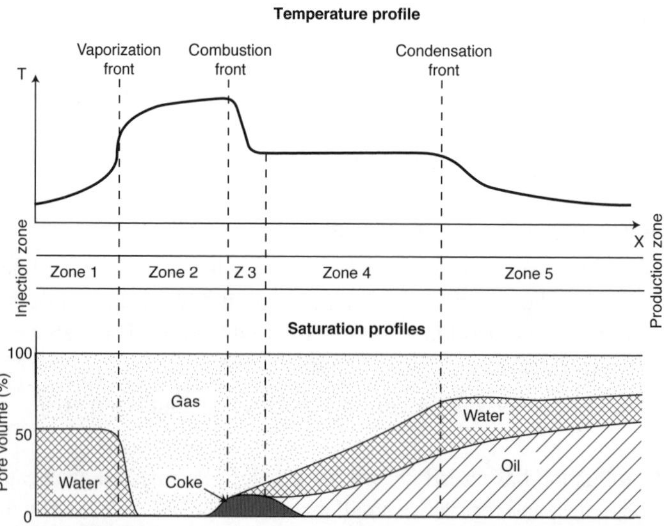

Figure 8.13

Wet Forward Combustion: Temperature and Saturation Profiles.
Source: Taken from [Bavière M *et al.*, 1991].

- Zone 3. Cokefaction zone: heavy fractions that have not been displaced or vaporized are pyrolyzed in this zone. This pyrolysis is combined with oxidation if all the oxygen has not been consumed in the combustion zone.
- Zone 4. When the temperature is sufficiently low, no further chemical changes are observed provided that the flowing gas does not contain unconsumed oxygen. In the region closest to the reaction zones, successive vaporization and recondensation of connate water and light fractions of oil take place. In the region where the temperature is lower than the condensation temperature of water, a water bank is observed in which water saturation is greater than the original saturation. Ahead of this zone an oil bank appears, with oil saturation greater than the original oil saturation. In any event, the two banks together constitute a zone with a high pressure drop. Plugging-up of the layer may even occur for very viscous oils. Beyond the oil bank, the porous medium gradually regains its initial characteristics.

8.2.3 ISC Application Design

A combustion field test is designed based on an estimate of the amount of fuel to be burned in order to propagate the combustion front, as well as an injection program that is appropriate for the characteristics of the candidate reservoir. Simplified evaluation methods have often been used for designing a combustion project. Three-dimensional numerical models are also used for predicting the propagation of the combustion process as a function of time [Petit H *et al.*, 1990].

8.2.4 ISC Field Applications

ISC has been applied and tested in a wide variety of fields and the experience gained has led to a valuable body of knowledge concerning this process. Over the years, some 200 *in situ* field tests have been performed and reported for most of these fields. Some of the fields have been commercially successful, while others were technically successful (they produced oil) and many "failed". One fact that stands out in the case of failures is that the process was often applied in the wrong type of reservoir, or the operating conditions were inappropriate. For many people involved in ISC operations – as in other EOR processes – proper geological reservoir characterization and selection is the key to project success. Although operational problems have been reported for most ISC field tests, it is unlikely that these problems have been responsible for the possible failures. Failure has often occurred because many operators view combustion as a last resort process and relegate it to fields where no other method has a chance of success.

Typical current operations are in Romania (Suplacu de Barcau, Balaria, East and West Videle fields), India (Balol, Santhal and Lanwa fields) and the United States in fields operated by Continental.

8.2.5 ISC: Favorable and Detrimental Factors

ISC has found less success than steam due to the difficulty of controlling the process. Fire fronts tend to advance far more erratically than steam fronts, probably because heterogeneities present in the reservoir are more detrimental, and it is much harder to modify operational parameters to obtain a homogeneous sweep of the reservoir.

The combustion process is reservoir-specific, and a reservoir whose rock and oil properties promote efficient oxygen utilization is more likely to succeed than one in which Low Temperature Oxidation (LTO) is promoted. It must be emphasized that ISC is a very complex process and there is no guarantee of success.

The most successful operations have been performed in sand or sandstone formations containing heavy- to medium-gravity crude oil (13-35°API). Moderate thickness (2-20 m [6-60 ft]), good permeability (greater than 100 mD), and a minimum oil content of $0.1 m^3/m^3$ are recommended. Depth is not crucial as for steam injection. Field applications range from 100 to 1,500 m (300-4,500 ft). The Romanian experience has shown that *In Situ* Combustion is particularly successful when designed in dipping reservoirs starting from the uppermost part of the structure and extending the process downward as the combustion front progresses down dip. The most significant reservoir characteristics in failed fireflood projects were: $\phi \times S_{oi} < 0.1$, $\phi < 0.15$ and $S_{oi} < 0.3$.

8.2.6 ISC Process: Operational Issues

The fact that the ISC process suffers from several real problems is widely observed and accepted. These include:

- High Gas Oil Ratios, and the concomitant high gas production rates cause a variety of mechanical problems, including well equipment erosion, gas-locking pumps, treating and separation, and pollution control.
- Producing wells may sand up due to increased water and gas production.
- Corrosion resulting from acids formed in the formation and high temperatures can take a heavy toll on well equipment. Acids and a host of low-temperature oxidation products lead to the formation of emulsions, the character of which may change from day to day. These chemicals may also induce precipitation of asphaltenes ("sludge") which can plug up the producers.
- Occasionally, a producing well may exhibit a high temperature that must be cooled *via* water injection with a corrosion inhibitor down the annulus, but must continue producing, since hot wells are often good producers.
- Other problems involve compressor breakdowns that sometimes lead to termination of field projects because the combustion zone has travelled far enough and ignition cannot be re-started by the time the compressor is finally repaired.

Although most of these problems can be solved, they make *In Situ* Combustion costly and labor-intensive. Cementing quality, sand control and ignition operations are also of major importance.

8.3 OTHER PROCESSES

8.3.1 Polymer Injection

Waterflooding of heavy oils is often plagued by poor sweep efficiency due to the severe contrast in water and oil viscosities. To alleviate or reduce this problem, a polymer must sometimes be added to the water in order to make it more viscous.

Polymer injection is thus usually envisioned as an EOR process for which the mobility ratio $M(\lambda_d/\lambda_o)$ between the displacing fluid (polymer solution) and displaced fluid (oil) is close to 1 (the mobilities of both fluids are defined as the ratio of their effective permeability k to their viscosity μ; $\lambda_d = k_d/\mu_d$; $\lambda_o = k_o/\mu_o$). Thus, polymer injection has been applied in reservoirs in which oil viscosity is generally less than roughly 100 cP, the general idea being that displacing much more viscous oil would require higher-viscosity polymer solution and thus high injection pressure (higher than the reservoir fracturing pressure). The process is sensitive to temperature (polymer degrades at high temperatures) and water salinity (high salinity reduces the polymer's viscosifying effect and thus has an impact on process economics). High permeability is also required in order to obtain sufficient injectivity for the viscous polymer solution.

However, with the advent of horizontal wells, this paradigm can be re-evaluated and it is likely that the polymer injection process could be applied in a more viscous oil reservoir where steam injection is not feasible (e.g. thin reservoirs). Recently, several pilot tests of polymer floods have been performed in the Pelican Lake field [Zaitoun A *et al.*, 1998] in Canada where the oil viscosity is roughly 1,600 cP under reservoir conditions at a depth of about 400 m (1,200 ft). Several parallel injection and production horizontal wells have been drilled and polymer solution injected below fracturing pressure. Excellent results have been obtained to date – far better than simple waterflooding in this reservoir [CNRL, 2007]. One reason for the success of this pilot test probably involves the very low effective relative permeability of the polymer solution, which was quite lower than would have been expected. The company operating the field has decided to extend the polymer flood to field scale.

Polymer flooding should thus be considered an "improved waterflooding" for heavy oil and rather than aim for a mobility ratio of 1, the target mobility ratio should be dictated only by economics.

One of the most successful polymer flood applications has been the giant Daqing field in China, where polymer flooding has been used to recover more than 48 million cu. m (300 million barrels) of oil [Demin W *et al.*, 2002] following the implementation of two successful pilots [Delamaide E *et al.*, 1994] (see Chapter 10).

8.3.2 CO_2 Injection

CO_2 is known for its high solubility in oils, with consequent mechanisms of oil swelling and viscosity reduction [Bavière M *et al.*, 1991]. Therefore, CO_2 flooding has the potential to recover heavy oil, as is the case in the Bati Raman field in Turkey. On another hand, CO_2

injection can also be implemented as a single well stimulation process similar to steam Huff & Puff, as described previously. The main issue in achieving profitable applications of CO_2 flooding is its very high mobility. The lower density and viscosity of CO_2 as compared to those of reservoir oils are responsible for gravity tonguing and viscous fingering, which are more severe than in waterflooding. To limit this effect, several solutions have been investigated, such as alternating CO_2-water injection and the addition of foaming agents along with CO_2. With the advent of horizontal wells, it is likely that new well architectures will be suitable for the use of CO_2 as a good displacing agent, with the advantage of keeping a large amount of this GreenHouse Gas (GHG) *in situ*. The issue of asphaltene precipitation for certain oils and in given thermodynamic conditions will have to be considered and carefully assessed to prevent plugging of the porous media, especially near the well bore region.

8.3.3 VAPEX

As mentioned previously, thermal recovery processes using steam suffer from high heat losses, high water requirements, the need for extensive facilities and adverse environmental impacts. After his research on SAGD, Dr. R. Butler thus suggested using light hydrocarbon vapor instead of steam to recover extra-heavy oils or bitumen and patented a new non-thermal vapor extraction process: VAPEX [Butler RM and Mokrys IJ, 1991].

The VAPEX process is closely related to SAGD. However, in the VAPEX process, the steam chamber is replaced with a chamber containing light hydrocarbon vapor close to its dew point at the reservoir pressure. The mechanism for oil viscosity reduction is no longer the increased temperature linked to heat supplied by steam, but rather dilution by molecular diffusion of the solvent in the oil. Diluted oil or bitumen – driven by gravity – drains to the horizontal well located below the horizontal well in which the solvent is injected. If the pressure used is close to the bubble point pressure of hydrocarbons, de-asphalting may in fact occur in the reservoir, leading to *in situ* upgrading of the oil with a substantial reduction in viscosity and heavy metal content.

A pilot test of the VAPEX process has been performed in recent years. The results of this pilot have not been officially released, but it seems that it did not perform as expected and additional research is needed to improve the process and take it to the commercial level.

Other hydrocarbons are sometimes used, in particular hot naphtha which acts *via* molecular and thermal diffusion, and allows for transport of the bitumen to the upgrader.

8.3.4 THAI and CAPRI

The Toe-to-Heel Air Injection (THAI) process [Greaves M and Xia TX, 1998] is very similar to *in situ* combustion. Its concept is to recover extra-heavy oil or bitumen using a vertical injector and a horizontal producer. The vertical injector is opened through the full thickness of the reservoir. Its shoe is placed close to the toe of the horizontal producer, located at the bottom of the reservoir. The combustion front generated from the vertical well should be perpendicular to the direction of the horizontal producer, and propagate through the reservoir from the toe to the heel of this well. Ahead of the combustion front, temperature is high and oil viscosity very low. Therefore, the oil flows by gravity to the horizontal producer where it can be recovered. THAI efficiency is estimated by its inventors to be very high – up to 80 percent of the original oil in place – while partially upgrading the crude oil *in situ* by thermal cracking. By placing an annular layer of catalyst on the outside of the perforated horizontal well, CAPRI – an extension of THAI – has also been proposed to enhance overall upgrading of the produced crude.

THAI has many potential benefits as compared to other *in situ* recovery methods, e.g. SAGD. These benefits include higher resource recovery, lower capital and production costs, minimal use of natural gas and fresh water, upgraded crude oil produced, reduced diluent requirements for transport and lower greenhouse gas emissions. THAI and CAPRI have been tested at the laboratory scale, but have not yet been field-tested. A THAI field test by Petrobank is currently underway at Whitesands in Canada. Results could determine the viability of this process and its possible field-scale implementation.

8.4 CONCLUSIONS

Production of heavy and extra-heavy oils as well as bitumen is technically and economically feasible *via* various Enhanced Oil Recovery processes. The processes most widely applied at field scale are cyclic steam stimulation, continuous steam injection and SAGD. However, due to the cost of these processes and their environmental impact, improvements – or even new processes such as VAPEX or THAI – must be investigated, studied and applied to recover a larger portion of the huge resources of these unconventional oils present in many locations throughout the world.

REFERENCES

Bavière M *et al.* (1991) Basic Concepts in Enhanced Oil Recovery Processes. Critical Reports on *Applied Chemistry*, **33**, Elsevier Applied Science.

Boberg TC, Lantz RB (1966) Calculation of the Production Rate of a Thermally Stimulated Well, *JPT*, Dec., pp 1613-1623.

Borregales CJ (1977) Steam Soak on the Bolivar Coast. (Presented at the *Canada-Venezuela Oil Sands Symposium*, Edmonton, 30 May-4 June, 1977. Published as The Oil Sands of Canada-Venezuela 1977, CIM Special volume 17).

Burger J, Souriaux P, Combarnous M (1985) Thermal Methods of Oil Recovery. Gulf Publishing Co., Houston. Editions Technip.

Butler RM (1991) Thermal Recovery of Oil and Bitumen. Edited by GravDrain Inc.

Butler RM and Mokrys IJ (1991) A New Process (VAPEX) for Recovering Heavy Oils Using Hot Water and Hydrocarbon Vapour. *JCPT*, **30**, 1, pp 97-106.

Chakrabarty C, Renard G, Fossey JP, Gadelle C (1998) SAGD Process in the East Senlac Field: From Reservoir Characterization to Field Application. (Presented at the 7th *UNITAR International Conference on Heavy Crude and Tar Sands*, Beijing, 27-30 Oct., 1998).

Chu C, Trimble AE (1975) Digital Simulation of Steam Displacement – Field Performance Applications. *JPT*, June, pp 765-776.

CNRL (2007) Canadian Natural Resources Corporate Presentation, Feb. 2007. www.cnrl.com/client/media/1242/1413/w_corppresfeb.pdf, pp 23.

Delamaide E, Corlay Ph, Demin W (1994) Daqing Oil Field: The Success of Two Pilots Initiates First Extension of Polymer Injection in a Giant Oil Field", SPE 27819. (Presented at the *1994 SPE/DOE IOR Symposium*, Tulsa, 17-20 April).

Delamaide E, Kalaydjian F (1996) A Technical and Economical Evaluation of Steam Foam Injection Based on a Critical Analysis of Field Applications. SPE 35692. (Presented at the *1996 SPE Western Regional Meeting*, Anchorage, 22-24 May).

Demin W, Jiecheng C, Junzheng W, Gang, W (2002) Experiences Learned After Production of More Than 300 Million Barrels of Oil by Polymer Flooding in Daqing Oil Field. SPE 77693. (Presented at the *2002 SPE ATCE Conference*, San Antonio, 29 Sept. – 2 Oct.).

Edmunds N (1999) On the Difficult Birth of SAGD (Steam Assisted Gravity Drainage). *JCPT*, **38**, 1, Jan., pp 14-17.

Egermann P, Renard G, Delamaide E (2001) SAGD Performance Optimization Through Digital Simulations: Methodology and Field Case Example. SPE 69690. (Presented at the *SPE ITOHOS Symposium*, Margarita Island, March 2001).

Giusti LE (1974) CSV Makes Steam Soak Work in Venezuela Field. *Oil and Gas Journal*, Nov., pp 88-93.

Greaves M, Xia TX (1998) Preserving Downhole Thermal Upgrading Using 'Toe-to Heel' ISC – Horizontal Wells Process (presented at the *7th UNITAR International Conference on Heavy Crude and Tar Sands*, Beijing, 27-30 Oct., 1998).

Gupta S, Gittins S, Picherack P (2003) Field Implementation of Solvent Aided Process", CIM 2002-299. (Presented at the *2003 CIM Annual Meeting*, Calgary, 11-13 June).

Hanzlik EJ, Mims DS (2003) Forty Years of Steam Injection in California – The Evolution of Heat Management. Paper SPE-84848-MS presented at the SPE International Improved Oil Recovery Conference, Kuala Lumpur, Malaysia, 20-21 October, 2003.

Hong KC (1999) Recent Advances in Steamflood Technology. SPE 54078. (Presented at the *1999 ITOHOS Conference*, Bakersfield, 17-19 March).

Jespersen PJ, Fontaine T (1993) The Tangleflags North Pilot – A Horizontal Well Steamflood. *JCPT*, **32**, 5, May 1993, pp 52-57.

Lacroix S, Renard G, Lemonnier P (2003) Enhanced Digital Simulations through Dynamical Sub-Gridding, CIM 2003-87 (presented at the *2003 CIM Annual Meeting*, Calgary, 10 June).

Marx JW, Langenheim RN (1959) Reservoir Heating by Hot Fluid Injection. *Pet. Trans. AIME*, 216, pp 312-315.

Myhill NA, Stegemeier GL (1978) Steam-Drive Correlation and Prediction, *JPT*, Feb., pp 173-182.

Neuman CH (1985) A Gravity Override Model of Steamdrive. *JPT*, Jan., pp 163-169.

Palmgren C, Renard G (1995) Screening Criteria for the Application of Steam Injection and Horizontal Wells. (Presented at the *1995 European IOR Symposium*, Vienna, May).

Petit H, Le Thiez P, Lemonnier P (1990) History Matching of a Heavy Oil Combustion Pilot in Romania. SPE/DOE 20249 (presented at *7th SPE/DOE IOR Symposium*, Tulsa, April, 1990).

Prats M (1977) Aquifer Plugging Steam Soak for Layered Reservoir. US Patent 4,064,942.

Prats M (1982) Thermal Recovery. SPE monograph Volume 7, SPE of AIME, Dallas.

Sarathi PS, Olsen DK (1992) Practical Aspects of Steam Injection Processes – A Handbook for Independent Operators. NIPER-580 (DE92001070), IIT Research Institute, Bartlesville.

Shin H, Polikar M (2004) Review of Reservoir Parameters to Optimize SAGD and Fast-SAGD Operating Conditions. CIM 2004-221 (presented at the *2004 CIM Annual Meeting*, Calgary, 8-10 June).

Vogel JV (1984) Simplified Heat Calculations for Steamfloods. *JPT*, July, pp 1127-1136.

Zahedi A, Johnson R, Rueda C (2004) Heat Management in Coalinga – New Insight to Manage Heat in an Old Field. Paper SPE-86984-MS presented at the SPE International Thermal Operations and Heavy Oil Symposium and Western Regional Meeting, Bakersfield, California, 16-18 March, 2004.

Zaitoun A, Tabary R, Fossey JP, Boyle T (1998) Implementing a Heavy-Oil Horizontal-Well Polymer Flood in Western Canada. (presented at the *7th UNITAR International Conference on Heavy Crude and Tar Sands*, Beijing, 27-30 Oct, 1998.

9 Heavy Oil Production: Pumping Systems

C. Wittrisch

Over the last decade, commercial interest in heavy oil production has received a boost with the advent of improved drainage architecture using horizontal and multilateral wells, as well as new recovery processes based on downhole technologies and involving cold or thermal production and improved artificial lift. Generally, downhole artificial lifts are used when the pressure in an oil reservoir has fallen to the point where a well will not produce at its economical rate *via* natural energy. This is particularly relevant for heavy oil and bitumen reservoirs, in which high viscosity combined with low reservoir pressure prevent their production *via* natural pressure drive.

The most important factors involved in producing heavy oil with downhole pumps or *via* artificial lifts include:

- Geological formation characteristics.
- Fluid characteristics such as temperature, pressure, density, viscosity, content of free or dissolved gases, presence of CO_2 and H_2S, and presence of chemical products (aromatics, etc.).
- Well architecture from vertical to horizontal.
- Well completion and sand control with wired screen or slotted liner.

9.1 COLD PRODUCTION OF HEAVY OILS

9.1.1 Natural Flow

Heavy oils with densities in the range of 12° to 20°API and high viscosities of 1,000 cP to 10,000 cP or more, along with low downhole pressures at the reservoir level, do not allow for production *via* natural flow from downhole to surface. Natural flow exists in some specific cases, but at a very low rate (10 bopd), with recovery limited to only 5-6% of the oil in place.

In some specific cases – e.g. in Canada (Alberta) or Venezuela – enhanced cold production is possible from unconsolidated sandstone reservoirs containing high-viscosity oil in the 12°API range *via* the use of horizontal wells and standard artificial lifts, in particular using Progressive Cavity Pumps (PCP) (see 9.3.3).

Several techniques have been developed to improve the per-day rate and recovery of heavy oils, and these are described in the following sections.

9.1.2 Downhole Injection of a Diluent

The cold production recovery process (described in Chapter 7) makes use of various types of diluents injected downhole to reduce heavy-oil viscosity. Diluents such as light oil, kerosene or naphtha are injected into the drain and naturally mixed with heavy oil. The use of diluents enables heavy-oil viscosity to be reduced from 100,000 cP to 1,000 cP or less. The diluted mixture is then lifted to the surface by a type of downhole pump, e.g. high-volume Progressive Cavity Pumps (PCP).

The Petrocedeno (ex Sincor) Project (see Chapter 10.2.2) is an interesting example of production *via* dilution of extra-heavy oil of 7.5-9°API (heavy oil upgraded to high-quality commercial 32°API Syncrude oil containing less than 0.1% sulfur).

The extra-heavy oil is produced downhole from 1,400 meters-long horizontal drains *via* downhole PCP or ESP pumps. The wells have 9 5/8" production casing with a 7" slotted liner or sand screen in the horizontal producing zone.

Well completion – as described in Figure 9.1 – involves 5 1/2" production tubing string from surface to the PCP pump. The inlet of the pump communicates with the annulus through a perforated tube. The diluent injection line of coiled tubing or small tubing string (1.6-inch OD) is located in the annulus between tubing and casing from surface to the pump level. The injection line is extended underneath the PCP pump *via* 2 3/8" tubing located inside the slotted liner. The 2 3/8" tubing arrives near the end of the slotted liner (or sand screen). The production string is lowered down in a single operation including downhole-to-surface tubing injection, PCP pump production tubing and injection line. The production rate can be up to 1,000 m^3/day (6,300 bopd) of diluted oil.

The diluent used is a 47°API naphtha. Injected at the bottom of the slotted liner, the naphtha moves slowly in the horizontal section of the drain *via* the effect of the pressure differential generated by pump operation. The heavy oil progressively transits from the reservoir to the liner, is naturally mixed with the diluent, and arrives at the pump inlet with a reduced viscosity that is acceptable for the pump. For example, under downhole conditions of 10,000 cP heavy oil at 50°C mixed with 20% naphtha, the viscosity is reduced to 200 cP. The viscosity of the mixed pumped fluid can be easily adjusted *via* the rate of diluent injected downhole through the injection line.

9.1.3 Downhole Injection of Water and Additives

The technology of forming an emulsion downhole is still at the experimental stage. Tests are underway to form an emulsion *via* energized mixing with water and additives. Jet pumps could be an option, with the driving fluid being water or a light water-oil emulsion injected through the nozzle in a convergent-divergent jet pump. The high mixing energy transferred between the heavy oil and power fluid would basically allow for the creation of a stable

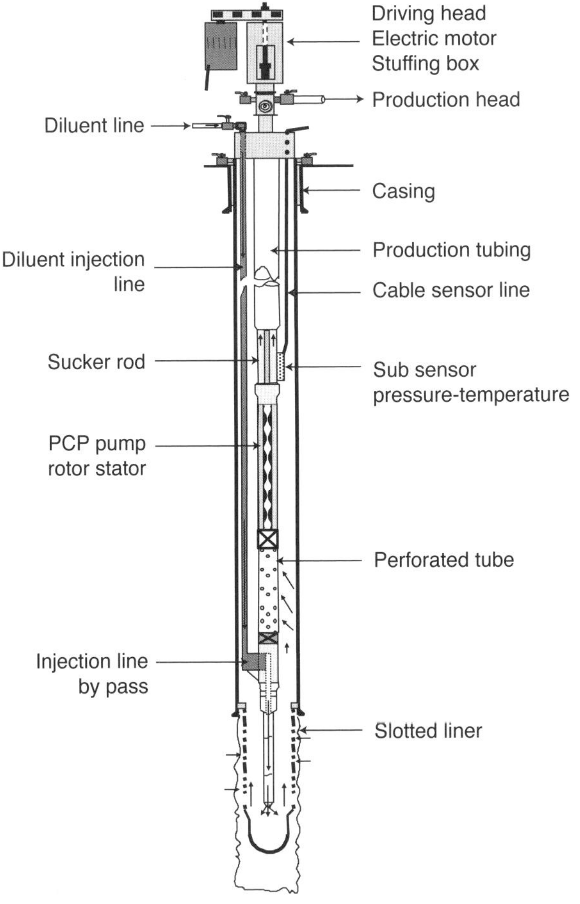

Figure 9.1

Cold Production with Diluent Injection.

emulsion. A PCP or ESP pump can be added to lift the emulsion. The efficiency of a jet pump is quite low: 25 to 30%, as compared to 90% for the PCP pump. The volume of power fluid will be higher than the volume of the oil produced; the result is a large volume of fluid transfer from surface to downhole and vice versa.

9.2 PRODUCTION *VIA* THERMAL RECOVERY PROCESSES

Thermal production recovery processes are another option to produce heavy oil (as described in Chapter 8). Hot fluid (e.g. steam) is injected to reduce downhole oil viscosity.

Artificial lift and completion components depend on the downhole thermal recovery process used, as well as conditions and constraints such as temperature and fluids.

9.2.1 Steam in Injection Well and Gas or Artificial Lift in Production Well

Standard assisted-lift means are used – e.g. gas lift or downhole pumps – depending on constraints. Gas lift is commonly used with the SAGD process. High-pressure natural gas is continuously injected at the base of the production tubing. The two-phase fluid – gas and a mixture of oil and water at high-temperature – rises to the surface. Gas and liquids are separated. The gas is totally or partially reused in a fired boiler to generate steam.

The artificial lift can also be a high-temperature pump (e.g. ESP or PCP). The main problem concerns reliability *versus* operation time of downhole pumps working with high-temperature fluid pumps involving interaction with elastomer components, electric motors and shaft bearings.

9.2.2 Other Processes

Chapter 8 describes other processes for enhanced heavy oil recovery, particularly miscible flooding with CO_2 injection. Standard artificial lift can be used but may involve corrosion problems associated with the presence of CO_2.

9.3 REVIEW OF ARTIFICIAL LIFT SYSTEMS

Well artificial lift design is a key aspect in the production of heavy oils.

The main issue involves the selection of optimized lifting technology to be used, specifically in terms of operating life of downhole and surface equipment, maintenance, environmental considerations and cost.

For this purpose, various configurations of downhole oil pump systems comprising pumps and drivers are described below.

9.3.1 Beam Pumps

A motor-driven surface system changes the rotation motion of the prime mover into an alternative linear pumping motion. A sucker-rod string transmits the surface motion and power to the downhole reciprocating pump, called a subsurface sucker-rod pump. Figure 9.2 shows typical beam pumps.

Figure 9.2

View of Beam Pumps.

9.3.2 Hydraulic Pumping Systems or Piston Pumps

A hydraulic high-pressure power fluid is pumped from the surface in order to actuate a downhole piston coupled with a reciprocating piston pump (Figure 9.3). The power fluid acts on a piston like a steam engine, but the power fluid is oil or water. The fluid transmits its power to the piston and returns to surface through another conduit if a closed loop power fluid (CPF) is used. In the open power fluid (OPF) design, the power fluid is mixed with the production when flowing back to the surface.

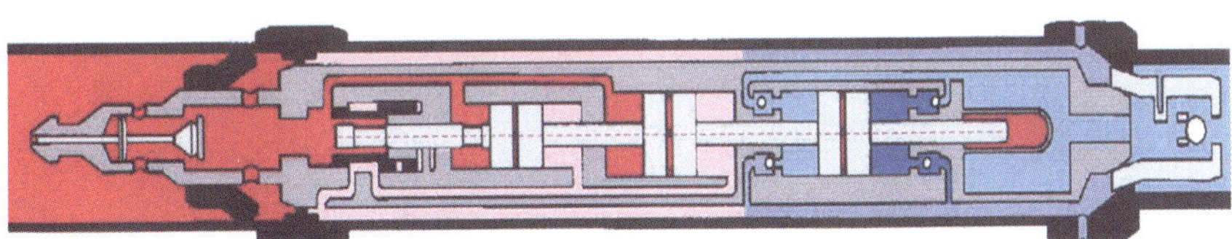

Figure 9.3

ALP type Piston Pump from Weatherford and Downhole Assembly Piston Pump [Cholet H, 2004].

9.3.3 PCP (Progressive Cavity Pump) Systems

Progressive Cavity Pumps (PCP) are now the most popular method of producing heavy oil in any type of well: vertical, deviated or horizontal. "Un peu d'histoire": Progressive Cavity Pumps were invented in 1930 by a French engineer, René Moineau, and developed for the industry by the company "Pompes & Compresseurs Mécaniques" (PCM). This company still manufactures and sells PCP pumps for the oil industry and oil production.

A PCP is essentially constituted of two helical gears, one inside the other (Figure 9.4). The metallic rotor is a single helical "rotating" inside the stator based on a double helical elastomer-lined nitrile in most cases. The external gear or stator has a *double* helical shape, one more than the internal *single* helical gear rotor. When the rotor is rotating, the fluid moves along the pump axis inside the cavities existing between the rotor and stator. The flow rate depends on various parameters such as: rotor diameter, pump eccentricity, length of the stator pitch and rotation speed. By modifying the various parameters, manufacturers can design and offer a large catalogue of pumps adapted to a wide range of well conditions in terms of flow rates, pressure gains (or pressure heads) and fluid types, from low to high viscosity.

Despite the range from manufacturers, the use of conventional progressive cavity pumps presents several drawbacks, e.g. in the case of horizontal or highly-deviated well configurations, where malfunction can result in tubing leaks caused by wear and tear or failure of the sucker-rod drive shaft.

Heavy, viscous oil leads to a frictional pressure drop in the tubing and viscosity increases the resistant torque of the drive string. Wells with low water cuts and/or high sand production induce high-torque loads on the rod string and surface wellhead. This is especially true if the pump unit is positioned in the build-up section or near the liner on top, close to a 90-deg. deviation.

Utilizing PCPs in thermal recovery processes can also be problematic due to the limited resistance of the stator lining (elastomer nitrile) in high-temperature conditions.

Consequently, oil producers applying thermal recovery processes are currently seeking a better artificial lift solution for their projects.

PCPs are generally driven from the surface, but they can be also driven by a downhole electrical submersible motor.

When PCPs are driven from surface (Figure 9.5), the stator is screwed at the tubing extremity and the rotor is fixed to the drive string of sucker-rods. On the surface, the drive head, absorbing the force of the sucker-rods, is operated by an electric motor and a speed

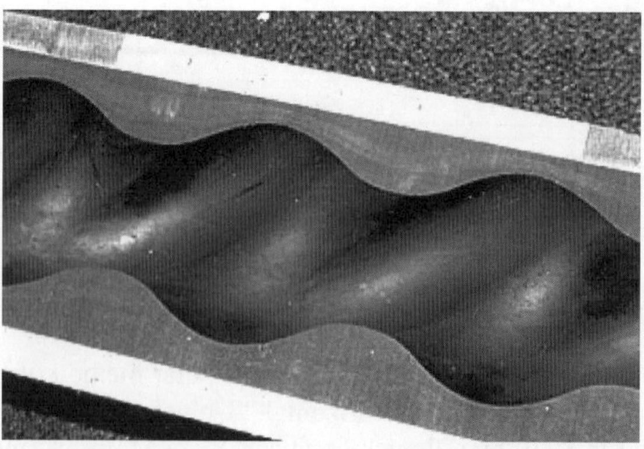

Figure 9.4

View of the Stator. Each full sinusoid represents one pressure stage of a pump.

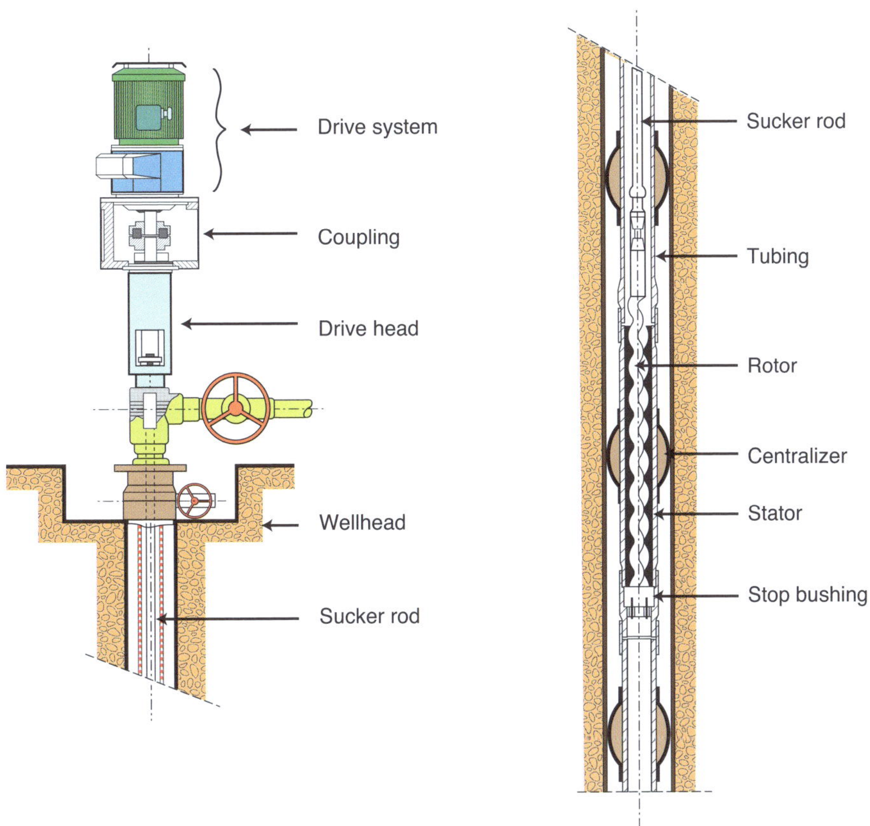

Figure 9.5

PCP Pump, Well Configuration of PCM Driven from Surface by Sucker-Rod [Cholet H, 2004].

reducer. Most downhole applications of PCPs are driven in this manner. However, the rotor can also be driven by an electrical submersible motor. Figure 9.6 provides a schematic view of such an Electrical Submersible Pump-PCP configuration.

9.3.4 Electrical Submersible Pump (ESP)

An Electrical Submersible Pump (ESP) is a multi-staged centrifugal pump driven by a downhole electrical motor (Figure 9.7). ESP pumps are used for high flow rates: from 64 m^3/d (400 bopd) to 12,000 m^3/d (75,000 bopd) depending on size and pressure gain.

ESP pumps are not suited for lifting highly viscous untreated oil, but they can be used to lift cold production diluted fluid with reduced viscosity obtained after injection of diluents (light oil, kerosene, naphtha or water with additives).

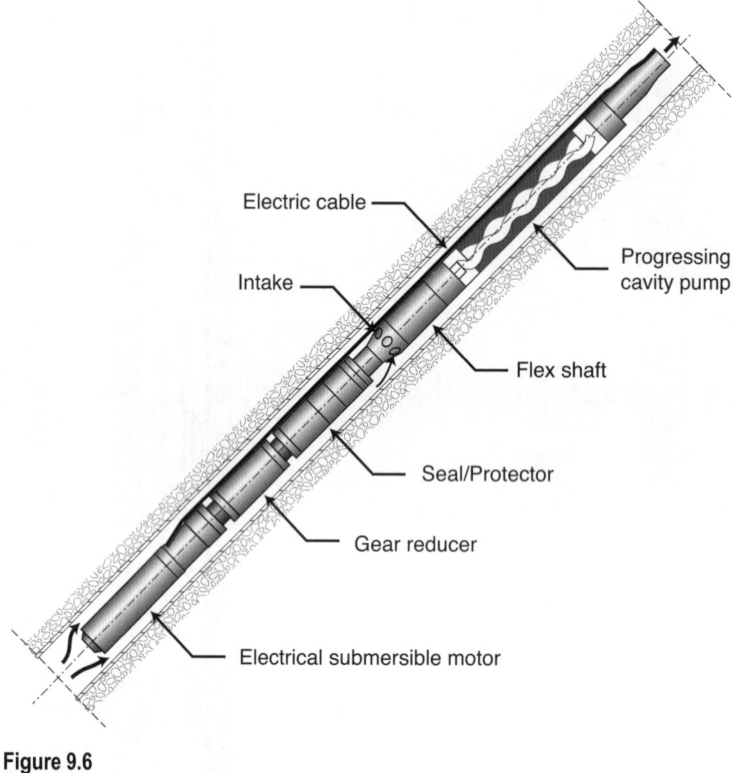

Figure 9.6

ESP-PCP Principle [Cholet H, 2004].

Thermal recovery presents several constraints for ESP artificial lift due to the high temperature applied on the electric motor, electric parts, power cable and pump components like bearings and seals. ESP manufacturers offer special designs for high temperatures, such as "hot line production" equipment with operating temperature ratings up to 288°C (550°F) for the power cable and motor.

9.3.5 Jet Pump Powered with High Pressure Fluid

High pressure power fluid generated on the surface by a pump is transferred downhole by dedicated tubing or through annular space casing-tubing. The power fluid activates a jet pump.

The power fluid enters into the pump (Figure 9.8) and passes through a nozzle, where nearly the entire pressure of the power fluid is converted to a velocity head if no loss occurs. The cross section of the ejector-diffuser throat annulus is determined by the specified inlet and outlet pressure and flow rate fluid characteristics. The high speed power jet fluid flows up to 70 m/s from the nozzle discharge into the production inlet chamber connected to the pump intake. The production fluid is entrained by the power fluid, and the combined fluids

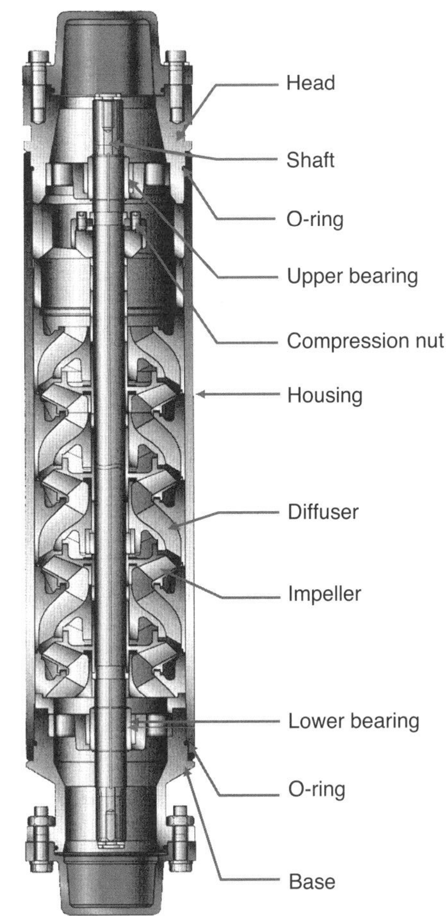

Figure 9.7

Multistage Centrifugal Pump Stage Geometry: Radial Flow or Mixed Flow (from www.esp.weatherford.com).

enter from the convergent to the throat of the pump, resulting in an increase of mixed fluid velocity. In the confines of the throat – which is always of larger diameter than the nozzle – complete mixing of the power fluid and production fluid takes place. During this process, the power fluid loses momentum and energy and the mixed fluid arrives in the divergent diffuser section that converts the velocity head to a static pressure change at the pump outlet. The commingled production fluid and power fluid return to the surface through the production tubing.

Hydraulic efficiencies of jet pumps are low: in the range of 20-32%. This depends on pump design, involving many geometrical parameters and the interaction with downhole operating conditions such as inlet pressures of aspirated and power fluid, outlet pressure of mixed fluids and the characteristics of both fluids.

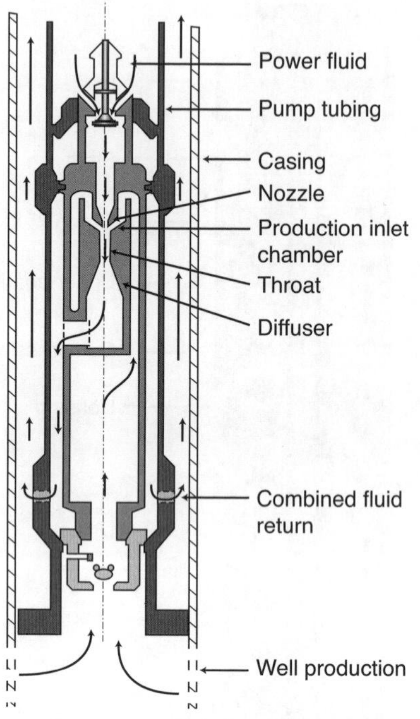

Figure 9.8

Downhole Jet Pump [Cholet H, 2004].

In standard circulation, the power fluid is pumped down the inner tubing and produced fluid is returned up the well annulus. In a reverse circulation jet pump, the power fluid is pumped down the annulus, and well production is returned up *via* the inner tubing.

In two-phase pumping, jet pumps can handle from 100 percent liquid to 100 percent gas with a large decrease in efficiency.

Jet pumps can be used in thermal production since they are not directly affected by the high temperature fluids, having no moving parts, elastomer or electric components. The main constraint of the jet pump is low efficiency and the significant volume transfer of power fluid needed to activate the pump's closed volume, in the range of 1:1.

9.3.6 Gas Lift

Natural gas under high pressure is injected through dedicated tubing or through the annular between casing and production tubing into the base of the production string or tubing. The injected gas makes the density of the fluid in the production string lighter and allows the two-phase mixture to rise up to the surface.

Gas lift can be injected continuously or intermittently in the production tubing at various depths through calibrated gas-lift valves. Gas compressor rates and pressure parameters are adapted to the gas lift requirement.

Gas lift is commonly used with SAGD heavy oil production in Canada. Natural gas under high pressure is continuously injected into the base of the production tubing. The two-phase fluid – gas and a mixture of oil and water at high-temperature – rises to the surface. Gas and liquids are separated, and the gas is then totally or partially reused in a fired boiler to generate steam.

9.3.7 Pump Monitoring

Pump monitoring is a key factor in extending the operating life of all types of downhole pumps (ESP, PCP and ES-PCP).

Where possible, downhole and surface sensors measuring pressure and temperature are linked to data controllers to reduce the risk of "pump off": a lack of fluid to lubricate the pump, allowing heat to build up which would destroy the elastomer stator of a PCP pump, multistage centrifugal or ESP pump. Alarm-controlled flow can set parameters to decrease risk, extend pump operating life and increase the total oil produced.

9.4 ONGOING AND FUTURE DEVELOPMENTS

9.4.1 PCP Pump with a Metallic Stator

The PCP pump with a metallic stator has a fixed positive clearance around the single helical rotor. This eliminates wear and basically increases pump life, but it allows leakage or slip-back due to suction, which decreases the net output flow rate for a given pressure rise. Conventional elastomeric stators have a negative to zero clearance to reduce leakage.

Bauquin J-L *et al.* [Bauquin, 2005] describe a special design for a PCP with metallic stator wherein the stator is made of two parts. The first part, in contact with the rotor, is made of a thin tube formed from a material of low elasticity like metal, and the second is designed to apply and maintain stress exerted by the first part on the rotor providing controlled zero to small clearance for the required pumped fluid pressure gain. Current difficulties involve manufacturing the metallic stator sleeve with tight and constant tolerance. For thermal production recovery, the objective is to use the PCP with metallic stator to prevent the chemical and thermal degradation observed on current nitrile-lined elastomer stators. Prototype PCP with metallic stator was tested by PDVSA and by PCM (Pompes & Compresseurs Mécaniques company). Patents on PCP metallic stator concepts are from IFP (now *IFP Energies nouvelles*) (FR 2794498 priority June.7, 1999, US 6336796 Jan.8, 2002) and from PCM (FR2826407 priority June 21, 2001, US 6872061 Mar.29, 2005).

More recently:

- PCM Oil & Gas has developed under the name Vulcain^{TM} a PCP pump with a metallic stator and a metallic rotor adapted to high temperature operations (400°C) and high pressure (100 bar), used to produce heavy oil with injection SAGD and cyclic steam stimulation processes [Parise, 2009].
- C-Fer Technologies has presented new development of metal-to-metal PCPs pumps [Guerra, 2009].
- Robbins Myers has extended the application range of PCP with the PCP Moyno HT660 (349°C-660°F) Metal-to-Metal rotor/stator.

9.4.2 PCP with Composite Material Stator from Coated Rotor

This concept is a PCP with stator made of a hard composite material such as a reinforcement fiber in a polymer resin matrix, and a rotor made of steel or composite material and coated with an even thickness of soft and durable polyurethane. The Precisionwall pump is supplied by EXOKO composite company.

9.4.3 Downhole Screw Pumps

CAN-K Artificial Lift Systems Inc. (Canada) manufactures a positive displacement pump with axial flow pattern and two metal screws. The principle is described in section 14.4 on rotary pumps, but difficulty was encountered in adapting this for downhole artificial lift. The Can-K is rotation-driven downhole by an ES motor or from the surface with sucker rods and a driving head.

9.4.4 PCP Driven with Coil Tubing and Injection of Diluents

The PCP rotor is driven from the surface by a continuous tube – referred to as coil tubing – rather than sucker rods. The coil tubing makes it possible to inject fluids (e.g. diluents) from the surface. The PCM metallic rotor can allow fluid diluent to flow through. In such cases, diluent can be mixed with heavy oil near the suction of the pump.

Another concept patented by IFP (now *IFP Energies nouvelles*) (IFP Patented US 7.290.608B2 Nov, 6, 2007) is the addition of a tube extension to the rotor. In such cases, the diluents are mixed with heavy oil directly in the horizontal section of the drain producer.

Table 9.1 gives a recap of downhole artificial lift processes for heavy oil production with their range of maximum rates and applicability in terms of maximum temperature.

Table 9.1 Recapitulation of Downhole Artificial Lift Processes for Heavy Oil Production.

	Type of Pumps		Rate (m^3/day) Depends Upon Conditions	For Heavy Oil	Pressure Gain (bar)	Temperature (°C)
Down Hole Pumps	Gas lift		8-4,000	acceptable	low	high possible
	Beam pump		100 max	acceptable	low	high possible
	Hydraulic pump	Piston	15-800	good	medium	high possible
		Jet pump	15-1,500	good	low	high possible
	Positive displacement	PCP	5-1,000	very good	medium	145-170
		ES-PCP	5-1,000	very good	medium	145-170
	Rotary centrifugal	ESP	40-6,000	good	medium	145-170

REFERENCES

Bauquin J-L, Boireau C, Lemay L, Seince L (2005) Development Status of a Metal Progressing Cavity Pump for Heavy-Oil and Hot-Production Wells. SPE 97796 (presented at the *2005 SPE/PS-CIM/CHOA International Thermal Operations and Heavy Oil Symposium*, Calgary, 1-3 Nov.).

Cholet H (1997) Progressing Cavity Pumps. Editions Technip, Paris.

Cholet H (2004) Well Production Practical Handbook. Editor Institut Français du Pétrole Publications, Editions Technip, Paris.

Guerra E, Sanchez A, Matthews C (2009) Field Implementation Experience With Metal PCP Technology in Cuban Heavy-Oil. SPE 120645 prepared for *2009 SPE Production and Operations Symposium*, Oklahoma City, April.

Gülich JF (1999a) Kreiselpumpen, Ein Handbuch für Entwicklung, Anlagenplanung und Betrieb. Springer, Berlin.

Gülich JF (1999b) Pumping Highly Viscous Fluids with Centrifugal Pumps, Parts 1 & 2, *World Pumps*, 1999, 395, pp 9-11.

Hydraulic Institute (1983) Standards for Centrifugal, Rotary and Reciprocating Pumps. 14th ed, Cleveland.

Klein ST (2002) Developments of Composite Progressive Cavity Pumps. SPE 78705 (presented at the *2002 SPE Eastern Regional Meeting*, Lexington, 23-26 Oct.).

Olivet A, Gamboa J, Kenyery F (2002) Understanding the Performance of a Progressive Cavity Pump with Metallic Stator. SPE 77730 (presented at the *2002 SPE ATCE Conference*, San Antonio, 29 Sept. – 2 Oct.)

Parise N, Lehman M, Amara AB (2009) PCM Vulcain™ Metal-to-metal PCP Artificial Lift Systems. Presented at the *2009 MEALF Conference*, Bahrain.

Stepanoff AJ (1940) Pumping Viscous Oils with Centrifugal Pumps, *Oil and Gas Journal*, **4**, pp 123-126.

10 | Examples of Large Heavy Oil Projects

**E. Delamaide, A. Kamp,
G. Renard, T. Rouaud,
R. Kasprik**

10.1 MINING PROJECTS

With regard to bitumen mining in Canada, there were three ongoing projects in 2007 and five others in the planning or construction phase. We will review these projects in the following sections. For the sake of clarity, use of the word "bitumen" refers to the raw product before upgrading. Rates and volumes correspond to the finished "oil" or "synthetic crude oil" product, unless specified otherwise.

10.1.1 Ongoing Projects

10.1.1.1 Overview

Table 10.1 provides an overview of existing projects in 2007: Suncor, Syncrude and Albian Sand Energy.

Figure 10.1 shows the production history of these three main projects.

10.1.1.2 Suncor

As shown in Table 10.1, Suncor was the first oil sand mining project with an initial capacity of 7,150 m^3/d (45,000 bopd) of oil, operating one upgrader. Nearly 30 years later, in 1996, the operation had produced 79.5 million m^3 (500 million barrels) of oil. At that time, as indicated in the previous figure, daily production had increased to roughly 12,100 m^3/d (76,000 bopd).

In 1998, the upgrader was expanded to accommodate the additional production of the Steepbank Mine. This resulted in a total production of 16,800 m^3/d (105,600 bopd) in 1999.

In 2001, a second upgrader was built to start the Millenium Mine, which resulted in doubling of the production capacity.

In 2005, production was reduced because of a fire at the oil sands facilities.

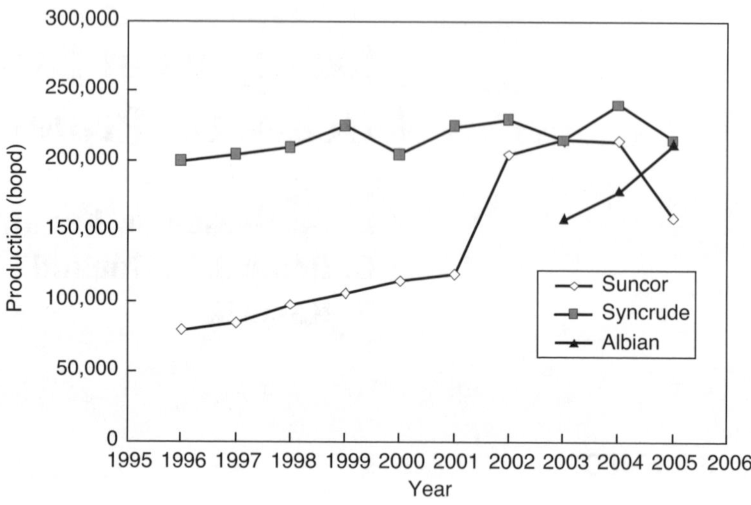

Figure 10.1

Production Profile *versus* Time for the Three Main Projects.

Table 10.1 Overview of Mining Projects in Canada in 2007.

	Suncor [a]		**Syncrude** [b]		**Albian Sands Energy** [c]	
Proved + Probable reserves (end 2005)	365.7 million cu. m (2.3 billion barrels) [d]		795 million cu. m (5.0 billion barrels) [e]		246.5 million cu. m (1.6 billion barrels)	
Starting date	1967		1978		2002	
Date/Mine	1967	Base Mine	1978	Mildred Lake	2002	Muskeg River Mine
	1998	Steepbank Mine	2000	Aurora Mine		
	2001	Millenium Mine				
Production as of 2005 — m^3/d / bopd	25,000 155,000		33,400 210,000		33,400 210,000	
Planned extensions (year/rate (m^3/d) or (bopd))	2008	55,700 350,000	2006	41,300 260,000	2008	15,900 100,000
	2010	79,500 500,000	2015	79,500 500,000		

[a] *[Suncor website]*
[b] *[Syncrude website]*
[c] *[Albian Oilsands website]*
[d] *[Suncor 2005]*
[e] *[Canadian Oil Sands Trust website]*

10.1.1.3 Syncrude

Syncrude was the second oil sands mine to start operations in 1978, at Mildred Lake. The hundred millionth barrel (15.9 million m^3) was produced 4 years later in July 1982, and in April 1998 the milestone of 1 billion barrels (159 million m^3) produced was reached.

Figure 10.1 shows the very steady production history of Syncrude from 1996 to 2005.

10.1.1.4 Albian Sands Energy Inc.

Albian Sands Energy Inc. is the operator of the Athabasca Oil Sands Project (AOSP), which is a joint venture between Shell Canada Limited (60%), Chevron Canada Limited (20%) and Western Oil Sands LLP (20%). The project includes two parts: the Muskeg River Mine operated by Albian Sands Energy Inc. and the Scotford Upgrader operated by Shell Canada Limited.

The Muskeg River mine stands on one of Shell's leases. It produced its first ore in August 2002, with bitumen production starting in December 2002 and synthetic crude production in 2003. Average production was 33,900 m^3/d (213,000 bopd) in 2005, as indicated in Figure 10.1.

10.1.2 Future Projects

A number of other projects are currently in various stages of planning and development. They are briefly reviewed below.

10.1.2.1 Canadian Natural Resources Limited (CNRL)

CNRL is the sole owner and operator of the Horizon project [CNRL website], which will be developed in three phases. Construction of the first phase started in early 2005 and first oil is targeted for mid-2008, with a production of 17,500 m^3/d (110,000 bopd) of synthetic crude oil (after ramp-up). The second phase plans to start producing in 2011 and should increase production to 24,600 m^3/d (155,000 bopd), and finally Phase 3 should start in 2012, bringing total production for the project to 36,900 m^3/d (232,000 bopd). CNRL is also evaluating the option of combining Phases 2 and 3.

Proved + Probable reserves were estimated at 461.1 million m^3 (2.9 billion barrels) at the end of December 2005.

10.1.2.2 Total E&P Canada Ltd. (formerly Deer Creek Energy Ltd.)

Total E&P Canada Ltd. [Total Canada website] took over Deer Creek Energy in August 2005, and thus became the owner of 84% of the Joslyn Creek Mine Project (the rest is owned by Enerplus Resources Fund).

The mining project is still in the planning stage, and may be modified now that Total is the operator. Plans thus far were for the mine to be developed in four phases of 8,000 m^3/d (50,000 bopd) each, with the first phase starting in 2013 and the following phases at three-year intervals afterward.

No upgrader is planned for the first phase, but one is being considered and could be constructed for the second phase of development.

10.1.2.3 Imperial Oil Limited/ExxonMobil Canada Ltd.

Imperial Oil [Imperial Oil website] and ExxonMobil are partners (70/30%) in the development of the Kearl Lake oil sands mining project.

Current plans consist of a first 16,000 m^3/d (100,000 bopd) phase to come on-stream in 2010, followed by a second phase of another 16,000 m^3/d (100,000 bopd) in 2012. A third 16,000 m^3/d (100,000 bopd) phase is also planned for 2018/2019.

The project does not include on-site upgrading; thus, the bitumen will have to be mixed with diluent to be shipped. The partners are currently evaluating a number of options regarding both the need for an upgrader and potential sources of diluent supply.

10.1.2.4 UTS Energy/Petro-Canada/Teck Cominco

The Fort Hills Project has been in the works for quite some time now, with many changes from year to year. The partners in 2007 were Petro-Canada (55%), UTS Energy (30%) and Teck Cominco (15%).

The development plan calls for 27,000 m^3/d (170,000 bopd) production by 2011, followed by an upgrader two years later.

10.1.2.5 Synenco/Sinopec

Synenco's Northern Lights project is the least-advanced of the mining projects [Synenco website]. The regulatory application was filed in May 2006. Two phases of 8,000 m^3/d (50,000 bopd) of production are planned, with a potential timing of 2009/2010 for start-up.

In 2005, Synenco sold 40% of its interest in the project to SinoCanada, a subsidiary of Sinopec, a large Chinese oil & gas company. The partners are also planning to build an upgrader.

10.2 *IN SITU* PRODUCTION PROJECTS

As indicated in Figure 10.2, there are several heavy oil basins throughout the world, located on all continents. This section includes a detailed review of specific heavy oil basins or fields produced in recent decades.

Of course, there are plenty of other fields producing heavy oils, but a selection was necessary. Those selected have the advantage of being well documented in the literature and representative of the various recovery processes used to produce these difficult reserves.

The projects are presented by recovery processes, from steam injection, widely used and beneficing from a long history, to the more recently implemented technologies.

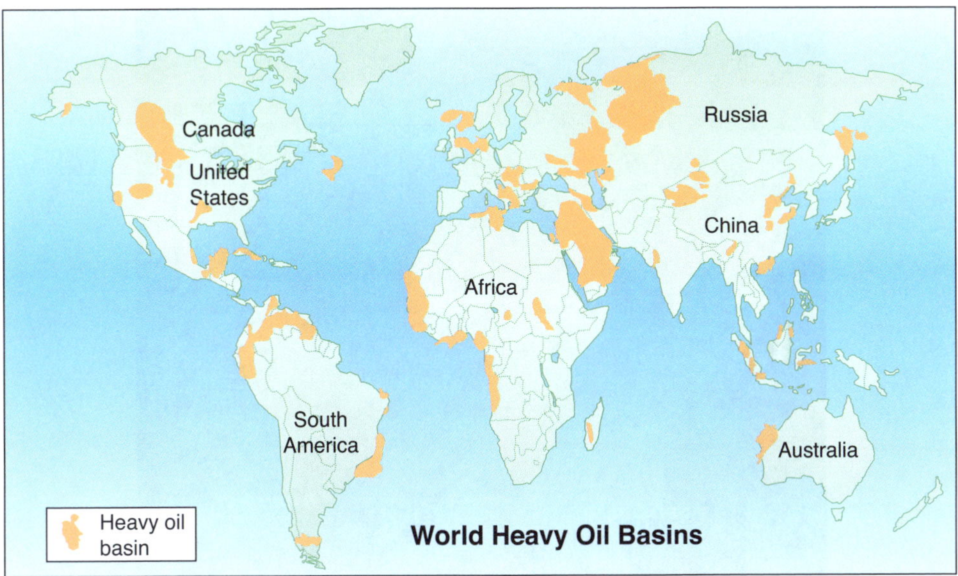

Figure 10.2

Location of Main Heavy Oil Basins throughout the World.

10.2.1 Steam Injection

10.2.1.1 California Oil Basin (US)

There are several heavy oil reservoirs in California, mainly located in the San Joaquin Valley (Figure 10.3). Their main characteristic is that they are quite shallow, explaining why most of them were discovered at the end of the nineteenth or beginning of the twentieth century, e.g. Coalinga in 1887, Midway Sunset in 1894, Kern River in 1899 and South Belridge in 1911. However, San Ardo – located in the Salinas Valley – was discovered later, in 1947.

Low depth favored the first drilling of near-surface seeps of heavy oil and tar by settlers and prospectors. Shortly after their discovery, in the early 1900s, primary production of the various fields peaked at several thousand barrels per day, after which production steadily declined. Because of the high oil viscosity at reservoir temperature, the primary production recovery factor proved to be low: between 5 and 12% of OOIP. In absence of heavy oil refining techniques, the demand for heavy oil on the market was also low. This changed in the early 1950s when progress was made on refining techniques. In the mid-1950s, a production increase was achieved using bottom-hole heaters. Some years later, in the early 1960s, successful experiments were done with steam injection and the process of Cyclic Steam Stimulation (CSS) and steam flood were applied to most of the Californian heavy oil

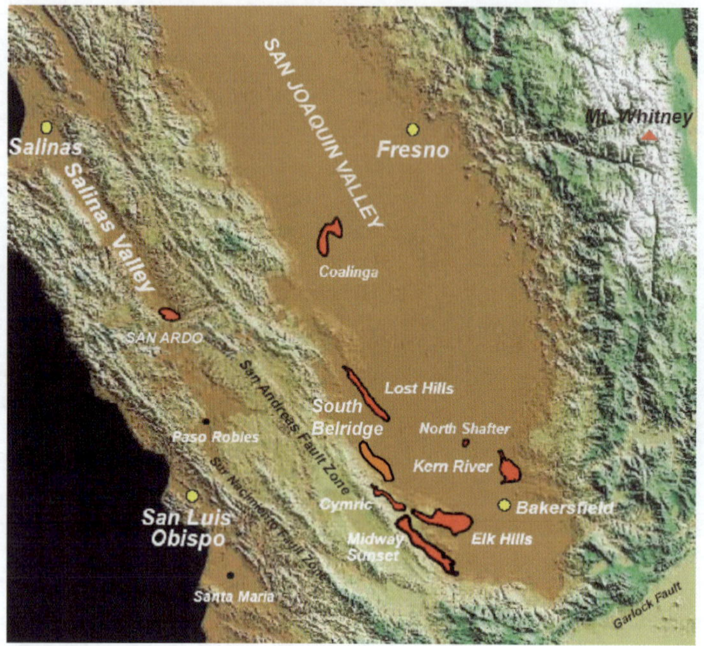

Figure 10.3

Location of Main Californian Heavy Oil Reservoirs.

fields. More recently, horizontal wells have been applied in some of the fields. This reduced the production decline. Kern River, Midway-Sunset and South Belridge have each already produced more than 1 billion barrels (160 million m^3) of oil.

Table 10.2 shows the main characteristics of five of the largest Californian heavy oil fields.

Figure 10.4 shows the number of projects and daily production rates in Californian heavy oil fields *versus* time, between the years 1978 and 2006 [Moritis G, 2006]. The relatively high number of projects is a result of the enormous field sizes and the fact that they were operated by numerous companies. In recent years mergers have taken place, thus decreasing the number of companies and projects.

As indicated in Figure 10.4, production from these thermal EOR projects peaked in 1986 at 76,315 m^3/d (480,000 bopd) and has declined in recent years to 45,470 m^3/d (286,000 bopd) in 2006.

In 2006, Kern River field – operated by Chevron Corp. – remained the largest single EOR thermal project in the United States, producing about 13,670 m^3/d (86,000 bopd), but this is a sizable decrease from 15,900 m^3/d (100,000 bopd) in 2004. Figure 10.5 shows the production profile of the Kern River field since its discovery. The large impact of thermal EOR methods is noticeable.

Table 10.2 Main Characteristics of Five of the Largest Heavy Oil Fields in California.

		Coalinga	Kern River	Midway Sunset	San Ardo	South Belridge
Formation		Temblor	Kern River	Monarch & Potter	Lombardi & Aurignac	Tulare
Area/Dimension	hectare km × km	16,800 8 × 21	6,400 10 × 6.4	19,200 4.8 × 40	1,500 7 × 2.2	4,610 14.4 × 3.2
	acre mile × mile	41,500 5 × 13	15,800 6.2 × 4	47,440 3 × 25	3,800 4.3 × 1.4	11,390 9 × 2
Depth	(m)	150-915	30-438	400	610-700	122-366
	(ft)	500-3,000	100-1,600	1,300	2,000-2,300	400-1,200
Thickness	(m)	15	107	90	60	40
	(ft)	50	350	300	200	130
Porosity	(%)	33	31	33	34	35
Permeability	(D)	0.7-3	1-10	0.5	1-7	0.5-5
°API		12-14	10-15	14	13	13
Specific gravity	g/cm^3	0.986-0.973	1-0.966	0.9725	0.979	0.979
Viscosity (@ res. cond.)	cP	50-5,000	500-10,000	1,500	2,500-40,000	1,875
Original Oil In Place (OOIP)	(10^6 m^3)	715	640	1,200	208	1,300
	(10^9 bbl)	4.5	4	7.5	1.3	8.3
Recovery factor (2004)	%	22	51	46	37	23

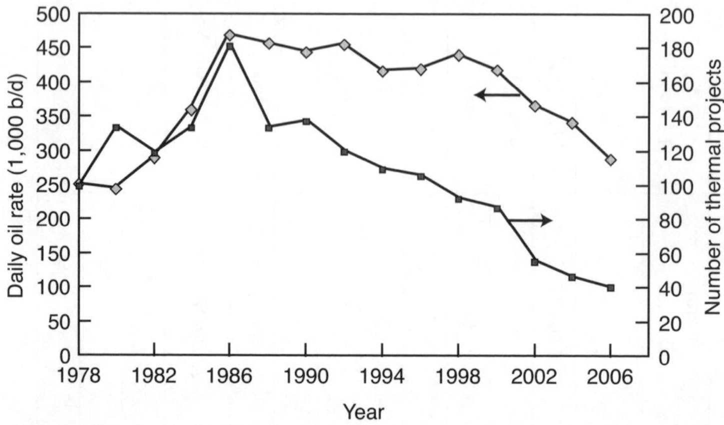

Figure 10.4

Californian Heavy Oil Production and Projects *versus* Time [Moritis G, 2006].

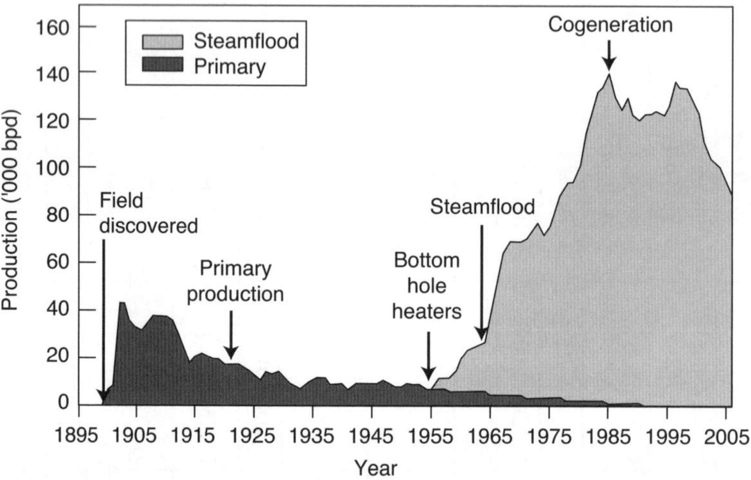

Figure 10.5

Production History of the Kern River Field (California).

From the year 2000 on, Chevron Corp. [Ormerod L *et al.*, 2006] has developed a real-time field surveillance and well services management system in its San Joaquin Valley Business Unit (SJVBU) that operates several heavy oil fields: Coalinga, Cymric, Kern River, Lost Hills, Midway Sunset, McKittrick and San Ardo. Chevron's thermal EOR projects in California produced 25,280 m^3/d (159,000 bopd) in 2006 and there were approximately 15,000 active wells in the BU, yielding an average oil production of 2.1 m^3/d (13 bopd) per well. Maintaining this very large number of wells for optimal production in the management of these fields was a key operational challenge. Operators must also bear the cost of injecting the equivalent of hundreds of thousands of barrels of steam per day *via* thousands of injection wells, and disposing of hundreds of thousands of barrels of produced water per day. Steam generation and treatment of produced fluids cost US $4 to $7 per barrel of produced oil. Cogeneration – i.e. the simultaneous production of electricity and steam or hot water in the same plant – has helped some heavy oil projects to remain economic during low oil price periods at the end of the 1990s and beginning of the current decade.

In 2006, Aera Energy LLC, a joint venture between ExxonMobil Corp. and Shell Inc., had the second largest thermal EOR production, producing 17,000 m^3/d (107,000 bopd) from 16 projects, but this also was a decrease from the 22,400 m^3/d (141,000 bopd) produced in 2004.

10.2.1.2 Indonesia (Duri Field)

The Duri field, operated by P.T. Caltex Pacific Indonesia (CPI) is located in the Riau province in central Sumatra Island (Figure 10.6). It has been the largest single thermal EOR project in the world for many years, producing more than 31,800 m^3/d (200,000 bopd). The field is approximately 18 km (11.2 miles) long by 8 km (5 miles) wide with developable areas of about 8,100 ha (20,000 acres).

The field was discovered in 1941, but was not put into production until the construction of a pipeline to the port of Dumai in 1958.

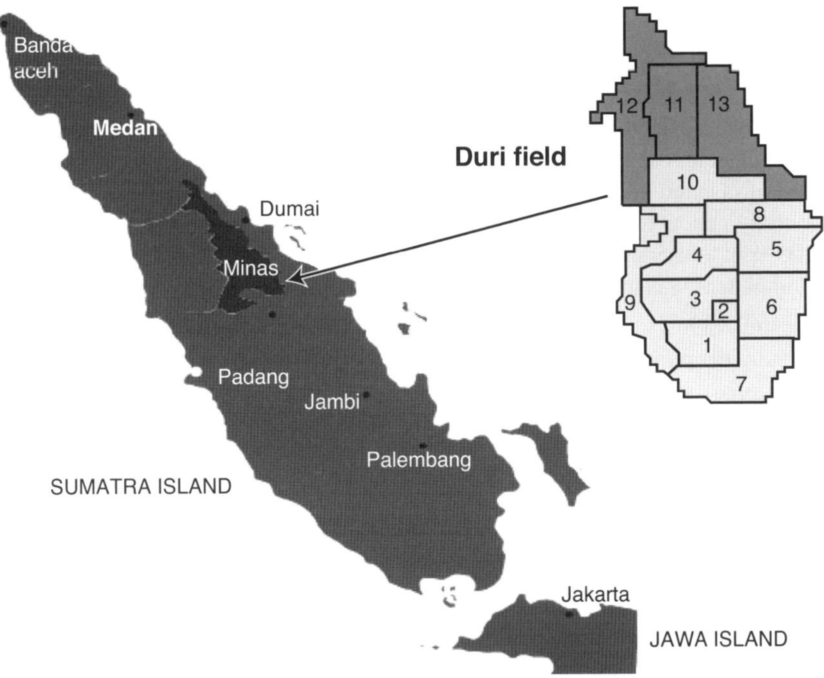

Figure 10.6

Location of the Duri Field (Sumatra Island).

Structurally, the Duri field is a faulted, asymmetric anticline. The Pertama/Kedua (P/K) and Rindu reservoirs, of early Miocene age, represent the majority of the reserves in the Duri field. Their main characteristics are the following: average depth 152 m (500 ft), average net pay thickness 37 m (120 ft), average permeability 1,500 mD, average porosity 34%, oil gravity 20°API, and oil viscosity 330 cP at reservoir temperature [38°C (100°F)].

Initially the field was developed under primary production, which peaked in the mid 1960s at 10,335 m^3/d (65,000 bopd). The recovery mechanisms were principally solution gas drive and compaction, leading to a recovery factor of 7% of OOIP.

Cyclic steam injection was tested first in the P/K reservoirs in the 1970s. Then a steamflood pilot consisting of 16 inverted 5-spot patterns on 6.2 ha (15 1/2 acre) spacing was initiated in 1975. The pilot was a success, recovering approximately 30% of OOIP, and led to the initiation of a major development, with the first project in 1985 in Area 1 located south-westward of the Duri field. The area was developed with inverted 7-spot patterns on 4.7 ha (11.625 acre) spacing. After 14 years on steamflood, recovery in Area 1 was about 64.5% of OOIP.

The second area (Area 2) started in 1986. The area was developed with inverted 9-spot pattern configuration on 6.3 ha (15.5 acre) spacing (250 m × 250 m [76 ft × 76 ft]). Steam injection was stopped in 1996, as the area was considered already mature. After 13 years on steamflood, it is considered that the Area 2 recovery factor had reached about 64.8% of OOIP.

Area 3, which was started in 1988, was developed with the combination of inverted 7-spot and inverted 9-spot patterns. After 10 years on steamflood, the Area-3 recovery factor had reached about 67.6% of OOIP.

Areas 4 and 5 were developed using 11 acre, 7-spot patterns (217 m × 217 m [66 ft × 66 ft]) in the mid-1990s with the P/K target sands. Although this design has been successful from both technical and economic perspectives, operational problems associated with the balancing of injected and produced fluids have necessitated the search for further optimization. A study conducted in 1989 recommended a "hybrid" combination of 6.3 ha (15 1/2 acre) of 9-spot and 5-spot patterns to help solve this problem. The first implementation of the hybrid pattern was in Area 6, followed by Areas 7, 8 and 9.

In the mid-1990s, CPI decided to evaluate the potential of steamflood in the Rindu reservoir and started a pilot with cyclic steam stimulation in most of the wells.

The production of the Duri field peaked at about 47,700 m^3/d (300,000 bopd) in the years 1994 to 1998 (Figure 10.7). Early in the current decade, Duri produced nearly 36,570 m^3/d (230,000 bopd) of oil *via* the injection of 151,000 m^3/d of CWE steam (950,000 bpd). About 20% of the oil was burned as fuel to generate steam. In more recent years, gas was used for this purpose. As indicated in Figure 10.7, cogeneration has been used since the early 1990s to optimize heat consumption.

Monitoring techniques to optimize steam efficiency were used in the Duri field (essentially the same techniques as those used in California fields). Radioactive tracers have been injected into injection wells to obtain the steam injection profile. High-temperature spinners have also been used. Wellhead temperatures, pressures and flow rates (at well head and locally using spinners) are measured in production wells. Such measurements allow detection of steam breakthrough. Other techniques include fiber-optic temperature surveys and oil

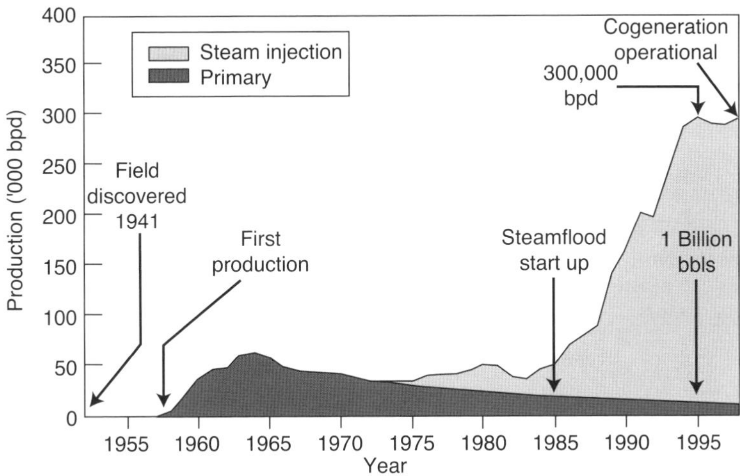

Figure 10.7

Production History of the Duri Field.

fingerprinting. Time-lapse 3-D seismic (4-D seismic) was also carried out in most areas to monitor steam growth over time and manage steam injection in order to optimize recovery efficiency.

The main producing mechanism in the Duri SteamFlood (DSF) after steam breakthrough is gravity drainage. Steam that is injected after steam breakthrough is used to maintain the downward growth of the expanding steam chest and compensate for heat losses to the formation and produced fluids.

Ultimate tertiary recovery from this US $1.8 billion project is estimated to be 300 million m^3 (2 billion barrels) of oil. The number of wells peaked at about 3,000 producers and 1,400 steam injectors.

10.2.1.3 Canada (Cold Lake Field)

Canada has enormous potential heavy or extra-heavy oils and bitumen (see section 2.2.1) resources, mainly located in the province of Alberta and Saskatchewan. The Alberta Oil Sands are located in three main deposits (Figure 10.8): Athabasca, Cold Lake and Peace River (from largest to smallest). The Athabasca deposit contains approximately 213 billion m^3 (1,345 billion barrels) of oil sands and is the world's largest known petroleum resource. The Cold Lake deposit has about 31 billion m^3 (195 billion barrels) of oil sands, and the Peace River deposit has about 20.5 billion m^3 (129 billion barrels). As of the end of 2005, the oil sands contained an estimated 27.7 billion m^3 (174 billion barrels) of recoverable bitumen and as of June 2004, 28 petroleum companies with 81 oil sands projects were offering to develop this resource.

Cold Lake is located 300 km (186 miles) northeast of Edmonton, mainly in the province of Alberta, and in Saskatchewan. It was discovered at the end of the 1950s. In the Cold Lake

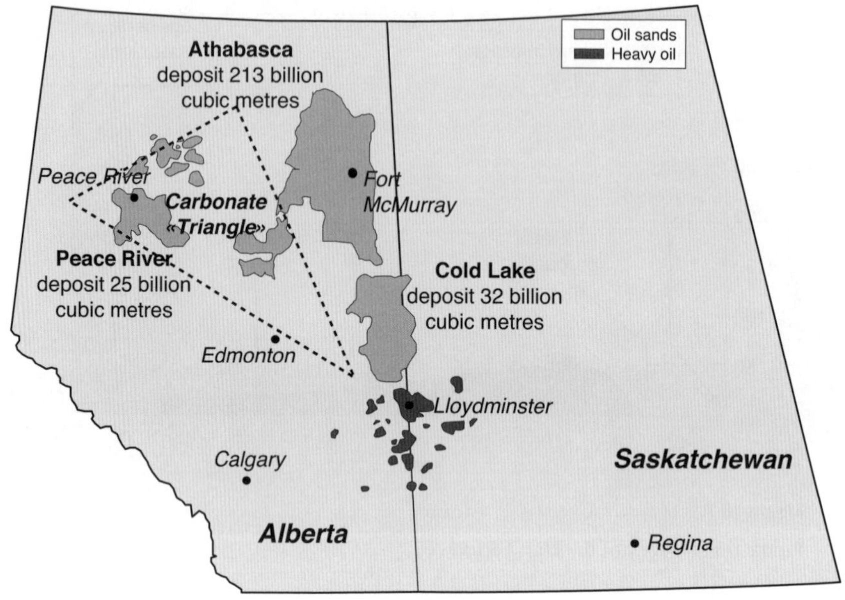

Figure 10.8

Main Canadian Oil Sands and Heavy Oil Deposits showing Cold Lake Area.

area, the three formations of the Mannville Group (McMurray, Clearwater and Grand Rapids) of Lower Cretaceous Age are all oil-bearing, but the thickest reservoir with the best petrophysical characteristics is the Clearwater formation. The reservoir is a thick sand deposited in an estuarine/deltaic environment at a depth of about 400-450 m (1,310-1,480 ft). Average pay thickness is 15 m (45 ft), but can reach over 40 m (130 ft) in some places. Reservoir quality is excellent with a porosity of 30 to 35% and a permeability of several darcies.

Due to their depth, Cold Lake's oil sands deposits cannot be developed using the open pit mines found further north in the Athabasca region of Alberta (see Chapter 6). Consequently, Imperial Oil Ltd (IOL) adopted Cyclic Steam Stimulation (CSS) as the most efficient process for the recovery of bitumen at Cold Lake. Cold Lake is the largest steam project in Canada and one of the largest thermal operations in the world. IOL operates in properties covering a huge area of 78,000 ha (31,565 acres) that holds over 2.4 billion m^3 (15 billion barrels) of oil in place. Bitumen produced is both very heavy (11°API) and very viscous (75,000 cP).

IOL began pilot experimentation with CSS at Cold Lake in 1964 and followed this pilot phase with commercial application in 1985 [Gallant RJ, 1993]. Since then, an extensive amount of experimental data and operating experience has been acquired, resulting in the continual evolution of steaming and operating strategies for CSS. The CSS process at Cold Lake involves periods of steaming (usually four to six weeks), followed by periods of "soaking" (four to eight weeks), followed by increasingly long periods of production (several months).

Operating CSS in Cold Lake offers some particular challenges. In order to inject steam to an economical level (about 200 m^3/d [1,260 bspd] cold water equivalent per well), sufficiently high pressure (9-10 MPa [1,305-1,450 psi]) must be achieved to mechanically fracture the formation. This results in complex fracturing and reservoir deformation behavior. The steam is injected at about 300ºC (572°F) into the fractures that develop. A high degree of interwell communication occurs when steam is injected, further complicating steam conformance and well behavior.

Furthermore, in the Cold Lake CSS process, the recovery mechanisms are complex and include compaction drive, solution gas drive, steam flashing and gravity drainage; also, the more unconventional bitumen mobilization processes of foaming and emulsification interact, particularly in the mature stages of depletion. To simulate these various and complex processes, IOL developed a mechanistic model which attributes most of the effectiveness of the CSS process to the formation – during production cycles – of an effective "foamy" impedance zone cushioned between cold and heated reservoir zones [Batycky JP, 1997]. The impedance or transition zone allows efficient exploitation of native solution gas drive to displace live bitumen at below steaming temperatures for a considerable portion of the production cycle. The heated zone closer to the wellbore acts as a thermally stimulated conduit for the displacement of bitumen into the wellbore. The cold reservoir supplies the compaction and solution gas drive energy. Therefore, CSS at Cold Lake is a surprisingly effective solution gas drive process.

To develop the Cold Lake area, IOL has used a phased development approach. Net recoverable reserves for the total phases have been estimated at 0.2 billion m^3 (1.2 billion barrels). More than 3,000 wells were operated in 2006 to produce 23,850 m^3/d (150,000 bopd) of crude bitumen, which corresponds to over 30% of Canada's total *in situ* production and over 8.3% of its total oil production. Vertical, directional and horizontal wellbores have been used. Recovery of 23% of Original Oil in Place (OOIP) is expected at a cumulative steam-oil ratio of 3.3. Figure 10.9 shows Cold Lake bitumen production *versus* time. Production in 2008 is expected to grow to 28,620 m^3/d (180,000 bopd).

Because of its high viscosity, Cold Lake bitumen must be blended with a diluent to meet pipeline specifications. To transport the huge quantities of blended bitumen, IOL and other companies in the Cold Lake area use a pipeline to the Hardisty terminal (Alberta). After recovery of the diluent, a pipeline is used to transport it from Hardisty to Cold Lake. Capacities of the two lines are on the order of 111,000 and 31,800 m^3/d (700,000 and 200,000 bopd), respectively.

In the year 2000, IOL started a cyclic-steam pilot of an enhanced process termed LASER, for "Liquid Addition to Steam for Enhancing Recovery" [Leaute RP and Carey BS, 2005], in which a diluent is injected together with steam. The overall performance of the pilot has been encouraging, and with a diluent recovery of 80%, has exceeded original expectations.

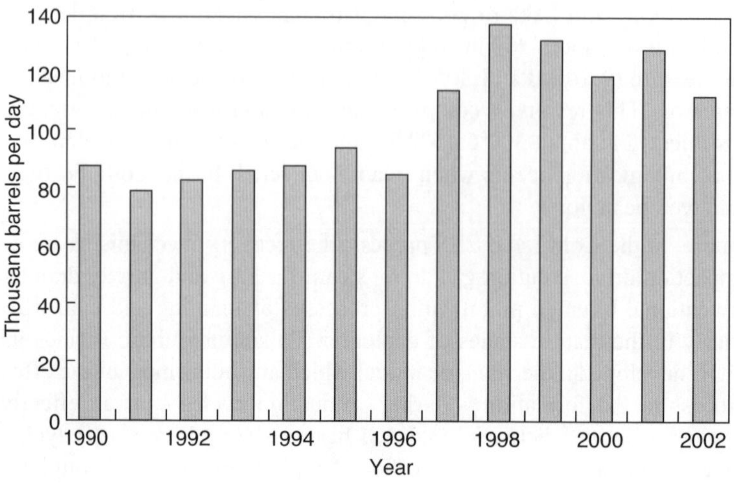

Figure 10.9

IOL Cold Lake Bitumen Production.

10.2.1.4 Canada (SAGD Projects)

In 2005, the Canadian petroleum industry produced an average of around 1,060,000 b/d of bitumen. Of this, 59% was mined while 41% was extracted *in situ*. *In situ* production of crude bitumen in Canada is split evenly between CSS and SAGD technologies due to the historic role of CSS in the development of *in situ* methods, but new projects are predominantly anticipating the use of SAGD technologies. In this section, the main SAGD projects currently underway in Canada are listed and briefly reviewed (see Table 10.3).

In 2004, there were 13 approved commercial SAGD projects in Alberta. A total of 93 producing wells accounted for 4.1 million m^3 (25.8 million barrels) of bitumen, at a cumulative average rate of 11,000 m^3/d (69,200 bopd), i.e. the average well production rate was 120 m^3/d/well (755 bopd/well) [Sadler K and Davis P, 2005].

A brief description of some of these projects is given below. Additional details can be found on the well-documented website of the Alberta Energy Utility Board [www.eub.ca] or the websites of the operators.

Senlac SAGD Project:

- The Senlac SAGD Thermal Project is the first field implementation of the SAGD process. It started in 1995, initially with a 3 well pilot.
- Senlac is a deep reservoir at a depth of about 750 m (2,460 ft). It is of excellent quality, with an average porosity of 33% and permeability varying from 5 to 10 Darcies. The medium-heavy oil, 12°API, has a viscosity of 5,000 cP at reservoir temperature 29°C (84.2°F). The net pay is between 8 and 15 m (26 to 49 ft). There is a bottom aquifer varying in thickness from 0 to 10 m (0 to 32.8 ft).

Table 10.3 Major SAGD Projects in the Alberta Province (Canada) as of Dec. 2006.

Project name	Operator	Date of start	Location	OOIP (million bbl)	2004 prod (bopd)	No. of well pairs	Cumul. Steam Injected (Mbbl)	Cumul. Oil Produced (Mbbl)	Lowest CSOR	CSOR all well pairs	Planned future production (bopd)
Senlac	Encana	Oct. 1995	Lloydminster	111	2,230	10	20,013	7,500	1.18	2.67	
MacKay River	Petro-Canada	June 1996	Athabasca	350	17,735	32	60,788	24,874	1.33	2.44	73,000 by 2020
Surmont*	Conoco-Phillips	Oct. 1997	Athabasca	20,000	435	3	6,413	1,978	1.38	3.24	27,000 by 2007 100,000 by 2012
Hilda Lake*	Blackrock Ventures	Oct. 1997	Cold Lake		500	2	6,036	1,524	2.76	3.96	34,000 by 2020
Hangingstone	Jacos	July 1999	Athabasca	500	7,100	15	30,331	10,284	1.77	2.95	60,000 by 2012
Foster Creek	Encana	May 2001	Cold Lake	29,000	30,000	44	108,874	41,415	1.42	2.63	150,000 by 2020
Christina Lake	Encana	Apr. 2002	Athabasca	15,000	4,500	4	11,896	4,742	1.72	2,51	250,000 by 2020
Burnt Lake	CNRL	Sept. 2002	Cold Lake			3				3.50	120,000 by 2020
Long Lake*	Nexen / Opti	May 2003	Athabasca	5,500	600	3	5,419	933	4.51	5.81	72,000 by 2007 291,000 by 2020
Firebag	Suncor	Jan. 2004	Athabasca	1,500	11,000	23	61,984	15,588	2.79	3.87	376,000 by 2020
Jocelyn*	Total (Deer Creek)	Sept. 2004	Athabasca	11,000	120	1	507	114	4.09	4,44	42,000 by 2020
Overall (all well pairs)						136	312,261	108,952	1.18	2,87	

* *Pilot*

- Five phases of SAGD operations have been performed from 1995 to 2006.
- Oil recovery varied for each well pair from 11 to 67% of OOIP, with an average of 32% of OOIP.
- Steam chamber monitoring has been performed using repeated time-lapse seismic (4D-seismic).
- A Solvent Added Process (SAP) test has been successful in a few of the wells.

MacKay River SAGD Project:

- The target reservoir is the McMurray formation at a depth of 80-135 m (262-443 ft) from the surface. There are no extensive underlying water or gas over bitumen issues.
- First steam and first oil of phase 1 were respectively in Sept. and Nov. 2002. Twenty-five well pairs were operated initially. Four 63 MW (215 MM Btu/hr) steam generators provided 9,500 t/d steam at 80% quality.
- Performance of Phase 1 as of end of Aug. 2006 was 3.6 million m^3 (22.7 million barrels) of bitumen produced at an average CSOR of 2.4. Recovery factor at that time was 29% of OOIP.
- Phase 2 started in Jan. 2006 with 16 additional well pairs.

Foster Creek SAGD Project:

- The reservoir target is located between 180 and 225 m subsea (590 and 740 ft) and has a thickness of 15 to 30 m (49 to 98 ft). Its permeability and initial oil saturation are respectively 6 darcies and 0.8 (0.5 in the transition zone).
- Phase 1A, 1B and 1C were successively in operation, each with a production capacity of 1,590 m^3/d (10,000 bopd) of bitumen. They included 28, 10 and 12 well pairs. Some wells experienced sand filling and were shut in. Three quarters of the producers were on ESP and the remaining on gas lift. The tendency has been to convert from gas lift to ESPs.
- A test of the VAPEX process was done in a well pair but a hydrate problem was encountered at operating conditions due to the simultaneous presence of methane (associated gas), propane and water.
- Phase 1D and 1E are planned to expand Foster Creek SAGD to 19,080 m^3/d (120,000 bopd) in 2008.

Christina Lake SAGD Project:

- McMurray formation is the production target of this SAGD project.
- Phase 1A SAGD operation in Christina Lake took place from 2002 to 2004 with a production capacity of 1,590 m^3/d (10,000 bopd) of bitumen. It comprised 6 SAGD pairs. Key points learned have been SAGD working in variable geology, optimization of start-up and completions, artificial gas lift performance, and bitumen depletion between top gas and bottom aquifer. Steam chambers were averaging 3,000 kPa (435 psi) at around 230°C [446°F]. The recovery factor in the 6 well pairs varied from 22 to 55% of OOIP. Injection of methane with steam in the well pairs (averaging 8 to 10 m^3 methane per ton of dry steam) helped to maintain pressure with no apparent penalty to oil recovery but with a positive impact on CSOR. 4D seismic was used to evaluate steam chamber growth along the well pairs.

– Extension of SAGD was planned to take place between 2006 and 2008 to gradually increase production to 20,350 m^3/d (128,000 bopd) (Phases 1B to 1E).

Long Lake SAGD Project:

– A SAGD pilot was started in April 2003 comprising 3 well pairs. After steam circulation, production began in Aug. 2003, but in Feb. 2004 the operator had to decrease the pressure injection in response to steam losses, creating economic concerns. Steam was thought to be escaping through a lean zone. Gas injection into the mobile zone was done with some success to reduce the problem.
– A solvent injection was also tried using a condensate as the solvent. However, it was stopped after a few weeks since the benzene concentration level was above permissible levels. Laboratory work was planned to select a more appropriate solvent.
– The drilling and completion of 78 SAGD well pairs was completed at the end of 2006 with an expectation for the start of the full project in 2007-2008.
– The Long Lake oil sands project is the first to integrate SAGD with an on-site upgrader. Production was planned to start in 2006, with the upgrader starting up in 2007. The plans include the SAGD wells producing 11,450 m^3/d (72,000 bopd) that will be upgraded to about 9,540 m^3/d (60,000 bopd) of synthetic crude.

Firebag SAGD Project:

– The Firebag reservoir properties are: depth 320 m (1,050 ft), average net pay 35-40 m (115-131 ft), permeability 6-10 darcies, bitumen saturation 85%, bitumen viscosity 10 million cP.
– The Firebag project consists of an experimental solvent injection pilot and a commercial SAGD scheme. Ultimately, the project will supply 57,240 + m^3/d (360,000 + bopd) of bitumen feed for the bitumen upgrader.
– The SAGD portion itself is expected to provide up to 22,260 m^3/d (140,000 bopd) in four – 5,560 m^3/d (35,000 bopd) stages at a designed SOR of 2.
– First steam was injected in Sept. 2003 and first oil sent to upgrader in Jan. 2004. As of the end of Feb. 2006, 22 out of a total of 40 well SAGD producers had produced 2 million m^3 (12.5 million barrels) of bitumen with a CSOR of 4 (4.6 one year earlier). Higher CSOR than expected is attributed to the high steam injection rates required to restart the wells after prolonged shutdowns at different times.
– At the end of 2005, ES-SAGD had been implemented on a limited basis. Solvent proportion was 2% of the steam volume.

Figure 10.10 shows the mean CSOR (all wells for each project) *versus* time for the main SAGD projects from start of first steam injection [MacKay K and Delamaide E, 2006]. As can be seen, for most of the projects the CSOR is between 2.5 and 4.

Table 10.4 summarizes the new projects planned as of the end of 2006 in the Alberta province, with expected production levels. High production has been forecast.

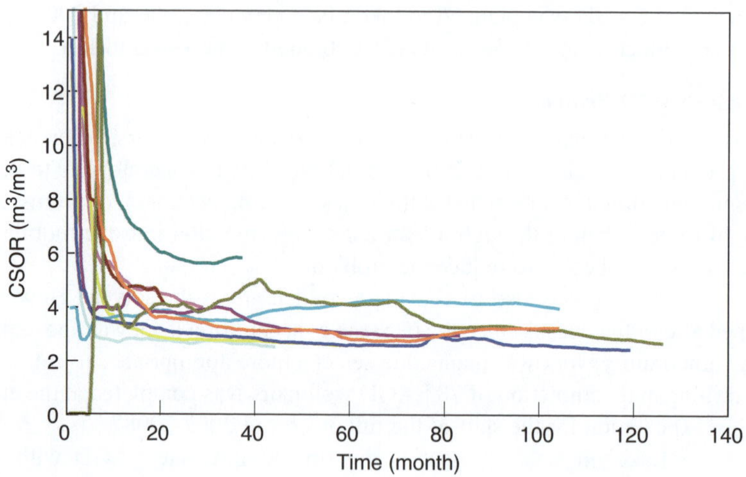

Figure 10.10

Evolution of CSOR *versus* Time for Major SAGD Projects of Alberta Province (time 0 at start of steam injection).

Table 10.4 Planned SAGD Projects in the Alberta Province as of Dec. 2006.

Project	Operator	Location	Start up Year	Planned Production (2020) (bbl/d)
Tucker Lake	Husky	Cold Lake	2006	30,000
Jackfish	Devon	Athabasca	2007	70,000
Hardy	MEG-Energy	Athabasca	2006	22,000
Kirby	CNRL	Athabasca		30,000
Meadow Creek	Petro-Canada	Athabasca		80,000
Lewis	Petro-Canada	Athabasca		80,000
Sunrise	Husky	Athabasca	2008	200,000
Birch Mountain	CNRL	Athabasca		60,000
Gregoire Lake	CNRL	Athabasca		120,000
Borealis	EnCana	Athabasca		100,000
Primrose East	CNRL	Cold Lake		30,000 (Start up with CSS)

10.2.1.5 Venezuela (Bolivar Coast)

Several oil field reservoirs (Figure 10.11) – i.e. Mene Grande, Bachaquero, Lagunillas (including Tia Juana) and Cabimas – are located along the east coast of the Maracaibo Lake of Venezuela, known as the Bolivar Coast [Borregales, 1977].

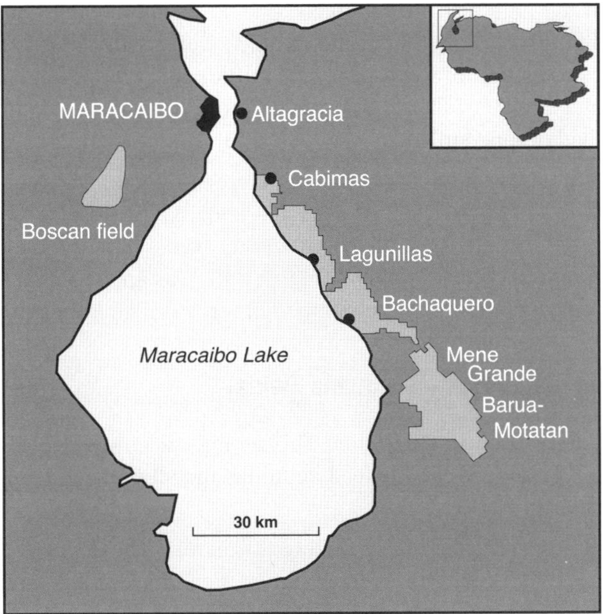

Figure 10.11

View of Main Venezuelan Heavy Oil Fields along the Bolivar Coast as well as the Boscán Field [Marquez L *et al.*, 2001].

Some specific aspects of the Bachaquero and Lagunillas (Tia Juana) fields now operated by PDVSA are outlined in the following section.

10.2.1.6 Venezuela (Bachaquero Field)

Bachaquero was discovered in 1934. Half of the reservoir is under the Maracaibo Lake and drilling/completion of the wells drilled from offshore have necessitated special attention and equipment.

Its main characteristics are the following: around 1,000 m (3,000 ft) deep, dipping 2 to 3°, STOOIP evaluated at 1.1 billion m^3 (7 billion barrels), 11.7°API gravity, *in situ* viscosity of 635 cP, reservoir horizontal permeability 2 darcies, porosity 33.5%, net oil sand thickness 61 m (200 ft). Initial pressure and temperature of 9,446 kPa (1,370 psia) and 53.3°C (128°F), respectively.

During the first 20 years, production was low since the field was poorly developed due to low well productivity (high oil viscosity) and high sand production.

Cold production began in 1960 with the development of new sand control techniques. The daily oil production rate increased from 795 m^3/d (5,000 bopd) in 1963 to 1,590 m^3/d (10,000 bopd) in 1965. In 1971, an intensive drilling program was initiated to complete the development under cold production with a normal spacing of 7.8 ha (19.3 acres) per well. In 1982, well spacing was reduced to 2.6 ha (6.4 acres) per well to increase primary recovery.

In 1971, part of the field was also produced under a massive cyclic steam injection system. In the year 2000, there were 370 cyclic-steamed wells with about 1,000 cycles associated with extra oil production of more than 12.7 million m^3 (80 million barrels), equivalent to only approximately 5.6% of OOIP at a rate of about 6,360 m^3/d (40,000 bopd). At that time, the pressure had dropped to 4,826 kPa (700 psia) from 9,446 kPa (1,370 psia) initially.

From 1995-1997, to stop production decline, three horizontal wells with horizontal sections of 390 to 475 m (1,280 to 1,560 ft) were put under CSS. Results were better as compared to vertical wells, but the cost of the horizontal wells made production with vertical wells more profitable. Injection of solvent or surfactant was applied with success, but there is no evidence that such a practice was continued in wells drilled after that period.

10.2.1.7 Venezuela (Lagunillas – Tia Juana Fields)

These fields were discovered in the 1930s. The geological unit which is the target for oil production is locally known as the Lower Lagunillas member of the Lagunillas formation, which is of Miocene age.

The fields cover a proven area of 3,560 ha (8,800 acres). The depth of top of sand is 457 to 610 m (1,500 to 2,000 ft), and the average net oil sand thickness is around 40 m (130 ft) [upper zone 24.4 m (80 ft) and lower zone 29.6 m (97 ft)]. Average porosity and horizontal permeability are 40% and 2 darcies, respectively.

Initial reserves were estimated at 440 million m^3 (2.75 billion barrels) with an API gravity ranging from 10 to 15. Oil viscosity at reservoir conditions [5,380 kPa, 43°C (780 psig, 110°F)] ranges between 100 and 10,000 cP.

Development from 1936-1963 was done by primary production. Ultimate primary recovery was estimated at 18.1% of OOIP, of which 12.5% was produced by the end of 1963 from some 540 vertical wells. Compaction was the main recovery mechanism.

In 1961, a steam drive project was operated in the northern part of the field. Seven 7-spot patterns with a central injector in each and a 132 m (435 ft) well spacing were used. With a cumulative OSR of 0.6 m^3/m^3 CWE and an increase in recovery of more than 20% of OOIP, steam drive has been an efficient displacement process in Tia Juana. This result was obtained despite a highly unsymmetrical flow pattern, with only one or two production wells responding per injection well. Although good results were obtained, the process was not considered attractive for large-scale application at the time since it was not commercially attractive under the prevailing economic conditions, and because it would suppress the highly effective compaction drive mechanism active in the Tia Juana field.

During the 1964-1966 period, a steam stimulation project was implemented in 73 vertical wells in the southern part of the reservoir. Injection rates per well varied between 190 and 480 m^3/d (1,200 and 3,000 bopd) (high injectivity due to good horizontal permeability of the pay zone). Oil recovery of 6% OOIP was produced, most of it after the first steam cycle. Oil rates of up to 240 m^3/d (1,500 bopd) were obtained in some wells, with a plateau production of 160 m^3/d (1,000 bopd) for several months in a few wells. Average OSR of 3 m^3/m^3 was observed (between 1.5 and 11 m^3/m^3). Compaction was still considered to be responsible for 40% of oil recovery under steam injection, with the remaining 60% due to thermal effects.

An enormous steam soak and steam drive operation in the Tia Juana field is well-documented in the literature. It concerns the M-6 project started in 1969 for steam stimulation and steam drive. The explanation regarding compaction is interesting, since it has been the main production mechanism acting on the Bolivar Coast reservoirs, with good recovery during primary production and steam soak. However, compaction has had a strong environmental impact since it has been responsible for continuous subsidence at ground level, necessitating the construction of dykes to prevent flooding of large areas of land. Theoretical considerations indicate that compaction is approximately equal to subsidence volumetrically for reservoirs which are large in relation to depth, as is typically the case for the reservoirs along the Bolivar Coast.

After 1990, several pilots were performed in the Tia Juana reservoir to test the performance of horizontal wells using various thermal processes such as CSS, SWSC (Single Well Steam Circulation) and SAGD. The most efficient process was SAGD, with a recovery factor greater than 50% of OOIP and a CSOR of 1.25 m^3/m^3 CWE.

Since the unification of Lagunillas and Tia Juana fields, steam flood operations have been performed in an offshore section, known as the LL-04 reservoir. This last operation is interesting in that it necessitated steam injection in offshore conditions that were not very shallow.

10.2.1.8 Venezuela (Boscán Field)

The giant Boscán field is located on-shore west of Maracaibo Lake, 40 km southwest of Maracaibo, in the Zulia state (Figure 10.11). It was discovered in 1946, and since 1947 has been producing a 10.5°API gravity asphaltic oil from the Eocene Misoa Formation, locally called the Boscán Formation.

The stratigraphic section at the Boscán field consists of sands and shales of Oligocene and Eocene ages deposited in a fluvial-deltaic depositional setting. The Boscán structure is a southwest-dipping monocline (flat, tilting surface) which contains oil over a very large depth range from 1,350 m (4,500 ft) to 2,860 m (9,500 ft) subsea. The field size is 20 × 35 square kilometers (12.4 × 21.7 square miles). The Original Oil in Place (OOIP) is estimated at 4.2 billion m^3 (26.4 billion barrels). Oil viscosity at reservoir conditions ranges between 130 and 500 cP. Reservoir permeability and porosity are high, respectively between 100 and 5,000 mD, and 10 to 26%.

Over the 58-year-life of the field, 797 wells have been drilled, with 522 of them still active as of the end of 2004. At that time, the daily production was approximately 18,000 m^3/d (113,000 bopd), and cumulative oil production was 190 million m^3 (1.2 billion barrels) or 4.5% of OOIP.

The reservoir drive is a combination of weak aquifer drive on the south flank, solution gas expansion and gravity drainage. As a result of the weak aquifer support, the field is pressure-depleted over large areas where there has been considerable fluid volume extracted. The field is dominantly on primary depletion, and was originally developed with a 1,000 m (3,280 ft) well spacing [Kumar M *et al.*, 2001]. After a short period of natural production, beam pumping was used to bring the oil to the surface. Later, well spacing was reduced to 600 m (1,970 ft). Despite this, ultimate recovery by primary production is expected to be low (less than 7-8% of OOIP).

Produced water has been re-injected into various parts of the reservoir since 1963. In the year 2000, Chevron and PDVSA began a managed water injection pilot project testing for enhanced oil recovery with pressure maintenance. Production data suggests that the oil decline rate was stopped as a result of water injection.

Because of the high initial oil in place and low primary production recovery, the field has considerable potential for improved oil recovery. However, the low API gravity of the Boscán crude and the reservoir depth make application of IOR methods difficult. A preliminary screening of possible IOR methods to increase production and recovery was done in the year 2000 [Marquez L *et al.*, 2001]. Cyclic steam injection and steam flood were considered as potential processes, but have not yet been implemented at pilot- or field-scale.

10.2.1.9 Russia

Deposits of heavy oils and bitumen have been found in almost every oil-bearing region within Russia; however, the greatest accumulations of these resources are concentrated in the Volga-Ural basin, Western and Eastern Siberian basins, and the Timano-Perchorsky basin. Estimates of Russia's heavy oil and bitumen range from 90 to 240 billion m^3 (575 to 1,500 billion barrels). This wide range of uncertainty is due to the fact that official statistics do not indicate the densities and viscosities involved.

The Gremikhinskoye field is one of the largest heavy oil fields in Russia. It has been developed for 20 years and in the year 2000, more than 900 wells were in production. Steam injection is used to improve overall recovery.

10.2.1.10 China

In 1995, China National Petroleum Corp. indicated it produced 24,170 m^3/d (152,000 bopd) of heavy crude from 7,100 thermal recovery wells. The four heavy oil production bases are Liaohe, Xinjiang, Shengli and Henan. Heavy oil production represented 5% of China's 1995 production.

10.2.2 Cold Production with New Well Architectures – Orinoco Belt (Venezuela)

Located in eastern Venezuela, the Orinoco oil belt (or Faja del Orinoco) covers an area of 54,000 square kilometres (20,900 square miles). It is one of the largest known accumulations of extra-heavy oil in the world, estimated to be around 190 billion m^3 (1,200 billion barrels) of hydrocarbons in place. These accumulations occur primarily in the Miocene Oficina and Freites formations. Reservoirs within these formations are on average 50 m thick (164 ft) and occur at a depth of 150 to 1,300 m (490 to 4,265 ft). Although the accumulations have an API gravity of 4-17°, they are mobile at reservoir temperatures – about 53°C (127°F) – with viscosities of 1,500 to 3,000 cP at reservoir temperature.

The oil belt was divided by the Venezuelan state-owned oil company PDVSA (Petroleos de Venezuela SA) into four sections: Machete, Zuata, Hamaca and Cerro Negro. In 1983, PDVSA conducted the first comprehensive evaluation studies of economically recoverable reserves which could be expected from the Orinoco belt. It concluded that 42.4 billion m^3 (267 billion barrels) or 22% of the resources could be recoverable for the whole Orinoco belt. However, the financial resources of PDVSA were limited and logically it placed a lower priority on exploration and development of these unconventional resources. In 1991-1992, Venezuela modified its policies and decided to open the door to foreign partners in selected areas of the hydrocarbon sector, including heavy oil activity.

Four integrated extra-heavy oil projects were then developed under 35-year-long "strategic association" joint ventures with PDVSA (Figure 10.12).

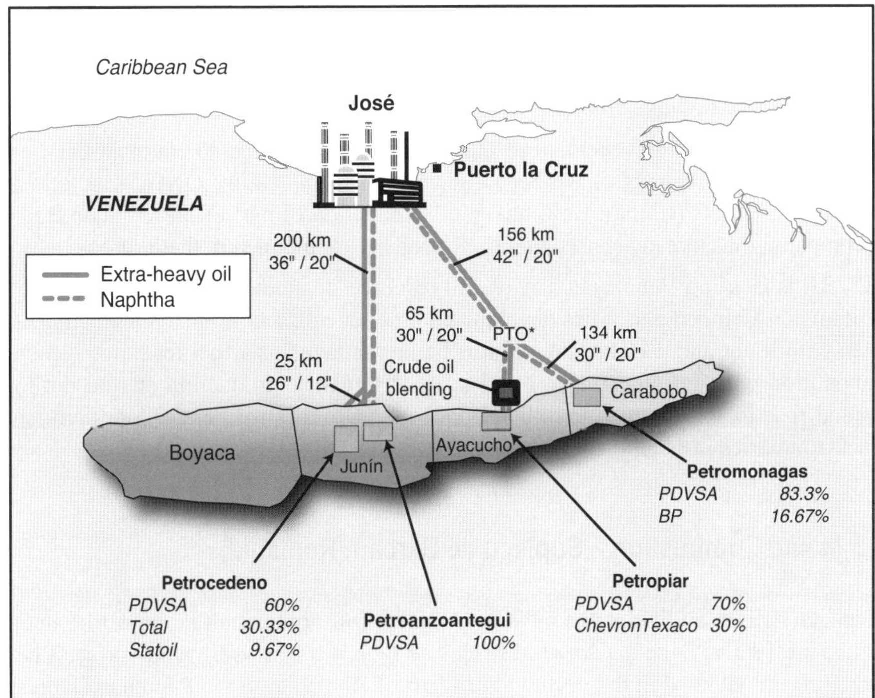

Figure 10.12

Orinoco Belt Integrated Heavy Oil Projects.

The four existing extra-heavy oil projects have a total capacity to produce around 95,400 m^3/d (600,000 bopd) of upgraded synthetic crude and are seeking to exploit 1.8 billion m^3 (11 billion barrels) of reserves at a combined total cost of US $15.9 billion (see Table 2.6).

All projects have been made possible and have reached their production plateau thanks to the use of horizontal and multilateral wells that have limited Capex and Opex expenses. These wells are drilled from clusters and are positioned in the reservoirs according to their best characteristics by integrated teams of geoscientist engineers (e.g. see Figure 7.8 in Chapter 7 for a multilateral well in the Petrozuata project).

The four projects use different upgrading methods to transform the extra-heavy crude into synthetic crude oil. The heavy crude is diluted at bottom-hole or at wellhead and is transported by pipeline to upgraders in San José on the coast. At the Jose Condominium, the heavy crudes are separated from their diluent and the major boiling point fraction is converted into lighter products. The level of upgrading depends on the project. In the Petrocedeno (ex Sincor) and Petropiar projects, the extra-heavy crude is converted into a 32 or 26°API Zuata sweet crude for export. In the Petroanzoategui and Petromonogas projects, the crude is only partially upgraded to a 14-22°API gravity crude that is exported to specific refineries in the United States.

Cold primary production is assumed for the four joint ventures. As indicated in Chapter 7 however, the recovery factor is limited to less than 10% of OOIP. The operators are therefore looking to EOR processes to increase recovery, e.g. by implementing cyclic steam or SAGD, or polymer flooding.

Since 2004, PDVSA has announced that it will open its assets to National Oil Companies. Cooperation agreements have been signed with NIOC (Iran), ONGC (India), Lukoil (Russia) and CNOOC (China). Cold heavy oil production of part of the Orinoco Belt with these companies has already started and EOR projects are planned in the future.

PDVSA is tackling Venezuela's heavy oil resource in another way. Through its subsidiary Bitumenes Orinoco SA, it has developed a product called Orimulsion – an emulsion of bitumen with water and a surfactant – that is being marketed as a fuel for power generation in competition with residual fuel oil and coal. Orimulsion production started during the 1980s and rose to 17,500 m^3/d (110,000 bopd) in 2001. It is sold under long-term supply contracts to utilities in several countries.

10.2.3 *In situ* Combustion – Suplacu de Barcau (Romania)

The Suplacu de Barcau field is the most well-known ISC application in the world, in operation since the 1960s [Gadelle C *et al.*, 1981]. The field is a heavy-oil, near-surface Pliocene clastic reservoir, situated in the northwestern part of Romania, in the Panonian Depression. The structure forms a monocline trending east-west, with a fault-dependent closure to the south (Figure 10.13).

Its main characteristics are: surface 1,700 ha (4,200 acres), depth of 50 to 200 m (164 to 656 ft), dip 5-8°, net pay thickness 4-24 m (13-79 ft), porosity 32%, permeability 2 darcies, reservoir oil viscosity 2,000 cP.

The reservoir was put into production in 1960, the main production mechanism being solution gas drive. Initial rates were in the range of 2 to 5 m^3/d/well but decreased very rapidly to 0.3 to 0.1 m^3/d/well. As a result, thermal methods were considered to increase oil production rates and recovery. Both ISC and steamflood were pilot-tested in the upper part of the structure during the period 1963-1970 in two zones at a semi-commercial size. Based on the results of these pilots, ISC was finally selected and commercially implemented in a line-drive exploitation by 1970. At the same time, steam injection was considered as a cyclic process to periodically stimulate production wells. Since the 1970s, the "linear" ISC front has propagated down-structure, parallel to the isobaths. The air injection wells are located along an east-west line of more than 10 km (6.2 miles); the distance between two adjacent injection (or production) wells is 100 m (300 ft). The position of the ISC front in 2000 is shown in Figure 10.14 [Paduraru R and Pantazi I, 2000].

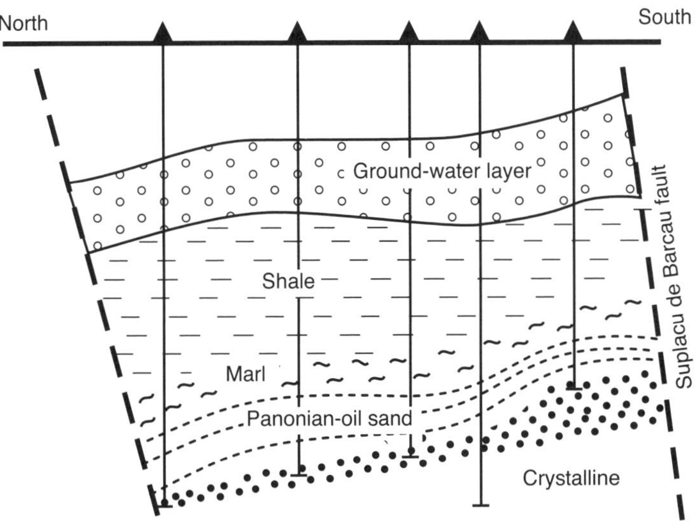

Figure 10.13

Panonian – Suplacu de Barcau field – Cross section. [Gadelle C *et al.*, 1981].

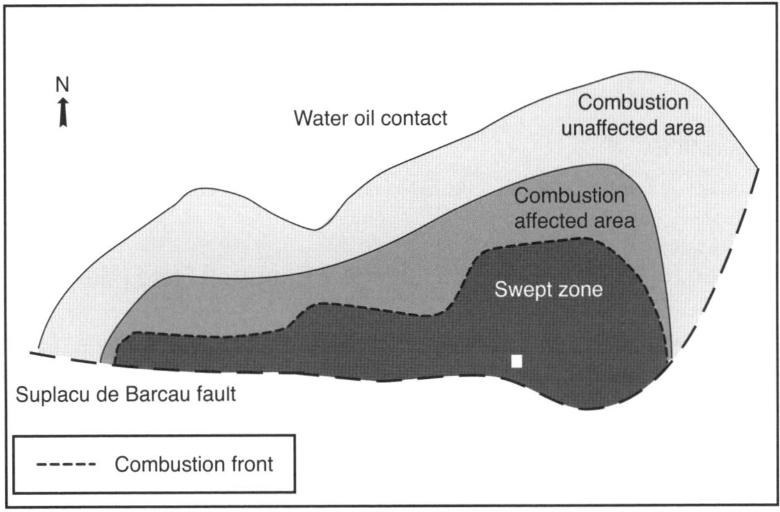

Figure 10.14

Location of the ISC front as of 2000 [Paduraru R and Pantazi I, 2000].

In 2006, there were 800 wells in production with a total production of about 1,200 t/d of crude oil. The cumulative crude oil extracted over roughly 40 years is about 18 million tons (Figure 10.15), which corresponds to a recovery factor of about 45.3% of OOIP [Panait-Patica A *et al.*, 2006]. Injection rates were ~ 2,000 thousand Sm^3/d of air through 90 wells and ~ 1,300 t/d steam through 24 wells.

The Romanian experience shows that *in situ* combustion is particularly successful when designed to start from the uppermost part of a structure and extending the process downward as the combustion front progresses down dip.

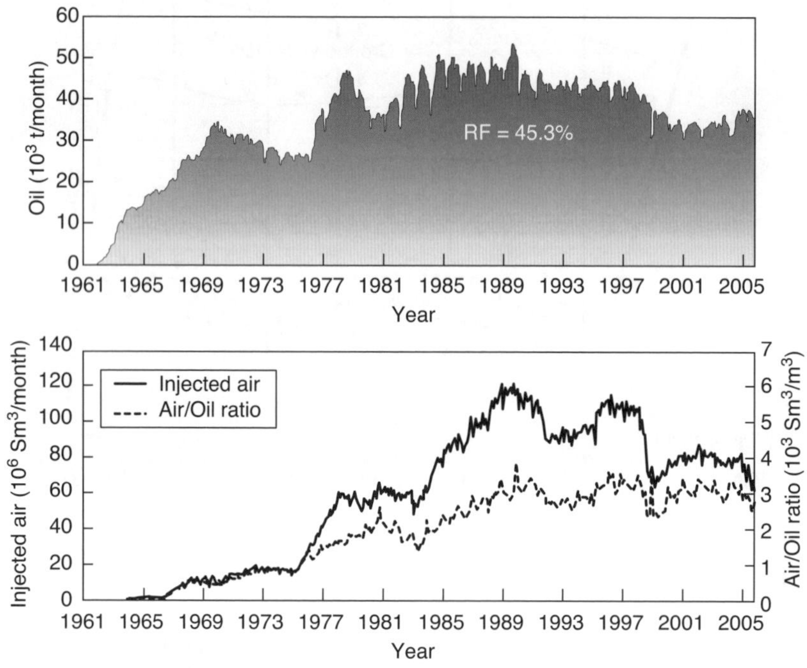

Figure 10.15

Suplacu de Barcau Production Profile as of 2006 [Panait-Patica A *et al.*, 2006].

10.2.4 Nitrogen Injection – Gravity Drainage – Cantarell Field (Mexico)

Cantarell, a complex of offshore naturally fractured carbonate fields, is located in the Bay of Campeche (Figure 10.16). It was discovered in 1976 by Pemex (Petróleos Mexicanos) and is considered the sixth largest oil field in the world. It is made up of five separate fields, namely Akal, Nohoch, Chac, Kutz, Ixtoc and Sihi. Akal is the largest with 5.1 billion m^3 (32 billion barrels) originally in place (86% of the complex) of a 19-22°API (0.9218 g/cm^3) heavy oil (2.3 cP in reservoir conditions) [Manceau E *et al.*, 2000].

Main pay zones in Cantarell are highly fractured-vuggy carbonate formations from Jurassic, Cretaceous and Lower Paleocene geological ages. Within Akal, formations are

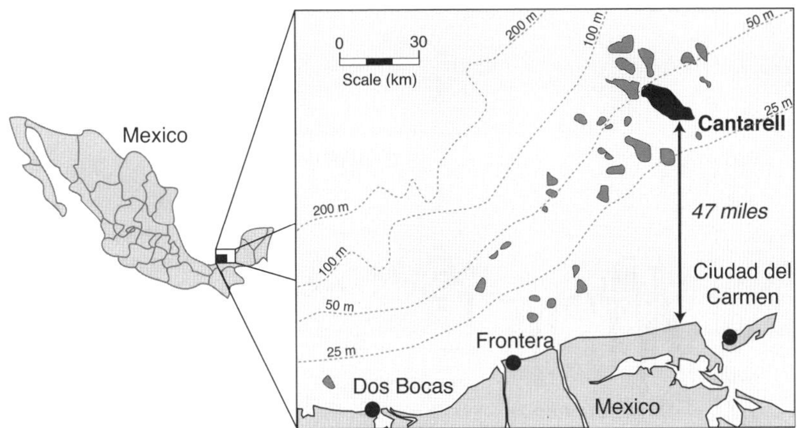

Figure 10.16

Offshore Location of the Cantarell Complex in México [Sanchez JL *et al.*, 2005].

hydraulically continuous and have an average thickness of about 1,220 m (4,000 ft) and structural relief of over 2,100 m (7,000 ft). Typical total porosity in the reservoir is 8% and it has been determined that up to 35% of it is secondary porosity, fractures and vugs. Typical absolute permeability in the matrix and fractures is 1 mD and 3,000 mD, respectively.

Production started in Akal in 1979 and by 1981 reached a peak of 180,000 m^3/d (1.16 million bopd), with an average production of 4,610 m^3/d (29,000 bopd) per well. Since then, up to 1995 the Cantarell complex produced at an almost constant rate of 159,000 m^3/d (1 million bopd). While pressure declined, new wells were drilled to maintain this production and gas-lift was required. From 1997 to 2004, production climbed due to launching of an aggressive long-term optimization program. This program consisted of infill drilling, modernization and expansion of production facilities and implementation of a pressure maintenance program by nitrogen injection in the year 2000.

The nitrogen injection project in the Cantarell Complex is the most ambitious pressure maintenance project in the world with regard to incremental oil, production rate, nitrogen injection rate, and investment of more than US $5 billion.

Nitrogen production facilities are located onshore and the nitrogen is sent offshore *via* pipeline for injection into the reservoir. The facilities doubled the world's current total output of nitrogen.

Figure 10.17 shows the production history of the Cantarell field, with the dramatic increase in oil production following the pressure maintenance by nitrogen injection. Nitrogen is injected into the gas cap through seven wells, drilled and completed at the top of Akal at a total injection rate of 33 million m^3/d (1.2 billion scf/d).

Figure 10.18 shows the oil production rate impact of the two main components of the Cantarell optimization project: pressure maintenance and additional drilling-expansion of production facilities [Moritis, 2006]. During the first four years of the pressure maintenance project, a cumulative 39.6 billion m^3 (1,400 billion scf) of nitrogen was injected into Akal.

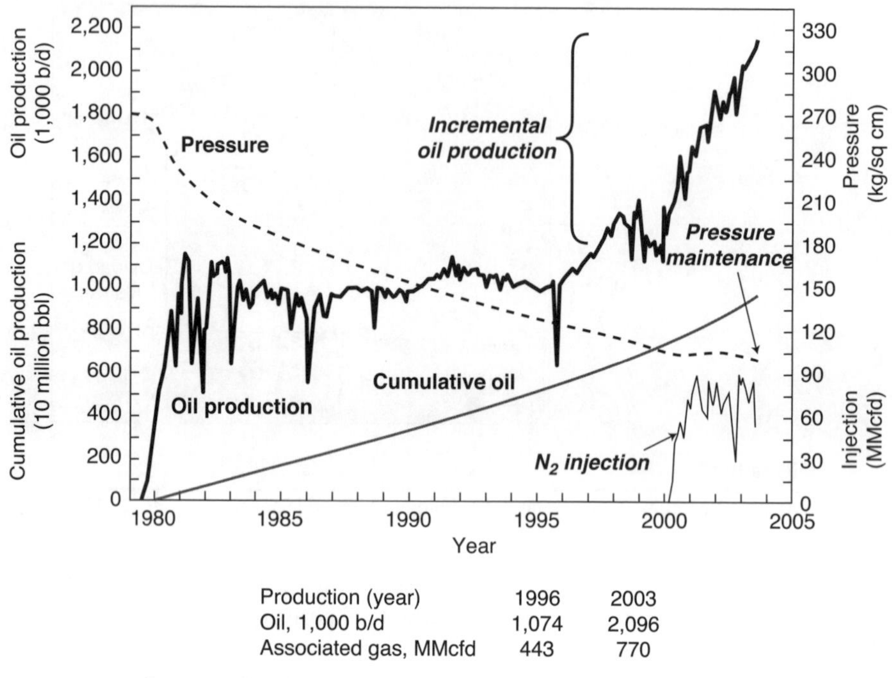

Production (year)	1996	2003
Oil, 1,000 b/d	1,074	2,096
Associated gas, MMcfd	443	770

Figure 10.17

Cantarell Field Production [Moritis, 2006].

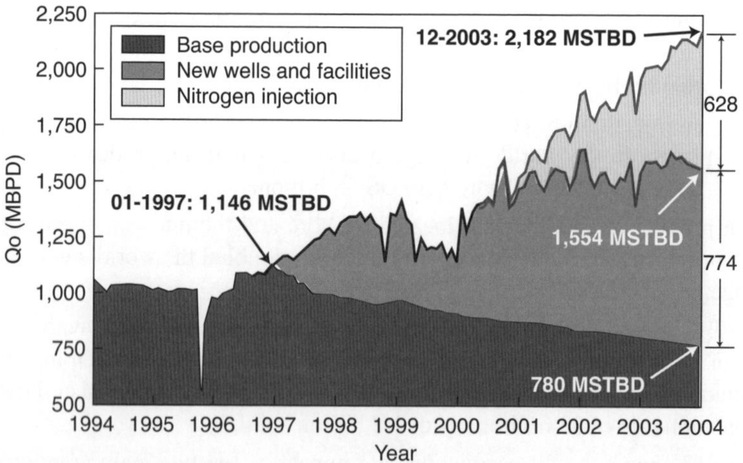

Figure 10.18

Pressure Maintenance, New Wells and Facilities Expansion: Impact on Oil Production – Cantarell field [Sanchez JL *et al.*, 2005].

The volume of gas in the secondary gas cap before injection was 41 billion m^3 (1,800 billion scf). The Gas Oil Contact has continued moving into a flat surface at a rate of 70 m (230 ft) per year during the injection process.

The project for pressure maintenance by nitrogen injection in Cantarell has proven to be very successful, both technically and economically. Gravity drainage of the oil by the secondary nitrogen gas cap has allowed a recovery of 1.74 billion m^3 (11 billion barrels) of oil (one-third of the original oil in place) as of 2005. An ultimate recovery of 2.5 billion m^3 (16 billion barrels) (50% OOIP) is expected at life-end. Cantarell oil production peaked in 2004 at 338,000 m^3/d (2.125 million bopd), itself representing about 55% of PEMEX total production. This made it the second-largest producing oil field in the world behind Ghawar (Saudi Arabia) with 715,000 m^3/d (4.5 million bopd), and ahead of Daqing (China) and Burgun (Kuwait) with around 159,000 m^3/d (1 million bopd). In 2005, PEMEX forecasted a fairly rapid decline in oil production – around 14% per year between 2007 and 2015 – due to water encroachment and gas coning in the wells.

10.2.5 Polymer Flooding – Daqing (China)

Daqing is the largest oil field in China and in 2006 produced more than 143,000 m^3/d (900,000 bopd). A unit of PetroChina Co. Ltd. operates the field, which is located 1,000 km (620 miles) north of Beijing in China's northern Heilongjiang province. The structure is a 145 km-long (90 miles), 10 km-wide (6.2 miles) and 700 to 1,200 m (2,300 to 6,560 ft) deep anticline trending N-NE/S-SW. There are three reservoirs with a very complex geology of fluvio-deltaic formation with several sand beds and great heterogeneities.

First commercial oil from the field was produced in 1959. Water injection was started very early thereafter and a relatively stable 1 million bopd production plateau was maintained for many years. However, the oil viscosity – 6-9 cP at reservoir conditions: 100 m (328 ft) and 45°C (113°F) – and degree of reservoir heterogeneity have resulted in moderate displacement efficiency and high water-cut in the production. In the 1980s, EOR techniques were evaluated to increase recovery and delay production decline. The initial screening showed that polymer flooding was the process best suited to reservoir conditions, and two pilots were implemented that demonstrated the potential of this process for the Daqing oil field [Demin W *et al.*, 2002].

In 1996, the Daqing Oil field started industrial scale polymer flooding in an area of 17,800 ha (44,000 acres), manufacturing its own polymer and annually injecting the largest quantity of polymer in the world. Annual production by polymer flooding in the year 2001 was more than 11.1 million m^3 (70 million barrels) and cumulative production was more than 47.7 million m^3 (300 million barrels), with more than 3,000 wells on polymer flooding (1,300 injectors, 1,700 producers) [Delamaide E *et al.*, 1994]. Incremental recovery by polymer flooding in the Daqing field is in the range of 12% ~ 15% OOIP. The increase in displacement efficiency and volumetric sweep efficiency each contributes about half of the total incremental recovery.

Polymer flooding is considered a great success in the Daqing field and the process is being considered for other heavy oil fields in China.

10.2.6 Offshore Fields: Special Issues

Most offshore heavy oil reservoirs are located in the Atlantic Ocean along the African Coast or Brazilian Coast. Today, the only applicable production processes for such reservoirs are primary recovery or water flood. Thermal methods cannot be used due to the huge loss of energy that would be observed when going through the sea water column. The use of polymer flooding where oil viscosity is not too high is being seriously considered for several fields. The main issue for the implementation of this process is the supply of huge quantities of polymer at the platforms.

REFERENCES

Albian Oilsands website: www.albiansands.com

Batycky JP, Leaute RP, Dawe BA (1997) A Mechanistic Model of Cyclic Steam Stimulation. SPE 37550 (presented at the *1997 ITOHOS Conference*, Bakersfield, 10-12 Feb.).

Borregales CJ (1977) Steam Soak on the Bolivar Coast. Canada-Venezuela Oil Sands Symposium, Edmonton, May 30-June 4. Published in *The Oil Sands of Canada-Venezuela 1977*, CIM Special volume 17.

Canadian Oil Sands Trust website: www.cos-trust.com

CNRL website: www.cnrl.com/horizon/.

Delamaide E, Corlay Ph, Demin W (1994) Daqing Oil Field: The Success of Two Pilots Initiates First Extension of Polymer Injection in a Giant Oil Field. SPE 27819 (presented at the *1994 SPE/DOE IOR Symposium*, Tulsa, 17-20 April).

Demin W, Jiecheng C, Junzheng W, Gang W (2002) Experiences Learned After Production of More Than 300 Million Barrels of Oil by Polymer Flooding in Daqing Oil Field. SPE 77693 (presented at the *2002 SPE ATCE Conference*, San Antonio, 29 Sept. – 2 Oct.).

Gadelle C, Burger J, Bardon C, Machedon V, Carcaona A (1981) Heavy Oil Recovery by *In Situ* Combustion. Trans. *AIME*, **271**, pp 2057-2066.

Gallant RJ, Stark SD, Taylor MD (1993) Steaming and Operating Strategies at a Midlife CSS Operation. SPE 25794 (presented at the *1993 ITOHOS Conference*, Bakersfield, 8-10 Feb.).

Imperial Oil website: www.imperialoil.ca

Kumar M, Sahni A, Alvarez JM, Heny C, Vaca P, Hoadley SF, Portillo M (2001) Evaluation of IOR Methods for the Boscán Field. SPE 69723 (presented at the *2001 SPE ITOHOS Symposium*, Margarita Island, Venezuela).

Leaute RP, Carey BS (2005) Liquid Addition to Steam for Enhancing Recovery (LASER) of Bitumen with CSS: Results from the First Pilot Cycle. CIM 2005-161 (presented at the *2005 Annual Technical Meeting of CIM*, Calgary, 7-9 June).

MacKay K, Delamaide E (2006) Review of Major SAGD Projects in the Province of Alberta (Canada). IFP Canada internal report.

Manceau E, Delamaide E, Sabathier JC, Jullian S, Kalaydjian F, Ladron de Guevara JE, Sanchez Bujanos JM, Rodriguez FD (2000) Implementing Convection in a Reservoir Simulator: A Key Feature in Adequately Modeling the Exploitation of the Cantarell Complex. SPE 59044 (presented at the *2000 SPE IPCEM Conference*, Villahermosa, Mexico, 1-3 Feb.).

Marquez L, Gonzalez M, Gamble S, Gomez E, Vivas MA, Bressler HM, Jones LS, Ali SM, Forrest GS (2001) Improved Reservoir Characterization of a Mature Field Through an Integrated Disciplinary Approach. LL-04 Reservoir, Tia Juana Field, Venezuela. SPE 71355 (presented at the *2001 SPE ATCE Conference*, New Orleans, 30 Sept.-3 Oct.).

Moritis G (2006) "2006 EOR Survey", *Oil and Gas Journal*, **104**, 15, 17 April 2006.

Ormerod L, Sardoff H, Wilkinson J, Erlendson B, Cox B, Stephenson G (2006) Real-Time Field Surveillance and Well Services Management in a Large Mature Onshore Field: Case Study. SPE 99949 (presented at the *2006 SPE Intelligent Energy Conference*, Amsterdam, 11-13 April).

Paduraru R, Pantazi I (2000) IOR/EOR – Over Six Decades of Romanian Experience. SPE 55169 (presented at the *2000 SPE European Petroleum Conference*, Paris, 24-25 Oct.).

Panait-Patica A, Serban D, Ilie N (2006) Suplacu de Barcau field – A Case History of a Successful *In Situ* Combustion Exploitation. SPE 100346 (presented at the *2006 SPE EAGE Annual Conference*, Vienna, 12-15 June).

Sadler K, Davis P (2005) *In Situ* Bitumen Overview and Activity Update in the Province of Alberta. SPE 97800 (presented at the *2005 ITOHOS Conference*, Calgary, 1-3 Nov.).

Sanchez JL, Astudillo A, Rodriguez F, Morales J, Rodriguez A (2005) Nitrogen Injection in the Cantarell Complex: Results After Four Years of Operation. SPE 97385 (presented at the *2005 SPE LACPEC Conference*, Rio de Janeiro, 20-23 June).

Suncor website: www.suncor.com

Suncor 2005 Annual Report.

Syncrude website: www.syncrude.com

Synenco website: www.synenco.com

Total Canada website: www.total-ep-canada.com

PART 3
Surface Transport

J-F. Argillier

This chapter focuses on transport issues of heavy oils, which are particularly challenging due to the high viscosity of this type of crude. At the surface level – due to their very high viscosity – *heavy crude oils* cannot be transported as such in pipelines and require additional treatments [Saniere *et al.,* 2004]. Using powerful pumps (see Chapter 14), these solutions consist in either reducing the viscosity by heating, dilution (see Chapter 11), oil-in-water emulsification (see Chapter 12) or upgrading (see Part 4), or lowering the friction in the pipe using core annular flow (see Chapter 13). Each of these is briefly described below.

Heat treatment: Since viscosity decreases very rapidly with increasing temperature, heating is an attractive method for improving the flow properties of *heavy crude oils* [Gerez *et al.,* 1996; Guevara *et al.,* 1998; Meyer *et al.,* 1998; Nunez *et al.,* 1998; Yaghi *et al.,* 2002]. A famous example is the Alyeska pipeline in Alaska which transports crude oil at approximately 50°C. However, designing a heated pipeline is not simple since it involves [Guevara *et al.,* 1998] many considerations: expansion of the pipelines, number of pumping/heating stations, heat losses etc. Other significant issues are the high costs and greater corrosion rate of the internal pipe due to the temperature. Moreover, a recent study [Evdokimov *et al.,* 2002] showed that heat treatment could induce changes in the colloidal structure of the crude oils and worsen their rheological properties.

Dilution: A classical method of enhancing *heavy crude oil* transport is by blending with a less-viscous hydrocarbon such as condensate, naphtha, kerosene or light crudes. According to Guevarra *et al.,* 1998, there is an exponential relationship between the resulting viscosity of the mixture and the volume fraction of the diluent, which makes dilution a very efficient method. However, in order to attain acceptable limits for transport, a fraction as high as 30% of diluent by volume is necessary and implies large pipeline capacity. Problems may also arise with regard to diluent availability [Crandall *et al.,* 1984]. This is forecasted to be a very crucial point in Alberta during the coming years. Diluent recycling could be a solution, but requires a large investment for the installation of an additional return pipeline. Use of drag reducing agent with diluted heavy oil can reduce significantly the frictional pressure loss under turbulent flow conditions, allowing a substantial increase in pipeline capacity.

Emulsion: The emulsion method consists in dispersing the *heavy crude oil* in water in the form of droplets stabilized by surfactants, leading to a significant reduction in viscosity. According to Rimmer *et al.,* 1992, a typical emulsion is composed of 70% crude oil, 30%

aqueous phase and 500-2,000 ppm of chemical additives. The resulting emulsion has a viscosity in the 50-200 cP range at pipeline operating conditions and is particularly stable. This method is applied in Venezuela for commercialization as "Orimulsion®" product [Layrisse *et al.,* 1998]. Orimulsion is an emulsion sold as fuel for electrical power plants. Recovering the crude oil for further processing involves breaking the emulsion, and such a process is not currently available. Issues related to water recycling must still be resolved.

Partial upgrading: This method consists in modifying the composition of heavy oils to make them less viscous. Upgrading technologies such as the hydrotreating processes traditionally used in refineries may be considered for this application. Suitable treatments of this kind have been developed, e.g. by ASVAHL, Association for the Valorization of Heavy Oils [Charlos *et al.,* 1988]. "Orimulsion is an emulsion sold..." (Solvahl de-asphalting process, Tervahl thermal treatment process and Hyvahl catalytic hydrotreatment processes). Recent studies aim to associate these various processes in order to optimize heavy crude conversion. These techniques will be described in further detail in Part 4. In Alberta, due to the lack of diluent, an increasing proportion of bitumen is transported as Synbit, which consists of bitumen diluted by upgraded Syncrude.

Core annular flow: The core annular flow mode can be an attractive method for the transport of viscous crude oil. In this transport technology, a water film surrounds the oil core and acts as a lubricant, such that the pumping pressure necessary for the lubricated flow is comparable to that needed for water alone. The water fractions are typically in the range of 10-30%. Many theoretical studies, laboratory and field tests have been carried out [Oliemans *et al.,* 1986; Joseph *et al.,* 1993 & 1997; Peysson *et al.,* 2007] and have shown that the configuration of core annular flow applied to heavy oils is stable. However, only two industrial examples of this technology are currently known. The main problem of the technology is that the oils tend to adhere to the wall, leading to restriction and an eventual blockage of the flow system. This kind of difficulty can be exacerbated during a shutdown operation, allowing stratification of oil and water phases and requiring a high restart pressure.

REFERENCES

Charlos J-C, Courdec J-L, Page J-F (1988) Heavy Oil Processing, a Synthesis of the ASVAHL Results in *E.C. 3rd European Community Symposium on New Technologies for the Exploration and Exploitation of Oil and Gas Resources*, March 22-24, Proceeding, **2**, E Millich *et al.* eds, Luxembourg.

Crandall G, Wise T(1984) Availability of Diluent May Inhibit Heavy Oil Exports, *Canadian Petroleum*, July-August.

Evdokimov IN, Eliseev DYu, ans Eliseev NYu (2002) Negative Viscosity Anomaly in Liquid Petroleum after Heat Treatment, *Chemistry and Technology of Fuels and Oils*, **38**, 3, pp 171-177.

Gerez JM (1996) Heavy Oil Transportation by Pipeline *International Pipeline Conference*, **2**, ASME.

Guevara E, Gonzalez J, Nuñez G (1998) Highly Viscous Oil Transportation Methods in the Venezuela Oil Industry, *Proceedings of the 15th World Petroleum Congress*, John Wiley and Sons, London, pp 495-501.

Joseph D, Bai R, Chen KP, Renardy R (1997) Core Annular Flows, *Annual Review of Fluid Mechanics*, **29**, pp 65-90.

Joseph D, Renardy Y (1993) *Fundamentals of Two Fluid Dynamics*, Springer Verlag, New York.

Layrisse R (1998) Viscous Hydrocarbon-in-Water Emulsions, *US Patent 4,795, 478.*

Márquez LJ, González M, Gamble S, Gómez E, Vivas MA, Bressler HM, Jones LS, Ali SM, Forrest GS (2001) Improved Reservoir Characterization of a Mature Field Through an Integrated Multi-Disciplinary Approach. LL-04 Reservoir, Tia Juana Field, Venezuela. Paper SPE-71355-MS presented at the Annual Technical Conference and Exhibition, New Orleans, Louisiana, 30 September-3 October.

Meyer RF (1998) World Heavy Crude Oil Resources, *Proceedings of the 15th World Petroleum Congress*, John Wiley and Sons, London, pp 459-471.

Nunez GA, Rivas HJ, Joseph D (1998) Drive to Produce Heavy Crude Prompts Variety of Transportation Methods, *Oil and Gas Journal*, Oct. 26, 1998, pp 59-67.

Oliemans R, Ooms G (1986) Core Annular Flow of Oil and Water Through a Pipeline in *Multiphase Science and Technology*, Hewitt GF, Delhaye JM and Zuber N editors, **2**, Hemisphere Publishing Corp., Washington.

Panait-Paticä A, Serban D, Llie N (2006) Suplacu de Barcau Field – A Case History of a Successful In-Situ Combustion. Paper SPE-100346-MS presented at the SPE Europec/EAGE Annual Conference and Exhibition, Vienna, Austria, 12-15 June.

Peysson Y, Bensakhria A, Antonini G, Argillier J-F (2007) Pipeline Lubrication of Heavy Oil: Experimental Investigation of Flow and Restart Problems, *SPE Production & Operations*, **22** (1), pp 135-140.

Rimmer D, Gregoli A, Hamshar J, Yildivim E (1992) Pipeline Emulsion Transportation for Heavy Oils in *Emulsions in the Petroleum Industry*, American Chemical Society, **8**, pp 295-312.

Sánchez JL, Astudillo A, Rodríguez F, Morales J, Rodríguez A (2005) Nitrogen Injection in the Cantarell Complex: Results After Four Years of Operation. Paper SPE-97385-MS presented at the SPE Latin American and Caribbean Petroleum Engineering Conference, Rio de Janeiro, Brazil, 20-23 June.

Saniere A, Hénaut I, Argillier JF (2004) Pipeline Transportation of Heavy Oils, a Strategic, Economic and Technological Challenge, *Oil & Gas Science & Technology*, **59** (5), pp 455-466.

Yaghi B, Al-Bemani A (2002) Heavy Crude Oil Viscosity Reduction for Pipeline Transportation, *Energy Sources*, **24**, pp 93-102.

11 Heavy Oil Dilution

P. Gateau, I. Hénaut, J-F. Argillier

Heavy or extra-heavy crude oils cannot be transported by pipeline without first reducing their viscosity. This is commonly performed by blending the oil with light hydrocarbons. In this case, the resulting viscosity of the mixture depends only on the dilution ratio, and on the respective viscosities and densities of the oil and the diluent. Classical diluents are light crudes, condensates and naphtha. Of course, diluent availability and its possible recycling need to be taken into account. Guidelines for solvent choice are mainly based on economical considerations such as cost and availability in the field.

11.1 HYDROCARBON SOLVENTS

Condensates were used until the end of the 1980s to transport nearly the entire production of Canadian crude [Crandall *et al.*, 1984; Urquhart, 1986; Todd, 1988]. Condensate is a very light oil obtained primarily from phase separation during the production of natural gas, so its availability depends on natural gas demand. According to forecasts made at the end of the 1980s [Todd, 1988] and since confirmed [Meyer, 1998], the production of condensates will not be able to satisfy market requirements since the demand is closely linked to the development of *heavy crude oil* production.

Moreover, condensates are poor solvents for asphaltenes. Asphaltene deposits may form and partially plug the lines [Todd, 1988; Burger *et al.*, 1984; Mehrotra, 1992; Anhorn *et al.*, 1994].

Blending with light crude oil can also reduce heavy oil viscosity. Light crude oils in the 35° to 42° API range can be used, although up to twice the volume of light crude as compared to condensate may be required to provide the same viscosity reduction [Urquhart, 1986]. This leads to a significant increase in the effluent volume, possibly resulting in additional pipeline capacity. As for condensates, supplies of light crude oils may fluctuate and their use as diluent may be limited since this would result in even less light crude for refinery supplies. Finally, due to their high saturates content, some light crude oils are poor solvents for asphaltenes and – like condensates – can induce asphaltene deposits.

Another substitute for gas condensates is available in the form of refinery naphtha. This product is a petroleum cut in the range of gasoline, with a carbon number ranging from C_5 to C_{11}. Because of its high API density (47.7°), naphtha is very efficient at diluting heavy oils

[Gonzalo-Rojas *et al.*, 1977]. It exhibits good compatibility with asphaltenes [Burger *et al.*, 1984; Wallace *et al.*, 1996] and is easily reusable. Dilution with naphtha associated with solvent recycling was used for Canadian crude from Cold Lake and from Lloydminster during the 1980s [Urquhart, 1986] and is currently the process used for the Petrocedeno (ex Sincor) project (Venezuela).

The typical evolution of viscosity as a function of the naphtha dilution rate for an extra-heavy oil (9°API) is shown in Figure 11.1 [Gateau *et al.*, 2004]. As can be seen, the viscosity of the blend decreases strongly with increasing diluent concentration. To obtain a viscosity of roughly 400 cP at 20°C, this particular 9°API extra-heavy oil would require roughly 25% naphtha by weight, i.e. roughly 29% by volume.

For mixtures of liquid hydrocarbons, including crude oils and fractions, the viscosity-composition curve is generally a monotonic, concave-upward function, and rarely goes through a minimum [Rahmes *et al.*, 1948] Adding a low-viscosity hydrocarbon to a heavy oil increases the intermolecular distance within the oil and decreases the viscosity and density [Gonzalo-Rojas *et al.*, 1977]. Generally speaking, the lower the viscosity of the diluent, the lower the viscosity of the diluted crude [Burger *et al.*, 1984; Gonzalo-Rojas *et al.*, 1977; Cabrera, 1982; Alcocer *et al.*, 1986]. However, since the viscosity ratio between the heavy oil and the solvent is very important, simple mixing rules do not apply. Empirical models based on multiple regression analysis have been developed to predict the viscosity of diluted heavy oils or bitumen. Lederer [Lederer, 1933] proposed a modified version of the classic Arrhenius expression to represent the mixture viscosity:

$$\log \mu = \left(\frac{\alpha V_O}{\alpha V_O + V_S} \right) \log \mu_O + \left(1 - \frac{\alpha V_O}{\alpha V_O + V_S} \right) \log \mu \qquad \text{Eq. 1}$$

where V_O, μ_O, V_S, μ_S are respectively the volume fraction and the viscosity for the oil (O) and the solvent (S) and where α is an empirical constant varying between 0 and 1. Shu [Shu, 1984] defined a generalized expression of α able to represent the viscosity of heavy oils or bitumen diluted with light hydrocarbons. He correlated this parameter with the viscosity ratio and the densities of solvent ρ_S and oil ρ_O:

$$\alpha = \frac{17.04 \left(\rho_O - \rho_S \right)^{0.5237} \rho_O^{3.2745} \rho_S^{1.6316}}{\ln \left(\frac{\mu_O}{\mu_S} \right)} \qquad \text{Eq. 2}$$

This correlation was based on data at temperatures from 297 to 301K and has been extended by Barrufet *et al.* [Barrufet *et al.*] up to 450K.

Note that the use of Drag Reducing Agent (DRA) with diluted heavy oil can reduce significantly the frictional pressure loss under turbulent flow conditions, allowing a substantial increase in pipeline capacity. The performance of DRA is known to depend on their own characteristics (molecular weight, structure, chemical composition) and on external parameters such as turbulence intensity, oil viscosity, etc. The mechanism by which DRAs interfere with turbulence is not fully understood and drag reduction is not easy to predict. Careful laboratory tests can help quantifying the potential efficiency of these additives [Darbouret *et al.*, 2009; Hénaut *et al.*, 2009].

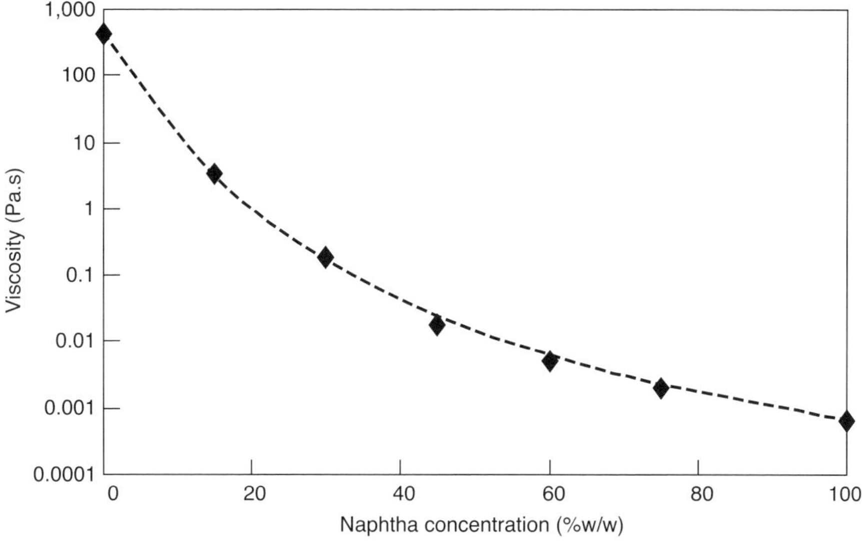

Figure 11.1

Naphtha Dilution of Oil A
Evolution of Viscosity as a Function of Dilution Rate.

11.2 NONHYDROCARBON SOLVENTS

The use of organic solvents other than hydrocarbons has been considered by some authors in experimental studies. Methyl Tert-Butyl Ether (MTBE) and Tert-Amyl Methyl Ether (TAME) have been studied in laboratory experiments as alternative thinners for heavy oils [Anhorn *et al.*, 1994]. These solvents are generally used in gasoline to improve octane number. No definitive comparison was made between hydrocarbons and these compounds. Alcohols – in particular pentanol – are at least twice as effective in reducing the viscosity of heavy oil as compared to kerosene [Storm *et al.*, 1999]. This suggests that the additional reduction of viscosity is due to interactions between the hydroxyl functions and some functionalities of the asphaltenes [Argillier *et al.*, 2001] To verify this hypothesis, some research experiments have been performed using solutions of asphaltenes in a hexanol/toluene mixture and comparatively measuring the radius of gyration of the asphaltenes and the relative viscosity of the solutions [Gateau *et al.*, 2004]. Both go through a minimum as shown in Figure 11.2.

This minimum is observed for a hexanol content of 10% in toluene. This confirms that hexanol has an effect on the organization of the asphaltenes in the diluted regime. By varying the concentration of asphaltenes in maltenes, Hénaut *et al.* [Hénaut *et al.*, 2001] have found the existence of a critical concentration above which the colloidal particles entangle. This structural change dramatically increases the viscosity of *heavy crude oils*. Limitation of

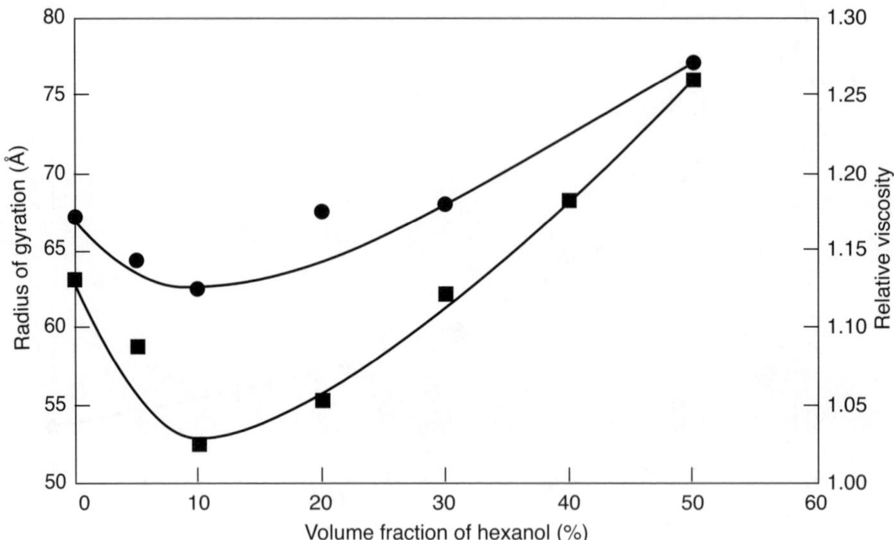

Figure 11.2

Influence of the Volume Fraction of Hexanol of 3% C5-asphaltenes in Hexanol/Toluene Mixture on
◆ the radius of gyration of the asphaltenes at 25°C
● the relative viscosity at 3°C.

entanglement will likely reduce the viscosity. This could be enhanced by increasing interactions between the solvent and the polar compounds of the crude oil (mainly the asphaltenes) and breaking the asphaltene-asphaltene interactions. Based on the previous results, polar solvents could assume this function. Polarity of the solvent can be characterized by the Hansen polar solubility parameter [Barton, 1992]. Mixing polar solvents with naphtha can result in a synthetic solvent of desired polarity. It has been shown that an increase in solvent polarity leads to a decrease in relative viscosity, defined as the ratio of the viscosity of the diluted crude oil to the viscosity of the solvent [Gateau *et al.*, 2004] (Figure 11.3). Among the tested products, MethylEthylKetone (MEK) is a good candidate. Using blends of naphtha and polar solvents can help in the transport of *heavy crude oils* by increasing the quantity of transported oil and keeping the pipeline flow rate constant. Obviously, the process must allow for recycling of the solvent and therefore, the boiling point of the co-solvent must fall within the boiling point range of the hydrocarbon cut of the mixture. This is the case for MethylEthylketone. One study has shown that MEK can present an economic advantage if losses are below 5 wt% [Argillier *et al.*, 2005]. Another potential interesting diluent in the use under light pressure of Dimethyl Ether [Héraud *et al.*, 2007].

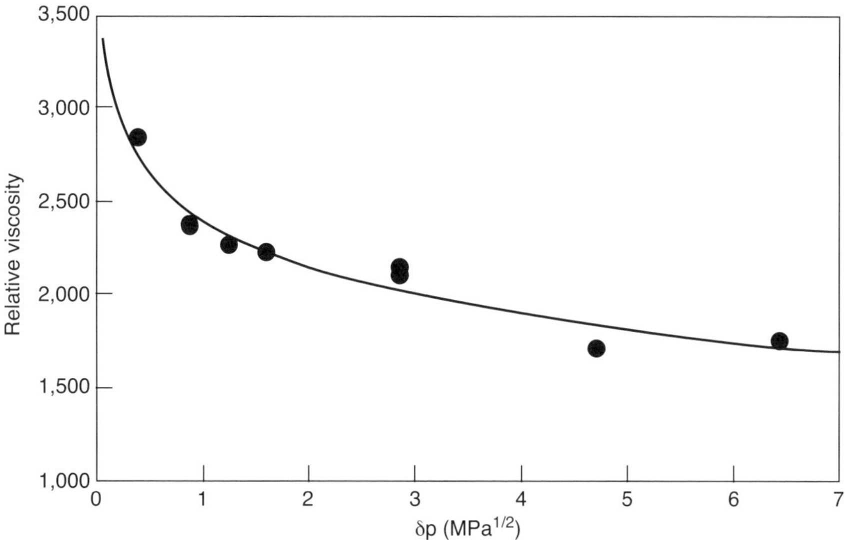

Figure 11.3

Evolution of the Relative Viscosity of Dilute Oil as a Function of Polarity of the Naphtha-Based Solvent.

REFERENCES

Alcocer CF, Menzie DE (1986) Development and Field Application of a Mathematical Model for Predicting the Kinematic Viscosity of Crude Oil/Diluter under Continuous Production Conditions., Proc. *61th SPE Annual Technical Conference*, New-Orleans, USA, pp 1-11.

Anhorn JL, Badakhshan A (1994) "Heavy Oil-Oxygenate Blends and Viscosity Model", Fuel, 73, 9, 1499-1503 and MTBE: A Carrier for Heavy Oil Transportation and Viscosity Mixing Rule Applicability, *J. Can. Pet. Technol*, **33** (4), pp 17-21.

Argillier J-F, Hénaut I, Gateau P, Héraud J-P, Glénat P (2005) Heavy Oil Dilution, *SPE/PS-CIM/CHOA International Thermal Operations and Heavy Oil Symposium*, 1-3 November, Calgary, Alberta, Canada, Paper 97763.

Argillier J-F, Barré L, Brucy F, Dournaux J-L, Hénaut I, Bouchard R (2001) Influence of Asphaltenes Content and Dilution on Heavy Oil Rheology, SPE 69711.

Barrufet MA, Setiadarma A (2003) Reliable Heavy Oil-Solvent Viscosity Mixing Rules for Viscosities Up to 450 K, Oil-Solvent Viscosity Ratios Up to $4x10^5$ and any Solvent Proportion. *Fluid Phase Equilibria*, **213**, pp 65-79.

Barton AFM (1992) Handbook of Solubility Parameters and Other Cohesion Parameters, 2nd Ed. CRC Press.

Burger J, Robin M (1984) Combinaisons de l'injection de solvant et des méthodes thermiques pour la production des huiles très lourdes., Proc. *11th World Pet. Cong.*, **3**, pp 251-260.

Cabrera H (1982) Production and Transport of Heavy Oil by Blending with Lighter Crudes, Proc. *Future Heavy Crude Tar Sands, Int. Conf.*, 2nd, pp 297-305.

Crandall GR, Wise TH (1984) Availability of Diluent may inhibit Heavy Oil Exports. *Can. Pet.*, **25**, pp 37-40.

Darbouret M, Hénaut I, Palermo P, Hurtevent Ch, Glénat P (2009) Experimental Methodology to Evaluate DRA Efficiency and Mutual Effects of Wax Deposit and DRA Action, presented at *14th International Conference on Multiphase Technology*, 17th-19th June 2009, Cannes, France.

Gateau P, Hénaut I, Barré L, Argillier J-F (2004) Heavy Oil Dilution, *Oil and Gas Science and Technology*, **59**(5), pp 503-509.

Gonzalo-Rojas G, Barrios T, Scudiero B, Ruiz J (1977) Rheological Behaviour of Extra Heavy Crude Oils from the Orinoco Belt, *Oil Sands*, pp 284-302.

Hénaut I, Barré, L, Argillier J-F, Brucy F, Bouchard R (2001) Rheological and Structural Properties of Heavy Crude Oils in Relation With Their Asphaltenes Content, SPE 65020.

Hénaut I, Darbouret M, Palermo T, Glénat P, Hurtevent C (April 2009) Experimental Methodology to Evaluate DRA: Effect of Water Content and Waxes on Their Efficiency, Proceedings of SPE International Symposium on Oilfield Chemistry, SPE 121 544, The Woodlands, Texas, USA.

Héraud J-P, Hénaut I, Forestière A, Argillier J-F (2007) Method of Optimizing Heavy Crude Transportation by Incorporation Under Pressure of Dimethyl Ether, *US Patent* US2007295642.

Lederer EL (1933) Proc. *World Pet. Cong.* (Lond.), **2**, pp 526-528.

Mehrotra AK (1992) "A Model for the Viscosity of Bitumen/Bitumen Fractions-Diluent Blends", *Journal of Canadian Petroleum Technology*, **30**(9), pp 28-32.

Meyer RF (1998) "World Heavy Crude Oil Resources", Proc. *15th World Pet. Congress*, pp 459-471.

Rahmes MH, Nelson WL (1948) Viscosity Blending Relationships of Heavy Petroleum Oils, *Analytical Chemistry*, **20**, pp 912-915.

Shu WR (1984) A Viscosity Correlation for Mixtures of Heavy Oil, Bitumen, and Petroleum Fractions, SPE 11280.

Storm DA *et al.* (1999) Drag Reduction in Heavy Oil, *Journal of Energy Resources Technology*, **121**, pp 145-148.

Todd CM (1988) Downstream Planning and Innovation for Heavy Oil Development – A Producer's Perspective. *J. Can. Pet. Technol.*, **27**, 1, pp 79-86.

Urquhart RD (1986) Heavy Oil Transportation: Present and Future. *J. Can. Pet. Technol.*, **25**, 2, pp 68-71.

Wallace D, Miadonye A, Henry D, Puttagunta VR (1996) Viscosity and Solubility of Mixtures of Bituminen and Solvent, *Fuel Science & Technology International*, **14**, N.3, 465-478.

12 Aqueous Emulsions

J-F. Argillier, I. Hénaut, D. Langevin

In this chapter, we will first review some of the main characteristics of crude oil emulsions and then discuss in more detail the transportation of heavy oil as an aqueous emulsion. As indicated in the introduction and reviewed in 2004 [Langevin *et al.*, 2004], one way to transport highly viscous hydrocarbons is to disperse the *heavy crude oil* in water as droplets.

These droplets are stabilized by surfactants, leading to a significant reduction in viscosity. According to Rimmer *et al.* [Rimmer *et al.*, 1992], a typical emulsion is composed of 70% crude oil, 30% aqueous phase and 500-2,000 ppm of chemical additives. The resulting emulsion has a viscosity in the 50-200 cP range at pipeline operating conditions (Figure 12.1) and is particularly stable.

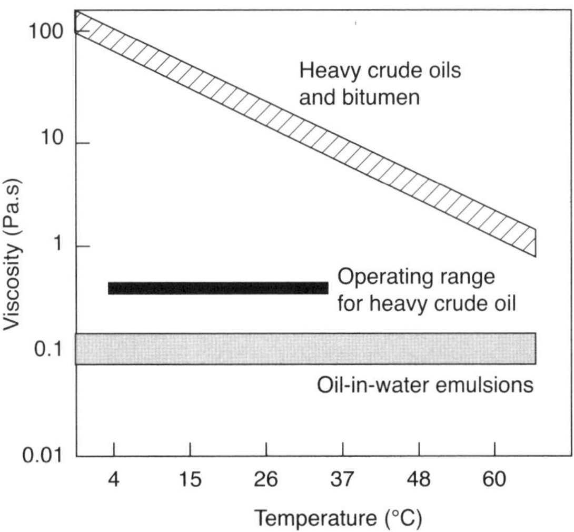

Figure 12.1

Reduction of Viscosities of *Heavy Crude Oils* and Bitumen by Conversion to Oil-In-Water Emulsions, adapted from [Rimmer *et al.*, 1992].

12.1 CRUDE OIL EMULSIONS

Emulsions of crude oil and water may be encountered at many stages during the drilling, producing, transporting and processing of crude oils, and in many locations such as hydrocarbon reservoirs, well bore tubings, surface facilities, transportation systems and refineries. Thorough knowledge of petroleum emulsions is necessary for controlling and improving processes at all stages. Many studies have been carried out during the last 40 years, leading to a better understanding of these complex systems. However, there are still many unanswered questions related to the peculiar behavior of these emulsions. The complexity comes mostly from the oil composition, in particular from the surface-active molecules contained in the crude. These molecules cover a large range of chemical structures, molecular weights, and HLB (hydrophilic-lypophilic balance) values; they can interact between themselves and/or reorganize at the water/oil interface. To make the system even more complex, these petroleum emulsions may also contain solids and gases.

Oil-water emulsions are fine dispersions of oil in water (O/W) or water in oil (W/O), with drop sizes usually in the micron range [Sjöblom, 2001]. In general, emulsions are stabilized by surfactants (Figure 12.2).

In some cases, multiple emulsions such as water in oil in water (W/O/W) or oil in water in oil (O/W/O) can be found. Emulsions can be stabilized by other species, provided that they adsorb at the oil-water interface and prevent drop growth and phase separation into the original oil and water phases. After adsorption, the surfaces become viscoelastic and the surface layers provide stability to the emulsion.

Crude oils – especially the heavy oils – contain large quantities of asphaltenes (high molecular weight polar components) that act as natural emulsifiers. Other crude oil components are also surface-active: resins, fatty acids such as naphthenic acids, porphyrins, wax crystals, etc. but generally they cannot by themselves produce stable emulsions. However, they can associate with asphaltenes and affect emulsion stability. Resins solubilize asphaltenes in oil, and remove them from the interface, therefore lowering emulsion stability.

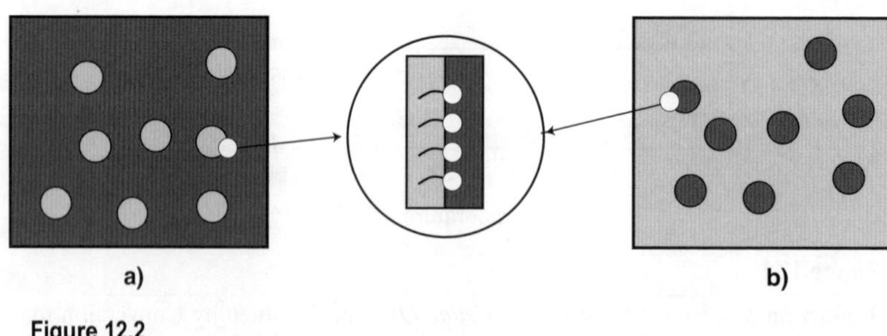

Figure 12.2

Schematic Representation of Emulsion Structures:
a) O/W Emulsion; **b)** W/O Emulsion. Encircled: Enlarged View of a Surfactant Monolayer Sitting at the Oil-Water Interface.

Waxes co-adsorb at the interface and enhance the stability. Naphthenic and other naturally-occurring fatty acids also do not seem to be capable of stabilizing emulsions alone. However, they are probably partially responsible for the significant dependence of emulsion stability upon water pH.

Particles such as silica, clay, iron oxides, etc. can be present in crude oils. These particles are naturally hydrophilic, but can become oil-wet (hydrophobic) due to long term exposure to the crude in the absence of water. A decrease in the size of oil-wet particles results in an increase in W/O emulsion stability. Emulsions with particles and asphaltenes combined can be much more stable than those stabilized by asphaltenes alone, provided that enough asphaltenes are present: all the adsorption sites on the particle surface need to be saturated by asphaltenes. Table 12.1 shows the resin and asphaltene content of various *heavy crude oils via* "heptane" asphaltene precipitation using SARA analysis.

One Venezuelan oil studied at *IFP Energies nouvelles* has a high density (9°API, Density 1.0). Its composition is shown in Table 12.2, which includes the difference induced by the asphaltene precipitating antisolvent (pentane or heptane).

Figure 12.3 shows examples of schematic molecular structures of asphaltenes, resins and naphthenic acids. The Venezuelan heavy oil in question has a high acid number, close to 4.3 (mg KOH/g crude needed for neutralization), due to its content in naphthenic acids.

Table 12.1 Resin and Asphaltene Content of Various *Heavy Crude Oils*.

Crude	° API	Resin wt%	Asphaltene wt%	Asph./Resin
Canada, Athabasca	8.3	14.0	15.0	1.07
Venezuela, Boscan	10.2	29.4	17.2	0.58
Canada, Cold Lake	10.2	25.0	13.0	0.52
Mexico, Panucon	11.7	26.0	12.5	0.48

Table 12.2 SARA Analysis and the Elementary Composition of Oil From Venezuela. The Extraction was Performed Using the ASTM 863-69 norm. According to [Jeribi M *et al.*, 2002].

(% weight)	SARA Pentane	SARA Heptane	(% weight) Heptane SARA Fractions				
			C	H	N	O	S
Asphaltenes	17	14.1	83.8	7.5	1.3	1.7	4.8
Resins	33	37.3	82.8	8.9	1.5	2.0	4.3
Aromatics	37	37.2	84.3	10	< 0.3	1.1	4.0
Saturated	12	11.4	86.6	13	< 0.3	< 0.2	< 0.1

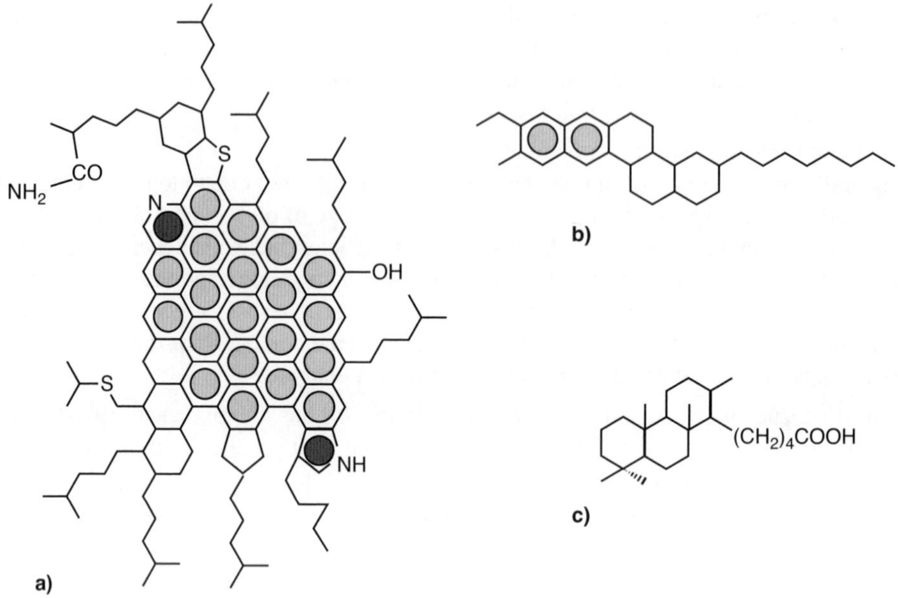

Figure 12.3

Examples of Schematic Molecular Structures: **a)** Asphaltene (adapted from proposal for 510C Residue of Venezuelan Crude, INTEVEP S.A. Tech. Rept., 1992); **b)** Resin (Athabasca tar sand bitumen); **c)** Naphthenic acid.

12.2 TRANSPORTATION OF HEAVY OIL AS AN AQUEOUS EMULSION

When mixing water and crude oil, a water-in-oil emulsion is generally created because the surface-active amphiphilic molecules in the crude are mostly oil-soluble due to their low polarity (depending on pH for fatty acids) and molecular weight. As the viscosity of an emulsion is always higher than the viscosity of the continuous phase, it is obvious that for transporting highly viscous heavy oil it is necessary to make an emulsion with water as the continuous phase [Ahmed *et al.*, 1999; Sun *et al.*, 1996; Hénaut *et al.*, 2001]. Making such an aqueous emulsion requires overcoming the natural tendency to form a water-in-oil emulsion. There are basically two methods for making a heavy oil in water emulsion [Elgibaly *et al.*, 1997]: the first is to add a high HLB (hydrophilic-lipophilic balance) natural or synthetic surfactant. In general, non-ionic emulsifiers are a good choice because they are not affected by the salinity of the produced water, are relatively cheap, and do not produce any undesirable organic residue that can affect the oil properties [Sharma *et al.*, 1998; Gillies *et al.*, 2000; Al-Sabagh *et al.*, 1997]. The second method is to "activate" the natural surfactants which are part of *heavy crude oil* composition: one can make these surfactants more hydrophilic by ionizing the acid functional groups carried by the fatty acids and asphaltenes (from low molecular weight species to carboxylic functions carried by asphaltenes) with a strong base like soda. Addition of strong bases is a well-known process in enhanced oil recovery by decreasing interfacial tension to very low values [Plegue *et al.*, 1989; Plegue *et al.*, 1986; Kessick *et al.*, 1982; Verzaro *et al.*, 2002; Wegener *et al.*; Sanchez *et al.*, 1994; Adewusi *et al.*, 1993]. The presence of natural saponifiable acids in some crudes may eliminate the need for expensive commercial surfactants. Emulsions suitable for pipeline transport – containing as much as 75% crude – have been made with some crudes mixed with alkaline water.

As shown in Figure 12.4, viscosity reduction can be very significant: more than 2 orders of magnitude [Zhang *et al.*, 2002]. For obvious economical reasons, oil operators wish to transport as much oil as possible and as little water as possible. To keep the viscosity of an aqueous heavy oil emulsion below the required specification value for pipeline transport (typically approx. 400 cP at ambient temperature), a maximum of 70% to 75% by volume of dispersed bitumen phase is acceptable. Above 70%, the viscosity becomes too high [Zhang *et al.*, 1991; Yaghi *et al.*, 2002; Zaki *et al.*, 2001; Zaki *et al.*, 2000].

The formation of oil-in-water emulsions to reduce the viscosity of *heavy crude oils* and bitumens has been under investigation to provide an alternative to the use of diluents or heat to reduce viscosity in pipelines [Zaki *et al.*, 2001; Zaki *et al.*, 2000; Plegue *et al.*, 1985; Layrisse *et al.*, 1990; Zaki *et al.*, 1997; Nunez *et al.*, 1996; Briant *et al.*, 1982; Fruman *et al.*, 1983]. Moreover, the viscosity of an emulsion is far less temperature-sensitive than that of a heavy oil [Zhang *et al.*, 1991; Yaghi *et al.*, 2002]. Simon *et al.* [Simon *et al.*, 1970] have also shown that restarting a pipeline after an emergency shutdown and re-emulsification of oil does not rise major problems.

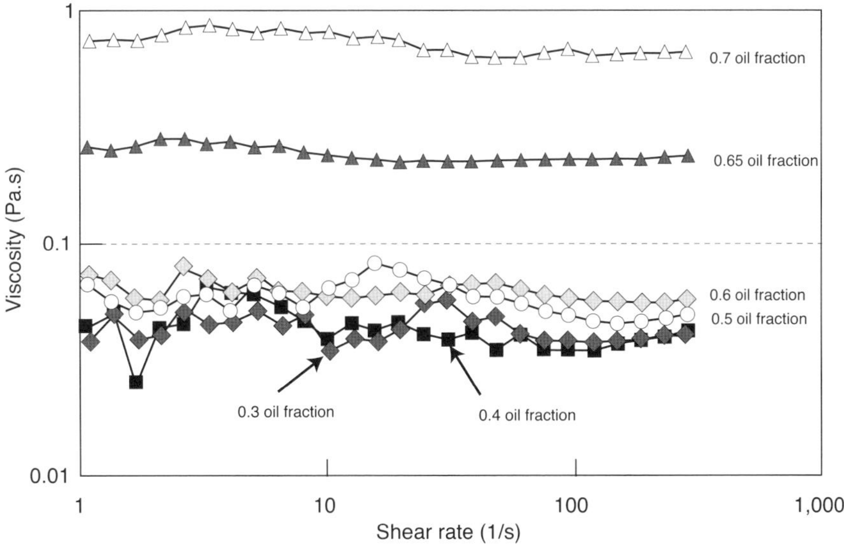

Figure 12.4

Viscosity of Oil/Water Emulsions at Different Shear Rates.

One of the major applications of heavy oil aqueous emulsions is the Orimulsion® process developed by PDVSA (Petroléos de Venezuela) in the 1980s. Orimulsion® is an aqueous bitumen emulsion, made of 30% water and 70% natural extra-heavy oil, which was directly used as a feedstock for power generation in thermo-electrical plants. The feasibility of this technology was clearly demonstrated by large-scale development of the commercial product. A good review of this process can be found in [Salager *et al.*, 2001]. Due to the current high price of crude, it would seem to be more economically interesting to separate the oil and water at the entry to a refinery, in order to refine the crude instead of burning it in power plants.

In addition to Orimulsion®, the technical viability of pipeline transportation of *heavy crude oil* as oil-in-water emulsions containing high fractions of oil was demonstrated in an Indonesian pipeline in 1963 and in a 13 mile long, 8 in. diameter pipeline in California [Plegue *et al.*, 1985].

As we have seen in the previous sections, emulsion stability depends on many parameters. Some of the main parameters include: oil composition in terms of surface-active molecules [Plegue *et al.*, 1985; Nunez *et al.*, 1996], water salinity and pH [Plegue *et al.*, 1986; Plegue *et al.*, 1985; Poteau *et al.*, 2005], oil/water volume ratio [Sun *et al.*, 1996], droplet size and polydispersity [Ahmed *et al.*, 1999; Sun *et al.*, 1996; Hénaut *et al.*, 2006 & 2009], temperature [Sun *et al.*, 1996; Sharma *et al.*, 1998], surfactant type and concentration [Sun *et al.*, 1996; Zaki *et al.*, 2000], dilution [Alvarez *et al.*, 2009] and mixing energy [Ahmed *et al.*, 1999; Sun *et al.*, 1996]. Many studies – mostly experimental in nature – have been carried out on oil-water emulsions [Ahmed *et al.*, 1999; Zaki, 1997]. However, the results of these studies have not always been consistent. The reason for this is that emulsions exhibit complex behavior which, as mentioned above, depends on many factors [Plegue *et al.*, 1989; Yaghi *et al.*, 2002]. A detailed review regarding the influence of the various parameters is beyond the scope of this article. An interesting review using formulation-composition maps of heavy hydrocarbon emulsions has been written by J-L. Salager *et al.* [Salager *et al.*, 2001].

Obviously, the rheology of these emulsions is an important issue for transportation. Some of the main parameters that control emulsion rheology are the oil dispersed volume fraction and droplet size distribution [Ahmed *et al.*, 1999; Lee, 1999]. Depending on conditions, heavy oil aqueous emulsions can present either Newtonian or shear thinning rheological behavior [Fruman *et al.*, 1983]. Droplet size distribution is a function of various parameters which are not independent of one another, in particular: surfactant type and concentration [Yaghi *et al.*, 2002; Zaki *et al.*, 2000], pH of the aqueous phase [Plegue *et al.*, 1985], surface tension [Ahmed *et al.*, 1999; Plegue *et al.*, 1989], mixing energy [Ahmed *et al.*, 1999; Zaki *et al.*, 2000], ionic strength [Ahmed *et al.*, 1999; Zaki *et al.*, 2000], temperature [Sharma *et al.*, 1998; Yaghi *et al.*, 2002; Chaudemanche *et al.*, 2009], chemistry of the natural surfactant of the crude oil [Al-Sabagh *et al.*, 1997; Plegue *et al.*, 1986], and pressure [Khan, 1996]. In addition to viscosity, the formation of a lubricating layer around the pipe wall resulting from migration of droplets away from the pipe wall due to the hydrodynamic conditions [Salager *et al.*, 2001] has been reported. This could explain the pressure drop during pipeline transportation, which is lower than that expected from rheological measurements [Nunez *et al.*, 1996; McKibben *et al.*, 2000]. Further investigation is needed on this topic.

As mentioned earlier, Orimulsion® was directly used as a feedstock for power generation in thermo-electrical plants. Another strategy for crude oil in water emulsions could be to separate the crude from the water after transportation – at the refinery entry – and subsequently upgrade the crude [Kessick *et al.*, 1982]. In this case, from a practical point of view, the emulsion properties required for a given dispersed oil volume fraction are as follows: 1) viscosity as low as possible, 2) sufficient stability in dynamic conditions during transportation and in static conditions in case of shutdown, 3) conditions which allow robust oil and water separation after transportation. One option for transporting heavy oil is to use NH_4OH as a base to saponify fatty acids so as to form the oil-in-water emulsion, and then break the emulsion after transport by evaporating the ammonia when heating the emulsion. The ammonia can then be recovered and recycled [Verzaro *et al.*, 2002]. Breaking these emulsions is a challenge for various reasons:

- The density of heavy oil is very close to the density of water. Without additional treatment, classical physical techniques using density difference (gravity separators, centrifuge systems, etc.) exhibit low efficiency for separating heavy crude and water. Raising the temperature of the emulsion, adding a low-density diluent or adding salt can increase the density difference and enhance the emulsion separation.
- Heavy oils contain a large amount of asphaltene, and more generally of amphiphilic molecules. As mentioned previously, these molecules interact and reorganize at the interface, forming strong films that give very stable emulsions.
- Inversion of the emulsion is difficult to avoid and is detrimental, leading to fine water droplets which are afterwards very difficult to separate, due to the high viscosity of the crude and the large amount of surface-active molecules.

Procedures for demulsifying *heavy crude oil* emulsions may combine different techniques: raising the temperature, adding a carefully chosen demulsifier system, adding solvents, adding salts, changing pH, applying an electrical field, stripping with gas, etc.

12.3 CONCLUSION

Emulsion behavior is largely controlled by the properties of the adsorbed layers that protect the oil-water surfaces. Knowledge of surface tension alone is not sufficient to understand emulsion properties, and surface viscoelasticity plays an important role in a variety of dynamic processes. Dispersing heavy oil or bitumen in water is a very efficient way to reduce the viscosity of the fluid by more than 2 orders of magnitude. It can be achieved by the addition of a hydrophilic surfactant or activation of the natural surfactants contained in the crude oil *via* a strong base. The stability and rheology of these emulsions depend on various parameters. The entire topic has received significant clarification in recent years, but some aspects deserve further investigation, in particular heavy oil emulsion rheology and flow properties, as well as interfacial properties of asphaltene/fatty acid mixed systems. This will help to achieve a better understanding of the relationships between microscopic interactions and macroscopic emulsion properties.

REFERENCES

Acevedo S, Escobar G, Gutierrez LB, Rivas H, Gutierrez X (1993) Interfacial Rheological Studies of Extra-Heavy Crude Oils and Asphaltenes – Role of the Dispersion Effect of Resins in *the Adsorption of Asphaltenes at the Interface of Water-In-Crude Oil-Emulsions. Colloids and Surfaces A-Physicochemical and Engineering Aspects*, **71**, pp 65-71.

Adewusi VA, Ogunsola AO (1993) Optimal Formulation of Caustic Systems for Emulsion Transportation and Dehydration of Heavy Oil. *Chemical Engineering Research & Design*, **71**, pp 62-68.

Ahmed NS, Nassar AM, Zaki NN, Gharieb HK (1999) Formation of Fluid Heavy Oil-In-Water Emulsions for Pipeline Transportation. *Fuel*, **78**, pp 593-600.

Ahmed NS, Nassar AM, Zaki NN, Gharieb KH (1999) Stability and Rheology of Heavy Crude Oil-In-Water Emulsion Stabilized by an Anionic-Nonionic Surfactant Mixture. *Petroleum Science and Technology*, **17**, pp 553-576.

Al-Sabagh AM, Zaki NN, Badawi AFM (1997) Effect of Binary Surfactant Mixtures on the Stability of Asphalt Emulsions. *Journal of Chemical Technology and Biotechnology*, **69**, pp 350-356.

Alvarez G, Poteau S, Argillier JF, Salager J-L, Langevin D (2009) Heavy Oil-Water Interfacial Properties and Emulsion Stability: Influence of Dilution, *Energy Fuels*, **23** (1), pp 294-299.

Asphaltene Deposition and its Control (2004).
http://tigger.uic.edu/~mansoori/Asphaltene.Deposition.and.Its. Control_html

Briant J, Fruman DH, Quemada D, Makria A (1982) Transport de pétroles bruts lourds sous forme d'émulsion huile dans eau. *Revue de l'Institut Français du Pétrole*, **37**, pp 809-821.

Chaudemanche C, Hénaut I, Argillier J-F (2009) Combinedd Effect of Pressure and Temperature on Rheological Properties of Water in Crude Oil Emulsions, *Applied Rheology*, **19**, 62210.

Elgibaly AAM, Nashawi IS, Tantawy MA (1997) Rheological Characterization of Kuwaiti Oil-Lakes Oils and their Emulsions. SPE 37259, pp 493-508.

Fruman DH, Briant J (2000) Investigation of the Rheological Characteristics of Heavy Crude Oil-In-Water Emulsions. International Conference on the Physical Modelling of Multi-Phase Flow, Coventry, England Paper J2. 19-4-1983. Gillies RG, Sun R, Shook CA. Laboratory Investigation of Inversion of Heavy Oil Emulsions. *Canadian Journal of Chemical Engineering*, **78**, pp 757-763.

Hénaut I, Barré L, Argillier J-F, Brucy F, Bouchard R (2001) Rheological and Structural Properties of Heavy Crude Oils in Relation with their Asphaltenes Content. SPE 65020.

Hénaut I, Saint-Michel F, Argillier J-F (2006) The Rheology of Concentrated Emulsion: the Influence of Drop Size Distribution, *4th World Congress on Emulsion*, N399, 3-6 October, Lyon, France.

Hénaut I, Courbaron AC, Argillier JF (2009) Viscosity of Concentrated Emulsions: Relative Effect of Granulometry and Multiphase Morphology, *Petroleum Science and Technology*, **27** (2), pp 182-196.

Jeribi M, Almir-Assad B, Langevin D, Hénaut I, Argillier JF (2002) Adsorption Kinetics of Asphaltenes at Liquid Interfaces. *Journal of Colloid and Interface Science*, **256**, pp 268-272.

Kessick MA, Denis ESt (10-8-1982) Pipeline transport of heavy crude oil. 157940(4 343 323). US.

Khan MR (1996) Rheological Properties of Heavy Oils and Heavy Oil Emulsions. *Energy Sources* **18**, pp 385-391.

Langevin D, Poteau S, Hénaut I, Argillier J-F (2004) Crude Oil Emulsion Properties and their Application to Heavy Oil Transportation, *Oil & Gas Science and Technology*, **59** (5), pp 511-521.

Layrisse IA *et al.* (8-5-1990)Viscous Hydrocarbon-In-Water Emulsions. 263896(4 923 483). US.

Lee RF (1999) Agents Which Promote and Stabilize Water-In-Oil Emulsions. *Spill Science and Technology Bulletin*, **5**, pp 117-126.

McKibben MJ, Gillies RG, Shook CA (2000) A Laboratory Investigation of HorizonTal Well Heavy Oil-Water Flows. *Canadian Journal of Chemical Engineering*, **78**, pp 743-751.

Nunez G, Briceno M, Mata C, Rivas H, Joseph D (1996) Flow Characteristics of Concentrated Emulsions of Very Viscous Oil In Water. *Journal of Rheology*, **40**, pp 405-423.

Plegue TH, Frank SG, Fruman DH, Zakin JL (1989) Studies of Water-Continuous Emulsions of Heavy Crude Oils. Prepared by Alkali Treatment. SPE 18516.

Plegue TH, Zakin JL, Frank SG, Fruman DH (1985) Studies of Water Continuous Emulsions of Heavy Crude Oils. SPE 15792.

Plegue TH, Frank S, Fruman D, Zakin JL (1986) Viscosity and Colloidal Properties of Concentrated Crude Oil In Water Emulsions. *Journal of Colloid and Interface Science*, **114**, pp 88-105.

Poteau S, Argillier JF, Langevin D, Pincet F, Perez E (2005) Influence of pH on Stability and Dynamic Properties of Asphaltenes and Other Amphiphilic Molecules at the Oil-Water Interface, *Energy & Fuels*, **19** (4), pp 1337-1341.

Rimmer D, Gregoli A, Hamshar J, Yildivim E (1992) Pipeline Emulsion Transportation for Heavy Oils in Emulsions in the Petroleum Industry, American Chemical Society, **8**, pp 295-312.

Salager JL, Briceno MI, Brancho CL (2001) Encyclopedic Handbook of Emulsion Technology. Sjöblom J (ed.). Marcel Dekker, New York, pp 455-495.

Sanchez LE, Zakin JL (1994) Transport of Viscous Crudes As Concentrated Oil-In-Water Emulsions. *Industrial & Engineering Chemistry Research*, **33**, pp 3256-3261.

Sharma K, Saxena VK, Kumar A, Ghildiyal HC, Anurada A, Sharma ND, Sharma BK, Dinesh RS (1998) Pipeline Transportation of Heavy Viscous Crude Oil as Water Continuous Emulsion in North Cambay basin (India). SPE 39537.

Simon R, Poynter WG (1970) Pipelining Oil/Water Mixtures, US Patent 3519006.

Sjöblom J (2001) Encyclopedic Handbook of Emulsion Technology. Marcel Dekker, New York.

Sun R, Shook CA (1996) Inversion of Heavy Crude Oil-In-Brine Emulsions. *Journal of Petroleum Science and Engineering*, **14**, pp 169-182.

Verzaro F, Bourrel M, Garnier O, Zhou HG, Argillier J-F (2002) Heavy Acidic Oil Transportation by Emulsion In Water. SPE 78959.

Wegener DC, Zornes DR, Maloney DR, Vienot ME, Fraim ML (28-8-2001) Heavy Oil Viscosity Reduction and Production, US Patent 6279653, Application Number 201925.

Yaghi BM, Al Bemani A (2002) Heavy Crude Oil Viscosity Reduction for Pipeline Transportation. *Energy Sources*, **24**, pp 93-102.

Zaki NN (1997) Surfactant Stabilized Crude Oil-In-Water Emulsions for Pipeline Transportation of Viscous Crude Oils. Colloids and Surfaces A-Physicochemical and Engineering Aspects, **125**, pp 19-25.

Zaki NN, Ahmed NS, Nassar AM (2000) Sodium Lignin Sulfonate to Stabilize Heavy Crude Oil-In-Water Emulsions for Pipeline Transportation. *Petroleum Science and Technology*, **18**, pp 1175-1193.

Zaki NN, Butz T, Kessel D (2001) Rheology, Particle Size Distribution, and Asphaltene Deposition of Viscous Asphaltic Crude Oil-In-Water Emulsions for Pipeline Transportation. *Petroleum Science and Technology*, **19**, pp 425-435.

Zhang J, Chen D, Yan D, Yang X (1991) Pipelining of Heavy Crude Oil as Oil-In-Water Emulsions. SPE 21733.

13 | Core Annular Flow

Y. Peysson

Oil flow as a single phase in a pipeline is a distributed method of transporting hydrocarbon resources over very long distances. The relatively low viscosity of typical crude oil is a key issue for such a method of transport due to its flow resistance. For example, the Brent viscosity is on the order of 3 cP under ambient conditions. For higher viscosity, an increase in the pipe diameter and a possible decrease in the mean flow velocity are possible options, but for very high viscosity, large pipes and low velocities rapidly become uneconomical.

Multiphase flow is also now widely used since it has the advantage of mixing gas, water and oil all together in a single pipe. However, flow conditions are more difficult to predict because there are various phase configurations. Core Annular Flow (CAF) is one particular two-phase flow regime where the oil phase is in the center of the pipe and water flows near the wall surface. A very attractive characteristic of this flow is that it is stable for an acceptable range of velocities, and the pressure drop is very small and only slightly depends on the heavy oil viscosity. Moreover, it is well-suited for heavy oil because the density is close to water density, implying a limited stratification tendency and the viscosity is high, which reduces the core deformation and limits modification of the flow regime.

These remarkable properties have been observed for a long time now: its industrial interest was noticed one hundred years ago. A 1904 patent of Isaacs and Speed [Isaacs *et al.*, 1904] in the United States first mentioned the ability to transport viscous product through "water lubrication". Despite this early interest, large-scale industrial pipelines for heavy oil are scarce and the first one was only built in the 1970s. This Shell pipeline near Bakersfield in California was 38 km long for a tube diameter of 15 cm. For more than ten years, a viscous crude oil was transported in the line with a flow rate of 24,000 bbl/d in a water-lubricated regime.

Since then, several studies have been dedicated to the Core Annular Flow regime and different reviews of the published work have been written [Oliemans *et al.*, 1986; Joseph *et al.*, 1997].

13.1 FLOW REGIME

Several flow regime configurations are possible when flowing a mixture of immiscible water and oil in a pipe. Multiphase flows of liquid-liquid and gas-liquid are crucial in the oil and gas industry and a large amount of work is available on flow regime characterization. A complete overview on multiphase flow was written in the 1970s [Govier *et al.*, 1972], and two-fluid dynamics were summarized more recently [Bai *et al.*, 1992].

Figure 13.1 shows various flow regimes for heavy oil and water in vertical and horizontal flows. Mean injection velocities are key parameters for flow regime determination. The configurations are also dependent on fluid properties such as density ratio and surface tension.

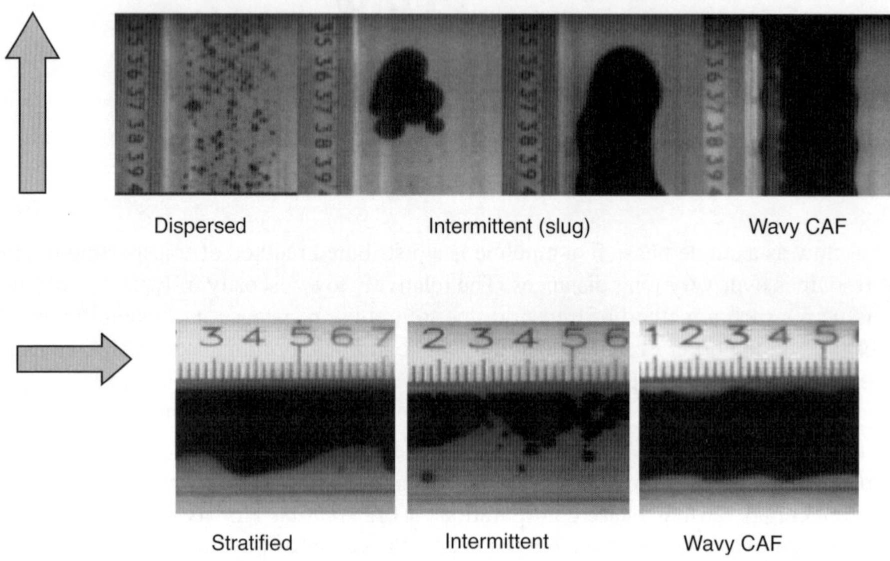

Figure 13.1

Various Flow Regimes of Viscous Oil in Water in Vertical and Horizontal Flow [Bannwart *et al.*, 2004].

13.2 CORE ANNULAR FLOW

Core Annular Flow is one of the flow regimes observed in two-phase immiscible flow. But a very specific property of it is that the pressure drop for the system is the lowest of all the flow regimes for equivalent liquid and water flow rates.

In this flow regime, water is at the pipe surface and lubricates the oil core. Perfect Core Annular Flow is shown in Figure 13.2. A small water layer is sheared and the velocity field is approximately linear if the difference in viscosity between the oil and water is large. In this case, the oil core is nearly a plug flow. Very weak deformations take place.

Perfect CAF appears to be very rare and can exist only for density-matched fluid. Several experimental observations have shown that waves are created at the water and oil interface [Bai *et al.*, 1992], leading to Wavy Core Annular Flow (WCAF). WCAF seems to be the type observed in operational situations. Indeed, for fluids with a difference in density, a buoyancy force will give a radial movement of the core. If no counter-balancing force is applied, the buoyant effect will push the core to the upper wall of the pipe. It has been shown that waves at the interface with a specific form are necessary to create sufficient lubrication force capable of counterbalancing the buoyancy force [Ooms *et al.*, 1984].

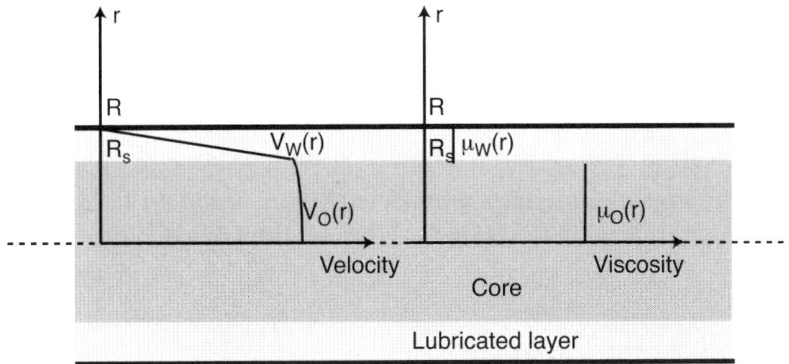

Figure 13.2

Flow of Oil In Water in a Perfect Core Annular Flow Regime. Schematic Representation of the Velocity and Viscosity Profiles.

13.3 STABILITY

Many authors have investigated the issue of the stability of such a system. Because of its simplicity, Perfect CAF without a density difference was first analyzed. It has been shown that stability in this case is achieved only for a narrow range of controlling parameters [Prezioki *et al.*, 1989]. For a fixed volume ratio between oil and water, the core annular flow is not stable at low velocity. Capillary instability due to surface tension appears and breaks the core. But the increase in velocity stabilizes the capillary instability and the flow regime can then be observed. However, for a larger velocity, PCAF flow again becomes unstable due to interfacial friction, and undulation of the flow arises, leading to Wavy CAF. But this configuration can re-stabilize. Only much higher velocities can give rise to break-up of the core or emulsification process.

The particular range of stability raised two main problems for industrial development. First, operating conditions (pump power, pipe geometry, etc.) must be compatible with the flow stability and second, if CAF is not stable at low velocity, the stop and restart of the flow must be addressed specifically.

13.4 PRESSURE DROP

Despite the problems mentioned above, a major advantage of CAF is that the associated pressure drop is the smallest of the available two-phase flow regimes. The transport of very viscous crude oil is thus possible with the pressure drop of water. Guevara *et al.* [Guevara *et al.*, 1998; Bensakhria *et al.*, 2004; Peysson *et al.*, 2007] compared different processes to transport heavy oil – e.g. heating, dilution or emulsion – and showed that CAF induces the smallest pressure drop in the pipe, actually not much different from the behavior of pure water transport.

Pressure drop measurement and friction factor estimation have been proposed in the literature. Arney *et al.* [Arney *et al.*, 1993] experimentally and theoretically analyzed the friction factor variation with a specific Reynolds number in the case of concentric CAF. Eccentricity of the core in the case of density difference has been studied [Huang *et al.*, 1994], and friction factors were estimated for different positions of the core in the pipe.

Operational oil companies are highly impacted by the heavy oil transport problem, and the very low pressure gradient involved in CAF has aroused much interest.

13.5 INDUSTRIAL DEVELOPMENT

The Shell pipeline near Bakersfield, California was discussed previously, but the literature also includes a description of another pipeline used to transport the bitumen produced in the Orinoco Belt of Venezuela. It is a 55 km-long pipeline transporting highly viscous oil (1.5 Pa.s). This commercial line was developed to carefully study the ability of this technique to transport heavy oil in Venezuela [Guevara *et al.*, 1991].

These two lines, as well as others detailed in various articles [Joseph *et al.*, 1999] demonstrated the capacity of this technique to transport heavy oil at an industrial scale despite the various related problems.

Indeed, restart conditions are not met in annular flow regime and require restart pressure, which could be high. A second observation is adhesion of the oil onto the wall in flow conditions. The pipe section can be restricted and an increase in pressure drop and even blockage can be observed.

Various solutions have been proposed to overcome these difficulties, e.g. cement treatment of the pipe inner surface [Arney *et al.*, 1996]. CAF obviously requires additional operational-scale demonstrations and currently is still not a widely-used technique in the heavy oil industry.

REFERENCES

Arney MS, Bai R, Guevara E, Joseph DD, Lui K (1993) Friction factor and Holdup Studies for Lubricated Pipelining – Experiments and Correlations, *Int. J. Multiphase Flow*, **19**, 6, pp 1061-1076.

Arney MS, Ribeiro GS, Guevara E, Bai R, Joseph DD (1996) Cement-Lined for Water Lubricated Transport of Heavy Oil, *Int. Journal of Multiphase Flow*, **22**, 2, pp 207.

Bai R, Chen K, Joseph DD (1992) Lubricated Pipelining: Stability of Core-Annular Flow. Part5. Experiments and Comparison with Theory, *Journal of Fluid Mech.*, **240**, pp 97-132.

Bensakhria A, Peysson Y, Antonini G (2004) Experimental Study of the Pipeline Lubrication for Heavy Oil Transport, *Oil & Gas Science and Technology*, **59** (4), pp 523-533.

Govier GW, Aziz K (1972) The Flow of Complex Mixture in Pipes, Von Nostrand Reinhold Company.

Guevara E, Gonzales SA, Nunez G (1998) Highly Viscous Oil Transportation Methods in the Venezueal Oil Industry, Proceedings of the *15th World Petroleum Congress.*

Guevara E, Zubillaga V, Zagustin K, Zamora G (1991) Core Annular Flow: Preliminary Operational Experience and Measurements in a 152 mm*55 km Pipeline, Proc. *5th Unitar Heavy Crude and Tar Sands International Conference*, **5**, Caracas, Venezuela.

Huang A, Christodoulou C, Joseph DD (1994) Friction Factor and Holdup Studies for Lubricated Pipelining – Laminar and k-e Modes of Eccentric Core Flow *Int. J. Multiphase Flow,* **20**, 3, pp 481-491.

Isaacs JD, Speed JB (1904) Method of Piping Fluids, US Patent N° 759374.

Joseph DD, Bai R, Mata C, Sury K, Grant C (1999) Self-Lubricated Transport of Bitumen Froth, *Journal of Fluid Mech.,* **386**, pp 127.

Joseph DD, Chen KP, Renardy YY (1997) Core Annular Flows, *Ann. Rev. Fluid Mech.,* **29**, pp 65.

Oliemans RVA, Ooms G (1986) Core-Annular Flow of Oil and Water Through a Pipeline, in *Multiphase Science and Technology*, Hewitt GF, Delhaye JM, Zuber N, **2**, Washington Hemisphere

Ooms G, Segal A, Van der Wees AJ, Meerhof R, Oliemans RVA (1984) A Theoretical Model for Core-Annular Flow of a Very Viscous Oil Core and a Water Annulus Through a Horizontal Pipe, *Int. Journal of Multiphase Flow*, **10**, 1, pp 41.

Peysson Y, Bensakhria A, Antonini G, Argillier J-F (2007) Pipeline Lubrication of Heavy Oil: Experimental Investigation of Flow and Restart Problems, *SPE Production & Operations*, **22** (1), pp 135-140.

Prezioki L, Chen K, Joseph DD (1989) Lubricated Pipelining: Stability of Core-Annular Flow, *Journal of Fluid Mech.,* **201**, pp 323-356.

14 Surface Pumps for Transport: Selection and Limitations

J. Falcimaigne

Pumping systems are required at surface for several purposes, such as:

- Short-distance transfers of produced oil from wells to a central processing facility or export hub.
- Handling of fluids between the various stages of a processing facility.
- Long-distance transport by pipeline.

The most suitable type of pump is chosen by considering the required duties (capacity and pressure rise), as well as the general characteristics of the fluid that must be handled: fluid viscosity is a crucial parameter to take into account. Two other important factors to consider are the presence of solid particles and sometimes the presence of gas in liquids. For heavy oils transport, pumps are selected based on their performance and energy expenses.

14.1 FLUID VISCOSITY

Fluid viscosity affects all pumps, but somewhat differently according to their type. Two main types of pumps are usually used at surface with heavy oils, and defined by their physical working principle: volumetric (or positive displacement) and centrifugal.

In general, heavy oil with viscosity as high as 100,000 cP can be pumped with positive displacement pumps. In practice, fluids of very high viscosity cannot be transported realistically over long distances. Pressure losses in the pipeline would become excessive and would require large and uneconomical energy expenses. Reasonably, the maximum viscosity is set at roughly 1,000 cP for transporting heavy oils by pipeline. Various techniques, discussed in the previous chapters, are used to reduce the viscosity to an adequate level.

Due to their general advantages, centrifugal pumps are used with viscous fluids, but within some limits. The upper economical limit is ordinarily about 100 to 150 cP. Above this value, positive displacement pumps are more attractive. However, centrifugal pumps can be used up to 500 cP and even above for high flow rates or for specific needs, with special design arrangements. In many cases, centrifugal pumps remain competitive for boosting heavy oil in pipeline transportation.

14.2 PRESENCE OF SOLID PARTICLES (TRANSPORT OF SLURRIES)

The possible presence of solid particles is the second point which must be considered when pumping heavy oils. Production of tar sands, for instance, requires handling fluids that are highly laden with solid particles. Such fluids are called slurries. The presence of solid particles increases the apparent viscosity of the mixture well above the viscosity of the oil. However, the main problem of pumping slurries is related to the abrasive wear created by solid particles. In some cases, the operating life of a pump may be reduced down to few months. In this respect, low velocity pumps are also better suited for handling slurries.

14.3 PRESENCE OF GAS IN LIQUID (MULTIPHASE FLOW)

Separation of gas from *heavy crude oils* is difficult due to the high liquid viscosity, and frequently some gas bubbles remain in the liquid flow after separation. This can affect pump performance, particularly with centrifugal pumps.

The processing of tar sands involves a flotation step and the boosting of a froth, which is a mixture of hot water, air, bitumen and sand particles. The air content can reach 10 to 30% and multiphase flow problems are then added to the effects of viscosity and solid particles. The performance of conventional pumps may be strongly de-rated by gas if contents are above a few percent.

Multiphase flow pumps are also used in Canada to surface boost the production from heavy oil fields with steam recovery. In these cases, fluids include condensates, gas and steam at very low pressure at pump suction, at very low liquid fractions, and consequently the pumps are more working as wet gas compressors. The differential pressure through the pump is not very high (around 10 bar) but the compression ratio reaches 9, leading to high temperatures at discharge, which requires efficient cooling.

14.4 PUMPING HEAVY OILS WITH POSITIVE DISPLACEMENT PUMPS

Volumetric pumps – also called positive displacement pumps – transfer a finite volume of fluid from a low-pressure side to a high-pressure side and are subdivided into reciprocating pumps (mainly piston pumps) and rotary pumps (e.g. PCP Progressing Cavity Pumps and screw pumps).

The discharge pressure is not determined by the characteristics of the pump, but by the resistance of the downstream piping system to the imposed flow rate. If the resistance is high, e.g. if a downstream valve is accidentally closed, the discharge pressure can become excessive. In such a case, the discharge pressure is only limited by the available power of the pump driver, or by the strength of pump and piping system components. For this reason, a positive displacement pump must always be used with a pressure relief valve enabling fluid release at discharge.

A minimum pressure can be required at pump suction to insure that the fluid properly fills the inlet chamber. If the suction pressure is too low, the fluid can partially vaporize. The volumetric efficiency drops and the pump generates vibration and noise. The minimum pressure is influenced by the pump rotation speed and by the fluid viscosity. The minimum pressure is often converted into head of the fluid, and sometimes called NPSH (Net Positive Suction Head or pressure provided by the pump) by analogy with a centrifugal pump

In estimating the required power, the flow rate which must be considered to calculate the compression work is the theoretical flow rate rather than the actual flow rate. This is because the fluid backflows are first compressed and then the corresponding energy is dissipated in leaks. To determine the power, the viscous friction losses and mechanical losses in timing gears, bearings and seals must be added to the fluid compression work. The necessary information is usually supplied by the pump manufacturer.

14.4.1 Piston Pumps

Alternative positive displacement pumps – which operate by the back-and-forth motion of a piston in a cylinder – deliver a constant volume of fluid for each piston stroke or each shaft revolution.

Piston pumps generally present very good efficiency and are well-adapted to high discharge pressures, but they have some drawbacks. First, they deliver a pulsed flow at the frequency of the piston strokes. To alleviate this problem, pumps with several cylinders are sometimes used (duplex, triplex and sometimes more). Accumulators are mounted on the discharge piping to smooth out flow pulsations. Another drawback is that valves and sliding seals between pistons and cylinders are not tolerant to solid particles or deposits. Consequently, piston pumps are not suited to handle slurry flows.

14.4.2 Rotary Pumps

In contrast to alternative pumps, rotary pumps transfer fluids *via* a rotational motion, which has the advantage of a continuous delivery of fluids without pulsations. They have good volumetric efficiency with viscous fluids, which is their principal domain of application. However, they are relatively heavy and expensive.

Two main rotary pump designs are widely used in the oil industry**: Progressing Cavity Pumps (PCPs**) and **screw pumps** (with twin or triple screws).

- Progressing cavity pumps are limited to capacities of approximately 200m^3/h and moderate pressure rises. They are mainly used for in-field transfers with small diameter flow lines. PCP principles and a system used for downhole artificial lift are described in Section 9.3.3.
- Screw pumps have good flow capacity – up to 1,000 m^3/h or more. Screw pumps are used in the industry for more than 70 years, mainly with viscous fluids or at low suction pressure or low available NPSH (Net Positive Suction Head or pressure provided by the pump) when centrifugal and other pumps are impractical. In the oil industry, they are used with *heavy crude oil* or as cargo pumps.

A screw pump is composed of two or three parallel helical meshing screws (Figure 14.1). The screws and their housing (or liner) delimit small chambers (or locks) filled by the fluid. The chambers move continuously along the rotation axes when the screws rotate and they transfer the fluid from suction side to discharge side. The theoretical volumetric flow rate is governed by the volume of chambers, i.e. the size and pitch of screws, and the rotation speed of screws. The differential pressure generated by the pump creates internal leakages through the housing and screws, or backflows, which reduce the theoretical capacity of the pump. With an incompressible fluid, the total differential pressure is distributed evenly between every chamber. The volumetric efficiency of the pump is directly related to the length of the screw (number of pitches), the size of gaps and – as the flow is essentially laminar in clearances – the fluid viscosity. A high viscosity reduces backflows and ensures lubrication between the screws and the housing in case of contact.

One shaft is directly coupled to the driver; the second shaft (or the other two shafts in the case of three screws) is driven through timing gears which transfer the torque. The screws usually rotate between 1,500 and 2,400 rpm. Typically, a pair of screws is arranged back to back on each shaft to balance the axial thrust produced by the pressure. In this configuration, the high pressure discharge is placed in the middle of the shaft and the low pressure suction is placed on the two sides to reduce the differential pressure on the seals.

Several other types of rotary pumps are also used with viscous fluids, e.g. gear pumps, lobe pumps, vane pumps, etc. Their description, advantages and limitations may be found in any textbook on pumps.

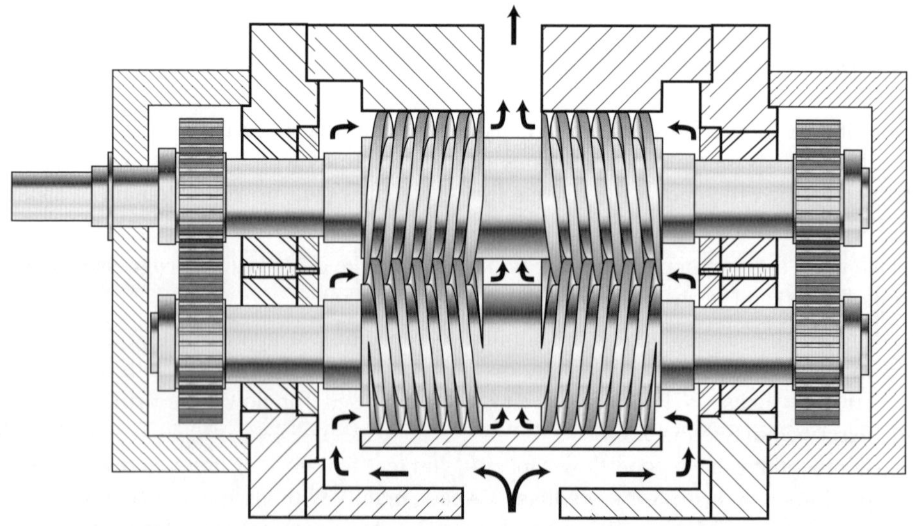

Figure 14.1

Diagram of a Typical Twin Screw Pump.

14.4.3 Fluid Viscosity: Impact on Positive Displacement Pumps

Fluid viscosity has an impact on several operating characteristics of positive displacement pumps (piston pumps or rotary screw pumps).

Power on shaft is increased because of higher shears and frictions in clearances: the pump efficiency declines slightly for this reason.

Outlet pressure is not influenced by effects of viscosity within the pump, but increased frictions in the piping system involve a greater resistance to flow and thus a greater pressure at pump discharge.

Flow rate is barely influenced: a slight improvement is generally observed due to a reduction of internal leakages (better volumetric efficiency).

Minimum suction pressure increases with viscosity.

Consideration of the second and forth points above usually leads manufacturers to derate the normal rotation speed of a given pump when it is operated with a highly-viscous fluid. The following table shows typical rotation speeds *versus* viscosity for rotary pumps, as a percentage of normal speed in water.

Kinematic Viscosity ν (cSt)	100	300	500	1,000	3,000
Rotation Speed (% normal)	98%	95%	90%	80%	60%

Consequently, the size of the screws must be increased to maintain a given duty flow rate with a lower rotation speed. In normal practice, the size of a rotary pump selected for a high-viscosity fluid is larger than the size used for a low-viscosity fluid and it is operated at a lower speed.

Viscosity is greatly influenced by the fluid temperature. For this reason, positive displacement pumps are usually designed to operate over a range of viscosities, with the minimum viscosity corresponding to the maximum fluid temperature and vice versa (e.g. cold or thixotropic product at start-up conditions). The pump flow rate is selected with minimum viscosity (lower volumetric efficiency). Efficiencies and drive power are determined for the maximum viscosity.

The presence of solid particles increases the risk of wear and abrasion. The rate of abrasion is extremely variable, depending on particle size, shape, hardness and content in slurry. Here again, the normal practice is to reduce the pump operation speed. The use of speeds down to 25%-40% of normal speed is frequent.

With very high viscosity, it is often necessary to heat the fluid to reduce the viscosity as much as possible. In such cases, the screw pump casing can be also adapted to heat the pump *via* a hot fluid.

In cases of transport of slurry with risk of wear and abrasion, a removable liner can be used in the screw housings, and the pump is preferably designed to ease the liner and screw exchange. Typical pressure relief valves are not well-suited to slurries; alternatively, rupture disks can be used.

14.5 CENTRIFUGAL PUMPS

14.5.1 Fundamentals

Centrifugal pumps are also called rotodynamic pumps or turbomachines. They impart kinetic energy to the fluid in a rotating part (called the impeller) and transform this energy into potential energy in a static part (called the diffuser). This category can be also divided into several sub-categories according to the overall shape of the impeller (axial, radial and mixed-flow). A centrifugal pump may be composed of a single impeller, or can be multi-staged with several impellers assembled on a common shaft and separated by fixed intermediate diffusers.

Roughly 80% of the pumps used in industry are centrifugal pumps, since for a given duty they are generally lighter, less cumbersome and less expensive than positive displacement pumps. They are also mechanically simpler and do not require safety valves. Jet pumps are rarely used at surface since they have a low efficiency.

According to their principle, centrifugal pumps involve high flow velocities and high shear rates. Significant internal losses of energy and lower efficiency are produced with highly-viscous fluids. In general, positive displacement pumps are better suited for boosting viscous fluids.

The specific work transferred to the fluid ΔW(energy per unit of mass flow) depends on the pump geometry, its rotation speed and the fluid flow rate, but is independent of fluid density. This specific work is usually characterized by the head $H = \Delta W/g$ homogeneous to a length unit.

The difference of discharge and suction pressures is not a characteristic of a pump: it depends on the fluid density and can be determined from the head by using the Bernoulli equation (conservation of mechanical energy) $H = \Delta\left[z + \frac{p}{\rho g} + V^2/2g\right]$.

As indicated above, the head depends on the flow rate. The characteristic curves of a pump describe the variation of head and efficiency *versus* flow rate (Figure 14.2). For a given rotation speed, optimum efficiency in energy conversion is obtained for a certain flow rate, call Best Efficiency Point (BEP). The system curve describes the head which must be balanced for a given flow rate in the piping system. It is often composed of a constant term which corresponds to the difference of level between the upstream and downstream ends of the piping system and a term proportional to the square of flow rate which corresponds to viscous losses. The operating point of the pump is established at the intersection of the pump curve and the system curve.

Best efficiencies are obtained when the general shape of the impeller is suited to the pump duty. For instance, axial flow impellers are best suited to high flow rates and radial flow impellers to lower flow rates with high heads.

When the fluid flows through the pump and enters the impeller, the velocity increases and the pressure drops, according to the Bernoulli equation. Various hydraulic energy losses

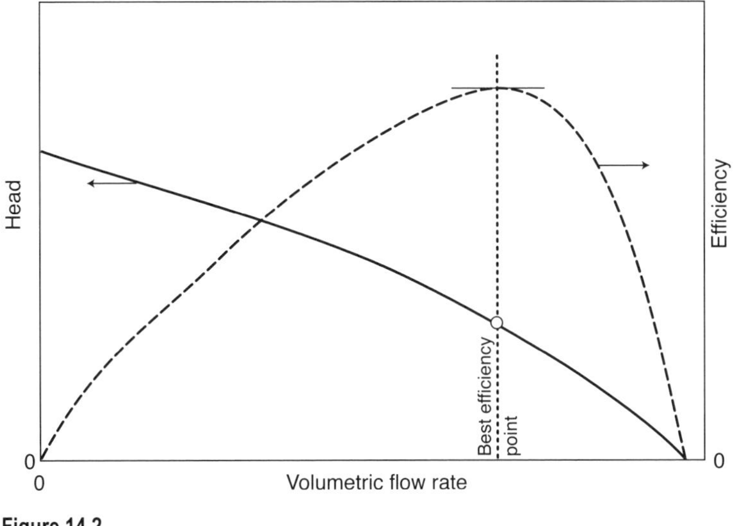

Figure 14.2

Typical Characteristic Curves.

also contribute to lower the pressure. To avoid the phenomenon of cavitations, the fluid must not vaporize in the pump. This requires having a minimum fluid head at the pump inlet. The minimum head required to avoid vaporization is a characteristic of the pump (NPSH). Like the head provided by the pump, it varies with the speed (with the same similarity laws) but is independent of fluid density.

The NPSH is determined by the pump manufacturer by tests in water at a standard temperature. The designer of the flow system must ensure that the available NPSH of fluid at the pump inlet is greater than the required NPSH of the pump, with adequate margins to account for the true vapor pressure of fluid and friction losses in inlet piping (both converted into head unit with the Bernoulli equation).

14.5.2 Influence of Viscosity

The characteristic curves of a centrifugal pump are usually established with water. Dynamic viscosity has a negligible effect on pump performance as long as it does not significantly exceed the viscosity of water, e.g. below 5 to 10 cP to provide an order of magnitude. At higher viscosities, high velocities and high shear rates increase internal friction and energy losses.

Consequently, the delivered head and efficiency are reduced. Simultaneously, the best efficiency point is displaced towards lower flow rates. Following the reduction of efficiency, the required power on shaft is increased. Figure 14.3 shows the evolution of the pump characteristic curves as the viscosity is increased. The required NPSH is also increased.

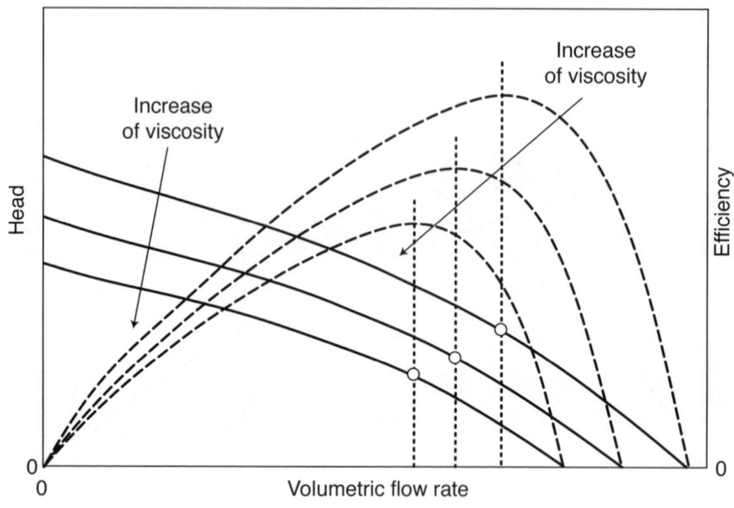

Figure 14.3

Influence of Viscosity on Head and Efficiency Curves (correspondence with y axes are similar than in Figure 14.2).

The performance reduction of a given centrifugal pump must be estimated in order to select a pump to operate with a viscous fluid. This is usually done through experimental viscosity correction factors applied to head, flow rate and efficiency.

The head and efficiency curves obtained when pumping a viscous fluid are usually derived from the corresponding curves in water by applying correction factors to flow rates, heads and efficiencies:

$$Q_v = f_q Q_w \qquad H_v = f_H H_w \qquad \eta_v = f_\eta \eta_w$$

The factors depend on viscosity, flow rate and head in water. Stepanoff [1940] and many others have published experimental data. It is usually presented in the form of correction charts. The type of impeller geometry is not considered in these charts. The more widely-used charts have been published by the U.S. Hydraulic Institute [2004]. They can also be found in several textbooks on centrifugal pumps. The correction factors were updated in 2004 and are now an ANSI Standard, still based on empirical data but with a theoretical background. Based on comparison to test data, the order of accuracy of the method is between +10% and –20% for kinematic viscosities up to 3,000 cSt.

Recently, an improved method has been proposed based on an analysis of the relevant losses [Gülich, 1999 a & b]. The proposed method is based on an analysis of the influence of fluid viscosity on different types of losses. A fraction of losses is virtually uninfluenced by viscosity, e.g. those resulting from internal backflows through clearances or gaps. In contrast, other losses are related to fluid frictions and increase with viscosity.

To evaluate their evolution, the method takes into account some basic parameters related to the impeller, e.g. specific speed and some characteristic geometrical data like diameter and blade length. In contrast to conventional correction charts, the approach introduces a differentiation with the geometric shape of the impeller. Nevertheless, it remains simple to apply since the geometry is described by a few selected parameters which govern the evolution of losses though non-dimensional Reynolds numbers.

The step-by-step calculation procedure is detailed in the listed references. According to published data, the accuracy of the method is ±5% for head and ±10% for efficiency.

14.6 SPECIFIC DESIGN PROVISIONS FOR SLURRIES

With positive displacement pumps, the reduction of rotation speed is the first measure taken to boost slurries. Rotation speeds may be reduced to roughly 1,200 rpm. Other current arrangements consist in designing pumps with the same channel size at discharge as at suction (ordinary discharge is smaller), and with a lined casing that is removable in case of wear, protecting bearings and seals from the intrusion of solid particles.

Special pumps are sometimes used with slurries: they are based on the formation of a vortex which boosts the fluid. The impeller is located in a recess to protect the blades from the main path of solid particles. These are called recessed impeller pumps, vortex pumps or torque flow pumps.

REFERENCES

Gülich JF (1999) *Kreiselpumpen, Ein Handbuch für Entwicklung, Anlagenplanung und Betrieb.* Springer, Berlin – English version: *Centrifugal Pumps*, Springer, September 2007 (ISBN-13: 978-3540736943).

Gülich JF (1999) Pumping Highly Viscous Fluids with Centrifugal Pumps, Parts 1 & 2, *World Pumps*, **395**, pp 9-11.

Hydraulic Institute (2004) *Effects of Liquid Viscosity on Rotodynamic (Centrifugal and Vertical) Pump Performance*, ANSI/HI 9.6.7 (ISBN: 1-880952-59-9).

Stepanoff AJ (1940) Pumping Viscous Oils with Centrifugal Pumps, *Oil and Gas Journal*, **4**, pp 123-126.

PART 4

Upgrading

A. Quignard

The oil industry has always adapted its refining tool to the evolution of crude oil characteristics and availability, and to the fuel – and petroleum-derived product market. This is especially true today with the high price reached for crude oils over the period 2007 to mid-2008, partly due to the growing demand of Asian countries, as well as the limited discovery of large new fields of conventional oils.

As shown on the Table 1, oil demand and oil consumption by product has significantly changed since the 1980's. From 1980 to 2006, statistics indicate total consumption growth of 57% with only a 10% increase in refining capacity, and a dramatic shift from heavy fuel oil to light (mainly gasoline) and middle distillates (jet fuel, diesel and heating oil).

Table 1 Evolution of the World Oil Market and Refining.

Year	1980	1985	1990	1995	2000	2005	2006
Word petroleum products Consumption (10^3 b/d)	53,236	50,849	58,248	69,830	76,280	83,080	83,719
Word oil refinery capacity (10^3 b/d)	79,158	72,920	74,526	76,186	82,265	85,929	87,238
Ratio world refinery capacity v/s World demand	67%	70%	78%	92%	93%	97%	96%
World consumption volumic ratio							
Light distillates	29.0%	31.2%	31.4%	30.8%	31.1%	31.1%	31.2%
Middle distillates	30.4%	32.7%	33.0%	33.9%	34.9%	35.9%	36.2%
Fuel oil	24.6%	17.9%	17.3%	16.2%	13.8%	12.2%	11.8%
Others	16.1%	18.1%	18.3%	19.1%	20.3%	20.8%	20.8%

Source: BP Statistical Review of World Energy 2007.

Consequently, refineries were working at full capacity in the years 2006-2008 and the need for deep conversion of heavy ends into light transportation fuels has been growing quite rapidly since the mid-1990s.

A second factor to be highlighted is that the development of heavy and extra-heavy crude oils is financially profitable and politically necessary to meet current and future fuel demand.

There is growing demand for naphtha in petrochemical uses as well as gasoline, jet fuel and diesel for ultra-low sulfur transportation fuels. On the other hand, with a very significant and irreversible decrease in heavy fuel oils and stabilization of heavy end use as bunker fuels and as asphalts for paving, the oil industry faces a major challenge regarding the processing of heavy and extra-heavy crude oils. There are two main reasons for this.

The first reason (Figure 1 and Table 2, see distillate yields) is related to the significant differences in the ratio of heavy ends to light ends between heavy or extra-heavy crude oils (i.e. Venezuelan heavy and extra-heavy crude oil) and conventional crude oils (i.e. West Texas Intermediate/WTI, Arabian Light or Ural). As show in Figure 1, it is mandatory to perform deep conversion of *heavy crude oils* and ultra-deep conversion of extra-heavy crude oils into a synthetic crude with a product distribution close to market requirements, using dedicated refining processes (one or several steps) as described in this chapter.

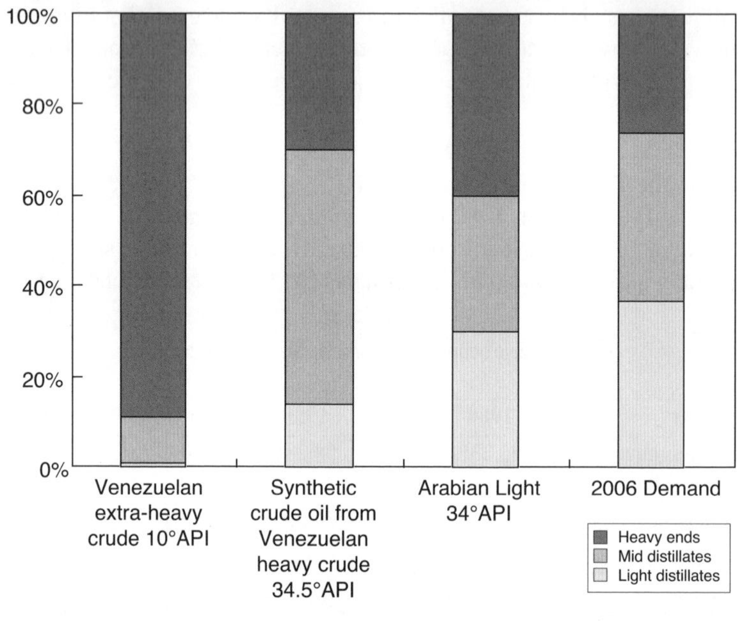

Figure 1

Example of Crude Oil Yield *versus* Product Demand.

The second reason is that the bulk composition of heavy and extra-heavy crude oils is far away from the specifications required by the market for petroleum-derived products (see Table 2 and Figure 2). The detrimental properties include high density (low API gravity) and viscosity level, very low hydrogen content and high content of high molecular weight compounds (resins and asphaltenes). This high content of heavy molecules, partially related to Conradson Carbon (a standardized method to measure the non-reactive polycondensed organic residue after applying pyrolysis conditions) and very closely related to asphaltene content, makes the heavy and extra-heavy crude oils very rich in hetero-elements such as sulfur, nitrogen and metals (mainly nickel and vanadium).

Table 2 Typical Properties of Crude Oils Based on Type and Origin.

Crude Oil	Unit	WTI	Ural	Forcados	Arabian Heavy	Duri	Maya	Cold Lake	Zuata	Boscan
Origin		USA	Russia	Nigeria	Saudi Arabia	Indonesia	Mexico	Canada	Venezuela	Venezuela
Type		extra-light	light	medium	medium	heavy	heavy	heavy	heavy	extra-heavy
API Gravity		39.8	31.3	29.5	27.3	20.8	21.2	21.2	14.8	8.9
Specific Gravity d25/25		0.8260	0.8692	0.8789	0.8911	0.9291	0.9267	0.9267	0.9672	1.0078
Elemenrary analyses										
Sulfur	wt%	0.33	1.39	0.18	2.89	0.20	3.35	3.69	2.81	5.61
Nitrogen	ppm	0.10	0.21	0.12	0.18	0.35	0.38	0.35	0.48	0.66
Hydrogen	wt%	13.59	12.60	12.70	12.25	12.25	11.58		11.10	9.90
Flow properties										
Pour Point	°C	-18	4	-9	-51	10	-29	-48	-21	13
Viscosity at 25°C	cSt	4.54	12.97	10.53	37.1	779	240	133	4,349	69,543
Viscosity at 50°C	cSt	2.50	7.40	5.20	16.9	175	66.0	40.1	305	4,190
Related to heavy ends										
Conradson Carbon	wt%	1.1	3.9	1.1	8.8	6.8	12.3	11.5	11.4	16.2
Asphaltenes	wt%	0.09	1.00	0.00	5.3	0.05	10.6	7.7	11.4	13.7
Nickel + Vanadium	ppm	4	62	5	79	33	334	218	356	1,409
Distillate yield										
Distillated at 175°C	wt%	28.9	7.9	13.4	16.5	2.9	12.5	17.7	5.5	0.6
Distillated at 370°C	wt%	67.3	51.4	64.0	44.6	25.4	37.6	36.0	28.4	16.0
Distillated at 550°C	wt%	90.0	78.3	90.6	66.5	52.4	60.0	58.7	57.1	40.0

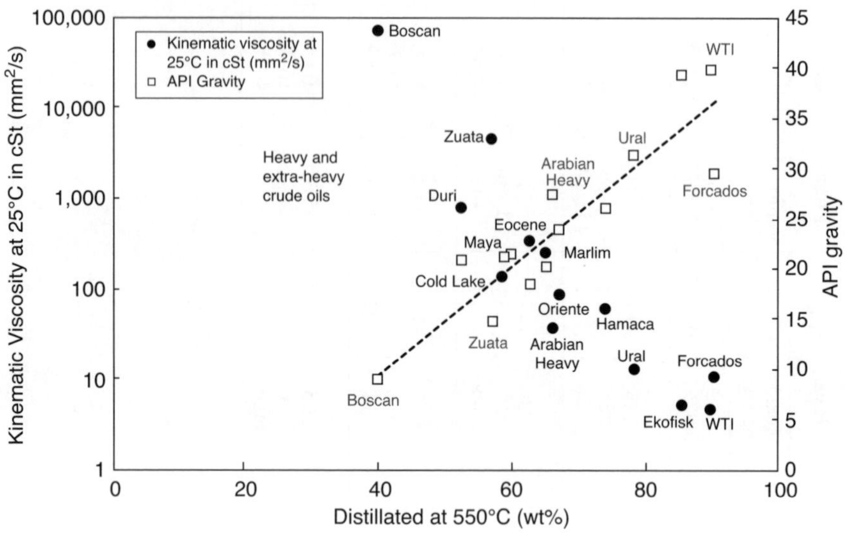

Figure 2

Crude Oil Classification According to API, Viscosity and Distillate Yield.

Subsequently, these types of crude oils need to be substantially upgraded through refinery processes.

These processes rely on two main principles:

1) Separation of the valuable part of the crude by physical processes such as solvent de-asphalting.
2) Transformation of the crude composition by deep thermal and/or catalytic processes with the contribution of added hydrogen (visbreaking, coking, hydroconversion processes, etc.).

The de-asphalting process (Chapter 15) corresponds to precipitation of the heaviest part of the oil (asphaltenes and a significant part of the resins) by the use of paraffinic solvents, which destabilize the asphaltene/resin aggregates. Since the heavy resins and asphaltenes with a high aromaticity (heavy polyaromatic structures) – and highly enriched with heteroatomic compounds such as sulfur, nitrogen and metals – are concentrated in the residual tar fraction (also called asphalt), the resulting raffinate (de-asphalted oil) is enriched in hydrogen, with saturates, naphthenes and aromatic hydrocarbons exhibiting a very low metal content and, to some extent a lower sulfur and nitrogen content. As a result this fraction is much more valuable. As far as hydrocarbon purification is concerned, this process, which is based on a chemical structure fractionation, is far more specific than classical distillation processes based on the boiling point of the compounds. By adjusting operational parameters such as the type of solvent, solvent-to-feed ratio and temperature, the quality and quantity of de-asphalted oil and residual tar can be adjusted by the operator. This makes the process a very flexible one.

Visbreaking is a mature, non-catalytic thermal process using steam injection, in which *heavy crude oils* are cracked under rather mild temperature conditions in order to increase distillate yield and reduce the viscosity of the distillation residues (Chapter 16). The thermal cracking involved in the visbreaking modifies the structure of the oil compounds. These changes imply a decrease in the size of the molecules and a subsequent decrease in the viscosity, but at the same time, aromatization of the asphaltenes and a correlative increase in the light saturate fraction of maltene that may decrease its solubility power for the asphaltenes. Such a situation potentially leads to destabilization and flocculation of the asphaltenes in the residual fraction. The engineering challenge of visbreaking is to adjust the thermal conditions in order to maximize the viscosity reduction and to avoid the onset of asphaltenes precipitation, as well as to limit the formation of gas and coke. Other thermal processes have been developed to increase the distillate yield without destabilizing the residual fraction. They may use hydrogen, hydrogen donor solvent or conditions close to or above the supercritical state of water. None of them are currently used on an industrial scale.

There are two types of visbreaking technology: the coil cracker process and the soaker process. These can be used separately or combined.

In the coil process, conversion of the feedstock is performed in a furnace at high temperature (470-500°C) during a short period of time. The soaker visbreaker achieves the conversion in a drum at a lower temperature (430-450°C) for a longer period of time. The difference of approaches relies on the kinetic nature of the cracking reaction.

Coking is a very severe non-catalytic thermal cracking process (Chapter 17). The objective is to use a disproportion reaction to recover as much hydrogen-enriched distillate as possible, and on the other hand a solid residue with a much lower hydrogen content: the coke. Moreover, the coke contains the unwanted metals and ashes together with a substantial part of the sulfur and nitrogen compounds. For commercial use, the distillates require further severe treatment or refinement in order to meet specifications, e.g. olefins, nitrogen, sulfur and aromatic content.

The two main processes that use the coking method are the delayed coking and the fluid/flexicoking.

The term "delayed" refers to the fact that the operating conditions of this semi-continuous process are designed to avoid coke formation in the main cracking furnace, and instead promote its production in large coke drums following the furnace. These drums, where the coke is deposited in large quantities, are alternatively in operation, in order to allow coke removal and collection.

Fluid coking is a continuous coke-making process based on fluid bed technology. In this process, circulating coke brings heat to the reactor and serves as reaction sites for the cracking of the heavy feedstock into lighter products. Fluid coking is very flexible and can handle any type of feedstock. It runs at a higher temperature (510°C to 540°C) and shorter residence time than delayed coking, as well as producing less coke. Liquid product yields are quite similar to those obtained in delayed coking, although the products are of lower quality.

Flexicoking is similar to fluid coking, with continuous gasification of the produced coke into low heating value fuel gas. It is very well adapted to extra-heavy oil, bitumen and tar upgrading with high sulfur content. Yields and product qualities are similar to those obtained in fluid coking. The main difference is that it virtually eliminates all solid coke to form fuel gas. This is an interesting process where there is a market for large amounts of low heating value gas.

The specifications and market demand for highly-refined light and middle distillates led to the development of catalytic hydroconversion (Chapter 18). The objective of using catalysts in the presence of high hydrogen pressure is to remove the unwanted heteroatoms (sulfur, metals and nitrogen) as well as promote hydrogenation of the product leading to higher transportation fuel and heating oil yields without coke rejection. For this purpose, various processes have been designed in order to function separately or in combination: removal of sulfur (hydrodesulfurization), removal of metals (hydrodemetallization), removal of nitrogen (hydrodenitrogenation), improvement in the hydrogen content of products (hydrogenation of aromatics and heterocyclics, secondary cracking with formation of light fraction and coke residue). These processes are operated using various technologies, mainly differing in the design for the feed and catalyst bed contact: fixed bed, moving bed, ebullated bed and slurry (entrained) bed. The latter is probably the most adapted for the heaviest feedstocks and is still at the pilot demonstration stage, but will be implemented industrially in the very near future.

Figure 3 summarizes the area of interest when using these conversion processes for upgrading heavy and extra-heavy ends, depending on the metal content of the feedstock and the conversion level to be achieved. The de-asphalting process may be combined with all of these processes, either upstream or downstream, except with the coking process, where it can only used upstream.

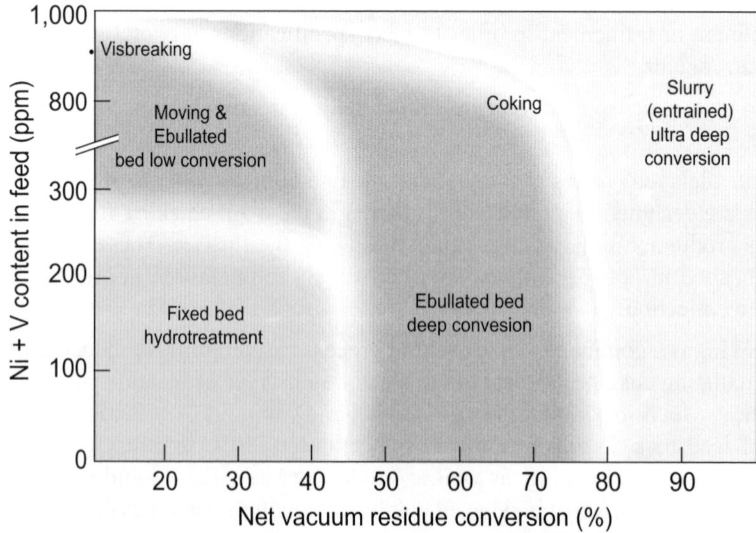

Figure 3

Refining Processes: Areas of Use for Heavy/Extra-Heavy End Conversion.

15 De-asphalting with Heavy Paraffinic Solvents

A. Quignard

The purpose of the Solvent De-Asphalting (SDA) process is to separate out a De-Asphalted Oil (DAO) concentrating the saturates, aromatics and lightest resins from a pitch, called asphalt, enriched in asphaltenes and in heaviest resins. The latter is contained in heavy petroleum feedstocks obtained after distillation of the crude oil, such as vacuum or atmospheric residua. The solvents used in the process are always light paraffinic solvents, including propane (C3), butane (C4), pentane (C5), hexane (C6) or even light straight run naphtha (C5 to C6).

The first industrial units were developed in the 1930s, based on a propane C3 SDA process for the production of high viscosity base oils to produce lubricants. Although far more commonly used than SDA with heavier solvents for conversion purposes due to the need for lubricant manufacturing, C3 SDA for base oil is beyond the scope of this book.

We will focus on SDA used as an intermediate process in the upgrading of heavy and extra-heavy oils. For this purpose, it must be stressed that the SDA process is not a conversion process in and of itself. It should always be considered as a separation process integrated with other refining processes used for the conversion of heavy ends, such as Visbreaking (Chapter 16), Coking (Chapter 17), Fluid Catalytic Cracking (FCC), Partial Oxidation/Gasification and Hydrocracking or Hydrotreatment/Hydroconversion of residua (Chapter 18).

When compared to distillation, the main separation process used in refining, the most relevant feature of the SDA process is that it allows for separation based on structural families rather than boiling point. This is a major difference with regard to reactivity *versus* the aforementioned conversion processes. Table 15.1 shows a comparison between vacuum distillation and C3 to C5 SDA with regard to yields and product quality.

Useful information on this process can be found in the references listed at the end of this chapter.

Table 15.1 Yield and Quality of Vacuum Distillation and SDA Products from Athabasca Heavy Crude (AR = Atmospheric Residue, DAO = De-Asphalted Oil, VGO = Vacuum Gas Oil).

Products	Unit	AR 375°C+	C3 DAO	C4 DAO	C5 DAO	VGO 343-524°C	VGO 375-500°C
Position on crude	wt%	21.5-100				18.3-50.9	
Yield on crude	wt%	78.5	23.0	40.3	74.6	31.7	20.00
Density at 15°C	kg/l	1.038	0.961	0.984	1.005	0.977	0.980
Viscosity at 100°C	mm^2/s	1,300	30	54	150	14.2	15
Conradson Carbon	wt%	16.7	1.5	3.9	8.9	0.4	0.2
C5 Asphaltene	wt%	18.0	0.05	0.05	0.76		
C7 Asphaltene	wt%	10.2	0.05	0.05	0.08	0.02	< 0.02
Nickel	wt ppm	101	1	4	34	< 2	< 2
Vanadium	wt ppm	280	2	6	101	< 2	< 2
Nitrogen	wt ppm	5,700	2,700	3,000	3,067	1,864	2,500
Sulfur	wt%	4.94	2.10	3.40	4.55	3.47	3.60

15.1 SOLVENT DE-ASPHALTING (SDA): PURPOSE AND PRINCIPLE

The Solvent De-Asphalting (SDA) operation is carried out in an extractor where the feed is placed in contact with a solvent that promotes precipitation of the asphaltic fraction. The efficiency of the separation between the oil and the asphaltic fractions is increased by operating with a temperature gradient between the top and bottom of the extractor. Light saturated hydrocarbons like propane, butane, pentane or hexane, are commonly used to precipitate the asphaltic fraction, called asphalt or pitch. This asphalt concentrates most of the impurities of the feedstock. Consequently, impurities in the oil – called De-Asphalted Oil (DAO) – are significantly reduced.

C3 Solvent De-Asphalting (SDA) is well-suited when the highest quality De-Asphalted Oil (DAO) is required on extra-heavy residua, but as explained in Chapter 15.2, DAO yield will always be very low. C4 SDA is used to combine a high-quality DAO with a more acceptable yield. Finally, C5, C6 or light gasoline SDA is preferred to maximize DAO yield but with a higher impurity level, such as metal content or residual carbon (also known as Conradson Carbon). It is also possible to combine two solvents to get intermediate DAO quality and yield, such as C3 + C4. The combination of heavier solvents, such as C5 + C6, is also technically and commercially feasible, but with minor differences *versus* DAO quality to yield.

15.1.1 Brief Description

A simplified flowsheet is shown in Figure 15.1. It indicates the de-asphalting step, as well as the separation steps to recover the solvent from the DAO, the asphalt product and solvent recycling. This solvent recovery step is very important from an economic standpoint, as explained in Section 15.4

As previously discussed, the SDA process should always be considered as a separation process, integrated with other refining processes used for the conversion of heavy or extra heavy residua. An example of integration is given in Figure 15.2. The SDA process can be used to purify the heavy or extra-heavy residua in order to blend the DAO with the straight run vacuum distillate before catalytic conversion processes, such as Fluid Catalytic Cracking (FCC), Residue Fluid Catalytic Cracking (RFCC) or Hydrocracking. The DAO may also be directly fed to a fixed or ebullated bed residue hydrotreatment or to any thermal cracking processes, such as Visbreaking or Hydrovisbreaking, with a much higher reactivity than the corresponding residue before SDA. The produced asphalt may typically be used as an extra-heavy fuel oil for combustion or in power plants. One upgrading route is to use it as a solid fuel to replace petroleum coke or coal with a much higher hydrogen content, as a feed to a gasifier for syngas ($CO + H_2$) or hydrogen production.

Further details regarding SDA integration in a refinery scheme will be discussed in Section 15.4.

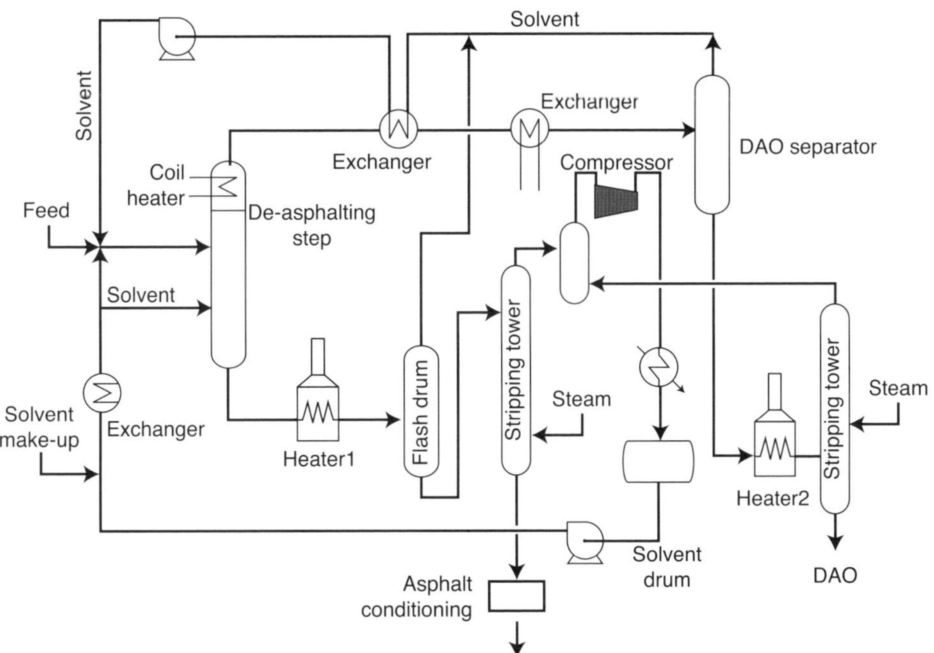

Figure 15.1

Simplified Scheme of SDA (Solvahl) Process.

Source: Resid and Heavy Oil Processing, Lepage JF *et al.*, Editions Technip, 1992.

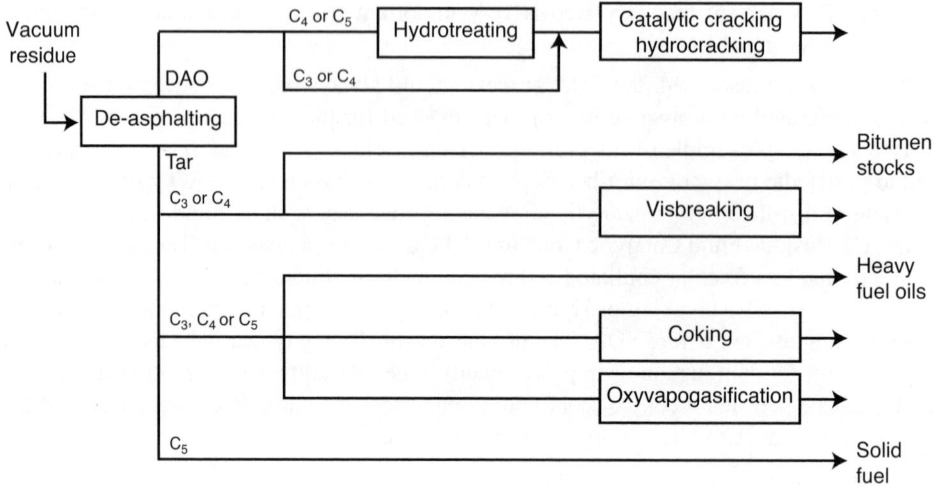

Figure 15.2

Example of SDA Incorporation in a Conversion Refining Scheme (Solvahl Process).

Source: : Resid and Heavy Oil Processing, Lepage JF *et al.*, Editions Technip, 1992.

15.1.2 Feed

Feed for de-asphalting is usually made up of a mixture of overflash and residue from the vacuum distillation. All residues may be considered as a colloidal system of high molecular weight compounds, classified into three categories:

- The first category – the oil medium – is the lightest part including paraffinic and cycloparaffinic structures (saturated compounds) with aromatic structures plus sulfur species, but without other impurities such as nitrogen and metals. These molecules are typically made up of 40-60 carbon atoms with a molecular weight of roughly 700.
- The second category is resins. They include much more condensed aromatic molecules with rather long aliphatic side chains, making them soluble in the oil medium. They also include sulfur and nitrogen species and metals (mainly nickel and vanadium). They typically have a molecular weight of approximately 1,000 and are made up of more than 60-80 carbon atoms. Because of their high aromaticity, these compounds are essential in the stability of the colloidal system resulting from the interaction between the oil medium and the asphaltenes (peptization of asphaltenes).
- The oil medium and the resin compounds are generally grouped together under the heading of maltenes.
- Asphaltenes have even more condensed aromatic structures than resins, including 6 to 20 aromatic rings with rather short paraffinic and cycloparaffinic side chains. Their average molecular weight is higher than 1,000. They concentrate the heteroatoms, such as sulfur, nitrogen, oxygen and collect most of the metals, such as nickel and

vanadium. As a consequence, they are not soluble in the oil medium. They are present in the form of tiny particles (several angstroms) and are "peptized" by the resins, which give them pseudo-solubility in the oil medium. This is the why resins are fundamental in creating a stable or unstable colloidal system.

A Straight Run residue is always stable because the asphaltenes are all "naturally" peptized by the resins. On a macroscopic scale, it looks like a perfectly homogeneous solution. But if the structures of the resins or the asphaltenes – or both – are modified by a thermal or catalytic treatment, or as a result of operating conditions such as pressure and temperature, this equilibrium may be broken and the asphaltenes can precipitate. These asphaltenes are said to be flocculated, in the form of large agglomerations of particles that can reach the size of several microns.

A typical composition of main chemical families (saturates: paraffins and naphthenes, aromatics, resins and C7 asphaltenes) *versus* boiling point is shown in Figure 15.3. Typical atmospheric residue (AR 350°C$^+$) and vacuum residue (VR 550°C$^+$) windows are also given.

This figure shows that the content of saturates and aromatics decreases very rapidly in the residua zone, with a drastic increase in the content of heavy hetero-polyaromatic structures: resins and asphaltenes. These heavy hetero-polyaromatic structures concentrate impurities such as nitrogen and metals.

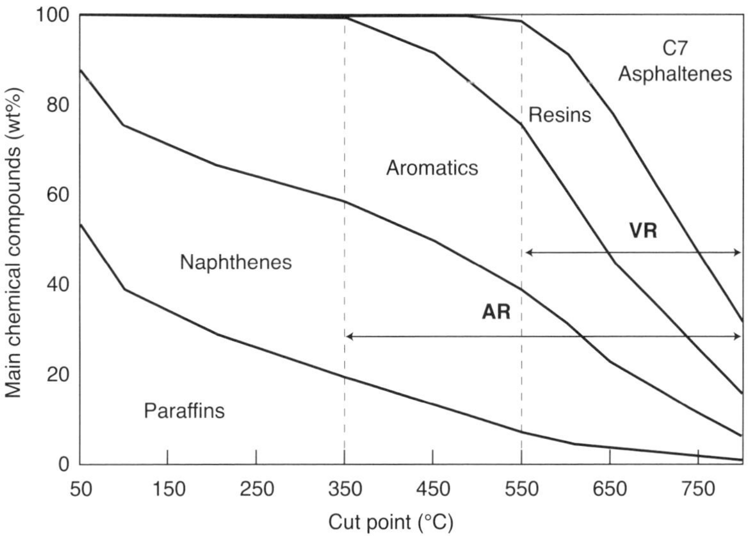

Figure 15.3

Typical Composition of a Crude Oil (AR = Atmospheric Residue VR = Vacuum Residue).

Examples of typical SDA feedstocks are given in Table 15.2 for Straight Run (SR) conventional and heavy oils from the Middle East, and in Table 15.3 for extra-heavy oils from Venezuela and Canada.

Table 15.2 Typical SDA Feedstocks (vacuum residua) from Conventional Heavy Oils.

Feedstock	Unit	Murban	Arabian Light	Kuwait export	Arabian Heavy	Buzurgan (Irakian Heavy)
Origin		UAE	Saudi Arabia	Kuwait	Saudi Arabia	Irak
Position on crude	wt%	90-100	79-100	72-100	67-100	48-100
Density at 15°C	kg/l	0.982	1.003	1.021	1.035	1.051
Viscosity at 100°C	mm^2/s	195	345	1,105	4,280	3,355
Conradson Carbon	wt%	12.9	16.4	15.4	22.5	22.5
C7 Asphaltene	wt%	1.2	4.2	5.2	12.2	18.4
Nickel	ppm	17	19	33	45	76
Vanadium	ppm	26	61	87	151	233
Nitrogen	ppm	3,180	2,875	3,850	4,980	4,500
Sulfur	wt%	3.03	4.05	5.35	5.28	6.20

Table 15.3 Typical SDA Feedstocks (atmospheric residua AR & vacuum residua VR) from Extra-Heavy Oil.

Feedstock	Unit	Cold Lake	Athabasca	Athabasca	Cerro-Negro	Morichal	Boscan
Origin		Alberta Canada	Alberta Canada	Alberta Canada	Venezuela Faja Orinoco	Venezuela Faja Orinoco	Venezuela Anzoategui
Type		AR	AR	VR	AR	AR	AR
Density at 15°C	kg/l	1.033	1.037	1.070	1.032	1.035	1.037
Viscosity at 100°C	mm^2/s	3,508	2,300	30,640	3,385	3,602	5,250
Conradson Carbon	wt%	18.3	16.7	21.9	20.4	19.3	18.0
C7 Asphaltene	wt%	14.2	7.3	14.9	10.9	13.2	15.3
Nickel	ppm	90	93	137	157	113	133
Vanadium	ppm	190	290	337	510	545	1,264
Nitrogen	ppm	4,600	6,010	6,000	8,240	nd	7,880
Sulfur	wt%	5.04	4.87	5.40	4.20	4.00	5.90

15.1.3 SDA Principle

In the oil, asphaltenes may be present in different forms as isolated tiny particles solvated by resins, as clusters of a few molecules, or as colloidal "large" micelles resulting from agglomeration of the clusters. The asphaltenes are in a metastable equilibrium with the surrounding maltene environment. The gradual addition of a saturated hydrocarbon, e.g. a light paraffinic solvent, breaks up this equilibrium and precipitates the asphalt phase. Under industrial SDA conditions (solvent-to-feed ratio > 2, temperature close to the critical limit), precipitation phenomena are very fast.

Operating parameters such as temperature and solvent-to-feed ratio, as well as solvent type, are the main factors influencing this precipitation and the quality of the separation between the extracted oil and the asphaltic phases. The lighter the solvent, the higher the asphalt yield; thus, the better the quality of the produced extracted oil or DAO. For a given solvent and solvent-to-feed ratio, a high extraction temperature increases DAO quality but decreases its yield. Variations in the oil-phase yield are even more sensitive, as the critical temperature of the solvent/oil mixture is approached. Pressure is not considered to be an operational variable, because it is related to the solvent type in order to achieve a liquid phase under extractor operating conditions and ensures that work is performed outside the critical point.

The influence of all these parameters is illustrated in Figure 15.4:

- Figure 15.4a illustrates the effect of solvent molecular weight on the theoretical extraction level (infinite dilution): the higher the molecular weight of the solvent, the lower the asphalt yield and the heavier the molecular weight of the DAO, with higher Conradson Carbon and more impurities concentrated in the resin fraction, e.g. sulfur, nitrogen or metals.
- Figure 15.4b illustrates the effect of temperature in the extractor on the theoretical extraction level: the higher the temperature, the lower the DAO yield, but the lower the impurity level as well.
- Figure 15.4c illustrates the effect of the solvent-to-feed ratio using a given solvent: the higher the ratio, the better the separation between asphaltenes and maltenes respectively in the asphalt and DAO fractions. At an infinite dilution ratio the separation is perfect.

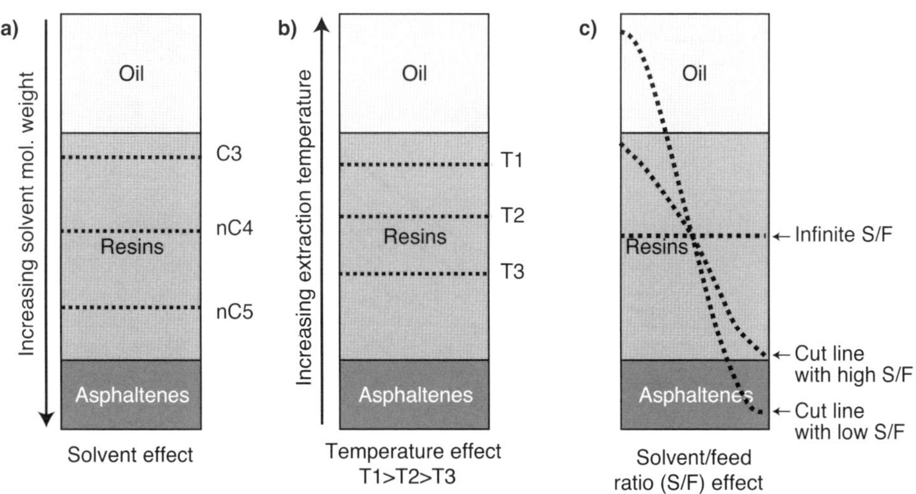

Figure 15.4

Influence of Operating Conditions on DAO/Asphalt Cut Point.

15.1.4 SDA Compared to Vacuum Distillation

Fractionation during an SDA operation occurs on the basis of structure rather than temperature or molecular weight. This main difference compared with a distillation operation is the advantage of SDA. In Table 15.4 and Figure 15.5, we provide a comparison of the properties of cuts obtained by vacuum distillation and pentane SDA of a Kirkuk conventional vacuum residue.

For a given yield (e.g. 22%), the vacuum distillate has higher Conradson Carbon (4%) and far more impurities (6 ppm metals) than the corresponding de-asphalted oil obtained at the same yield (no metals and less than 1% Conradson Carbon).

Table 15.4 Comparison between Straight Run Vacuum Distillates (VGO) and De-Asphalted Oil (DAO) from a Kirkuk Vacuum Residue.

Feedstock	Unit	VR	VG01	VG02	VG03	DA01	DA02	DA03
Yield on VR	wt%	100.0	21.9	35.2	48.3	21.6	48.6	88.2
Cut Point/SDA Solvent	°C	530°C$^+$	500-600	500-650	500-700	C3	C4	C5
Density at 15°C	kg/l	1.019	0.966	0.968	0.975	0.910	0.946	0.970
Viscosity at 100°C	mm^2/s	600	33	44	65	22	51	104
Conradson Carbon	wt%	18.0	4.0	5.6	6.6	0.8	3.4	7.8
Nickel	ppm	50.0	2.0	8.0	17.0	0.2	2.8	13.0
Vanadium	ppm	118.0	4.0	16.0	32.0	0.1	2.6	1.0

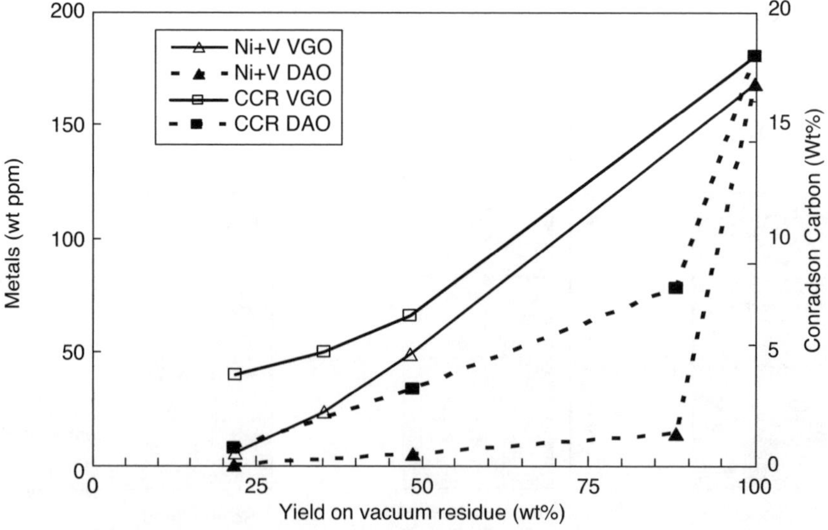

Figure 15.5

Comparison of Metal Content (Ni + V) on Vacuum Distillates (triangle) and DAO (square) from a Kirkuk Vacuum Residue.

For the same Conradson Carbon and impurities level (metals), the yield in de-asphalted oil is more than double (roughly 50%) the corresponding vacuum distillate (see Table 15.4).

We present a comparison of structure analyses obtained by C^{13} NMR of an Arabian Light butane de-asphalted oil and an Arabian Light vacuum distillate in Table 15.5. It can be seen that de-asphalted oil concentrates the aliphatic carbon in the DAO along with hydrogen, the content of which becomes even higher than that of the vacuum distillate, which is nonetheless much lighter.

The same comparison held for Athabasca extra-heavy Atmospheric Residue (AR) between the SR VGO C3, C4 and C5 DAO in Table 15.6. It can be seen again that the C3 and C4 de-asphalted oils saturated content is higher than in the Straight Run (SR) Vacuum Gas Oil (VGO) with higher hydrogen content and higher yields on crude (twice the SR VGO yield on the C4 DAO). The C5 DAO is slightly more aromatic than the SR VGO, but with a yield on crude three times higher.

In addition to high selectivity and as compared to distillation, the SDA process results demonstrate its ability to produce purified oils that can be more easily value-enhanced by conversion processes.

Table 15.5 Comparison of Structure of a Vacuum Distillate and a C4 DAO from an Arabian Light Vacuum Residue.

Products	Unit	VR	VGO	C4 DAO
Yield on crude	wt%	19	25	10
Density at 15°C	kg/l	1.019	0.930	0.945
Viscosity at 100°C	mm^2/s	600	9.8	55
Hydrogen	wt%	10.4	11.9	12.1
Saturated C (C13 NMR)	wt%	65.8	79.8	82.0
Aromatic C (C13 NMR)	wt%	34.2	20.2	18.0

Table 15.6 Comparison of Structure of a Vacuum Distillate and a C4 DAO from an Athabasca Atmospheric Residue.

Products	Unit	AR 375°C$^+$	VGO 375-500°C	C3 DAO	C4 DAO	C5 DAO
Yield on crude	wt%	78.5	20	23.0	40.3	59.5
Yield on AR	wt%	100	40	29.3	51.3	75.8
Density at 15°C	kg/l	1.038	0.980	0.961	0.984	1.005
Viscosity at 100°C	mm^2/s	1,300	15.0	30.0	54.0	150
Conradson Carbon	wt%	16.7	< 0.1	1.5	3.9	8.9
C5 Asphaltene	wt%	18.00	0.1	< 0.05	< 0.05	0.80
Saturates	wt%	20.0	28.5	40.3	30.2	24.5
Aromatics	wt%	36.0	52.0	49.5	49.3	47.7
Resins	wt%	33.8	19.5	10.2	20.5	27.8
C7 Asphaltene	wt%	10.2	< 0.05	< 0.05	< 0.05	0.08
Metals (Nickel + Vanadium)	wt ppm	381	0.7	3	10	135

15.1.5 Solvent Effect

The most important process parameter is the nature of the SDA solvent. This is what primarily determines the yield and qualities of the extracted oil phase. The heavier the paraffinic solvent, the higher the yield in maltenic phase, at the expense of its quality as evidenced by parameters such as metals, Conradson Carbon, density, hydrogen content and viscosity. The influence of solvent type on the oil/resins/asphaltenes separation is schematically shown in Figure 15.6. The heavier the molecular weight of the solvent; the heavier the molecular weight of the resin fraction in the de-asphalted oil, and the higher the asphaltene concentration in the asphalt.

For example, propane, butane and pentane exhibit solvent power for soft and hard resins, while ethane precipitates not only asphaltenes and resins but also a large fraction of the oil medium. Propane proves to be particularly well-adapted to the production of lubricants, butane for the production of a very high-quality DAO directly suitable as a Residue Fluid Catalytic Cracking or Hydrocracking feedstock, and pentane for the production of a less-reactive DAO with more impurities but with a very high yield, especially well-suited for any conversion refining scheme including a preliminary hydrotreatment step to reduce metals and Conradson content (Table 15.7).

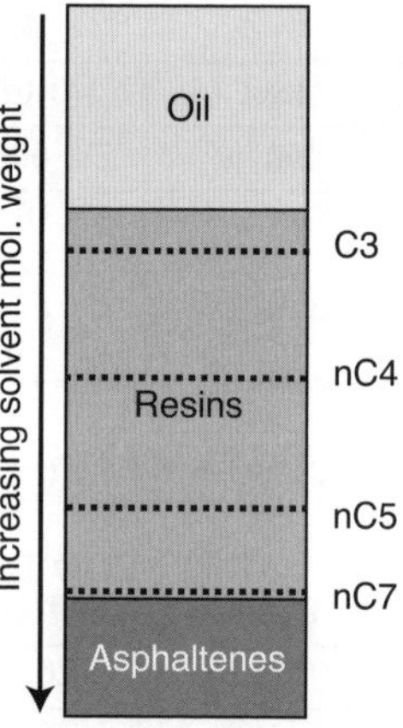

Figure 15.6

Influence of Solvent Type on SDA Cut Point.

The higher the molecular weight of the aliphatic solvent, the higher the extraction of heavy polar resins, resulting in a dramatic increase in DAO yield with a more polyaromatic overall structure, and with higher Conradson Carbon, higher metal content and, to some extent, greater sulfur and nitrogen content.

This solvent type effect on the extracted maltenic phase yield is quantified in Figure 15.7 for different feedstocks: Straight Run (SR) Poso Creek atmospheric residue, SR Arabian

Table 15.7 Effect of Solvent Type on an Athabasca Atmospheric Residue: Comparison to DAO and Vacuum GasOil (VGO) Qualities and Yields.

Products	Unit	AR 375°C+	VGO 375-500°C	C3 DAO	C4 DAO	C5 DAO
Yield on crude	wt%	78.5	20	23.0	40.3	59.5
Yield on AR	wt%	100	40	29.3	51.3	75.8
Density at 15°C	kg/l	1.038	0.980	0.961	0.984	1.005
Viscosity at 100°C	mm^2/s	1,300	15.0	30.0	54.0	150
Conradson Carbon	wt%	16.7	< 0.1	1.5	3.9	8.9
C5 Asphaltene	wt%	18.00	0.1	< 0.05	< 0.05	0.80
Saturates	wt%	20.0	28.5	40.3	30.2	24.5
Aromatics	wt%	36.0	52.0	49.5	49.3	47.7
Resins	wt%	33.8	19.5	10.2	20.5	27.8
C7 Asphaltene	wt%	10.2	< 0.05	< 0.05	< 0.05	0.08
Metals (Nickel + Vanadium)	wt ppm	381	0.7	3	10	135
Nickel	wt ppm	101	0.5	1	4	34
Vanadium	wt ppm	280	0.2	2	6	101
Nitrogen	wt ppm	5,700	2,500	2,700	3,000	3,067
Sulfur	wt%	4.94	3.6	2.1	3.4	4.55
Carbon	wt%	85.2	85.9	86.0	85.7	85.2
Hydrogen	wt%	9.3	10.3	11.6	10.6	9.9

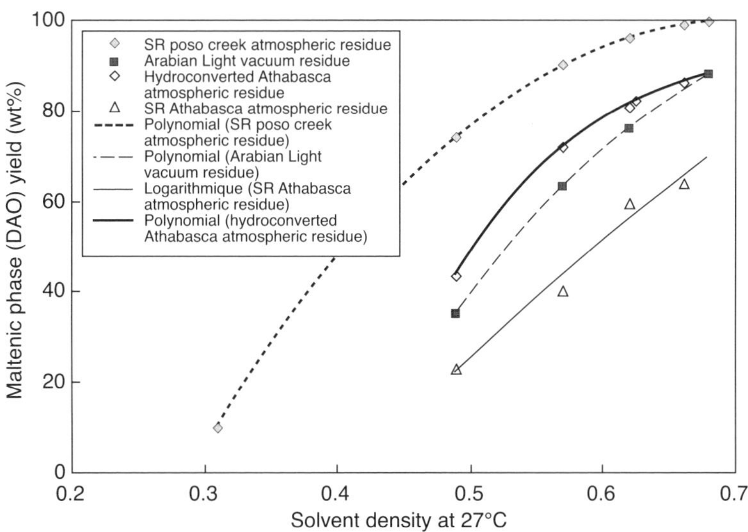

Figure 15.7

Influence of Solvent Type on Extraction of the Maltenic Fraction (DAO) for Various Feedstocks.

Light vacuum residue, hydroconverted Athabasca atmospheric residue and SR Athabasca atmospheric residue. The overall trend is the same, regardless of the feedstock.

15.2 DE-ASPHALTING PROCESS

15.2.1 Extraction Principle

There are three steps in the SDA extraction process, corresponding to three physical zones in the extractor (Figure 15.8):

- Precipitating the asphaltic phase (resins + asphalt) in the medium (feeding) zone.
- Deresining (light resins at the top with the DAO, entrained heavy resins precipitated down to the bottom with the asphalt) in the top zone.
- Settling of the asphalt plus re-extraction of light resins in the bottom zone.

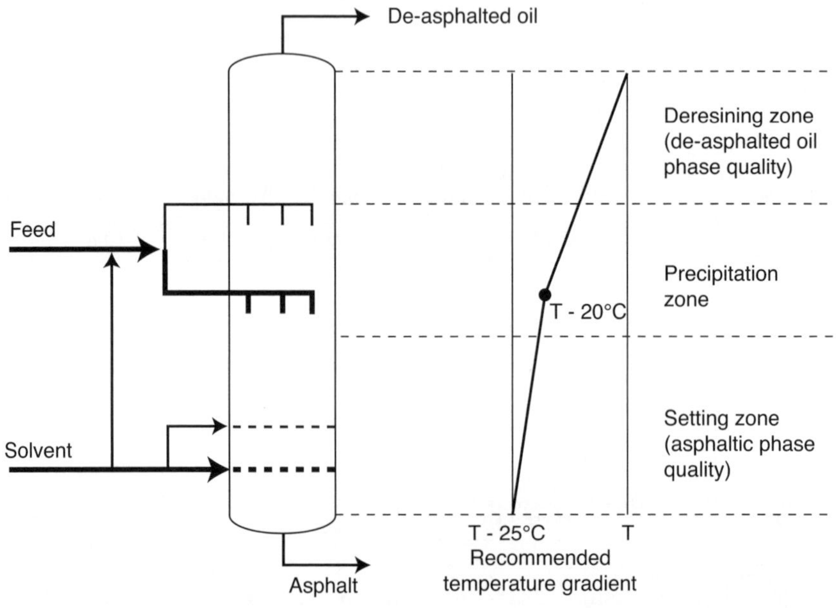

Figure 15.8

Principle of SDA Extractor Operation in Three Zones.

Source: Petroleum Refining, Vol. 2 Separation Processes, Wauquier JP, Editions Technip, 2000.

15.2.1.1 Precipitation Step

Precipitating the asphalt means breaking the equilibrium within the colloidal system between maltene and asphaltene media, as described previously.

The most important variables in this precipitation are:

- Type of solvent (molecular weight).
- Solvent-to-feed ratio.
- Temperatures (gradient plus absolute levels).

The influence of these parameters is discussed in the next paragraph. The operating pressure is always chosen so that it is higher than the solvent's critical pressure. As discussed, this is not an operating parameter than can actually be adjusted.

In an industrial de-asphalting process, precipitation of the asphalt fraction does not induce an actual flocculation of the asphaltenes because the resin content remains high and the remaining resins have high peptizing power for asphaltenic moities. Hence, the asphaltenes remain pseudo-peptized.

The precipitation zone is generally made up of:

- Pre-dilution zone, upstream from the extractor, whose main objectives are to reduce feed viscosity and approach feed saturation with the solvent.
- Feed inlet to the extractor, where the separation takes place between the DAO and the pitch (asphaltic phase) in contact with the solvent fraction, which is introduced at the bottom of the extractor and tends to rise because of its much lower specific gravity.

15.2.1.2 Deresining Step

Deresining consists in generating an internal reflux that improves the separation between the oil medium and the resins.

The solvent/oil mixture is heated at the top of the extractor by steam coils. The resinous fractions – at the least the heaviest – are precipitated downwards in the extractor, where the rising counter-current flow of solvent/oil mixture re-dissolves the lightest resins at a lower temperature. The higher the temperature in the deresining zone (resulting in lower viscosity), the better the separation between oil and resins, resulting in higher quality DAO but with a lower yield. The highest temperature limit is determined by the critical temperature of the solvent. The temperature gradient set in the deresining zone allows the precipitation-redissolution cycle to be maintained. Settling of resinous particles is also promoted by an increase in the solvent-to-oil ratio. The theoretical fundamentals are described in the literature [Wauquier, 2000].

The design of this deresining zone is very important to obtain good DAO quality. It must account for:

- Minimum residence time for the solvent + oil + resins medium to allow even the smallest resinous particles to separate out.
- Maximum rising velocity, depending of the geometry of this zone and use or internals, for the solvent + oil mixture to allow even the smallest resinous particles to settle out.

Along with the constraints of maximum extraction column diameter (up to 7 m), the last two rules limit de-asphalting capacity to roughly 1.5 Mt/year (30,000 BPSD: Solvahl process).

15.2.1.3 Settling Step

Settling the asphalt consists in washing the asphalt emulsion in the solvent + oil mixture with a counter-current flow of pure solvent.

The settling zone is located between the feed inlet and the bottom of the extractor. Stripping of the asphalt is promoted by an increase in the solvent ratio (i.e. replacing the solvent + oil medium with pure solvent), at the lowest possible temperature (increasing the solvent specific gravity and consequently increasing the solubility of the resinous fractions in the solvent).

A typical recommended temperature gradient is indicated in Figure 15.8. It must allow the overhead reflux in the column to be maximized in order to improve DAO quality and promote stripping of the asphalt in the bottom of the reactor.

The minimum temperature at the bottom of the settling zone is the result of the maximum admissible viscosity for the pitch + solvent to be pumped outside the extractor. The efficiency of asphalt stripping zone can be significantly modified, depending on the choice of contacting internal equipment.

The absolute temperature level varies depending on the solvent type, but is within the 50/60°C to 220/230°C range.

As previously mentioned, the main operating parameters for good separation are temperature level and gradient, as well as solvent/feed ratio.

15.2.2 Operating Variables

15.2.2.1 Feed Type

A. Effect of Cut Point

For a given set of operating conditions (temperature gradient in the extractor, solvent type and ratio), an increase in light fraction content in the feedstock residual fraction results in increased DAO yield at the expense of quality. This is because the relative volume of solvent *versus* resins and asphaltenes decreases for a given solvent ratio. Table 15.8 shows that Cold Lake, a Canadian extra-heavy feedstock that resembles Athabasca crude, illustrates such an effect between de-asphalting of the crude and of the atmospheric residue with C3 and C5 solvent at 8: 1 solvent to feedstock ratio. The demetallization rate is higher for the DAO from the atmospheric residue, and despite dilution of the DAO with the atmospheric distillate fraction in the DAO from the crude, the other purification levels (sulfur, nitrogen, Conradson Carbon and asphaltenes) are not much better than for the DAO from the atmospheric residue.

B. Effect of Feed Type

Table 15.9 for C3 solvent and Table 15.10 for C5 solvent illustrate these effects, based on the feedstocks previously presented in Table 15.2. Except for sulfur, metals and nitrogen content, the characteristics of the DAO are very similar, regardless of feedstock origin. The main differences lie in DAO yields since they are closely related to feedstock composition: saturates, aromatics, resins and asphaltenes.

Table 15.8 Effect of Cut Point on SDA Performance for a Cold Lake Feedstock.

Products	Unit	Crude	AR °C+	C3 DAO from crude	C3 DAO from AR	C5 DAO from crude	C5 DAO from AR
Yield on crude	wt%	100		51.4		74.0	
Yield on AR	wt%		100.0		48.0		72.6
Density at 15°C	kg/l	0.992	1.035	0.949	0.951	0.970	0.980
Viscosity at 100°C	mm^2/s	95	187	15.5	20.1	31.9	53.7
Conradson Carbon	wt%	11.0	13.5	1.6	1.8	4.6	6.8
C7 Asphaltene	wt%	7.5	7.8	0.2	0.4	0.18	0.12
Metals (Nickel + Vanadium)	wt ppm	204	293	6.7	4	51	44
Nitrogen	wt ppm	3,900	4,815	1,850	1,870	3,200	3,600
Sulfur	wt%	4.19	4.36	3.16	3.08	3.82	3.98
Purification levels (wt% vls feed)							
Metals (Nickel + Vanadium)				98%	99%	75%	85%
Nitrogen				62%	61%	18%	25%
Sulfur				28%	29%	9%	9%

Table 15.9 Effect of Feedstock Type on C3 SDA Performance.

Feedstock	Unit	Murban	Arabian Light	Kuwait export	Buzurgan (Irakian Heavy)	Boscan
Yield on crude (feedstock)	wt%	10.0	21.0	28.0	52.0	78.0
C3 DAO yield on feedstock	wt%	54.0	45.0	29.0	31.0	33
Density at 15°C	kg/l	0.924	0.933	0.937	0.945	0.953
Viscosity at 100°C	mm^2/s	38.0	34.9	46.0	30.0	33.0
Conradson Carbon	wt%	2.0	1.7	1.9	2.0	1.8
C7 Asphaltene	wt%	< 0.05	< 0.05	< 0.05	< 0.05	< 0.05
Nickel	ppm	< 1	1	1	2	6
Vanadium	ppm	< 1	1.4	2	4	10
Nitrogen	ppm	800	1,200	1,300	930	1,600
Sulfur	wt%	1.80	2.55	2.70	3.60	4.70

15.2.2.2 Solvent Type

As mentioned earlier in Paragraph 15.1.5, the heavier the de-asphalting solvent, the higher the DAO yield at the expense of its quality. In Paragraph 15.2.2.1 (Tables 15.9 and 15.10), we discussed typical properties including Conradson Carbon, sulfur, nitrogen, and metal content associated with the type of feed.

Table 15.10 Effect of Feedstock Type on C5 SDA performance.

Feedstock	Unit	Murban	Arabian Light	Kuwait Export	Buzurgan (Irakian Heavy)	Boscan
Yield on crude (feedstock)	wt%	10.0	21.0	28.0	52.0	78.0
C3 DAO yield on feedstock	wt%	93.0	85.5	78.0	64.2	64.0
Density at 15°C	kg/l	0.970	0.974	0.986	0.985	0.987
Viscosity at 100°C	mm^2/s	124	105	201	78	135
Conradson Carbon	wt%	9.0	7.9	8.9	7.4	6.4
C7 Asphaltene	wt%	< 0.05	< 0.05	< 0.05	< 0.05	< 0.05
Nickel	ppm	8	7	12	12	36
Vanadium	ppm	9	15.5	28	26	270
Nitrogen	ppm	1,560	2,170	2,070	2,350	4,330
Sulfur	wt%	2.85	3.65	4.70	4.80	5.30

The values of these parameters *versus* the solvent type – from C3 to C5 – are illustrated with an Arabian Heavy vacuum residue feedstock in Figure 15.9. The overall trend is always the same regardless of the feed. Moreover, the heavier the solvent, the harder the produced asphalt, as exemplified with an Arabian Heavy vacuum residue in Figure 15.10, which shows how the asphalt softening point [1] changes with solvent type and DAO yield. When solvent molecular weight is increased, the asphaltenes-to-resins ratio in the asphalt product also increases, resulting in a harder asphalt (also see explanation in Paragraph 15.1.5, Figure 15.6).

Solvent type is the main process parameter used to control DAO yield and purification level. It also dictates the minimum operating pressure, which is slightly higher than the solvent's critical pressure, i.e. 30-40 bars for C3/C4 and 20-40 bar for C5/C6, which is not an operating variable since it is fixed (little effect on separation selectivity).

15.2.2.3 Solvent-to-Feed Ratio

SDA operation selectivity can be significantly improved by increasing the solvent ratio: this is the main variable allowing for improved DAO quality. In general, when the solvent ratio is increased, the DAO yield decreases to a value compatible with a settling step, then grows until it reaches an asymptotic value (Figure 15.11). An increase in DAO yield causes harder asphalt to be produced, as discussed previously.

Figure 15.12 shows how the solvent ratio effect influences extraction selectivity.

1. The softening point is the temperature at which a metallic ball passes through a calibrated ring. The higher the temperature, the harder the asphalt. The softening point is closely related to viscosity. It is used when viscosity cannot be measured because the temperature is too high.

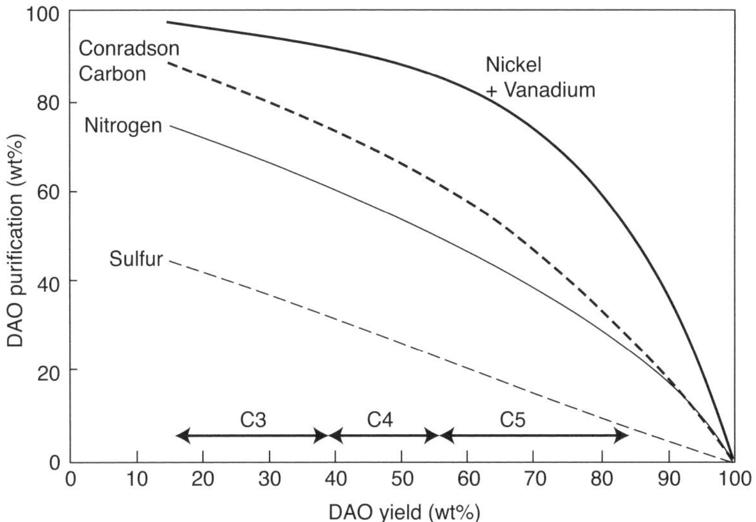

Figure 15.9

Typical Influence of Solvent Type on the Purification Level of a DAO from Arabian Heavy Vacuum Residue.

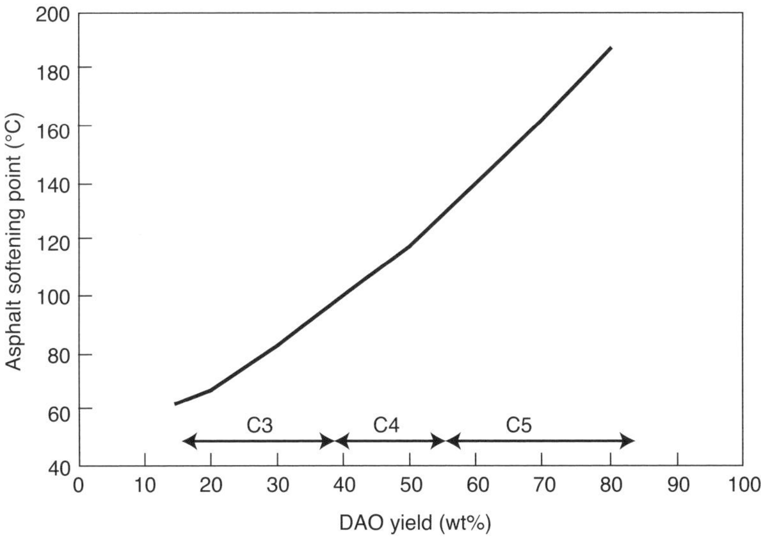

Figure 15.10

Typical Influence of Solvent Type on Asphalt Softening Point for Arabian Heavy Vacuum Residue.

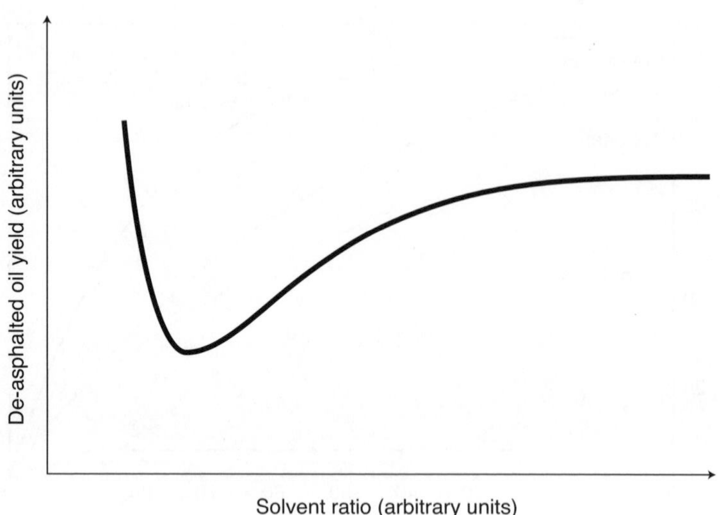

Figure 15.11

Effect of Increased Solvent Ratio on DAO Yield.

Source: Petroleum Refining, Vol. 2 Separation Processes, Wauquier JP, Editions Technip, 2000.

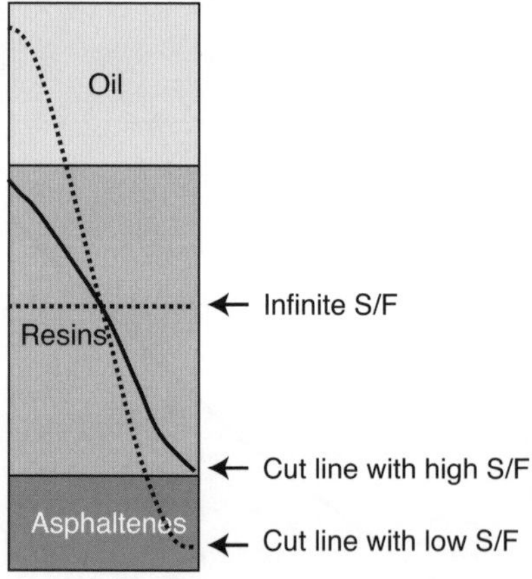

Figure 15.12

Influence of Solvent-to-Feed Ratio (S/F) on Extraction Selectivity.

Thus, the solvent ratio has a significant influence on purification levels, especially for asphaltenes and metals, as shown on Figures 15.13 and 15.14 for an Arabian Heavy DAO. To achieve a high purification level for asphaltenes, which is compatible with further refining process upgrading such as Hydrotreatment or Hydrocracking (i.e. less than 0.05% C7 asphaltenes), a minimum Solvent-to-Feed (S/F) ratio of 4: 1 is required, and a typical industrial level is 5: 1.

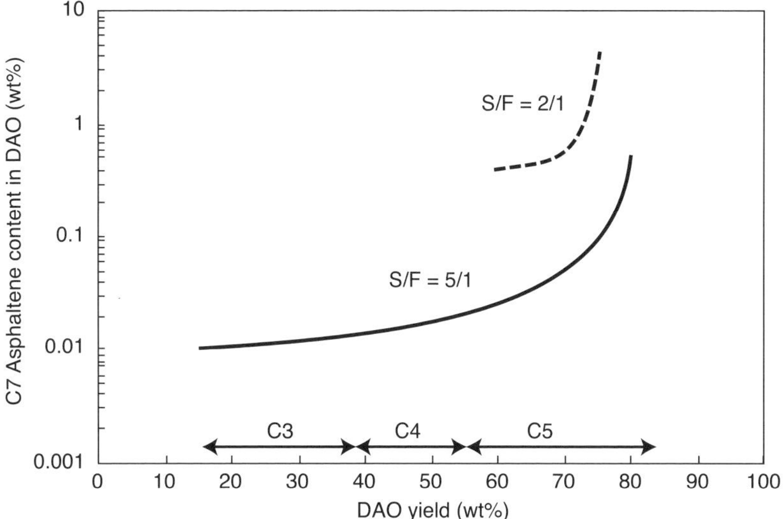

Figure 15.13

Solvent-to-Feed Ratio (S/F): Effect on C7 Asphaltene Content of a DAO from Arabian Heavy Vacuum Residue.

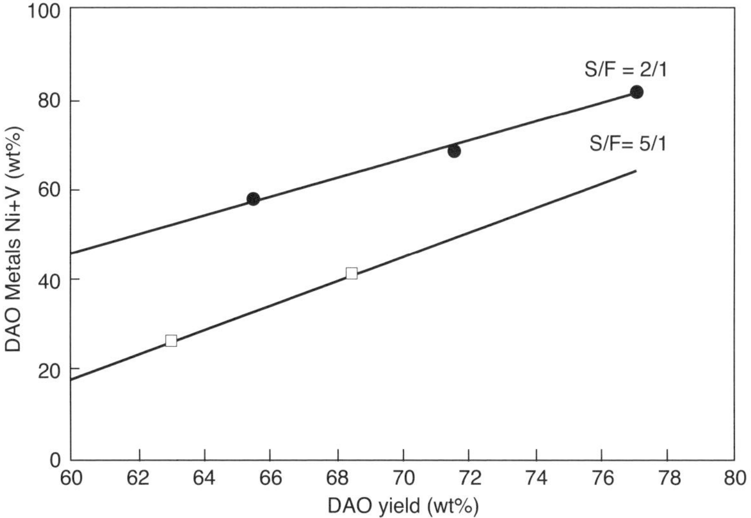

Figure 15.14

Solvent-to-Feed Ratio (S/F): Effect on DAO yield on an Arabian Heavy Vacuum Residue.

15.2.2.4 Temperature Effect

The effect of operating temperature on extraction selectivity is a result of the change in solvent power. Regardless of the types of feeds and solvents, a higher operating temperature causes a decrease in the solvent's surface tension and solvent power (as schematically shown in Figure 15.15).

For a given solvent and a given feed, an increase in operating temperature causes softer pitch to be produced (precipitation of increasingly light resins) along with lighter (more purified) and lower-yield DAO. As the supercritical temperature is approached, complete precipitation of the oil is about to occur. The solvent supercritical temperature is therefore the upper limit for a given solvent. As previously discussed, optimization of extraction selectivity means adopting a temperature gradient to set a significant internal reflux in the deresining zone (to increase DAO quality without decreasing yield, a difference of > 20°C between the feed and the top-zones is recommended), with a minimum temperature in the settling zone (a difference of < 5°C between the feed and the bottom zone is recommended).

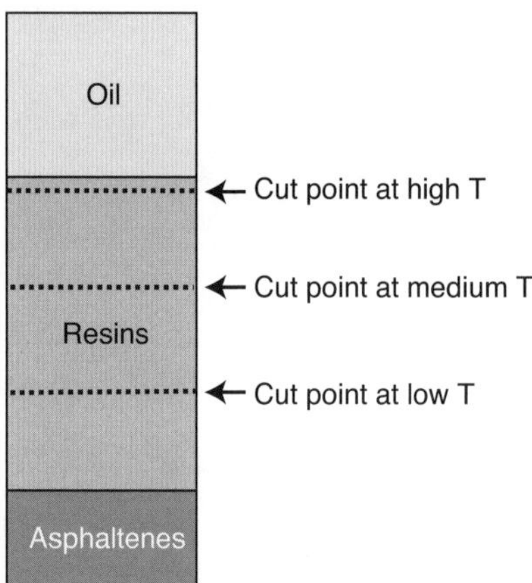

Figure 15.15

Schematic of Temperature Influence on Extraction Selectivity.

Table 15.11 illustrates the effect of operating temperature on the extraction selectivity of a very asphaltenic extra-heavy atmospheric residue from the Adriatic – Rospomare oil – using C5 solvent. The higher the temperature level, the higher the purification level on the DAO at the expense of its yield, and the lower the asphalt softening point. Thus, temperature is another main parameter that can be used to adjust DAO yield and qualities to some extent.

Table 15.11 Effect of Operating Temperature on Extraction Selectivity for an Atmospheric Residue of Rospomare oil with C5 solvent (T is the basis temperature; variations related to T are shown in °C).

Operating Conditions	Unit	Feed			
Top temperature in extractor	°C		T	T-10	T-20
Bottom temperature in extractor	°C		T-40	T-50	T-60
DAO					
Yield on feed	wt%	100	54,7	56,7	58,5
Density at 15°C	kg/l	1.055	0.995	0.999	1.000
Viscosity at 100°C	mm^2/s	4,180	138	147	149
Sulfur	wt%	8.10	6.04	6.20	6.24
Nickel	wt ppm	76	7	8	10
Vanadium	wt ppm	274	50	59	61
C7 asphaltenes	wt%	25.2	0.05	0.06	0.11
Tar (asphalt)					
Density	kg/l		1.143	1.145	1.149
Softening point	°C		159	167	169

15.2.2.5 Other Variables

Feed flow rate: the feed flow rate affects yields and qualities of de-asphalted products only to the extent that the extractor's maximum capacity is exceeded (flooding with carryover to the DAO).

Operating pressure: this parameter must be kept at a value higher than the critical pressure of the solvent used. Therefore, it is set by the solvent type and consequently is not an adjustable parameter.

Solvent injection: feed pre-dilution is an effective means for reducing viscosity and approaching saturation in the feed. Injection position in the extractor can be optimized, depending on the type of feed processed.

15.2.2.6 Summary of the Effect of Operating Variables

Table 15.12 shows the usual domain of SDA operating conditions *versus* solvent type.

Table 15.13 recapitulates the influence of solvent ratio and extraction temperature on the produced de-asphalted oil and asphalt (or tar). DAO quality is related to metals, nitrogen, sulfur, asphaltene content and Conradson Carbon, as well as density and viscosity.

15.3 INDUSTRIAL UNITS: BRIEF DESCRIPTION

The main equipment is the extractor. Basically there are two types of extractors. The first type is the column extractor (Figure 15.16a) with plates, baffles or any internal devices such as ring packing to increase liquid/liquid contact. This type of equipment is more suitable for high-quality DAO but has a limited feed capacity. Thus, it is more adapted to C3 or C4 solvent when high purity is required, such as for lube applications.

Table 15.12 Usual Domain of Operating Conditions in the SDA Process.

Solvent	Unit	C3	C4	C5	C6
Solvent/Feed ratio	vol/vol	6/1 to 10/1	5/1 to 8/1	3/1 to 6/1	3/1 to 6/1
Temperature (extractor)	°C	55/90	90/130	160/200	180/230
Pressure	MPa	3/4	3/4	2/4	2/4
N° of theoretical stages		> 3	2 to 3	1 to 2	1 to 2
Typical DAO yield	wt%	40	65	80	85

Table 15.13 Solvent Ratio and Extraction Temperature: Overall Trends for Product Quality.

Operating Conditions Tendency	DAO Yield	DAO Quality	Pitch Yield	Pitch Softening Point
MW ↗ from C3 to C6	↗	↘	↘	↗
S/F (Solvent/Feed) ↗	↗	↗	↘	↗
Temperature ↗	↘	↗	↗	↘

The second device is the horizontal extractor (Figure 15.16b). Its selectivity is a bit lower than that of the vertical extractor, but the feed capacity is higher. Thus it is well-adapted for a heavy ends conversion refining scheme with C4 to C6 solvents.

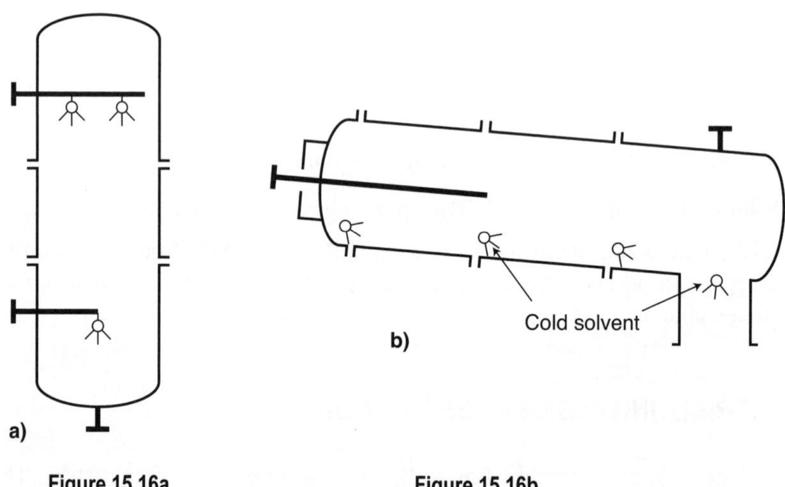

Figure 15.16a

Vertical extractor.

Figure 15.16b

Horizontal extractor.

15.3.1 Industrial Processes on the Market

The main industrial processes on the market are the ROSE™ (Residuum Oil Supercritical Extraction) process from KBR, initially developed by Kerr-Mc-Gee; the UOP Demex™ + FW'S LEDA, also called UOP/FW USA Solvent De-Asphalting process, resulting from the merging of UOP and Foster Wheeler USA in 1996; and the Solvahl® process from Axens.

15.3.2 Solvent Recovery Technique

All industrial processes use recycling of the solvent under supercritical conditions, resulting in a significant reduction of the vaporization energy, in order to reduce operating costs.

For example, the Solvahl® process solvent recovery is described in Figure 15.17.

This supercritical solvent recovery at low solvent/feed ratio was developed by *IFP Energies nouvelles* and referred to as the "Opticritical" process. The energy savings is typically 30%.

The Solvahl® process scheme has yet been displayed in paragraph 15.1.1 and on Figure 15.1. Selection of adequate pressure and temperature levels in the supercritical DAO separator results in the recovery of two phases:

- Upper phase which contains the majority of the solvent and a small portion of the DAO (< 0.6 w%). After cooling, this phase is directly recycled at the inlet of the unit without any compressor or pumping device.
- Lower phase contains the DAO with some remaining solvent (20 to 25 wt%), which is removed downstream within a conventional stripping column.

The solvent extraction and recovery zones are indicated in Figure 15.17.

The heavier the solvent, the more economically attractive the recovery under supercritical conditions. This is why it is especially suitable for C4, C5 or C5/C6 de-asphalting processes.

15.3.3 Economics

Table 15.14 shows a relative comparison of economics – Opex and Capex – based on conventional SDA with traditional solvent recovery conditions *versus* supercritical and Opticritical SDA Solvahl® solvent recovery conditions (Figure 15.17).

Table 15.14 Overall Economics of the SDA Processes.

	Conventional SDA	Supercritical SDA	Opticritical SDA Solvahl®
Investment	Base	Base × 0.9	Base × 0.7
Utilities			
Electricity	Base	Base × 1.1	Base × 0.9
Fuel	Base	Base × 0.7	Base × 0.6
Steam	Base	Base × 4	Base

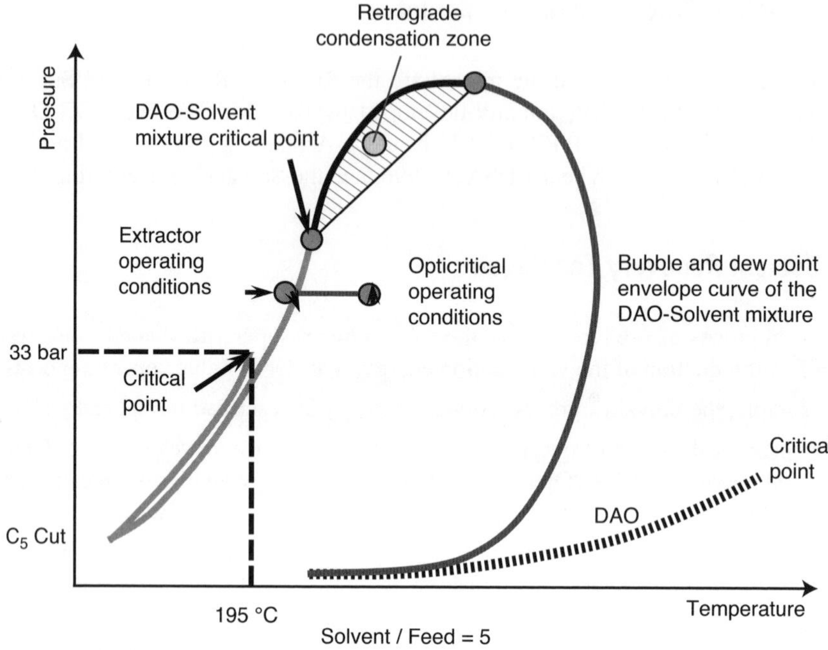

Figure 15.17

Solvent Recovery under Supercritical Solvent.

15.4 EXAMPLES OF PROCESS INTEGRATION WITHIN A REFINING SCHEME AND POSSIBLE USE OF DE-ASPHALTED OIL (DAO) AND ASPHALT

The Solvent De-Asphalting (SDA) process is very versatile and may be integrated – either in an existing or new refining scheme for heavy or extra-heavy oil – in many different ways, upstream or downstream from a conversion unit, in order to increase the overall conversion level. The use of an SDA unit makes it easier to fully convert heavy oil into transport fuels.

It can be integrated with virtually any conversion unit, such as a thermal conversion process (visbreaking, coking, partial oxidation POX or gasifier), catalytic conversion unit (VGO/DAO Fluid Catalytic Cracking FCC or DAO/Resid Resid Catalytic Cracking RCC), fixed bed hydroconversion unit (Residue hydrotreatment HDM or VGO/DAO HydroCracking HCK), or Ebullated Bed (EB) hydroconversion unit.

15.4.1 Examples of Integrated Schemes

Since there are many possible refining schemes that can include an SDA, we will provide only a few examples to illustrate how it can be combined upstream or downstream from one or several conversion units.

Figure 15.18: SDA + VGO/DAO Type EB illustrates the use of a C4 or C5 SDA before an ebullated hydroconversion unit dedicated to heavy VGO and DAO, like the H-OIL$_{DC}$ process. This association, plus recycling of the resid proceeding from the ebullated bed reactor, results in a high level of conversion into transport fuels of intermediate qualities. Nevertheless, for high-quality commercial products, a final hydrotreating to reduce the sulfur level and aromatic content is mandatory.

Figure 15.19: Resid Type EB + SDA shows the same refining scheme but with a resid ebullated bed (such as the LC fining or the H-OIL$_{RC}$ processes) before the SDA. Like the previous scheme, a final hydrotreating to reduce the sulfur level and aromatic content is mandatory.

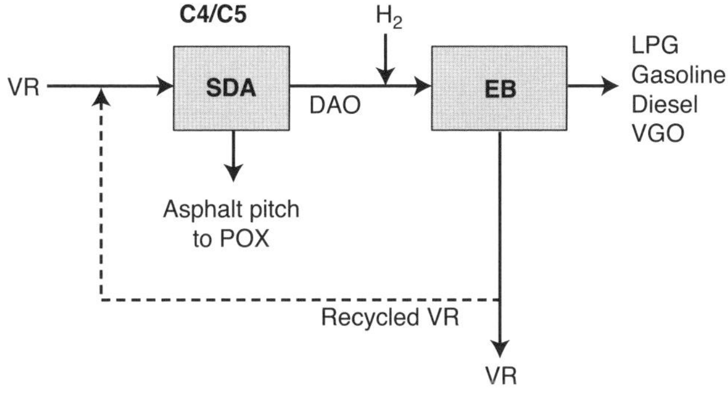

Figure 15.18

SDA + VGO/DAO Type EB.

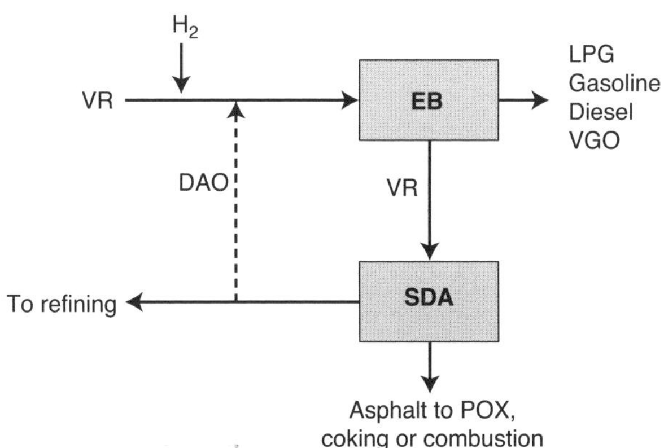

Figure 15.19

Resid Type EB + SDA.

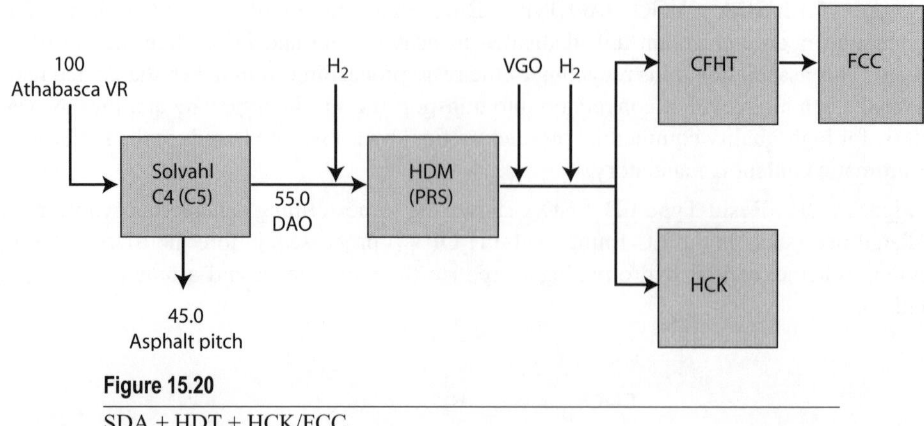

Figure 15.20

SDA + HDT + HCK/FCC.

Figure 15.20: SDA + HDT + HCK/FCC is similar to the previous scheme but dedicated to C5 SDA with two routes: the first for maximizing gasoline and the second for maximizing mid-distillates, jet fuel and diesel. Like the previous scheme, the final products are ultra-low sulfur products directly suitable for transport fuels.

In this scheme, the produced C5 DAO is purified in a hydrotreatment unit (HDM) to reduce its metal content and Conradson Carbon. In the first route, the demetallized C5 DAO is mixed with VGO and used as a feedstock for an FCC unit after a pretreatment (CFHT) or directly sent to an RCC. In the second route, the VGO and DAO are fed to an HCK unit.

This fourth scheme is the most complicated and involves a higher level of investment, but it is very flexible and can optimize the conversion of heavy oil into gasoline, jet fuel and diesel.

For all of these schemes, the pitch or asphalt produced can be used as a heavy fuel oil for combustion, as feed for a partial oxidation or gasification unit, or for a coker (see details in next paragraph).

15.4.2 Asphalt: Possible Uses

Thus, despite its high content of impurities such as sulfur, nitrogen and metals, asphalt is a better quality feedstock with a higher H/C ratio for combustion or POX than coal or petroleum coke. It can subsequently be used:

- As a high-sulfur solid fuel oil for combustion, exhibiting a higher heating value than petroleum coke (from delayed coking), or any type of coal due to its higher hydrogen content (7 to 9 wt%), as compared to 4 to 5% for coal and 3 to 4% for a petroleum coke; or as an extra-heavy liquid fuel, after blending with an aromatic cutter stock such as a light cycle oil from FCC to reduce viscosity.
- As a feedstock for a partial oxidation unit (POX) or a gasifier to transform the pitch into a syngas ($CO + H_2$) that can be used as an H_2 source, or gas for a turbine device to generate electricity, or transformed into methanol, DiMethyl Ether (DME) or high-quality transport fuel using the Fisher Tropsch reaction.

- As a coker feed to make additional transport fuel.
- For road paving, some specific qualities of asphalt are required that can be suitable for C3/C4 asphalt products, but not for C5 asphalt because it is too hard.

REFERENCES

Adewusi VA, Ademodi B, Oshinowo T (1991) Relative Efficiency of Phenol Water Ether Mixture and Nitrobenzene System for De-Asphalting Low-Asphaltic Crude-Oil Residue, *Fuel Process Technol.*, **27**, 1, pp 21-34.

Anon (1992) De-Asphalting, in *Refining Handbook 92*, Hydrocarbon Processing.

Bousquet J, Laboural T (1987) Deasphatable Oils Acceptable FCC Feed in Europe, *Oil and Gas Journal*, 20 April, 1987, pp 62.

Chang CPC, Murphy JR (1982) De-asphalting, in *Encyclopedia of Chemical Processing and Design*, McKetta JJ and Cunningham WA Eds., **14**, pp 149, Marcel Decker Inc., New York.

Chen S-L, Jia S-S, Luo Y-H, Zhao S-Q (1994) Mild Cracking Solvent De-Asphalting – A New Method for Upgrading Petroleum Residue, *Fuel*, **73**, 3, pp 439.

Curtis CW, Tsai K-J, Guin JA (1987) Effects of Solvent Composition on Coprocessing Coal with Petroleum Residua, *Fuel Process. Technol.*, **16**, pp 71.

Handbook of Petroleum Refining Processes, 3rd edition, McGraw Hill, 2003, pp 10.37-10.61.

Le Page *et al.* (1992) Resid and Heavy Oil Processing, Editions Technip, Paris.

Mohan S, Rana *et al.* (2007) A Review of Recent Advances on Process Technologies for Upgrading of Heavy Oils and Residua, *Fuel*, **86**, pp 1216-1231.

Peries JP *et al.* (1995) From Heavy Oils to Marketable Products: IFP Proposes New Schemes for Maximum Conversion, UNITAR International Conference on Heavy Crude and Tar Sands, Houston, Texas, 12-17 February, 1995.

Wauquier JP (2000) Petroleum Refining, **2**, Separation Processes.

Internet websites related to SDA licensors

AXENS: www.axens.net/html-gb/offer/offer_processes_94.html.php

KBR: www.theyesmen.org/agribusiness/halliburton/kbr/hydroChem/petroRefine/residuum Processing.html

FW/UOP: www.uop.com/search/cherche.asp?target=SDA&Submit = search

16 | Visbreaking

A. Quignard, S. Kressmann

The thermal decomposition of hydrocarbons was discovered accidentally in 1861 in a small refinery in New Jersey when a batch distillation unit was left unattended for a while. The increase in temperature in the distillation drum was such that the heavy residue broke down into a light gasoline fraction. Later, around 1910, batch thermal cracking processes were developed (temperature of roughly 400°C, 5-7 bar of pressure). The first continuous process was constructed in 1925.

The objective of this process is a mild cracking of the residues, and possibly crude oil, to produce distillates and a residual fuel. Historically, the initial motivation for building a visbreaker (VB) was to produce distillate and gasoline in a very economical way. It is now used in refineries to reduce the viscosity of residue-fuel fractions. The name "viscosity breaker" refers to this function. In addition, it allows for the production of a limited quantity of light distillates.

The fuel used in conventional power plants that produce electricity requires a low viscosity. This requirement is generally met by blending the residue with light material. Adjusting viscosity – which is done by blending the residue fraction with various cutter stocks – is expensive due to current demand for distillates. The objective of a visbreaker is to reduce this quantity of cutter stock by lowering the residue viscosity *via* the cracking of high molecular weight molecules in lighter cuts. Blending is therefore minimized, but might still require the use of low-value cutter stocks; for example, using light cycle oil (mainly 70 to 80% aromatic light distillate) exiting the Fluid Catalytic Cracking (FCC) unit is a current practice.

The light materials converted during the visbreaking phase require post-treatment to meet specifications due to their inherent content in olefins following the cracking operation. Generally, the overall mass-balance for the distillates produced and used for the blend is nearly equilibrated, resulting in cost savings and justifying implementation of VB. The heavy distillate can be fed to the FCC for gasoline production. This is the main reason for worldwide construction of this low-cost process as compared to others for a mild conversion residue. The market share is roughly 25% of worldwide residue processing capacity.

In summary, the aim is to decrease the viscosity of heavy residues in order to make them usable as heavy fuel oils. This applies to:

- Atmospheric residues – to produce converted gasoline and diesel oil while complying with residual fuel oil stability and viscosity specifications.
- Vacuum residues – this is the main application, to reduce viscosity to a level such that they can be used as industrial fuel oils after adding a light diluent (gasoil).

The current trend is to replace thermal processes on the atmospheric residue with Residue Fluid Catalytic Cracking (RFCC), in accordance with decreasing fuel oil demand, or with residue hydroprocessing units that produce high-quality products. However, these thermal operations are still used as mild cracking processes to inexpensively upgrade atmospheric and vacuum residue from conventional and heavy oils. These residues are not easy to process using fixed bed technologies or other technologies without a significant investment cost. VB could be used advantageously on this low-cost feed where the need to reduce viscosity is important. This market share will likely decrease in the future since sulfur fuel oil specifications will become increasingly severe, although this is highly-dependent on the regulations of each country. Some countries would like to achieve very low sulfur fuel oil to lower their power plant emissions, and this will constitute a future challenge for the new technologies.

Visbreaking can also be used in the field for extra-heavy oil production, to reduce viscosity and make this oil suitable for pipeline transportation to a dedicated refining site. It avoids or minimizes the use of solvent, such as Straight Run (SR) Naphtha, usually used to reduce viscosity in the pipe and then recycled to the exploration site.

16.1 TECHNICAL BACKGROUND

Visbreakers were designed to reduce viscosity. Some general publications present a good introduction to this subject [Hildebrand, 1916; Van Kerkvoort *et al.*, 1952; Martin *et al.*, 1954; Heithaus, 1962; Barton, 1955; Mellan, 1968].

Two approaches have been developed for reactor configuration on the market:

- Coil cracker process (most common).
- Soaker cracker process.

Based on these processes, several other thermal cracking processes have been developed to convert residues more deeply than that achieved in visbreaking without coke rejection. These include Deep Thermal Conversion (Shell/ABB Lummus Global) [Barton, 1983], High Conversion Soaker Cracking HSC (Toyo Engineering Corporation and Mitsui Zosen chemicals) [Hirschberg *et al.*, 1984; Lewis *et al.*, 1985], Hydrovisbreaking (Tervahl-H AXENS) [Berryman *et al.*, 1987] and Aquaconversion (INTEVEP) [Rovesli *et al.*, 1988; Lewis *et al.*, 1988; Wiehe, 1996]. Finally, a process named Eureka (Chyoda Co.) [Pauli *et al.*, 1996; Kassinger, 1997] is more related to delayed coking (see Chapter 17), since it produces a solid pitch in the form of flakes. These processes will be not described here since they are based on the same principles as VB, except that they include technology improvements and a catalyst booster to improve the conversion level without coke formation. A few of them are or have been in operation, but most have still not been fully demonstrated.

The coil of the visbreaker performs the conversion operation *via* high temperature (470-500°C) and a short residence time (from several tenths of seconds to a few minutes). The design of the furnace is critical and will be discussed later. The soaker visbreaker achieves the same level of conversion *via* the combination of a lower temperature (430-450°C) and a longer residence time (10-30 minutes). The soaker drum is combined with the furnace to complete the conversion (40-60% of the overall conversion), resulting in a longer operating time, lower operating cost (–15%) and an indirectly lower investment cost for the furnace and exchangers. The key element of the soaker is to minimize coking and increase run length before cleaning and decoking of the heater and cracker drums. The turnaround is typically scheduled on an annual base using the soaker configuration *versus* several months only (i.e. 6 months) using a coil visbreaker. Quenching the effluent is necessary to protect the downstream equipment from coke build-up, and to minimize fouling of the exchangers. The main drawback of the VB process – which limits on-stream time – is this cleaning of the exchangers and decoking of the reactor zone. Quenching efficiency, in addition to temperature and residence time control, are the key parameters. Steam is generally used in the reactors to adjust residence time and reduce coking tendency in the furnace (< 0.5 wt% feed).

Visbreaker conversion is limited by the stability of the unconverted fuel at the outlet of the cracking zone. Generally, the production of distillates and viscosity reduction provide incentives to increase the severity of the treatments, resulting in problems of fuel oil compatibility and storage stability. To overcome this major concern regarding stability and compatibility, the fuel oil is generally produced by blending the visbreaker bottoms with appropriate refinery streams, which results in a stable fuel oil and achieves fuel storage specifications. Several technical analyses have been developed to anticipate and control these stability issues. A short review will be provided regarding the various physico-chemical parameters involved in operating the visbreaker to achieve maximum viscosity reduction and conversion while minimizing gas and coke formation, and finally obtain a stable fuel. It should be noted that cracking at high temperature generates olefinic components in the converted products, which should be post-treated. Also, the removal of sulfur and nitrogen and the addition of hydrogen is necessary in the gasoline and distillate cuts. Hydrotreatment is generally performed on naphtha and diesel cuts. The vacuum gas oil is usually sent to the FCC. Visbreaker severity is limited due to this stability concern, which depends on feed composition. Several studies have been performed to understand the impact of operating conditions and feed effects. The following two paragraphs will focus on feed composition and chemistry of the visbreaker. This should be helpful in understanding its behavior and optimizing the use of this process in a refinery or directly in the field to reduce viscosity.

16.2 FEED COMPOSITION

In highly simplified terms, a residue is a colloidal system composed mainly of hydrocarbons of polydispersed molecular weight called maltenes and asphaltenes. These two types of compounds are arbitrarily defined by solubility: the asphaltenes are insoluble in n-C7 following a standard analysis (*ASTM-D6560*). The structure of the asphaltenes is not clearly defined and considered to be an association of complex molecules with a high molecular weight and a dominant aromatic feature, containing aliphatic chains, heteroatoms (S, N and O) and heavy metals (Ni, V) (See Chapter 4). Maltenes have a lower molecular weight than asphaltenes. They consist of paraffinic, naphthenic and aromatic molecules and also contain heteroatoms and heavy metals, but to a lesser extent than asphaltenes.

The maltenes are generally separated by polarity using liquid chromatography and provide Saturates, Aromatics and Resins (SAR). The resins are the most polar compounds and contain most of the nitrogen, but also oxygen, sulfur and metals. These families of compounds are fully polydispersed molecular weights and chemical composition is quite different from one feed to another.

The four fractions – Saturates, Aromatics, Resins and Asphaltenes (SARA) – are defined in terms of solubility, polarity and molecular weight. It is a continuum of species that, when combined, constitute a stable medium (residue) which feeds the visbreaker unit. The cracking reactions involved in the VB process will preferentially convert some fractions, leading to perturbation in the equilibrium of the colloidal system. The properties of the SARA fractions are highly-dependent on their chemical composition, which consequently influences their behavior during processing. In spite of their molecular complexity, these fractions are related to their crude oil origins, and organic matter types. They can be classified according to several different types of organic origins which are described in the Geochemistry chapter.

Generally, the asphaltenes are thought to contain large polynuclear aromatic rings (from 4-20 rings). The condensed aromatic rings can link together through alkyl chains and/or simply exhibit alkyl side-chains; the length of the alkyl chain can vary from 1 to 20 – and possibly 40 – carbons. This structure is related to the composition of the crude, depending on its geological origin and history. Another property of the asphaltenes is the risk of self-aggregation in the solution *via* stacking of the aromatic rings. In the petroleum residue, the asphaltenes are considered to be peptized by the resins in order to avoid the stacking. The formation of micelles is primarily due to the interaction between asphaltene and resin fractions, but the aromatics and saturates also play a key role. Currently, the nature of the intermolecular or intramolecular forces involved in the formation of asphaltene micelles is not clear. It has been suggested that a number of forces may be involved in forming asphaltene micelles, including Van der Waals attraction, dipole-dipole interaction, hydrogen bonding, electron transfer, or – bonding interaction between aromatic rings. This is the major reason why asphaltenes are considered the most unwanted molecules in refinery upgrading processes, since they result in coke and sludge formation, and catalyst deactivation in other processes. As we shall see later, the VB process doesn't convert the asphaltenes, but instead transforms their initial structure into more pericondensed aromatic structures (polyaromatic structures with higher condensed aromatic carbon ratio). When increasing the severity, we observe the creation of coke precursors – which are insoluble in C7 – at high temperatures.

Consequently, the effect of asphaltenes on the physical properties of heavy oils and bitumen has been studied extensively. It has been demonstrated that the viscosity of petroleum is significantly influenced by the presence of asphaltenes and their concentration and size. Other asphaltene properties that could be correlated to the insoluble (coke precursor) formation are: degree of condensation of polynuclear aromatics and degree of alkyl-substitution of polynuclear aromatics. The balance between resins and asphaltenes is one of the parameters to be monitored during the operation. The utilization of aromatics, especially condensed polyaromatics, has a positive effect on the peptization of asphaltenes, but the use of saturate solvent could destabilize the equilibrium between resins and asphaltenes.

Since the asphaltenes play a role in the rheological and physical properties of the residue, we must examine the visbreaking chemistry involved during the operation. Thermal cracking has an impact on SARA fraction distribution, which subsequently impacts the system's physical chemistry. This evolution could result from chemical modification of the internal chemical composition of each fraction, or simply from evolution of the proportion of each fraction *versus* visbreaking treatment severity.

16.3 VISBREAKING CHEMISTRY

16.3.1 Thermal Cracking

Thermal cracking occurs under the action of temperature as low as 200°C. But the evolution is very slow and may take hundreds, thousands or even millions of years. This slow thermal cracking occurs in the conversion of organic matter into crude oil under geological conditions. The chemistry involved at low and higher temperatures is the same: free radical mechanisms were discovered in 1933-1943 *via* the pyrolysis of light hydrocarbons, and have been extensively studied since this period. All of the reactions are endothermic. The transformations involved proceed by thermal activation and take place according to the following free radical steps:

a) Initiation – formation of free radicals through homolytic cleavage of the C-C, C-S, C-O and C-N bonds. Initiation through C-H is quite low, but could occur at high temperature (above 400°C).

b) Propagation – several types of reactions are occurring:

- Free radicals created previously can activate the molecules of the surrounding substrate *via* so-called H-transfer reactions or metathesis. This is the fastest reaction for transferring the radical from one molecule to other molecules.
- Decomposition of the free radicals in olefins, and creation of a new free radical of low molecular weight, ensuring the production of light compounds. This reaction is referred to as β-scission and results in α-olefins. For the alkyl-aromatic, the reaction is very fast and the α-olefins are generated quickly.
- Isomerization and cyclization reactions can occur depending on the type of free radical and configuration of the hydrocarbon molecules; creation of the naphthenic ring can be observed.
- Condensation reactions: subsequent free radical reactions of molecules containing aromatics could occur, resulting in the formation of high polyaromatic compounds. This reaction can lead to the formation of coke through secondary condensation reactions. The asphaltenes are mainly subjected to this reaction.
- Secondary reactions can occur on the products created during the first pass *via* all of the previous reactions; this is the so-called chain mechanism, and the number of loops could be very large. Reaction selectivity is determined by this chain mechanism of the propagation reactions.

c) Recombination: two free radicals will recombine to form a bond. This reaction slows down the transformation of the residue. This recombination takes place for all free radical saturate, aromatic or naphthenic compounds.

Thus, all SARA fractions participate in thermal cracking, and the lowest energy bonds (C-C, C-S, S-S) are cracked first. The asphaltenes, resins and pericondensed aromatics are dealkylated. The reactivity of high molecular weight species is higher than for low molecular weight species, e.g. the aliphatic short chain. Consequently, the cracking rate of the paraffins, alky-naphthenes and alkyl-monoaromatics is very fast. On the other hand, the rate of thermal cracking of the condensation reactions involving asphaltenes and pericondensed aromatics is slow as compared to previous reactions, and strongly depends on temperature, as indicated in Table 16.1.

Table 16.1 Atomic Bond Energy in Hydrocarbons.

Bonds	kcal/mole
C-C	83
C=C	146
C≡C	200
C-H	100
C-S	65
C-O	86
O-H	111
S-H	81
S-S	54
C-N	73
N-H	93

Source: www.cem.msu.edu/~reusch/VirtTxtJml/react2.htm

These condensation reactions lead to the pericondensation of aromatics in asphaltenes and resins, which are coke precursors. The behavior of all fractions combined is difficult to understand and to predict *via* a kinetic model. Strong competition exists between the cracking of saturate, naphthenes, resins and asphaltenes capturing the free radicals and leading to coke. All reactions are fully controlled by temperature and the initial chemical composition of the feed.

In addition, there are reactions involving the asphaltene heteroatoms (sulfur and oxygen):

- Formation of H, S thiophenes and mercaptans.
- Formation of phenol, if oxygenates are already present in the feed.

The metal impurities, mostly attached to nitrogen in porphyrin and non-porphyrin structures, remain in the residual fraction.

Since most of the chemical reactions during the visbreaking process are thermally driven, there is no selectivity in bond cleavage. Moreover, the conversion to lighter products is mainly thermally driven regardless of whether the reaction is carried out in the presence of hydrogen and/or a metal catalyst. These reactions play a major role and should be controlled to minimize coke occurrence.

At low to moderate severities, i.e. 400-430°C, 10-20wt% conversion can be achieved without major coke formation. Under these conditions, changes in the molecular structure of bitumen are relatively small since most of the C-C bonds remain intact. A significant MW reduction must take place before heavy cuts are converted to distillates.

16.3.2 Physico-Chemical Transformation

In a stable residue, the asphaltenes are kept in a colloidal solution in the surrounding maltenic medium by solvation with the resins and pericondensed aromatics present in the medium. Extensive studies can be found in the literature dealing with these considerations. In the visbreaking operation, the cracking and condensation reactions combine to break this equilibrium:

a) Cracking decreases the average molecular weight and solvent power of the substrate environment.

b) Dealkylation of the asphaltenes leads to a reduction in the H/C ratio, the asphaltenes become more compact and the tendency to condense will start.

c) Condensation reactions and hydrogen donations increase the aromaticity of the asphaltenic micelles and their incompatibility with the maltenes, which become more saturated than before. The asphaltenes can begin to stack together – increasing in size – and possibly flocculate.

The art of the visbreaking operation consists in approaching this flocculation threshold but not exceeding it. This is a compromise to be achieved and conversion is limited by this threshold. It is therefore important to be able to accurately anticipate whether the asphaltenes remain in the form of colloidal dispersions (peptisation) or begin to flocculate.

Several methods have been proposed to determine the peptised state of the asphaltenes:

a) Hot filtration tests to measure the actual or potential amount of flocculated asphaltenes (total sediments in residual fuels ASTM D4870-04).

b) Measurement of the tendency to flocculate *via* the addition of a binary solvent (O-xylene-isooctane or toluene-n-cetane, a combination of solvent and anti-solvent). Most of the relevant techniques are based on this principle. Martin and Bailey were the precursors in the 1950s (for example, see ASTM D7060-04 on Flocculation Ratio and Peptizing Power in Residual and Heavy Oils).

c) Measurement of flocculation *via* physical analysis (microscopy, laser, etc.)

d) Combination of methods b & c above.

Based on experience, it is difficult to extrapolate the threshold measurement (limit of asphaltene flocculation) because the measurements are generally performed under standard conditions (atmospheric pressure and low temperature) and often involve dilution. The precise state of asphaltene peptization under reaction conditions is currently unknown. Extensive work has been carried out *via* analytical and physical chemistry methods to characterize molecular weight, size and physical state. Currently, we do not have a clear understanding of this matter and most experience is based on expertise and extensive work on pilot, bench and batch reactors.

The addition of a binary solvent (method b) is the most common technique. It was developed in the 1950s in the studies of Martin & Bailey.

16.3.3 Reaction Kinetics

Most of the products obtained are the result of chain reactions by radicals. From a kinetic standpoint, the reaction rate can be represented by a first order expression:

$$V = \frac{dx}{dt} = k(1-x)\exp(-\frac{Ea}{RT}) \text{ or } k't = \ln(\frac{1}{1-x})$$

where x is the weight fraction of the converted feed corresponding to a cut point generally close to 350°C, V is the net conversion, k is the rate constant, t is the residence time and Ea is the apparent activation energy.

The activation energy (Ea) varies according to the nature and composition of the feed. This model is acceptable within a small range of temperature and conversion. Extrapolation is very unreliable. Considering the previous mechanism described, it is very fortunate that the reaction order is close to 1. Moreover, the activation energy at high temperature (450-500°C) should be the same as that at moderate temperature (400-450°C) due to the chain mechanism order type of free radical reaction. The activation energies for high molecular weight molecules are roughly 40 kcal/mole, increasing up to 50 kcal/mole for low molecular weight molecules. Condensation reactions are around the range of 60-90 kcal/mole, so they are largely favored by a temperature increase.

According to data from the literature, the atmospheric residue has an activation energy of 75 kcal/mole, which can shift to 55 kcal/mole for the vacuum residue, when removing the low molecular weight portion of the residue. If the asphaltenes are removed from the vacuum residue (De-Asphalted Oil or DAO, see Chapter 15), the activation energy decreases to 36 kcal/mole because condensation reactions are minimized. Most of the resins will crack very quickly due to the presence of heteroatoms, which facilitate the creation of free radicals.

Generally, the design of the coil soaker is based on the residence time, which depends on the rate constant of the feed. Lengthy experience is required to optimize this design in order to achieve the maximum conversion with a stable residue. This experience is mainly based on data obtained from pilot or batch reactors.

As previously explained, thermal cracking is not a selective process because all of the molecules react according to their own scheme and interact strongly with each other *via* H transfer reactions. Based on experience, as severity rises, the product slate favors light and heavy ends. At much higher severity, we reach the coking process field, and the yield of gas and coke increases greatly.

The recycling mode could be used to modify visbreaker selectivity. Unfortunately, this is an expensive operation and condensation reactions are increased by the increase in asphaltene concentration. Improvements have been investigated, including the addition of additives or catalysts in the reaction zone to increase conversion. These actions increase selectivity by controlling H-transfer and condensation reactions (e.g. aquaconversion process) but have not been fully demonstrated. Hydrovisbreaking was developed and successfully met the requirement, but no commercial process was built due to issues of cost. The slurry process has been investigated and described elsewhere in this book (see Chapter 18). Control of condensation reactions and management of H-transfer reactions are the key elements for success in increasing liquid conversion.

16.4 PERFORMANCE

The two visbreaker configurations (coil and soaker) provide different results and the main objectives of their utilization are:

- Maximum reduction of viscosity: coil configuration (Figure 16.1)
- Maximum production of distillate: soaker configuration (Figure 16.2)

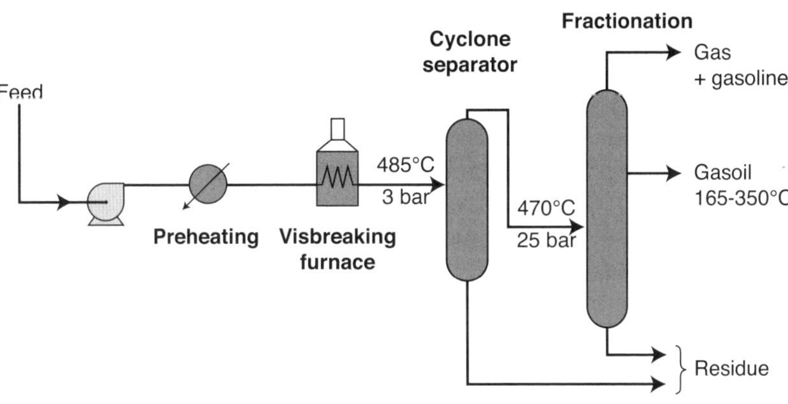

Figure 16.1

Visbreaking Coil Configuration.

Source: Petroleum Refining, Vol. 3 Conversion Processes, Leprince P, Editions Technip, 2001.

Decoking operations are easier for the coil configuration but run length is lower, ranging from 4 to 6 months. The soaker provides longer runs (typically one year), but the coke is more difficult to remove during the turnaround. A lower cost of operation (10-15%) is the main advantage of the soaker, and an up-flow stream provides increased benefit. The main advantage is avoiding over-cracking of distillate and light ends due to the high activation energy of these compounds. Selectivity is slightly improved and gas production is minimized with this configuration.

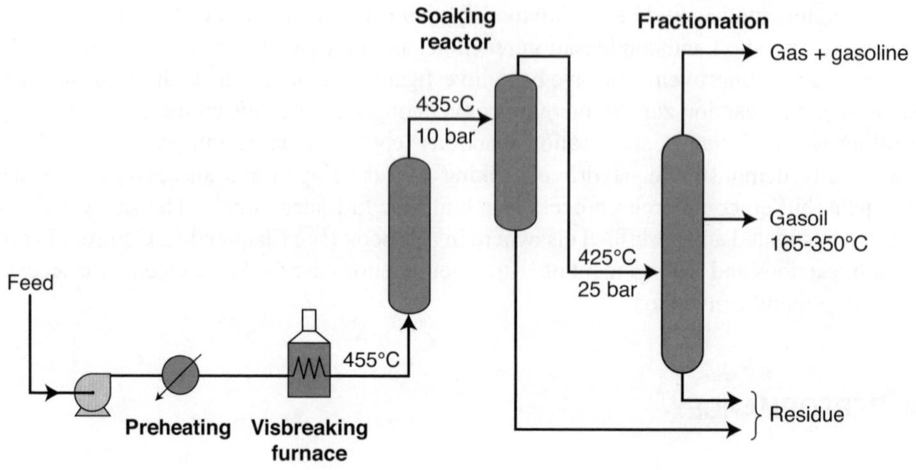

Figure 16.2

Visbreaking Soaker Configuration.

Source: Petroleum Refining, Vol. 3 Conversion Processes, Leprince P, Editions Technip, 2001.

Generally, four cuts are split at the outlet of the visbreaking process: a gaseous fraction (H_2S, NH_3, C1-C4), a gasoline (typically C5-165°C), a diesel (typically 165-350°C) and a residue (350°C $^+$). The yields and properties of these products depend on the type of feed and on the conversion achieved by an appropriate choice of operating conditions (see Table 16.2).

Net residue conversion can be defined as the difference between the 350°C $^+$ in feed minus the 350°C $^+$ in product *versus* the 350°C $^+$ in feed. It is within the range of 20 to 30%. It fully depends on the severity and the feed quality.

Table 16.2 Typical Yields Obtained by Visbreaking for Different Residues (AR= atmospheric residue VR= vacuum residue).

	Unit	Arabian Heavy AR	Arabian Heavy VR	Cerro Negro AR	Maya AR
Density 15/4	kg/l	0.997	1.039	1.037	1.008
Viscosity at 100°C	mm^2/s	206	4,122	3,475	311
Yield	wt%				
Gas		2.69	4.04	2	0.98
Gasoline		5.8	4.94	2.5	1.45
Gasoil		12.28	7.65	10.3	9.74
Residue		79.23	83.37	85.2	87.83

16.5 PRODUCT QUALITY

The quality of products is not very high (Table 16.3) and a post-treatment must be applied in order to use them in the pool of refinery products as finished product bases.

Table 16.3 Quality of Visbreaking Products for Different Residues (AR= Atmospheric Residue VR= Vacuum Residue).

	Unit	Cerro Negro AR	Maya AR	Arabian Heavy VR
Density 15/4	kg/l	1.037	1.008	1.039
CCR	wt%	21	18	24
Viscosity at 100°C	mm^2/s	3,475	311	4,122
Sulfur	wt%	4	4.39	5.32
C7 asphaltenes	wt%	13.9	15.6	13.1
Product Properties				
Gasoline				
Density 15/4	kg/l	0.729	0.754	0.737
Sulfur	wt%	1.12	1.81	1.46
Gasoil				
Density 15/4	kg/l	0.865	0.845	0.855
Sulfur	wt%	2.69	2.52	2.45
Residue				
Density 15/4	kg/l	1.074	1.02	1.051
Viscosity at 100°C	mm^2/s	2,350	268	2,237
Sulfur	wt%	3.6	4.41	5.7
C7 asphaltenes	wt%	21.9	17.3	20.9
CCR	wt%	25	20	30

- The gaseous fraction is mainly H_2S and C1 to C4 hydrocarbons. The H_2S must be eliminated by amine treatment before recovering the fuel gas (C1-C2) and LPG (C3-C4). The hydrocarbon fraction contains equal parts of unsaturated compounds (ethylene, propylene and butenes) and their saturated homologues.
- The gasoline (C5-165°C) is a low-quality motor fuel, with a Research Octane number (RON) in the range of 60-70. A typical PONA (Paraffins/Olefins/Naphthenes/Aromatics) is 35-45-15-9 wt%. The high olefin content (45%), presence of diolefins, and high content of sulfur and nitrogen compounds make an adapted hydrotreatment necessary. The diolefin content (Maleic Anhydride Value = 10) could – if dilution with other cuts is insufficient – result in the need for a specific hydrotreatment before the fraction can be used as feed for catalytic reforming to increase the octane number.
- The gasoil (165-350°C) has a moderate to acceptable cetane number (40-50 depending on the feedstock structure) and contains up to 2-3% wt sulfur. It has high olefin

content (bromine number close to 25g/g/100g). Stability problems during storage could occur due to gum formation and ASTM color (ASTM D1500) varies rapidly. Hydrotreatment in a blend with straight run material is mandatory.

- Some refineries have integrated the visbreaker with a downstream vacuum tower to recover the vacuum distillate, also called Vacuum Gas Oil (VGO). This product can be sent to the FCC, but its quality is poor, namely due to nitrogen and hydrogen content. The VGO should preferably be pre-treated before being sent to the FCC unit.
- The residue (350°C $^+$) has improved viscosity in comparison to the feed. However, a certain amount of fluxing agent must be added to meet the commercial standards of heavy fuels and the sulfur content, except for low or ultra low sulfur feedstocks, will more than exceed 1% (current usual specification for ultra heavy fuel oil).

  In Europe, where the major component of heavy fuels is a visbroken residue, problems may arise concerning the compatibility and stability of these fuels. This implies the use of blending rules. The use of LCO, as an aromatic gasoil, to adjust the viscosity is a great help to meet this stability requirement. It may also involve the use of specific additives to prevent or delay asphaltene flocculation, avoid corrosion or improve combustion properties in order to decrease soot emissions.

  The tendency of visbroken resids to flocculate and their influence on the stability of heavy fuels during storage is the subject of much attention today. Methods for evaluating this stability and the tendency to instability have been developed (see Physico-Chemical Transformation 16.3.2).

16.6 PROCESS

16.6.1 Operating Variables

In an industrial unit, several operating variables can be used to modify or adjust performance: temperature (furnace-coil and soaker), pressure, flow rate and steam injection.

16.6.1.1 Furnace Exit Temperature

This temperature is used as a criterion although the reactions occur in the furnace tubes. This temperature varies between 430 and 490°C, depending on the feed processed and the type of unit, since the unit can in fact include a soaking drum where the reaction continues after the furnace. The furnace exit temperature is selected depending on the desired conversion: an increase 10°C causes conversion to increase by a few%.

16.6.1.2 Feed Flow Rate

By increasing feed flow rate, residence time is shortened, but flow conditions in the furnace tubes and soaking drum are also modified. A 3°C increase in the furnace exit temperature is enough to offset a 10% increase in the feed flow rate for the same conversion.

16.6.1.3 Pressure

In units without a soaking drum, pressure of a few bars is sufficient to keep the feed from vaporizing. In units with a soaking drum, the pressure is chosen so that the desired products are vaporized and quickly leave the reaction zone, while the heavy products soak in the liquid phase. Five to eight bars is used for short residues (vacuum residues) and ten to twelve bars for long residues (atmospheric residues).

16.6.1.4 Steam Injection in the Furnace Tubes

Steam is injected to improve the hydrodynamic flow and consequently decrease the residence time. Thus, a drop in conversion is observed which must be offset by raising the temperature.

16.7 IMPLEMENTING THE VISBREAKING PROCESS

16.7.1 Process Flow Schemes

After pre-heating, the feed is raised to cracking temperature in a pipe still. At the exit of the pipe still, the reactions are stopped by injecting a cold gasoil or residue quench. The hydrocarbon mixture passes through a cyclone separator and then undergoes fractionation into gas, gasoline, gasoil (part of which is used for quenching) and residue (Figure 16.1). In some cases, the pipe still effluent is sent to a soaking drum that lengthens reaction time (Figure 16.2) and allows the pipe still to operate at a temperature that is approximately 50°C lower.

Finally, if the aim is to maximize gasoil production, visbreaking can be combined with a thermal cracking unit using the heavy distillate from the fractionation tower as a feed (Figure 16.3).

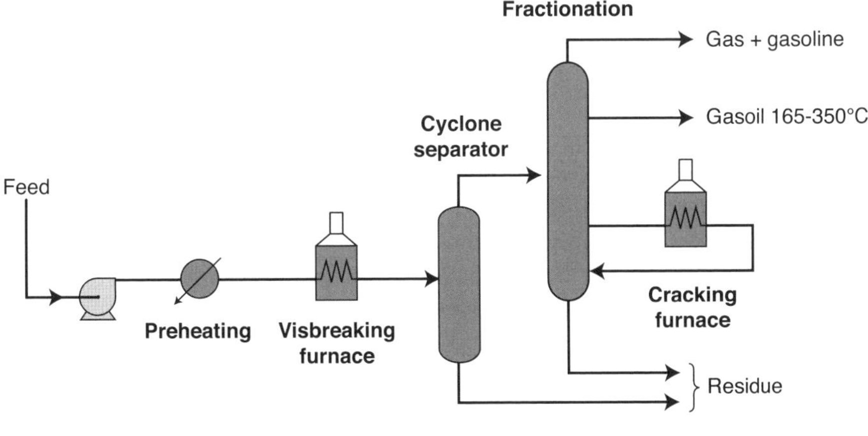

Figure 16.3

Visbreaking Combined with Thermal Cracking.

Source: Petroleum Refining, Vol. 3 Conversion Processes, Leprince P, Editions Technip, 2001.

Figure 16.4 shows the flow scheme of a visbreaking unit with the pipe still and its soaking drum. The effluent is then sent to an atmospheric distillation column which provides a gaseous fraction (H_2, H_2S, C_1, C_2), and LPG, a naphtha cut that is subsequently fractionated into light gasoline and heavy naphtha, a gasoil and an atmospheric distillate. The residue is routed to a vacuum column and is split into a VGO and a visbroken residue. The VGO, mixed with the atmospheric distillate, is sent to catalytic cracking, while the visbroken residue is mixed with gasoil, which acts as a fluxing agent to adjust its viscosity. The flow scheme (Figure 16.4) of the unit, typically operating with 135 t/h of feed, shows the yields of various products and the most common operating conditions (temperature and pressure) for the furnace, the soaking drum and the atmospheric distillation.

16.7.2 Specific Equipment

16.7.2.1 Furnace

The function of the furnace is to raise the feed to the reaction temperature and to keep it at a residence time long enough to accomplish the desired conversion (Figure 16.5).

Different types of furnaces can be used:

- Where heat is transferred:
  - By convection alone
  - By radiation alone
  - By a combination of both
- Some furnaces may have several in-series radiation cells, may include vertical or horizontal tubes and may be fired by gas or fuel oil.

However, they must meet specific requirements in order to function properly:

- Average heat flow must be 22 to 30 kW/m^2, and in no case should the heat flow exceed 60 to 70 kW/m2 locally. Beyond this, coke deposition will accelerate and become detrimental to proper furnace operation (increased pressure drop and increased tube skin temperatures).
- Tubes must be able to resist 650°C. Steel alloys (9 Cr-0.5 Mo) are generally selected.
- Steam or condensate injection devices must be installed in the zone where hydrocarbon vaporization is slight in order to increase the turbulence of the flow, thereby reducing coke deposits. Steam injection ratio of roughly 0.5% of the feed is common.
- Decoking equipment is indispensable. It includes an injection device for an air/steam mixture that allows furnaces to be decoked at roughly 550°C.

16.7.2.2 Soaking Drum

Located just downstream from the furnace, this drum is designed to lengthen the feed residence time so that the furnace can operate at a lower temperature (Table 16.4).

Approximately 40 to 60% of cracking occurs in the drum. Since these reactions are endothermic, the exit temperature is lower than the inlet temperature by 10 to 20°C.

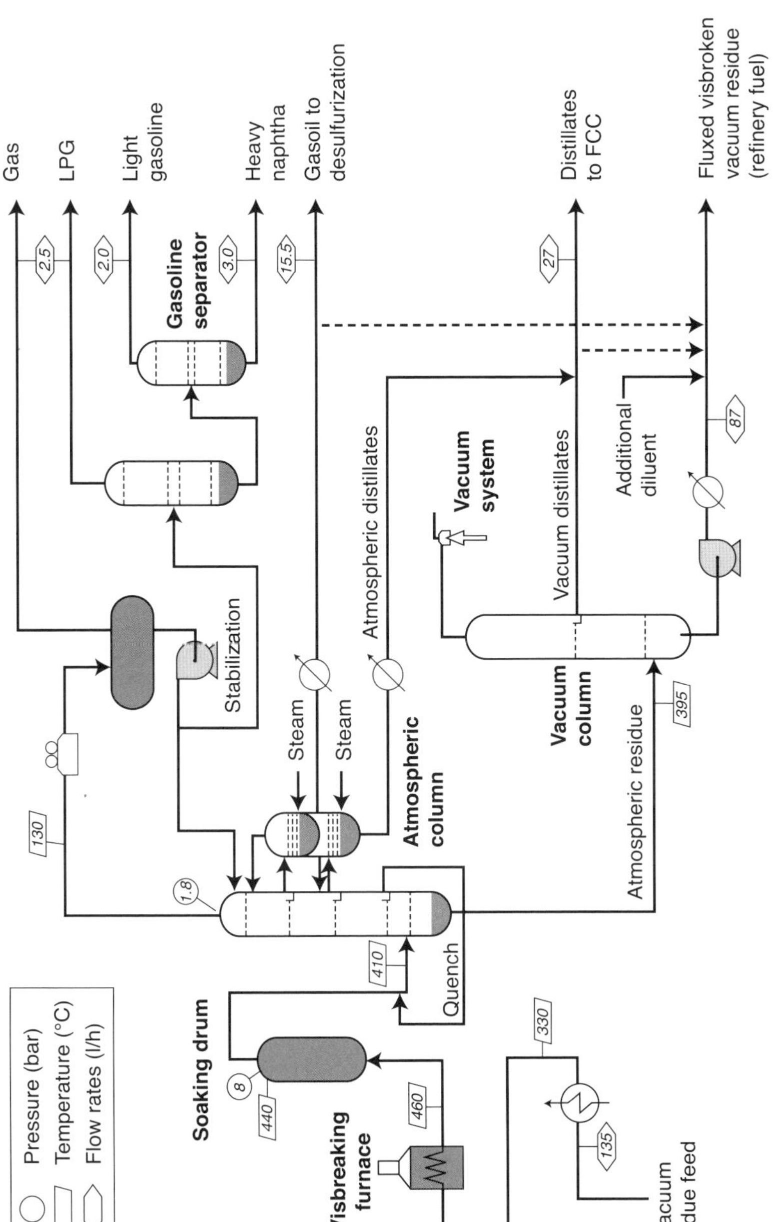

Figure 16.4

Typical Visbreak Flow Scheme.

Source: Petroleum Refining, Vol. 3 Conversion Processes, Leprince P, Editions Technip, 2001.

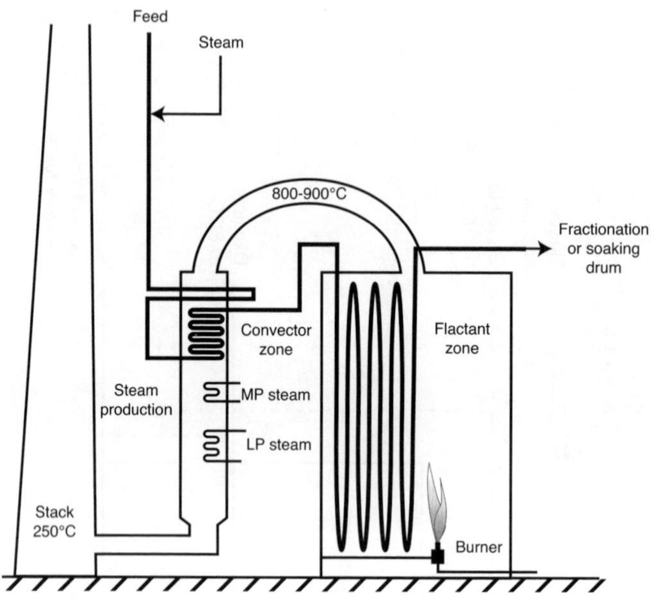

Figure 16.5

Visbreaking Fumace.

Source: Petroleum Refining, Vol. 3 Conversion Processes, Leprince P, Editions Technip, 2001.

Table 16.4 Temperature/Residence Time Relationship at 6% Gas + Gasoline Conversion.

Furnace Exit Temperature (°C)	Residence Time (mn)	Soaking Drum
410	32	yes
440	8	yes
455	4	yes
470	2	no
500	0.5	no

The soaking drum is a reactor with flow from bottom to top (up-flow mode), a volume of around 15 m3/1,000t/day of feed and a height-to-diameter ratio of 5 to 8. These dimensions correspond to a residence time of 15 to 20 min. The soaking reactor provides the following advantages:

- 15% reduction in fuel oil consumption due to the furnace exit temperature, which is 20 to 30°C lower.
- Longer running time between two decoking operations, since the coke deposit rate in the furnace is three to four times slower than in conventional units.

- Better selectivity due to the difference in activation energy between the production of light molecules (250 kJ/mol) and heavier molecules (230 kJ/mol). As a result, gasoil components contribute less to gas and gasoline production with a soaking drum and the conversion of heavy fractions is higher.
- Reduction in furnace and heat exchanger costs. Despite the cost of the chamber itself, overall investment is 10 to 15% lower.

A recent development in soaking drums is to equip them with perforated trays. The trays are designed to even out the residence times of various portions of the feed, thereby favoring high conversions.

16.7.2.3 Cyclone Separators

The function of this equipment is to separate the reaction products into a gaseous and a liquid phase without lowering the temperature. The flash occurs in the upper part of the separator, whose tangential inlet provides good gas/liquid separation. The liquid trickles down the walls and comes together in the lower part. Some of the liquid is cooled by exchange with the unit's feed and used as a quench in the lower compartment (Figure 16.6).

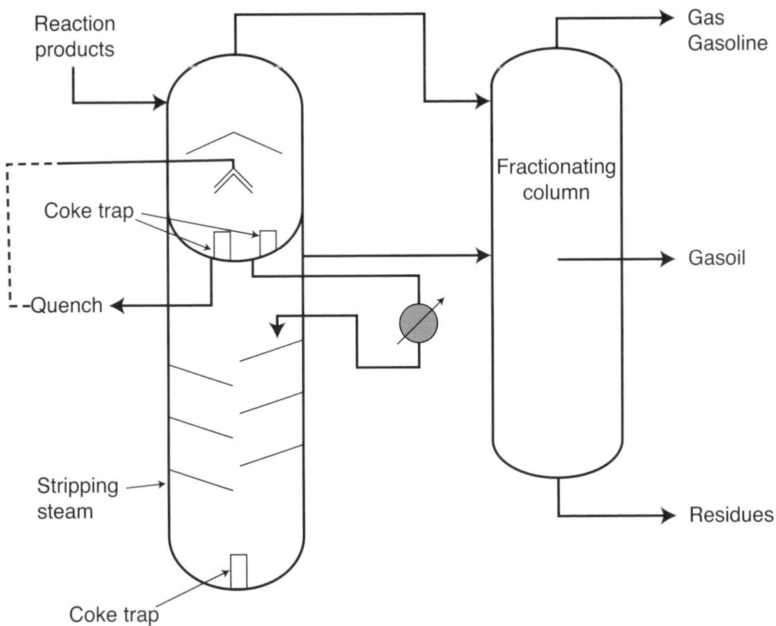

Figure 16.6

Diagram of a Cyclone Separator.

Source: Petroleum Refining, Vol. 3 Conversion Processes, Leprince P, Editions Technip, 2001.

16.7.2.4 Heat Exchangers

The thermal cracking residue has a strong tendency to foul the heat exchangers in which it circulates. Heat exchangers therefore need to be equipped with bypasses so that they can be cleaned without shutting down the unit. Additionally, pieces of coke that might plug exchanger tubes must be retrieved by coke traps and filters incorporated at the pump inlet.

16.7.2.5 Coke Traps and Filters

Coke traps are installed at the bottom of vessels operating at high temperature. They hold back pieces of coke larger than a few centimeters. Filters are of basket type installed upstream from the pumps and can be cleaned without stopping operation.

16.8 PROCESS WATER TREATMENT

Process water from the top of the fractionating column contains H_2S, mercaptans, phenols and thiophenols, organic acids and hydrocarbons. It must be stripped to eliminate the H_2S. The phenols remaining in solution are eliminated by biological treatment. During decoking operations, the furnace effluent is cooled and water is added to it to collect the fine particles of coke that it contains. The water recovered from the furnace must be filtered, settled and treated. The coke recovered (several tons) during maintenance shutdowns is a dry product that is incinerated.

16.9 ECONOMICS

The basic thermal cracking patents are now public. The enhanced process patents and intellectual property are held mainly by Shell Global Solutions International BV/ABB Lummus Global BV, Foster Wheeler/UOP LLC and Axens. About 130 visbreaking facilities have been built worldwide.

Energy consumption is approximately 1.5 to 2% of the feed for a visbreaking unit without soaking. When soaking is included in the process, energy consumption is decreased by 10%. When residue visbreaking is followed by cracking of the distillate, energy consumption reaches 2.5 to 5% of the feed, depending on the degree of thermal recovery.

Table 16.5 shows the distribution of visbreaking capacity by geographical zone, which amounted to 161.2 Mt/year in 2002 [Hydrocarbon Publishing, 2002]. Note that more than 50% of worldwide capacity is located in Europe, with 42% in Western Europe.

The InSide Battery Limits (ISBL) investment for a typical 20,000 bpd/1,100,000 t/y delayed coker unit is roughly US \$33-34 million on a year 2004/2005 basis (West Europe/ US Gulf Coast).

Table 16.5 Distribution of Visbreaking Capacity by Geographical Zone in 2002.

Region	Crude Capacity	Visbreaking Capacity	Ratio Visbreaking to Crude Capacity	% of Worldwide Visbreaking Capacity
	10^6 t/y	10^6 t/y	wt%	wt%
North America (USA + Canada)	888	5.8	0.7%	3.6%
Latin America/Carribean	384	18.9	4.9%	11.7%
Western Europe	661	67.8	10.3%	42.1%
CIS & Central/Eastern Europe	512	19.0	3.7%	11.8%
Middle East	326	23.6	7.3%	14.7%
Africa	154	4.4	2.8%	2.7%
Asia-Pacific	968	21.7	2.2%	13.5%
Wordwide	**3,893**	**161.2**	**4.1%**	**100.0%**

Source: Hydrocarbon Publishing, Q4 2002, www.hydrocarbon publishing.com.

REFERENCES

Barton Eg, AFM (1975) Solubility Parameters, *Chem. Review*, **75**, 6, pp 731-753.

Barton AFM (1983) *Handbook of Solubility Parameters and other Cohesive Parameters*, CRC Press, Boca Raton, Florida, USA.

Berryman TJ, Lewis CPG (1987) Cleanliness and Stability of Residual Fuels, *ASTM Symposium on Marine Fuels*, Orlando.

Catellanos J, Cano JL, Briones VM, Del Rosal R (1992) *Revista del Instituto Mexicano del Petroleo* XXIV, **3**, pp 68.

Douwes BA *et al.* (1999) (Shell International Oil Products), Thermal Conversion Technology in Modern Power Integrated Refining Schemes, Paper AM 99-24, *NPRA Annual Meeting San Antonio, Texas*, March 21-23, 1999.

Fukiyama H *et al.* (2000) (Toyo Engineering), Toyo Recent Commercial Experience and Development of Heavy Oil Upgrading, *Petroleum Refining Technology & Economics in Russia and the CIS*, Conference, Paris, France, October 12-13, 2000.

Handbook of Petroleum Refining Processes, 3rd Ed., McGraw Hill, 12.91-12.105, 2003.

Haples R, *Petroleum Refinery Process Economics*, Pennwell Books, Penwell Publishing Co., Tulsa.

Heithaus JJ (Feb. 1962) Measurement and Significance of Asphaltene Peptization, presented at the *Symposium on the Fundamental Nature of Asphalt* sponsored by the Petroleum Division of the ACS at the 138th ACS meeting in New York, Sept. 1960. Also published in J. Institute of Petroleum, **48**, pp 458.

Hildebrand HJ (1916) Solubility, *J. Am. Chem. Soc.,* **38**, pp 1452.

Hirschberg AW, de Jong LNJ, Schipper BA, Meyers JG (1984) *Soc. Pet. Eng. J*, **24**, pp 283.

Houde E (UOP LLC), Pereira P (PDVSA-INTEVEP) (2001) Aquaconversion *vs.* Thermal Processing of Heavy Crudes, *NCUT Symposium* Calgary, Alberta, Canada, Sept. 17-18, 2001.

Hydrocarbon Publishing, Q4 2002 – www.hydrocarbon publishing.com

Kassinger R (1997) Fuel Blending: How to Minimise the Risk of Incompatibility, *6th International Conference on Stability and Handling of Liquid Fuels*, Vancouver, Oct., 1997, pp 489-501.

Lewis CPG, De Jong F (1988) Effects of Storage on the Stability of Residual Fuel Oil, *3rd International Conference on Stability and Handling of Liquid Fuels*, London, Sept. 1988, pp 49-60.

Lewis CPG, Johnson EK, Johnson and Berryman TJ (1985) The Stability of Residual Fuels. Theory and Practice of the Shell Concept, Paper presented at the *16th CIMAC Conference* in Oslo.

Martin CWG, Bailey DR (1954) The Stability and Compatibility of Fuel Oil and Diesel Fuel, *J. Institute of Petroleum*, **40**, May 1954, pp 365.

Mellan I (1968) *Compatibility and Solubility*, Noyes, New Jersey.

Ogata Y *et al.* (Toyo Engineering) (1992) Well Defined HSC Process Proves Reliability, *International Symposium on Heavy Oil Upgrading and Utilization*, Fushung, Liaoning, China, May 5-8, 1992.

Pauli A (1996) The Automated Flocculation Titrimeter, *Preprint of the Am. Chem. Soc.*, Div of Fuel Chemistry, **41** (4), pp 1276-1281.

Pereira P *et al.* (PDVSA-INTEVEP) (2000) Aquaconversion Technology Offers Added Value to Venezuela Synthetic Crude Oil production, *Oil and Gas Journal,* Nov. 20, pp 56-57.

Pereira P *et al.* (PDVSA-INTEVEP) (1998/99) Increasing Resid Conversion at Curacao Refinery, *Petroleum Technology Quarterly*, Winter 1998/99, pp 29-37.

Redelius PG (2000) Solubility Parameters and Bitumen, *Fuel*, **79**, pp 27-35.

Refining Processes 2006 (2006) Deep Thermal Conversion (Shell Global Solutions International BV and ABB Lummus Global BV), *Hydrocarbon Processing.*

Refining Processes 2006 (2006) Visbreaking (Foster Wheeler/UOP LLC & Shell Global Solutions International B.V. and ABB Lummus Global B.V.), *Hydrocarbon Processing.*

Refining Processes 98 (1998), Visbreaking (IFP, now *IFP Energies nouvelles*), *Hydrocarbon Processing*, Nov., pp 111.

Rovesli WC, Anderson RP (1988) Prediction of Storage Stability and Compatibility of Residual Fuel Oils, *3rd International Conference on Stability and Handling of Liquid Fuels*, London, Sept., 1988, pp 24-35.

Scuster R (1995) *La Rivista di Combustibili*, **49**, 2, pp 45-49.

Tatasuka T *et al.* (Chiyoda Co.) (1994) Renewed Attention to Eureka Process, *Residue and Heavy Oil Workshop*, Dahran, Saudi Arabia, Dec. 4-9.

Van den Berg FGA (2000) Developments in Fuel Oil Blending, *7th International Conference on Stability and Handling of Liquid Fuels*, Graz, Austria, September 24-29, 2000.

Van den Berg FGA (2001) Case Studies on Fuel Instability, Paper Presented at the *Symposium on Stability and Compatibility During the Production, Transportation and Refining of Petroleum*, National Centre of Upgrading Technology, Calgary, Alberta, Canada, September 17-18, 2001.

Van Kerkvoort WJ, Nieuwstad AJJ, Van der Waarden M (1955) Le comportement des systèmes hydrocarbures-asphaltènes. Un facteur important pour la stabilité et la compatibilité des fuel-oils résiduels. IVe Congrès International du Chauffage Industriel, Group II, Section 22, Preprint n° 220, pp 1-8, Paris, 1952. Published in the *US Library of Congress* under n° PB 116765.

Vilhunen J, Quignard A, Pilviö O, Waldvogel J (2000) Experience in Use of Automatic Heavy Fuel Oil Stability Analyzer, *7th International Conference on Stability and Handling of Liquid Fuels*, Graz, Austria, September 24-29, 2000.

Watari R *et al.* (Chiyoda Co.) (1997) Eureka in Combination with Texaco Gasifier Improves Refinery Integration, Paper AM 97-52, *NPRA Annual Meeting San Antonio*, Texas, March 16-18, 1997.

Wiehe I (1996) Two-Dimensional Solubility Parameter Mapping of Heavy Oils, *Fuel Science and Techn. Int'l*, **14**, 1&2, pp 289-312.

17 | Coking

A. Quignard, S. Kressmann

The coking process is by far the most widely used technology for reducing the carbon-to-hydrogen ratio of petroleum feedstocks. Installed capacity represents roughly 31% (around 6 million b/d) of world refining capacity for treating petroleum residue. This thermal conversion process transforms residue into valuable liquids. It is operated at a higher temperature than the visbreaking process and is widely used in refineries, especially in the United States and Venezuela refining scheme. It can treat virtually any type of heavy end since there are no limitations on feedstock characteristics. The developed technology is based on two types: semi-continuous batch bed and fluidized bed.

The first technology, called delayed coking, consists of the thermal decomposition of hydrocarbons in an empty drum, where the coke produced in the reaction is deposited inside the reactor drum. At the same time, the light hydrocarbons exit the drum and are routed to a treatment plant. Several major companies (Foster Wheeler, UOP, Kellogg, Conoco-Phillips and Lummus), as well as others (Koa Oil, Sinopec, Nippon, Chiyoda, etc.) have the know-how required to build and operate this type of process.

In the second technology, the hydrocarbons are decomposed in a fluidized bed of coke particles, part of which is withdrawn. The most common version of this type of process is Fluidcoking, extended to Flexicoking, which was developed by Exxon Research and Engineering Co. Other processes have been also developed: the ART process, Eureka process, and LR coking/Satcon process (Lurgi). Some of these make use of a solid to help to continuously remove the coke built-up during the thermal reaction. These technologies are at the borderline of Residue Fluid Catalytic Cracking (RFCC) technology, wherein the solid used is an active catalyst involving catalytic reactions along with the thermal reactions.

The objectives of the coking process are to:

- Upgrade heavy, high-sulfur, low-quality bottoms (typically vacuum residue) by converting them to lighter products (naphtha + distillates) which should ultimately be post-treated using hydrotreatment processes. In addition, the process frees substantial amounts of premium distillate flux stock, which was previously used for fuel oil blending from the vacuum residue, without conversion processes.
- Upgrade heavy/extra-heavy, high-viscosity crudes in the production field to transform them into Synthetic Crude Oil (SCO), which can be easily transported and handled in a refinery.

Coking is a non-catalytic thermal cracking process based on the concept of "carbon rejection". The heaviest, hydrogen-deficient portions of the feed (i.e. asphaltenes and resins) are rejected as coke, which essentially contains all of the feed metals, ash, and a substantial portion of the feed sulfur and nitrogen.

Since coking is a thermal pyrolysis process, like the de-asphalting process (see Chapter 15) it is insensitive to feedstock quality, such as metals, nitrogen, asphaltenes or Conradson Carbon content. It can handle virtually any pumpable hydrocarbon (e.g. atmospheric and vacuum residue, tar sand bitumen, SDA pitch, heavy whole crudes, shale oil, Fluid Catalytic Cracking or Resid Catalytic Cracking heavy distillates, etc.). Liquid product yields mainly depend on the carbon content of the feed. The higher the carbon content, the lower the yield of cracked liquid.

Representative coking reactor yield data for vacuum residua include:

- 10-13 wt% gas product.
- 60-70 wt% total liquid product.
- 20-30 wt% reactor coke.

Downstream treatments of coker products depend on refining needs:

- Essentially complete removal of sulfur can be achieved in $C4^-$ gas. This material can then be used for refinery fuel, LPG recovery or H_2 manufacture. C3-C4 can be used as alkylation plant feed.
- More than 90% desulfurization can be achieved on coker naphtha *via* conventional hydrotreating. Hydrotreated naphtha is suitable as a component of the motor gasoline pool or as a petrochemical feedstock. After a severe hydrotreatment to hydrogenate the olefins and remove all sulfur and nitrogen species, it can also be sent to the reforming unit to increase the octane number.
- Coker gasoils can be directly converted to more valuable clean Fuels *via* Catalytic Cracking (FCC), hydrotreatment or hydrocracking processes.

17.1 DELAYED COKING

Delayed coking is the most widely used technology in the refining industry. In the United States or in Venezuela, where its use is very widespread, the main objective is to convert heavy residues into light products, with the coke being used as a solid fuel or for specific applications such as feeding thermal or cement plants. In Europe, Asia, the Far East and Middle Eastern countries, the objective is to produce a specific grade of coke that can be used – after calcination – to manufacture anodes for the aluminum industry (sponge coke [1]) or for the steel industry (needle coke [2]). Delayed coking is a semi-continuous process that can be adapted to a wide variety of feed types and produce different grades of coke depending on the feed structure and market needs.

This technology produces far more coke than any other residue conversion processes. The delayed coker produces less gas and total liquid product than the fluidized bed technology.

1. 80-85% of the calcinated petcoke market involves the production of consumable anodes for aluminium reduction (0.4-0.45% wt consumption vs. Al).
2. Used predominantly to produce graphite for electrodes used in steel manufacturing (9-10% of the calcinated petcoke market).

17.1.1 Process Description

In the delayed coking process (see Figure 17.1), the heavy residuum is pumped to the bottom of the combination tower, which acts as an accumulator for the feed to the coking furnace as well as a product fractionator.

The bottoms of the combination tower are pumped to the coking furnace where thermal cracking takes place at roughly 480°C to 515°C. Short residence times and high mass velocity (enhanced by steam injection near the furnace entrance) are required to avoid coke deposition in the furnace tubes.

The furnace outlet stream goes to an up-flow coke drum where the endothermic cracking reaction is completed. Coke is deposited in the drum at 415-465°C and 0.1 MPa to 0.4 MPa (1-4 bar), while the vapors are sent to the combination tower for separation into various product streams and recycled.

The upper portion of the combination tower is used to separate the cracked products from the coker unit.

Vapor from the top of the combination tower is condensed with a portion of the liquid naphtha returned to the tower as reflux, while the uncondensed vapor is sent to light ends.

The delayed coking process uses at least two (or more) coke drums which operate on sequenced 24-48 hour cycles so that one drum is in its coking cycle while the other drum is being decoked. When one coke drum is filled with coke, it is switched to a cooling and decoking cycle and an alternate drum is placed in coking service.

17.1.1.1 Feed Types

Delayed coking units generally process heavy residues such as:

- Vacuum residue from conventional and heavy crudes or atmospheric residue from heavy and extra-heavy crudes.
- Topped bitumen.
- Catalytic cracking decanted oil or slurry.
- Visbreaker residue or thermal cracking tar.
- C3 to C6 asphalt or lube oil extracts.
- Other.

Product yields and quality depend on four characteristics: the Conradson Carbon residue, sulfur content, metal content and distillation cut point or viscosity if the boiling point is too high to be measured.

17.1.1.2 Carbon Residue

There are two methods of determining this characteristic: the Conradson Carbon Residue (CCR) and the Ramsbottom methods [1]. The Conradson method is more frequently used, so

1. Residual carbon content corresponding to unconverted residue under standardized pyrolysis conditions.

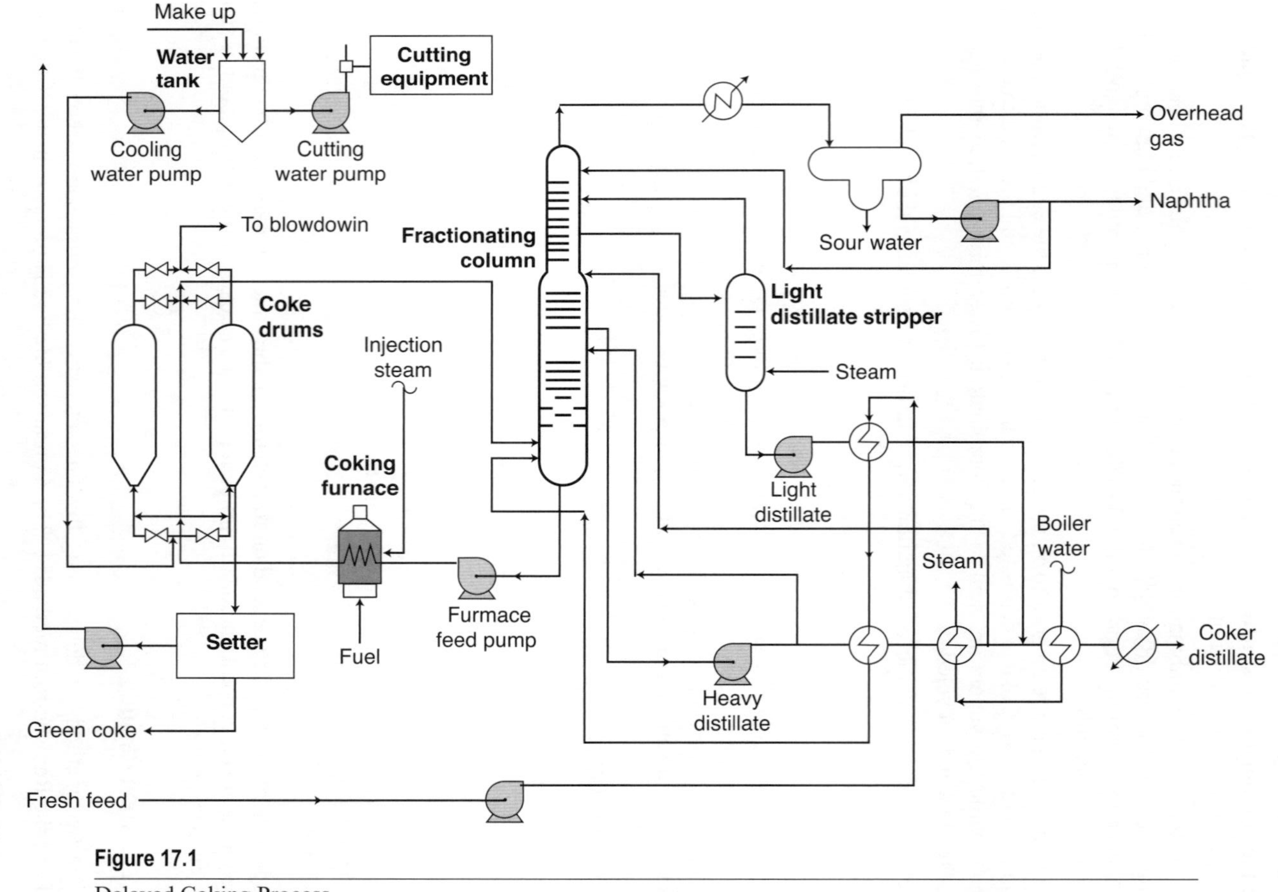

Figure 17.1

Delayed Coking Process.

Source: Petroleum Refining, Vol. 3 Conversion Processes, Leprince P, Editions Technip, 2001.

references herein will always refer to this value. The higher the CCR, the higher the coke yield at given operating conditions. Since the usual objective is to maximize the distillate yield – i.e. minimize coke production – the higher the CCR, the more difficult it is to achieve this objective. In the past, CCR values were normally lower than 10% weight in the atmospheric residues used as feed. Today, with the use of heavier residues from heavier crudes, CCR values in excess of 20% – or even 30% weight – can easily be found.

17.1.1.3 Sulfur and Metal Content

The sulfur in the residue tends to concentrate in the coke and in the heavy distillate. Similar to CCR, changes in feed types and origins have resulted in an increase in their sulfur content. Consequently, the sulfur content in coke is currently increasing.

Metals – mainly nickel and vanadium – also tend to be higher in feeds, and end up almost exclusively in the coke. A metal content of over 500 ppm weight or more can be found in coking unit feeds from heavy and extra-heavy crudes.

17.1.1.4 Distillation Cut Point

The distillation cut point for a typical Vacuum Residue (VR) is approximately 550°C for conventional crude oils, but it can vary depending on the origin of the crude and operation of the refinery vacuum distillation unit. For example, the cut point for VR from extra-heavy oils such as Alberta or Orinoco Belt, is reduced to approximately 500°C because of their very high viscosity. The distillation cut point is related to CCR, sulfur and metal content. The higher the cut point, the higher the impurity content. As such, it influences coking product yields and quality.

17.1.1.5 Chemical Reactions

The exact mechanism of coking reactions is relatively complex and this makes it rather difficult to provide a precise description of the reactions and changes taking place in the reactor. Two main mechanisms can be identified which co-exist for most of the feeds processed, and there are three separate and successive stages in the reactions that take place. The first mechanism involves dealkylation of cyclic compounds and cracking heteroatom reactions of high molecular weight asphaltenes and resins. The second mechanism consists of:

1) Partial vaporization and preliminary cracking of C-C bonds as the temperature is rising through the furnace to reactor temperature.
2) Cracking of the vapor phase as it passes through the coke drum.
3) Successive condensations and aggregation-precipitation of the liquid phase retained in the reactor until it is converted into coke.

These reactions are all endothermic and cause a drop in temperature of roughly 50°C in the reactor (coke drum). Since it is a purely thermal cracking, all the reactions described for the visbreaking process in Chapter 16 apply to the coker process (see Chapter 16 for further details).

17.1.2 Operating Conditions

Coking temperature has a direct impact on the Volatile Combustible Matter (VCM) of the coke produced. At higher temperatures, a larger proportion of the feed is vaporized at the reactor inlet, thereby causing less coke to be formed. It can therefore be asserted that coke yield decreases as temperature increases, at constant recycle ratio and pressure. However, at temperatures that are too high, the coke produced will be very hard and difficult to remove by hydraulic cutting systems. At temperatures that are too low, soft coke with a high VCM is produced.

At constant temperature and recycle ratio, an increase in pressure keeps more of the hydrocarbons in the liquid phase in the coke drum, thereby increasing coke and gas yields.

The recycle ratio has the same effect as pressure on product distribution: the higher the ratio, the higher the coke and gas yields. The recycle ratio controls the end point of the heavy distillate. The higher the ratio, the higher the quantity of heavy product that is recycled to the coking reactor where it is converted into coke and gases.

17.1.3 Typical Product Yields and Characteristics

Table 17.1 shows typical yields from various feeds, from light conventional to extra heavy oils (CCR from 5 to 22% and metals from 45.50 ppm to 900 ppm).

Table 17.1 Typical Yields at Constant Recycle Ratio vs. Feed Types.

Feedstock		**A**	**B**	**C**	**D**
Cut Point	**°C**	**485**	**485**	**540**	**540**
Density at 15°C	kg/l	0.983	0.955	1.015	1.047
Conradson Carbon	wt%	5.2	11.1	15.6	22.0
Sulfur	wt%	0.60	0.50	3.40	5.30
Metals (Ni + V)	ppm	50	44	90	910
Product					
Yields: $C4^-$	wt%	6.2	7.4	9.2	10.5
Yield: Naphtha C5-195°C	wt%	18.5	20.4	17.4	21.4
Density at 15°C	kg/l	0.757	0.733	0.748	0.762
Sulfur	wt%	0.10	0.20	0.50	0.90
Yield: Gasoil 195°C$^+$	wt%	65.3	54.5	48.5	33.0
Density at 15°C	kg/l	0.922	0.853	0.905	0.933
Yield: Coke	wt%	10.0	17.7	24.9	35.1
Sulfur	wt%	1.10	0.80	5.10	6.40
Metals (Ni + V)	ppm	500	249	361	2,592

Source: Petroleum Refining, Vol. 3 Conversion Processes, Leprince P, Editions Technip, 2001.

17.1.4 Products

17.1.4.1 Gas

The gas produced includes fuel gas used in the coking unit or sent to the refinery fuel gas network after desulfurization by amine washing, and LPG (C3-C4) with high olefin content.

17.1.4.2 Naphtha

After hydrotreating, the light naphtha is sent to the gasoline pool and, after a specific hydrotreatment required due to its high sulfur and nitrogen content, the heavy naphtha can be used as a catalytic reforming feed after deep hydrotreating, generally blended with SR naphtha. The direct route to the gasoline pool is not possible due to high diolefin content, which induces instability *via* gum formation, and high sulfur content.

17.1.4.3 Distillates

The light distillate (light coker gas oil) requires medium to high severity hydrotreating in order to stabilize it (color and olefin content), decrease its nitrogen and sulfur content and, in most cases, increase its cetane number before being sent to the refinery gasoil pool. The heavy distillate (heavy coker gas oil) can be used as feed for the FCC, mixed with straight run vacuum distillate. It can also be sent as feed (in a blend with a straight run vacuum gas oil called SR VGO) to a hydrocracker to boost the kerosene and gasoil yield from crude. It is a much more difficult feed to be hydrocracked than a SR VGO.

17.1.4.4 Coke

The type of coke produced – either electrode, metallurgical or fuel grade – has a significant effect on the overall economics of delayed coking. It is mandatory to have a long-term market for the coke in order to achieve acceptable profitability.

Coke produced in delayed coking contains Volatile Combustible Material (VCM) and moisture. Calcining is the treatment step where the moisture and VCM are reduced and the fixed carbon is raised to over 99 wt%. Graphitization is the process which converts calcinated coke into finished electrodes.

A. Coke for Anodes

This coke is called sponge coke because of its characteristic spongy appearance. Sponge coke has small, unconnected pores surrounded by very thick walls; it is the usual form of delayed coke. However, not all cokes of this type are necessarily of anode grade. To meet the specifications required for anode coke, the metal content must be low (below 200 wppm), and the same is true for the sulfur content (generally less than 1.8 to 2.5% weight). These two characteristics are directly related to the quality of the feed. However, this type of feed is often paraffinic and tends to produce low-density coke. Anode grade coke must be calcinated before being used. This often takes place in the refinery to reduce Volatile Combustible Material (VCM). The most important characteristic of this type of coke is its density after calcination, measured by VBD (Vibrated Bulk Density). It should be above 780 kg/m3. Green coke, i.e. rough coke before calcination, is characterized by its VCM (8-10 wt%) and its hardness. These two properties have a direct influence on the VBD of calcinated coke. Hardness is normally expressed by the HGI coefficient (Hardgrove Grindability Index), which should be roughly 80.

The choice of feed for a unit designed to produce anode coke is crucial, since the properties of the feed determine the characteristics of the coke produced. Feeds are usually residues from atmospheric or vacuum distillation of conventional low sulfur and relatively low metal contents. Other feedstocks can be the heavy distillates produced from Fluid Catalytic Cracking (FCC).

B. Needle Coke

Needle coke has large interconnected pores surrounded by thick walls. Needle coke is of a higher quality, produced by coking special aromatic tars such as decanted oil from Fluid Catalytic Cracking (FCC), thermal cracking tar, steam cracking residue and coal tar. Needle coke is used to manufacture graphite electrodes for electric arc steel mills. It must undergo calcination and other treatment before it attains its final characteristics. Its crystalline structure is in the form of small needles that are not only responsible for its name, but also its electric conductivity. High density and a particle size distribution exhibiting low fines content, as well as a low Coefficient of Thermal Expansion (CTE), are important properties in obtaining a high-quality calcinated coke.

C. Fuel Coke

Heating coke is by far the most commonly produced grade of coke worldwide. This type of coke is considered to be a byproduct of delayed coking operations. This grade of coke generally has high sulfur (3-7 wt%) and metal content (0.5-1 wt%). Physically, it looks like something between sponge coke and shot coke. Shot coke consists of ball bearing-like spheres of coke ranging from 0.5 to 1.5 cm in diameter. Shot coke usually has a low HGI and is generally used as fuel in cement plants. Fuel coke units are characterized by very low operating pressure coke drums that are enormous – up to 8.5 meters in diameter and up to 37 m in height.

17.1.5 Coking Cycle

A pair of coke drums is generally used in semi-continuous operation (from 2 to 6 depending on the capacity to be handled). While one of them is in the reaction stage, the other undergoes steam purging, cooling and coke removal by cutting with a high-pressure stream of water. These operations are normally carried out in a 24-hour cycle.

The decoking operation takes place first by a steam-out, then cooling which allows the reactor to be opened. This can be done manually or more often semi-automatically, but it can be entirely automated. Hydraulic cutting is the most common method of removing the coke from the drum. High-pressure jets of water are used to cut through the coke and remove it from the drum in layers. The mixture of coke and water coming out of the bottom is sent to the coke separation and handling section, in order to recover the water to be reused after removing most of the coke and fines.

17.1.6 Coke Calcination

In the step of calcining petroleum coke from a delayed coking unit, the green coke is subjected to very high temperatures (1,200 to 1,500°C) for 90 min. or more after it has been crushed. The goal is to remove the volatile compounds (VCM) from the green coke. This improves its physical characteristics, such as thermal conductivity and actual density. There are two calcination methods: rotary cylindrical furnace and rotary hearth.

17.1.7 Delayed Coking: Conclusion

Delayed coking is semi-continuous process for residue upgrading. On conventional heavy ends, the main reason to build a delay coker is to produce coke and achieve a conversion of residues into light products. Design is strongly related to the use of the coke. The advantage is to be able to produce green coke with lower sulfur content. The investment is moderate and the heat input is significant since the reactions are endothermic. The drawbacks of the delay coker are complex coke recovery, low on-stream factor due to the semi-continuous mode, and poor quality of products which forces the refinery to add a significant amount of hydrogen downstream of the coker process.

For heavy and extra-heavy oils, bitumens and tars, the goal is to achieve significant conversion into distillates. In this case, the coke can only be used as a solid fuel or as a gasification feed to produce hydrogen, synthetic gas, electricity and steam. An integrated coking process has been developed by Exxon to integrate the coking operation with coke combustion, and is called Flexicoking.

17.2 FLUID COKING – FLEXICOKING

Fluid Coking is a continuous process operating at more severe thermal cracking conditions. The feedstock is cracked in a continuous fluid solids process. The technology and process unit consists mainly of a fluidized bed reactor where coking takes place and a burner vessel where a portion of the coke is burned to provide heat for the process.

This process is extremely flexible. It will handle almost any residual feedstock that can be pumped and yields valuable distillate products, after hydrotreatment. The process operates at conditions similar to delayed coking, at essentially atmospheric pressure, but at a significantly higher thermal level: 510°C to 540°C.

The reaction produces small particles of petroleum coke, formed in the process itself, which circulate in a fluid state between the vessels and act as a heat transfer medium. Similar to the delayed coking process, the major disadvantage of fluid coking is the production of large quantities of coke product; it is subject to similar quality/marketing constraints. The main outlets for coke are the steel industry if the sulfur content is low, fuel used in steam/power plants and cement kilns. The normal outlet for coke is combustion in coke-fired steam boilers. A limitation on its use is that it requires the design of grassroots boilers with no severe environmental restrictions regarding sulfur emissions or a to add an expensive flue gas desufurization equipment.

Flexicoking is a residuum conversion process that combines coke gasification with conventional fluid coking. This process was licensed by Exxon in 1970s. It is very well-adapted for upgrading extra-heavy oils, bitumens and tars.

In the Flexicoking process, most of the coke produced – except for a small quantity containing high-ash and high-metal contents – is converted to low Btu fuel gas in the gasification reactor of the flexicoking unit. After a desulfurization step, the low Btu fuel gas contains less than 5% of the total sulfur contained in the gross coke, and its sulfur level is approximately equivalent to a fuel oil containing 0.25 wt% sulfur. Thus, particularly for high-sulfur feedstocks, flexicoking can be a very suitable conversion process since it produces marketable liquid products, plus a low Btu fuel gas that may be burned in areas to produce steam or possibly electricity. It virtually eliminates the constraint of marketing low-quality coke with a very high sulfur and high metal content.

The choice of flexicoking as a conversion process rests primarily on the ability of the refinery to effectively consume the large quantities of low Btu gas produced. Ideally, the low Btu gas should be used to replace higher-value petroleum products, such as low sulfur fuel oil or sales gas; however, if the disposition value of the coke gas is lower (e.g. power generation), it may be less costly to burn fluid coke directly and employ flue gas desulfurization to control sulfur and particulate emissions.

17.2.1 Flexicoking Process Description

A simplified diagram of a flexicoking unit is shown in Figure 17.2. The unit consists of a fluidized bed reactor, a series of cyclones and scrubbers at the top of the reactor, a heater where the coke circulating from the reactor is pre-heated by the gas and coke from gasification, a gasification reactor, a cooling system for the furnace overhead gas and a recovery system for coke fines.

The residue feed, pre-heated to approximately 250-350°C, enters the reactor where it is thermally cracked to produce a product containing gas, naphtha, distillates and coke that is laid down on particles of fluidized coke The overall heat of vaporization and the heat of reaction required by the residue are provided by a stream of hot coke coming from the furnace. The cracked products are cooled and washed. The light fractions rise to the top of the column and then go to a fractionator where they are separated into gas, naphtha and distillates for further treatment in the refinery. The coke from the coking reactor is sent to the furnace where it is heated by the coke and gas from the gasifier. Part of the circulating coke is routed from the furnace to gasification, where it is converted – in the presence of air and steam and at high temperatures (800 to 1,000°C) – into H_2, CO, N_2, $C0_2$, H_20, H_2S and a small amount of COS.

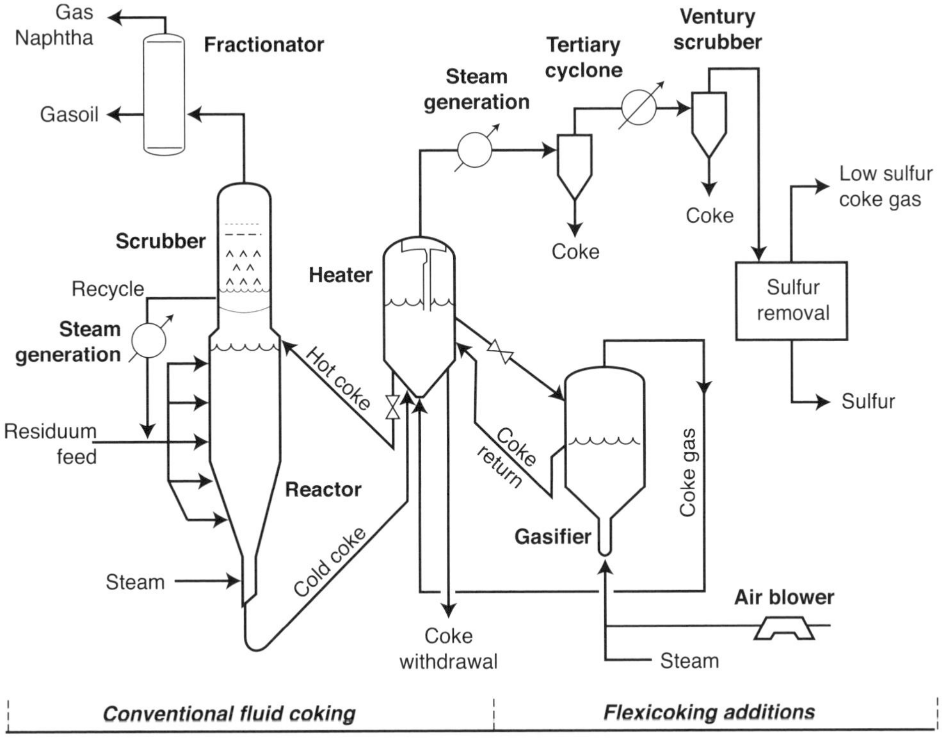

Figure 17.2

Flexicoking Process Simplified Flow Diagram.

Source: Petroleum Refining, Vol. 3 Conversion Processes, Leprince P, Editions Technip, 2001.

The gas produced by gasification, along with the particles of entrained coke, is sent back to the furnace. There it is cooled by contacting the coke from the reactor, thereby supplying part of the heat required in the reactor. A coke stream, sent from gasification to the furnace, provides the rest of the heat. The coke oven gas leaving this reaction section is used in a waste heat boiler to generate steam before going through cyclones and a washing unit, including a venturi and a washing column, to recover coke fines. The gas, minus its solids, is sent to the treatment section for H_2S removal.

On-stream factor: a normal limitation to run length has historically been cyclone snout coking. Average service factor over time is 91%. Recently, the factor has been improved to 95% for several units. Turnaround time for a fluid coker ranges from 10 to 30 days; flexi-cokers require up to 45 days.

17.2.2 Chemical Reactions

The coking and cracking reactions that occur in the fluidized bed are similar to those that take place in the delayed coking reactor.

Regarding the gasification reactions, standard oxidation reactions occur in the gasifier in the presence of air and steam. Two reaction rates are competing in the oxidation zone of the fluidized bed:

In the oxidation zone of the fluidized bed:

$$C + 1/2\ O_2 \rightarrow CO \qquad DH^0_{298} = -111\ \text{kJ/mol} \qquad (1)$$

$$CO + 1/2\ O_2 \rightarrow CO_2 \qquad DH^0_{298} = -283\ \text{kJ/mol} \qquad (2)$$

$$C + O_2 \rightarrow CO_2 \qquad DH^0_{298} = -394\ \text{kJ/mol} \qquad (3)$$

In the reduction zone:

$$C + CO_2 \leftrightarrow 2\ CO \qquad DH^0_{298} = 173\ \text{kJ/mol} \quad \text{(Boudouard reaction)} \qquad (4)$$

$$C + H_2O \leftrightarrow CO + H_2 \qquad DH^0_{298} = 131\ \text{kJ/mol} \quad \text{(water gas reaction)} \qquad (5)$$

$$CO + H_2O \leftrightarrow CO_2 + H_2 \qquad DH^0_{298} = -41\ \text{kJ/mol} \quad \text{(shift reaction)} \qquad (6)$$

Reactions 1 to 3 are very rapid and take place both on the surface of the coke and in the gaseous phase. Reactions 4 and 5 are much slower and determine the overall gasification reaction rate. Reaction 6 occurs exclusively in the gaseous phase and is an equilibrium reaction under the gasification conditions.

17.2.3 Typical Product Yields and Characteristics

Flexicoking yields of liquid hydrocarbon products are similar to yields obtained in delayed coking with one exception: there is no coke production and coke is transformed into gas. This is the main interest of flexicoking and fluid coking where there is no coke market. In the flexicoking process, approximately 97% of the initial coke is converted into gas with a low heating value (roughly 4,200 to 5,000 kJ/std m3). The unit produces a low Btu gas. This gas contains from 45 to 55% nitrogen. Its characteristics are very similar to those of blast furnace gas. Once the gas has been purified by amine treatment and solid particles have been removed, it becomes a safe and reliable fuel gas.

Because of the complexity of the flexicoking and fluid coking processes, the investment level is much higher than that for the delayed coking process (by a factor of 1.6; see section 17.3). This is the main reason why most of the installed coking units (see section 17.3) are based on the latter process.

The quality of the liquid products is lower than that of the delayed coker, due to the high severity of the flexicoker. Naphtha and distillates produced are processed in the refinery's hydrotreating units and then in the catalytic cracking or hydrocracking units, as well as in the desulfurization units which can recover 95% of the sulfur present in the feed.

An example of a flexicoking unit material balance is given for an Arabian Heavy feed in Table 17.2. It shows the estimated yields before gasification and the heating value of the gas from the gasification section. Product qualities are shown in Table 17.3.

Table 17.2 Flexicoking Product Yield on an Arabian Heavy Vacuum Residue.

	Units	Arabian Heavy
Feed Properties		
Cut range	°C	565+
Gravity	°API	3.0
Sulfur	wt%	6.0
Nitrogen	wppm	4,800
Conradson Carbon residue	wt%	27.7
Metals (Ni + V)	wppm	269
Yields Based on Fresh Feed	wt%	
H_2S		1.5
C_1-C_3		10.1
C_4		2.1
C_5-188°C		10.6
188-343°C		14.0
343-510°C		24.9
Purge coke		1.0
Gross coke		35.8
Gas Heat Content (LHV)		
Btu/scf		127
$X10^3$Btu/bbl fresh feed		1,370

17.2.3.1 Uses for Low Btu Gas

The gas with its H_2S and solids removed is a clean fuel that can replace natural gas and other fuels consumed in a refinery. After purification, the gas can be used as fuel in several ways: in specifically designed furnaces, or boilers with or without electricity generation by turbo-generators, or in gas turbines, although there has been no industrial experience as yet.

17.2.3.2 Use of Purge Coke

This coke comes from three different sources in the unit: bottom discharge from the fluidized coke bed, fines coming from the cyclones, and moist fines coming from the columns. It is fairly low in sulfur but high in metals, especially in the fines. The dry product is a relatively high-quality fuel. The moist fines can also be used as fuel or the metal they contain can be recovered in conventional facilities.

Table 17.3 Flexicoking Product Quality on an Arabian Heavy Vacuum Residue.

Cut (°C)	Unit	Naphtha C5-190°C	AGO 190-343°C	343-510°C
Density at 15°C	kg/l	0.750	0.895	0.987
Nitrogen	wt%	0.005	0.008	0.29
Sulfur	wt%			
Conradson Carbon	wt%			3.0
Bromine number	g/100g	120.0	60.0	30.0
Viscosity at 40°C	mm^2/s		3.0	
Viscosity at 100°C	mm^2/s			11.5
Pour point	°C		–23.0	28.0
Cetane Number			30.0	
PONA Analysis				
Paraffins	vol%	20.0	15.0	
Olefins	vol%	55.0	26.0	
Naphthenes	vol%	10.0	2.0	
Aromatics	vol%	15.0	57.0	
Low Btu Gas Composition (after treatment)				
CO	vol%	18.60		
CO_2	vol%	10.50		
H_2	vol%	21.00		
N_2	vol%	45.60		
CH_4	vol%	0.80		
COS	vol%	0.01		
H_20	vol%	3.50		

17.2.4 Flexicoking: Conclusion

The main disadvantages of flexicoking are the low Btu gas, poor product quality, coke fines outlet and high investment cost. The flexibility, high conversion level, high throughput (100,000 bpd) and heat generated are the main reasons to build a flexicoker. The high severity hydrotreatments or hydrocracking of coker liquid products is mandatory to produce clean fuel and fuel oil with the correct specification. The overall cost of the flexicoker and post-treatment is significant, but well-proven and efficient for a huge throughput associated with very difficult feedstocks containing high metal, sulfur and Conradson Carbon content.

17.3 HYDROTREATMENT OF COKER LIQUID PRODUCTS

As for the delayed coker, the flexicoking process provides the conversion of heavy bottoms to lighter distillates, but the liquid products must be further refined to meet stringent product demands.

Whatever is the process use, all coker products are very high in diolefin, olefin, nitrogen and sulfur content than any virgin or catalytically processed oils. Therefore severe hydrotreatment or hydrocracking of the liquid products is mandatory. By employing hydrotreating processing, impurities such as sulfur and nitrogen can be eliminated, and olefins plus diolefins can be saturated into more stable paraffinic and naphthenic structures. Multiple aromatic rings can be saturated to lower ring structures and to naphthenes, as required for downstream catalytic cracking feeds or for meeting diesel oil and jet fuel product specifications.

The three hydrotreating processes to be considered include:

- Naphtha Hydrotreatment unit.
- Diesel/Jet Hydrotreatment unit (aromatic saturation unit).
- Heavy Coker Gasoil Hydrotreatment or Hydrocracking unit.

17.3.1 Naphtha Hydrotreatment Unit

The objective is to produce a naphtha that is suitable to be processed in today's modern bimetallic catalytic reformers (usually < 0.1 vol% olefins and < 1.0 wppm of nitrogen and sulfur). A C_5-180/190°C coker naphtha feed is used. LPG treatment can also be done in this unit, if needed.

17.3.2 Jet Fuel/Diesel Hydrotreatment Unit

This unit can process the Jet Fuel/Diesel fraction 180/190°C-340/350°C alone or blended with other distillates, such as virgin distillates or catalytic cracking products. In addition to reducing olefin, sulfur and nitrogen content, this unit must also saturate aromatics to produce the required diesel cetane number for the diesel fraction and the required jet fuel smoke point. Proper selection of catalyst and operating conditions is required to attain the product quality goals. This unit is typified by relatively high to very high pressures, depending on the coker feedstock nature, and high hydrogen uptake. Thus, it must be properly designed to control the high exothermic heat of reaction of olefin and aromatic hydrogenation.

17.3.3 Heavy Coker Gasoil Hydrotreating or Hydrocracking Unit

The coker VGO severe hydrotreatment is mandatory before being used as a catalytic cracker feedstock. The unit produces a substantial quantity of middle distillate product and some naphtha as a result of the severe processing of heavy coker gas oil (81-87% HDN and 99% HDS).

This "conversion" of the feed to lighter products is primarily due to boiling point reduction associated with sulfur and nitrogen removal and hydrogenation of multi-ring aromatics. Some direct carbon-carbon bond breakage also occurs. While the additional distillate and naphtha products are of valuable economic benefit to the refiner, their quality might need to be improved if their disposition is into the automotive products market.

The distillate product, which is extremely low in sulfur and nitrogen relative to heating oil specifications, has a high aromatic content. This product would therefore not meet current specifications for jet and diesel fuels. Further hydrogenation of this material within the GO conventional HDS unit would not be efficient enough and is best carried out in a dedicated high hydrogen pressure unit.

The C5-180/190°C naphtha fraction produced in the GO hydrotreating unit is fairly high in organic nitrogen content, approximately 20 wppm. This material by itself would therefore not be a suitable feedstock for an existing virgin naphtha pre-treater upstream of a catalytic reformer. It can be easily upgraded in the coker naphtha hydrotreating unit. Alternatively, the naphtha could be treated in the Diesel/Jet hydrotreating unit or possibly simply blended with the virgin naphtha, depending on the final naphtha pool properties.

If a higher quality of naphtha, jet fuel and diesel is required, the preferred method is to send the coker VGO with virgin VGO to a high pressure hydrocracking unit with a specific hydrotreatment to reduce the high impurity level of the coker VGO to a very low level. Hydrogen consumption will be higher than with virgin VGO alone due to the aromaticity and impurities level.

17.4 CAPACITY, COKE PRODUCTION AND ECONOMICS FOR ALL TYPES OF COKING PROCESSES

Worldwide coking capacity, mainly based on the delayed coker process (roughly 90% of the total coking capacity, with 253 MMt/y in 2003), is shown in Table 17.4. It can be seen that roughly 70% of worldwide coking capacity is located in the Americas, with about 55% in North America. Capacity installed in Venezuela is also significant and is predicted to rapidly increase in the next 10 to 15 years.

Worldwide fluid and flexicoking capacity (Table 17.5) is much lower, with 31 MMt/y in 2003, and mainly concentrated in the Americas with 90% of the total capacity. Most of the worldwide capacity is concentrated in the United States of America, Canada and Venezuela.

Worldwide coking capacity (Table 17.6) accounted for 284 MMt/y in 2003, as shown in Table 17.6, with a total green petcoke production of 80.3 MMt/y. Worldwide coke production is forecast to be over 110 million t/y in 2010, up from roughly 90 million t/y in 2005.

By that time, it is estimated that roughly 30% of the world's petcoke production will be used as a carbon source in the metallurgical industry (electrodes, anodes and special metals). Nevertheless, the largest petcoke market is for combustion and energy production, since it is an attractive solid fuel with a higher heat value and low ash and moisture content (but a lower volatile content) as compared to coal. Currently, 65% of petcoke is co-fired with coal for power generation on boilers equipped with flue gas scrubbing devices to reduce SO_2 emissions. It is also directly burned as a kiln fuel in the cement industry. The latter application represents the main use for petcoke; a high sulfur content is desirable since sulfates are essential in this industry, and the use of high-sulfur coke reduces the calcium sulfate (gypsum) requirement.

It can also be used in gasification facilities for producing synthetic gas for the chemical industry (ammonia, methanol and DiMethyl Ether (DME)), or to produce hydrogen for the refining industry, plus electricity and steam. Currently, there are several gasification projects underway and this method of upgrading coke will undoubtedly increase in the future, based on high crude oil prices and development of the conversion of heavy and extra-heavy oils.

Due to the location of coke production, most of the petcoke used in the world is imported from the USA and Venezuela.

Table 17.4 Geographical Distribution of Delayed coking Capacity in 2008.

Region	Crude Capacity	Coking Capacity	Ratio Coking to Crude Capacity	% of Worldwide Coking
	k BPSD	k BPSD	%	%
North America	21,127	2,099	9.9%	54.6%
South America	6,725	452	6.7%	11.8%
Europe	16,933	405	2.4%	10.5%
CIS region	8,317	142	1.7%	3.7%
Middle East	6,993	83	1.2%	2.2%
Africa	3,358	79	2.3%	2.0%
Asia-Pacific	23,514	583	2.5%	15.2%
Worldwide	86,965	3,841	4.4%	100.0%

Source: IFP Energies nouvelles and Oil and Gas Journal.

Table 17.5 Geographical Distribution of Fluid + Flexicoking Capacity in 2008.

Region	Crude Capacity	Coking Capacity	Ratio Coking to Crude Capacity	% of Worldwide Coking
	k BPSD	k BPSD	%	%
North America	21,127	513	2.4%	53.2%
South America	6,725	77	1.1%	8.0%
Europe	16,933	82	0.5%	8.5%
CIS region	8,317	73	0.9%	7.6%
Middle East	6,993	0	0.0%	0.0%
Africa	3,358	0	0.0%	0.0%
Asia-Pacific	23,514	219	0.9%	22.7%
Worldwide	86,965	964	1.1%	100.0%

Source: IFP Energies nouvelles and Oil and Gas Journal.

Table 17.6 Geographical Distribution of Total Coking Capacity in 2008.

Region	Crude Capacity	Coking Capacity	Ratio Coking to Crude Capacity	% of Worldwide Coking	Coke Produced
	k BPSD	k BPSD	%	%	k t/d
North America	21,127	2,612	12.4%	54.4%	129
South America	6,725	529	7.9%	11.0%	25
Europe	16,933	487	2.9%	10.1%	16
CIS region	8,317	215	2.6%	4.5%	9
Middle East	6,993	83	1.2%	1.7%	3
Africa	3,358	79	2.3%	1.6%	2
Asia-Pacific	23,514	802	3.4%	16.7%	20
Worldwide	86,965	4,805	5.5%	100.0%	204

Source: IFP Energies nouvelles and Oil and Gas Journal.

A delayed coker requires approximately 2 m^3/year of antifoaming agent and 8 m^3/year of demulsifying agent.

For a conventional unit with two coke drums working on a 32 to 36-hour cycle (16-18 hours in production + 16-18 hours decoking), it is common practice to have four operators for the unit itself, plus three additional operators for the decoking crew if the decoking cycle is not automated. A third drum can be added to provide increased flexibility.

The limiting factor for a single train unit (two drums) is the size of the drum, with a maximum diameter of 30 feet and length of 120 feet, corresponding to 45,000 bpd capacity per drum, with coke production of roughly 2,500 t/d.

The Inside Battery Limits (ISBL) investment for a typical 20,000/25,000 bpd corresponding to a 1.1/1.4 MMt/y unit is roughly US $110-120 million for a delayed or a fluid coking unit and roughly US $170-190 million for a flexicoking unit, on a year 2006 US Gulf Coast basis.

REFERENCES

Aiba T, Kaji H (1983) *Chem. Eng. Progr.,* **79**, 1, pp 76.

Allan DE, Metvailer WJ, Wiechert S, King RC (1981) Advances in Fluid and Flexicoker Technology, *Chem. Engineering Progress,* **77**, pp 12.

Allan DE, Blaser DE, Lambert MM (1982) Flexicoking of Residue with Synthesis Gas Production, *NPRA Meeting*, San Antonio, Texas.

BP Statistical Review 2004.

SFA Pacific Phase 7 (2003) Upgrading Heavy Crude Oils and Residues to Transportation Fuels.

Deblase R, Elliot JD, Hartnett DE (1984) Delayed Coking Process Update, *188th Meeting Am. Chem. Soc.,* April 1984, St Louis.

Elliott JD (Foster Wheeler) (2002) Delayed Coking: Recycle to Zero Recycle Concepts & Considerations, paper AM-03-3, *NPRA Annual Meeting*, San Antonio, Texas, March 17-19, 2002.

Elliott JD (1990) Latest Coker Designs Increase Liquid Yields, Reduce Emissions, *Oil and Gas Journal,* Nov. 8.

Hammond DG (ExxonMobil) (1998) Fluid Coking and Flexicoking Fundamentals, Presentation at *AIChE Spring National Meeting*, New Orleans, Louisiana, March 8-12, 1998.

Janseen HR, leaman GL (1984) Improved Coking Design Can Up Liquid Yields, *Oil and Gas Journal,* June 25.

Logwinuk AK *et al.* (1982) *Hydrocarbon Processing,* **58**, 9, pp 119.

Mallik R, Hamilton G (2002) Delayed Coker Design Conditions and Project Execution, *NPRA Annual Meeting*, March 17-19, 2002.

Tatasuka T *et al.* (Chiyoda Co.) (1994) Renewed Attention to Eureka Process, *Residue and Heavy Oil Workshop*, Dahran, Saudi Arabia, Dec. 4-9, 1994.

Watari R *et al.* (Chiyoda Co.) (1997) Eureka in Combination with Texaco Gasifier Improves Refinery Integration, Paper AM 97-52, *NPRA Annual Meeting* San Antonio, Texas, March 16-18, 1997.

Weiss H (1985) Paper Presenting During a Meeting on New Lurgi Process Developments, Lurgi Head Offices, May 6-7, 1985.

Weiss H *et al.* (Lurgi) (1999) Low Cost Process for Residue Conversion, *Petroleum Quarterly*, Summer 1999, pp 79-83.

Yui S *et al.* (Syncrude Canada Ltd.) (2001) Processing Oils Sands Bitumen is Syncrude's R & D Focus, *Oil and Gas Journal,* April 23, pp 46-57.

18 Catalytic Hydrotreatment and Hydroconversion: Fixed Bed, Moving Bed, Ebullated Bed and Entrained Bed

J. Verstraete, D. Guillaume, M. Roy Auberger

18.1 CATALYTIC HYDROTREATMENT AND HYDROCONVERSION

As indicated by its name, catalytic hydrotreating consists in treating a feedstock in the presence of a catalyst under hydrogen pressure. The first industrial units for catalytic residue hydrotreating were built in the early 1960s with the aim of producing low-sulfur fuel oil. However, increasing market demand for light and middle distillates and decreased demand for heavy products led to the development of catalytic hydroconversion processes, whose main goal is to convert the atmospheric or vacuum residues into lighter, more valuable products. This intensive and continuing R&D effort has resulted in a large number of catalytic hydrotreating and hydroconversion processes. These processes can be classified into four categories (Figure 18.1), depending on the catalytic reactor technology used: fixed bed, moving bed (co- or counter-current), ebullated bed or entrained bed reactor.

Adding a catalyst to a thermal conversion process leads to a much more complex chemistry, since it promotes reactions that could not be carried out otherwise. Indeed, currently used catalysts initiate or accelerate a wide range of reactions that can occur consecutively or in parallel:

- Hydrogenation of aromatic cycles and heterocycles (HDAr).
- Hydrogenolysis of C-S and C-N bonds (HDS, HDN).
- Removal of metals (HDM).
- Disaggregation and cracking of asphaltenes (HDAs).
- Reduction of Conradson Carbon content (HDCCR).

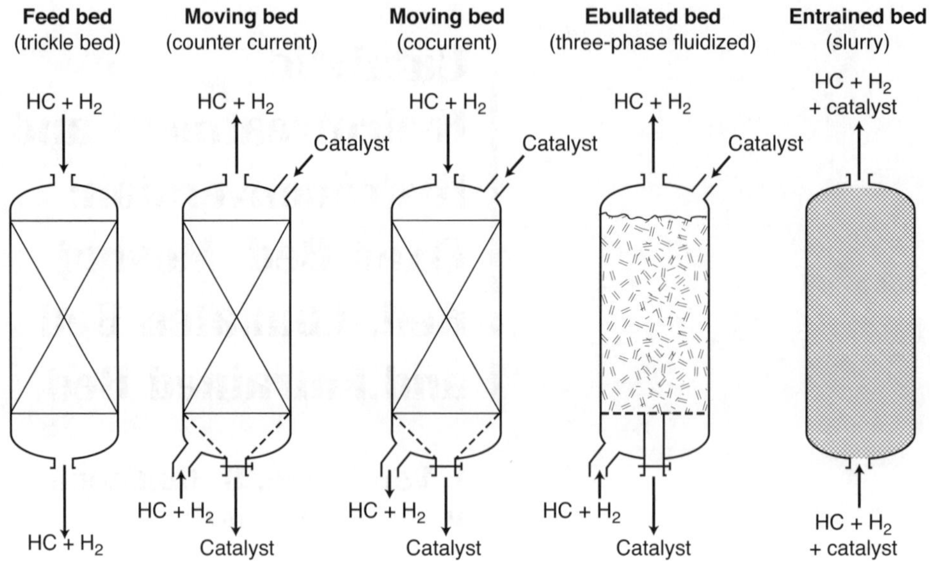

Figure 18.1

Types and Operation Modes of Catalytic Residue Hydrotreating and Hydroconversion Processes.

Hydroconversion reactions are therefore both thermal and catalytic in nature. Cracking reactions are mostly thermal reactions. They concern all molecules with sufficiently labile bonds and follow a radical formation mechanism. It should also be stressed that thermal reactions occur in the entire bed, i.e. including the void spaces in between the catalyst particles. Cracking reactions are always endothermic, but subsequent hydrogenation of the products is generally highly exothermic. Elimination reactions like those involved in HDS, HDM, etc. are mostly catalytic. The reaction sequence comprises the diffusion of the reactants inside the catalyst particles and their adsorption onto the catalytic surface. Once the reactions have taken place, the reaction products need to diffuse through the catalyst particle towards the bulk phase on the outside of the grain. These reactions are usually exothermic.

18.2 REACTION CHEMISTRY

18.2.1 Hydrodesulfurization (HDS)

This is one of the most important objectives of hydrotreatment. Specifications regarding sulfur content in both light and heavy fuels are becoming increasingly stringent. In addition, sulfur is also a contaminant for many downstream conversion processes.

Sulfur compounds may be present in oil residues in two forms [Speight, 2000]:

– Sulfides, which can be found in asphaltenes, where they connect naphtheno-aromatic rings through sulfide or disulfide bridges, can be catalytically decomposed as well as thermally decomposed, since their C–S bond is fragile (320 kJ/mol).

$$H_{59}C_{29}{-}S{-}CH_2{-}C_6H_4R \xrightarrow{H_2} C_{29}H_{60} + CH_3{-}C_6H_4R + H_2S$$

– Thiophenes, which can only be decomposed catalytically, may undergo two parallel reaction pathways that differ from each other at an intermediary hydrogenation before the hydrogenolysis step that removes the sulfur atom as H_2S.

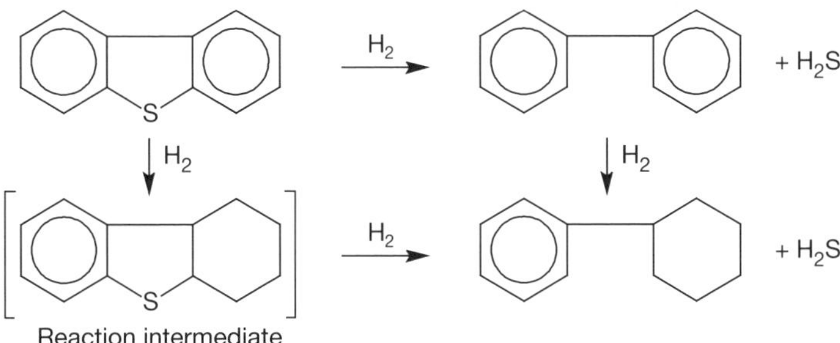

18.2.2 Hydrodemetallization (HDM)

In principle, all metals can be present in residues from the various crude oils, but those with the highest abundance are generally Ni and V. Elimination of metals from residues is an absolute necessity. Not only do metals poison the hydrodesulfurization catalysts, but they must also be removed if the hydrotreated residue is to be used as Low Sulfur Fuel Oil (LFSO). Unless metals are eliminated, they pose a major threat of corrosion as well as powder formation in combustion gases [Speight, 1991].

HDM reactions are known to be thermo-catalytic. Deposition of metals (mainly V and Ni) stems from the disaggregation of asphaltenes and resins and their partial hydrogenation [Toulhoat *et al.*, 1990]. Hydrogen sulfide also participates in demetallization reactions, where it leads to irreversible formation of metal sulfide deposits (V_3S_4 and Ni_3S_2). The metals that are not in porphyrin-type structures are considered to be more reactive, i.e. for metals included in porphyrins, the porphyrin structures need to be partially hydrogenated before metal removal can take place [Speight, 1991]. In general, the elimination of vanadium is generally faster than that of nickel.

It should also be mentioned that the sulfide layers deposited on the catalyst also have a catalytic effect on the HDM reactions, increasing the metal deposition rate [Toulhoat *et al.*, 2005]. As metal deposits continue to be laid down on the catalyst surface, they start to plug up the catalyst pores and eventually contribute – along with coking reactions – to catalyst deactivation. Depending on the structure of the catalyst, pore plugging may even occur at relatively low levels of metal deposition, since the metals are mainly deposited in the outer shell of the catalyst particles [Furimsky and Massoth, 1999; Verstraete *et al.*, 2007].

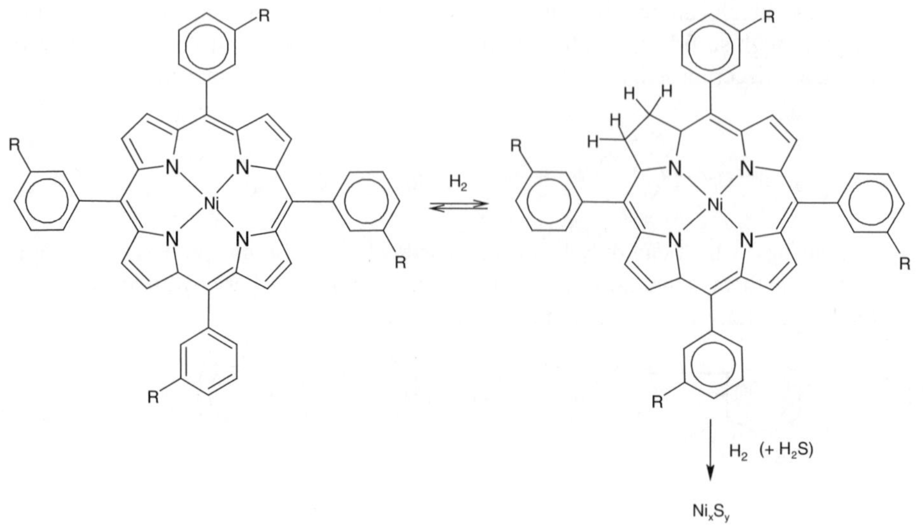

18.2.3 Hydrodenitrogenation (HDN)

Hydrodenitrogenation (HDN) has not been studied as intensively as HDS, mainly because in the past it was more important to remove sulfur from fuels than nitrogen, since the sulfur specifications were more severe. Furthermore, residue HDN is far more complex than HDS.

Nevertheless, HDN is very important since European regulations state that the amount of NOx in combustion gases should not exceed 450 mg/Nm^3. To achieve this goal, the nitrogen content in fuel oils must be as low as 0.2-0.4 wt%. Moreover, nitrogen is an extremely important contaminant for downstream conversion processes. Catalytic cracking catalysts used in FCC are strongly inhibited by nitrogen. The nitrogen content in FCC feedstocks should therefore preferably be lower than 0.2 wt%. With regard to feedstocks for hydrocracking units, the nitrogen content needs to be reduced to levels below 10 wtppm.

Nitrogen is present in residua in two main forms: non-basic heterocyclic structures (e.g. pyrroles, indoles, carbazoles, benzocarbazoles, etc.) and basic heterocyclic structures (e.g. pyridines, acridines, quinolines, benzoquinolines, etc.). The latter are less reactive and have a stronger tendency to be adsorbed on the catalyst, thereby inhibiting its activity.

HDN is catalytic in nature. It has been suggested that the first step in the HDN of nitrogen-containing aromatic molecules in oil fractions is the transformation of the strong aromatic C–N bonds into weaker aliphatic C–N bonds by hydrogenation of the heterocyclic aromatic molecules. Removal of the nitrogen atom from the resulting alkyl-amines can then take place by nucleophilic substitution and Hofmann -H elimination. Hence, contrary to thiophene HDS, there is no direct hydrogenolysis route for HDN. The intermediary hydrogenation step of the aromatic cycles is therefore always necessary. HDN will thus be more sensitive to hydrogen partial pressure than HDS, since nitrogen containing heterocycles must always be hydrogenated before hydrogenolysis of the C–N bond.

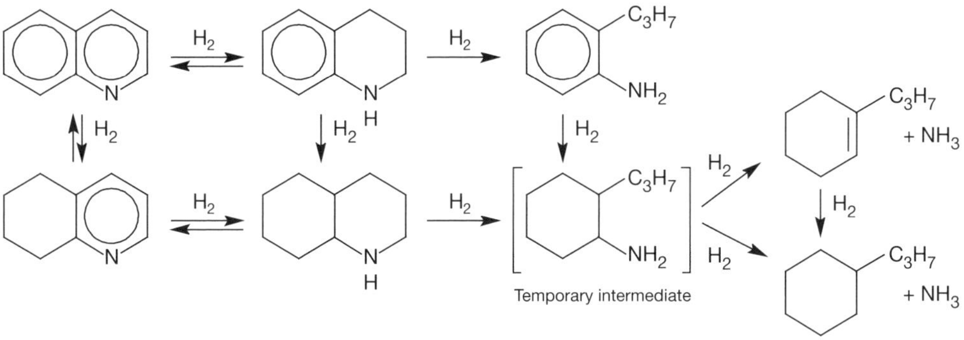

18.2.4 Hydrogenation of Aromatic Rings and Heterocyclic Compounds (HDAr, HDCCR)

Hydrogenation of aromatic rings and heterocyclic compounds leads to an increase in the H/C ratio of products and a decrease of Conradson Carbon by as much as 80%. The latter is important if the product is to be used as FCC feed, because the Conradson Carbon content of the feed to a resid FCC unit typically must be lower than 8-10 wt% [Ross *et al.*, 2000].

Hydrogenation increases the reactivity of heterocyclic compounds since they lose the stabilization provided by electronic resonance. It is therefore a crucial step in HDM and HDN and a very favorable one in HDS and ring-opening reactions. Polyaromatic compounds are more reactive than monoaromatic species.

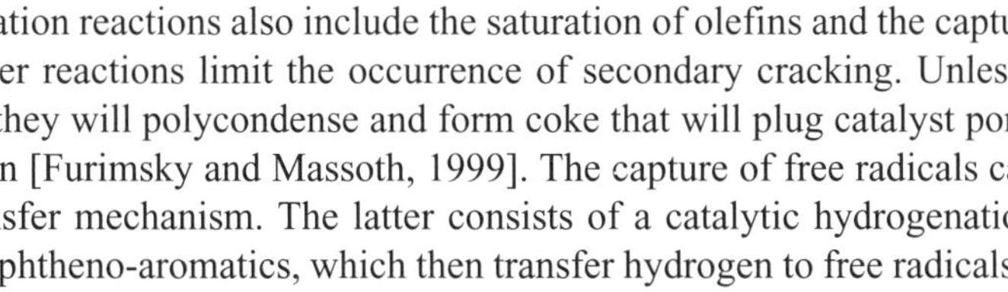

Hydrogenation reactions also include the saturation of olefins and the capture of free radicals. The latter reactions limit the occurrence of secondary cracking. Unless free radicals are captured, they will polycondense and form coke that will plug catalyst pores and lead to its deactivation [Furimsky and Massoth, 1999]. The capture of free radicals can be part of a hydrogen transfer mechanism. The latter consists of a catalytic hydrogenation of polyaromatics into naphtheno-aromatics, which then transfer hydrogen to free radicals present in the catalyst bed. During this step, naphtheno-aromatics revert to polyaromatics, which are then catalytically rehydrogenated. This hydrogen transfer mechanism constitutes the basis for the H-solvent donor type processes.

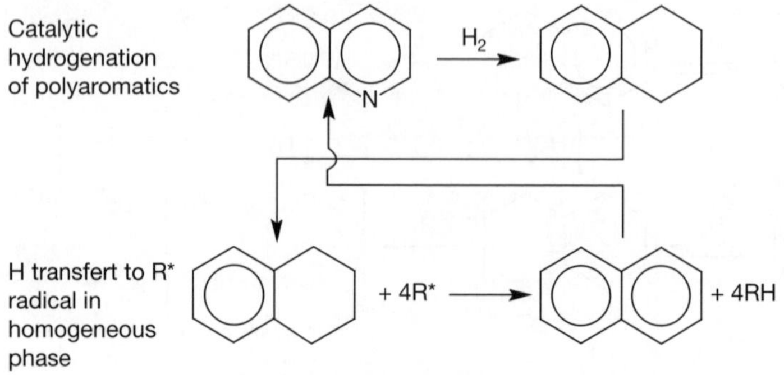

Catalytic hydrogenation of polyaromatics

H_2

H transfert to R* radical in homogeneous phase

+ 4R* → + 4RH

18.2.5 Cracking and Condensation Reactions

Cracking reactions are of tremendous importance since they are at the origin of the formation of light products with high added value such as gas and above all gasoline, as well as gasoil and vacuum distillates. The vacuum residue conversion can achieve up to 90 wt% under very severe conditions.

Although hydrogenation, hydrogenolysis and catalytic hydrocracking are three significant mechanisms to produce lighter fractions, when it comes to residues, thermal cracking at high temperatures is often considered to have the most important contribution. These thermal reactions follow a radical chain mechanism and take place in the entire reactor volume, including in the void spaces of the catalyst bed.

Radical chain mechanisms lead to the formation of both lighter fractions by secondary cracking of free radicals, and polyaromatic coke precursors by condensation of heavy radicals [Speight, 1991].

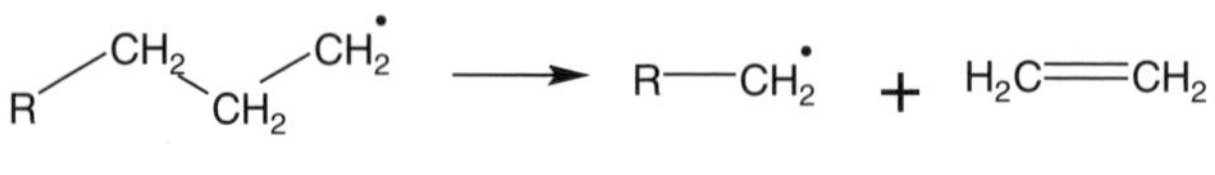

$$R{-}CH_2{-}CH_2{-}\dot{C}H_2 \longrightarrow R{-}\dot{C}H_2 + H_2C{=}CH_2$$

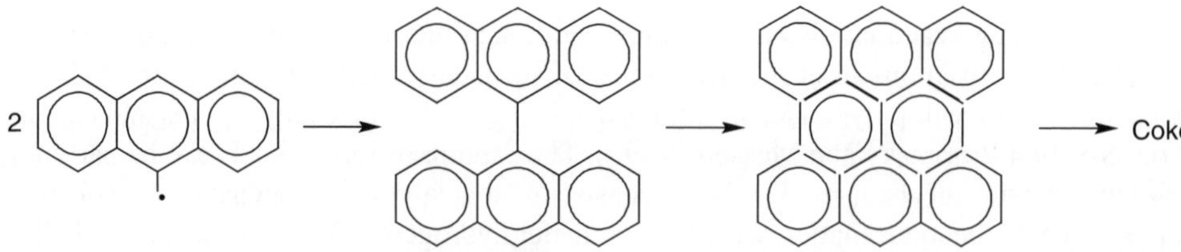

2 → → → Coke

These heavy radicals can be captured by activated hydrogen, terminating the chain mechanism and thus stopping their tendency to condense and form coke.

During these radical reactions, not all C–C bonds in the saturate, aromatic, resin and asphaltene molecules undergo thermal cracking to the same extent. Weak aliphatic C–C bonds are the preferential targets of thermal cracking. Despite this, naphthenic C–C bonds can also be cracked during the process, although to a lesser extent. The aromatic C–C bonds are stabilized by electronic resonance and are therefore less prone to thermal cracking.

Activation energies for thermal cracking and condensation reactions are roughly 160 kJ/mol and higher, while activation energies for hydrogenation and hydrogenolysis are typically in the region of 100 kJ/mol [Le Page *et al.*, 1992]. Such a difference may disturb the balance between catalytic and thermal reactions if the operating temperature is too high. Obviously, high temperatures will favor cracking and condensation reactions, and will therefore lead to significant coke and/or sediment formation [Fujita *et al.*, 2002].

18.3 CATALYTIC HYDROTREATING AND HYDROCONVERSION PROCESSES

As mentioned in the introduction, several types of reactor systems suitable for resid hydroprocessing have been developed and commercialized. Catalytic hydrotreating processes essentially differ in the configuration and type of reactor used. In each case, the main incentive was to find a way to extend the operation by avoiding frequent catalyst replacements. Reactors using granular catalysts can be fixed bed, moving bed or ebullated bed. Entrained bed operation is a different approach for extending cycle length. The various methods of operation have been schematically depicted in Figure 18.1.

Catalytic residue hydrotreating processes can hence be classified into four classes:

- Fixed bed processes.
- Moving bed processes.
- Ebullated bed processes.
- Entrained bed processes.

The main characteristics of these processes are listed in Table 18.1 and typical performances are displayed in Table 18.2.

Table 18.1 Main Characteristics of Residue Hydrotreating Processes.

Type of Process	Fixed Bed	Moving Bed	Ebullating Bed	Entrained Bed
Number of units (2007)	58	5	15	Demo only [a]
Max. feed Ni + V content (wtppm)	120-300	300-700	> 700	> 700
Tolerance for impurities	Low	Low	Average	High
Max. 550°C + conversion (wt%)	40	40	80	90+
Distillate quality [b]	Medium	Medium	Medium	Poor
Fuel oil stability	Yes	Yes	Borderline	No
Unit operability	Easy	Difficult	Difficult	Difficult

[a] *two industrial units are under construction and scheduled to be started in the near future*
[b] *rough distillate as it is without any hydrotreating, sequential hydrotreating or hydrocracking*

In practice, all resid hydroprocessing is carried out at a total pressure of 14 to 20 MPa (140 to 200 bar), temperature ranging from 380 to 480°C, hydrogen recycle of 300 to 1,500 Nm^3/m^3, and low space velocity, typically between 0.1 and 0.6 h^{-1}. In all processes, the reactors are therefore designed to operate at these high pressures and temperatures. For manufacturing and transportation reasons, their diameter does not typically exceed 5 to 5.5 m and they cannot be more than 25 to 30 m long. These constraints will therefore directly impact the maximum capacity per train. Hydroprocessing therefore involves much higher investment and operating costs than thermal processes, but provides greater selectivity for light products, which are also much better in quality.

Table 18.2 Typical Performance for Residue Hydrotreating Processes.

HDS	80-95%
HDM	80-99%
HDN	20-60%
HDCCR	50-80%
HDAs	50-90%
540°C + conversion	30-90%

18.3.1 Fixed Bed Processes

Fixed bed processes typically consist of a series of multi-bed adiabatic reactors. In each reactor, gas and liquid flow cocurrently from top to bottom in a trickle flow, with the gas phase as continuous phase. A typical process flowsheet is shown in Figure 18.2.

Since the reactor diameters are large, the quality of the fluid distribution at the top of the bed is very important. Indeed, poor initial distribution can result in non-irrigated and bypassed areas in the catalyst bed, reducing catalyst utilization. Moreover, these zones can rapidly become clogged, creating pressure drops in the catalyst bed that further enhance the poor fluid distribution.

To decrease the work that needs to be provided by the recycle compressor and to maintain a sufficiently high hydrogen partial pressure at the outlet of the last reactor, the pressure drop over the reactors must be minimized by [Le Page *et al.*, 1992]:

- Increasing the void fraction of the catalyst bed by optimizing the equivalent diameter of the catalyst grains.
- Decreasing the superficial linear velocities of the reaction fluids by reducing the L/D ratio of the total catalyst bed or using parallel trains.
- Decreasing the viscosity of the liquid feed, especially for very viscous vacuum residues, by diluting with a lighter product.

Minimizing reactor pressure drop at the start of a run does not suffice in and of itself. It is also necessary to ensure that the pressure drop evolves as slowly as possible up to the operating limit of the unit. Avoiding extremely high temperatures that cause polymerization and

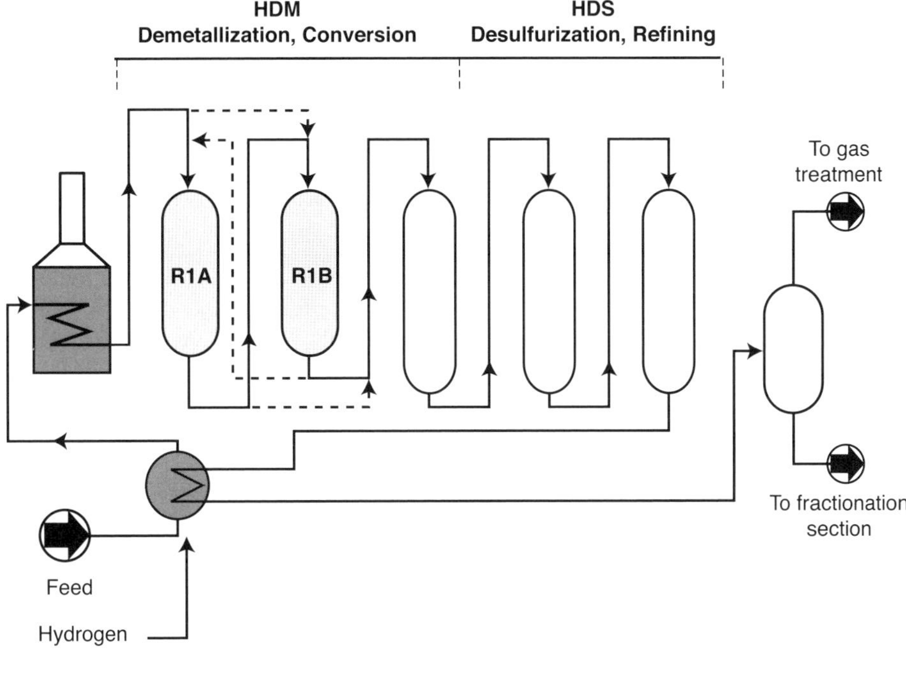

Figure 18.2

Example of a Typical Fixed Bed Process Flowsheet (Hyvahl Process).

condensation reactions, and maintaining a high hydrogen partial pressure are two ways to do this [Fujita *et al.*, 2002]. During the operation, solid impurities like salts and sediments may also plug the top of the catalyst bed leading to poor distribution of the reaction fluids and high pressure drops. Installation of a guard bed with adapted grading of large-size guard material and/or catalyst at the top of the first reactor helps to avoid the formation of a sediment crust [Furimsky, 1998]. In addition to grading the size and void fraction of the particles, associating catalysts with a grading of pore size and activity may also significantly increase the cycle length [Kressmann *et al.*, 1998]. The installation of filters and/or an additional desalting stage can also be a solution.

The temperatures and temperature profiles in the catalyst beds must be carefully controlled. Due to the exothermic character of the reactions, the temperature profile increases along the reactor. Temperature control is assured by quenches, i.e. injection of a cold fluid, such as recycled gas, either between two reactors or between two catalyst beds. After each quench, the reaction fluids must be redistributed.

Most fixed bed residue hydrotreating processes use at least two different catalyst types placed in series, each having specific and complementary roles. The first catalyst family is composed of one or more HDM catalysts, essentially designed to ensure that the feed arriving at the HDS catalyst section is demetallized. Its role is therefore to disaggregate asphaltenes and to remove the metals contained in the heavy molecules. The second catalyst family consists of the HDS catalysts. Their main function is to promote hydrorefining reactions through HDS, HDN and HDCCR. In many cases, an intermediate catalyst, in terms of HDM and HDS performance, is put between the two former catalysts. The proportions of each catalyst type must be adjusted depending on the feedstock and desired process performance [Furimsky, 1998]. Compared to the hydrodesulfurization catalyst alone, this combination of catalysts achieves higher demetallization levels for nearly the same desulfurization conversion.

The main limitation of using fixed bed reactors involves catalyst deactivation, which requires shut-down of the unit after a period of time [Rana *et al.*, 2007]. Fixed bed hydrotreating processes generally aim to achieve run lengths of up to one year. To maintain process performance during the entire cycle, the inlet temperatures of the catalyst beds are continually increased to compensate for catalyst deactivation. The maximum bed temperatures that can be allowed on fixed bed units depend on the hydrogen partial pressure, but are generally not above 420 to 430°C.

The metal content of the feed is one of the main parameters for selecting the catalyst types and configuration of the catalytic reactors. For feeds containing between 25 and 50 ppm, a dual catalyst system may be required, i.e., the front-end catalyst should possess a high metal tolerance, whereas the tail-end catalyst should have high catalytic activity. For metal contents ranging between 50 and 100 ppm, a triple catalyst system with a front-end HDM catalyst that has a high metal storage capacity may be required [Furimsky, 1998]. If guard beds are associated with the process, the feed metal content that can be tolerated is higher. When the feed metal content exceeds 200 ppm, two swing guard reactors may be necessary, with one in operation and one on standby. Alternatively, the guard reactors (Figure 18.2) can also be advantageously used in a permutable configuration [Kressmann *et al.*, 1998; Ross *et al.*, 2000]. In this case, the two guard reactors (A and B) are put on-stream in series at the unit start-up. When the lead reactor (A) can no longer maintain its performance or its operation, reactor A is bypassed and taken out of service. At this moment, reactor B is the only guard reactor in operation. After unloading reactor A and installing and conditioning the fresh catalyst, reactor A is put back into service in lag position. When reactor B can no longer maintain its performance or its operation, the scenario is repeated and the reactors are again permuted.

Above 300 to 350 ppm of metals in the feed, fixed bed hydrotreating processes can no longer be operated over 11-month cycles. In this case, the cycle length needs to be reduced or other technologies, such as moving bed or ebullated bed processes, must be employed.

How the configuration of catalytic reactors and catalyst types depends on feed metal content is shown in Figure 18.3.

Currently, several licensors offer fixed bed residue hydrotreating processes on the market [Rana *et al.*, 2007]. The main industrially-available processes are: RCD Unionfining process licensed by Universal Oil Products (UOP), RDS process licensed by Chevron Lummus Global (CLG), Hyvahl process licensed by Axens, RESIDfining process licensed by Kellogg Brown & Root (KBR) and ExxonMobil Research and Engineering (EMRE), and the Shell Residue Hydroconversion process licensed by Shell Global Solutions (SGS).

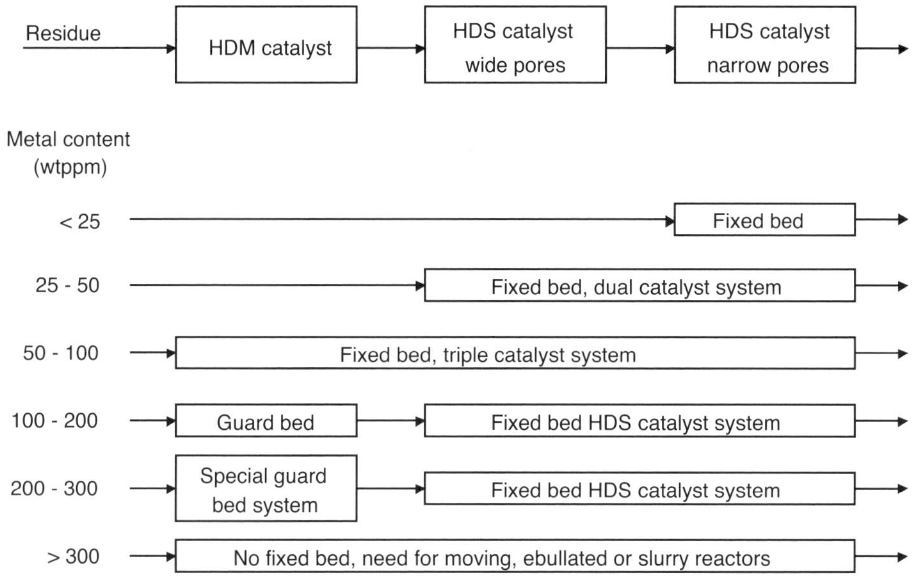

Figure 18.3

Configuration of Catalytic Reactors and Catalyst Types: Dependency on Feed Metal Content (adapted from [Furimsky, 1998] and [Moulijn *et al.*, 2001]).

18.3.2 Moving Bed Processes

Moving bed technology aims to increase the cycle length of a residue hydrotreating process, particularly for metal-rich feeds, by continuous catalyst renewal. In these processes, the catalyst moves slowly, by gravity, from the top to the bottom of a specially conceived reactor. The catalyst flow is close to plug flow, except at the inlet and outlet sections of the reactor.

In moving bed processes, the catalyst flow rate is much lower than the velocities of the fluids. Hence, the catalyst bed works in each point of the reactor as a fixed bed reactor, since the catalyst residence time is much higher than the fluid residence times. The main difference with the fixed bed process is that, for fixed conditions, a steady state can be reached in which the metal content on the catalyst in each point of the reactor remains constant, while in fixed bed processes, the metal content on the catalyst continually increases over time. This feature avoids annual shutdowns for catalyst replacement and allows operation at constant operating conditions, while achieving constant process performance and product quality.

In moving bed reactors, the use of spherical catalyst beads is recommended, as these particles flow better than extrudates due to the smaller internal angle of friction. Catalyst transfers from the fresh catalyst hoppers to the reactor and from the reactor to the spent catalyst hoppers are achieved by liquid lifts, while flow control is achieved by several series of special valves.

The catalyst and reactor fluids can circulate cocurrently in down-flow. This technology is used by Shell [Scheffer *et al.*, 1998] with Bunker-type reactors (Figure 18.4 and Figure 18.5). In the second variant of the moving bed process, the catalyst also circulates from top to bottom in the reactor, but the reaction fluids circulate from bottom to top counter-currently to the catalyst. The countercurrent moving bed technology is used by Chevron's OCR process (Figure 18.6) [Meyers, 2004]. The main advantage is that in this case, the sediments in the feed enter at the reactor bottom and are contacted by the most deactivated catalyst. Hence, the sediments are extracted from the reactor as soon as they are deposited. In the countercurrent moving bed reactor, the fluid velocity can be increased up to the point where the catalyst bed becomes slightly expanded. Hence, the observed pressure drop over the reactor is lower than for a fixed bed reactor or a co-current down-flow moving bed reactor, but this also requires that the bottom of the reactor is equipped with a proper gas-liquid distribution system. In this configuration, however, the linear velocities of the liquid and gas phases must remain relatively low to avoid lifting the catalyst bed, thus blocking the gravitational downward flow of the catalyst, and producing catalyst fines.

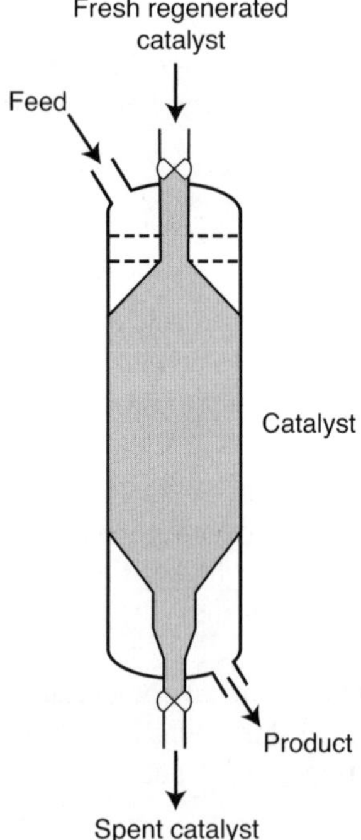

Figure 18.4

Shell Bunker Type Reactor.

Source: Resid and Heavy Oil Processing, Lepage JF *et al.*, Editions Technip, 1992.

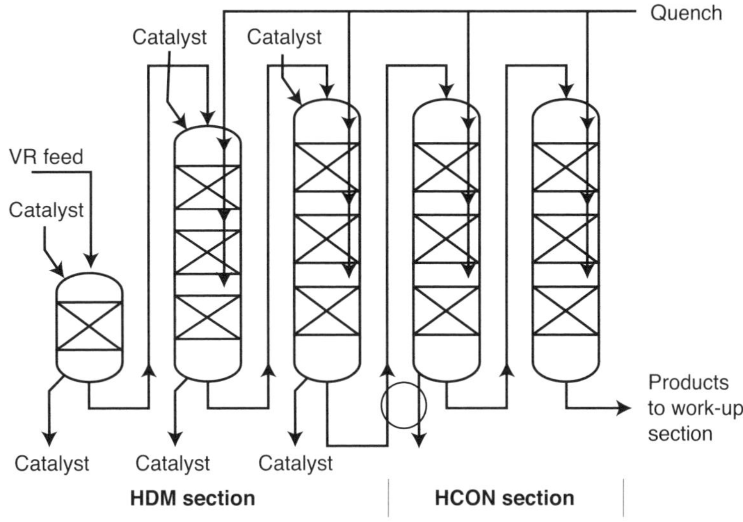

Figure 18.5

Shell Bunker Type Reactor System.
Source: Shell, Catalysis Today, vol. 43, pp. 217-224, 1998.

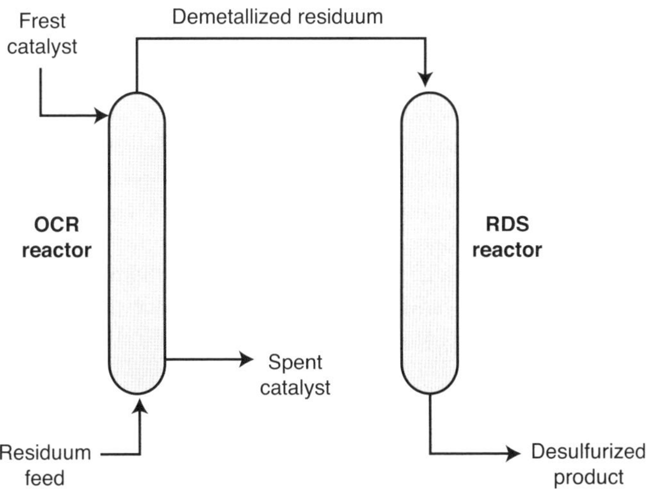

Figure 18.6

Chevron's OCR Type Reactor System.
Source: Chevron, 1994.

Even with metal-rich feeds, moving bed reactors are only used for the first reactor(s) in a hydrotreating process, where the initial conversion and most of the demetallization and asphaltene disaggregation is carried out. Behind the moving bed reactors, the other reactors generally use conventional fixed bed technologies.

As mentioned above, several moving bed processes are industrially available, e.g. the Onstream Catalyst Replacement (OCR) technology licensed by Chevron Lummus Global and the HYCON Bunker technology licensed by Shell Global Solutions.

18.3.3 Ebullated Bed Processes

The ebullated bed technology also aims to increase the time between two shut-downs of a residue hydrotreating process by continual catalyst renewal. To this end, the catalyst is held in a fluidized state by means of a significant liquid recycle. In this way, a fraction of the catalyst inside the reactor can be withdrawn and replaced by fresh catalyst on a daily basis [Wisdom and Colyar, 1996; Ross *et al.*, 2000]. After an initial stabilization period, a steady state will be reached in catalyst age and activity distribution, resulting in a so-called equilibrium catalyst [Fujita *et al.*, 2002]. Hence, an ebullated bed process allows operation at constant operating conditions with constant process performance and product quality over time. Therefore, the run length for the ebullated bed reactor system does not depend on the decline of catalyst activity – as in a fixed bed system – but is set by the refiner's inspection and turnaround schedule for the entire processing facility, which is typically 24 to 36 months.

A schematic representation of an ebullated bed reactor is shown in the Figure 18.7. In such a reactor, the catalyst is fluidized through the upward lift of liquid reactants (feed oil plus recycle oil coming from the ebullating pump) and gas (hydrogen feed and recycle) which enter the reactor plenum and are distributed across the bed by means of a distributor grid plate. The catalyst bed expansion level typically lies between 1.3 and 1.5 times the settled bed level. The height of the ebullated catalyst bed is measured by density detectors and controlled by the flow rate of recycled liquid, which is adjusted by varying the speed of the ebullating pump (a canned centrifugal pump). The ebullating pump recovers its recycle liquid from an internal vapor/liquid separator, called the recycle cup [Kressmann *et al.*, 2000]. This separation of vapor and liquid to provide feed to the ebullating pump can also be accomplished outside of the ebullated bed reactor in a hot high-pressure separator. A very important feature is the ability to add fresh catalyst to and withdraw spent catalyst from the reactor, thereby controlling the average catalyst age, and hence the activity of the equilibrium catalyst inside the reactor. Moreover, this also allows for changing the type of catalyst, in order to adjust for different feedstocks or levels of desired performance, without shutting down the reactor [Colyar and Wisdom, 1997].

Ebullated bed reactors have the characteristics of a CSTR [1] reactor and a fluidized bed reactor. Operation in an ebullated state therefore results in a back-mixed, nearly isothermal bed with a low and constant reactor pressure drop, because bed plugging and channeling are eliminated. These reactor characteristics allow for control of reaction exotherms and

1. Continuous Stirred Tank Reactor corresponding to a perfectly well mixed reactor.

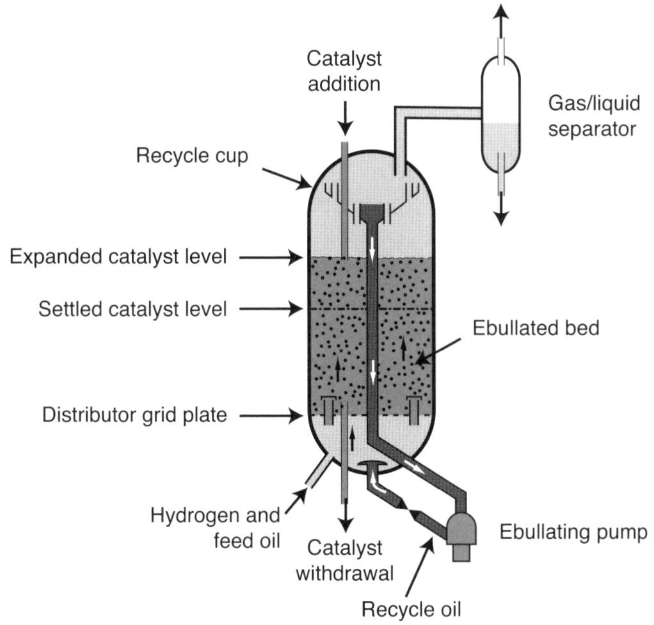

Figure 18.7

Ebullated Bed Reactor Technology (H-Oil$_{RC}$ process).

processing of feedstocks containing solids, while providing a more flexible way to change feedstocks and operating objectives than in a fixed bed reactor system. The nearly isothermal profile in the ebullated bed reactor also avoids temperature hotspots and therefore allows operation at higher temperatures than fixed bed reactors, thus achieving higher conversion levels. The drawbacks of a back-mixed reàctor is low hydrogenation efficiency compared to a near plug flow fixed bed reactor and the loss of reaction volume related to the 1.3 to 1.5 times expansion of the catalyst bed.

In an ebullated bed reactor, the catalyst reaches equilibrium activity after a relatively short period of time (typically around 1 month). The activity level of this equilibrium catalyst is determined by the level of feed contaminants, catalyst replacement rate for the reactor and operating conditions [Fujita *et al.*, 2002]. At this equilibrium catalyst activity, the ebullated bed reactors will therefore produce a product of constant quantity and quality. Because of this advantage, ebullated beds are well-suited for applications requiring a long, uninterrupted run length. Hence, ebullated beds are more suited for heavier feeds or for the most severe conversion requirements [Speight, 2000]. Indeed, the ebullated bed process is designed to handle the most problematic feeds, such as vacuum residues and heavy oils with a high content of asphaltenes and including sediments [Wisdom, 1995]. Their flexibility also allows them to be used for coal liquefaction and coprocessing mixtures of heavy feeds with coals and plastics [Furimsky, 1998].

A process scheme for an ebullated reactor process is shown in Figure 18.8. The process typically consists of 2 to 3 reactors in series, followed by a separation section and a compression section. In between the reactors, an Inter-stage Separation System can be provided to eliminate the lightest fraction before entering the next reactor [Gauthier *et al.*, 2007]. Because of back-mixing in the reactor section, temperature regulation of the exothermic hydrotreating reactions is accomplished by injecting low temperature feed oil into the unit. In order to maintain control of the reactor, a separate hydrogen heater and oil heater is necessary for safety reasons. Control of the temperature and quench oil of the reaction section can be ensured through hydrogen bypass of the reaction system. At the outlet of the reaction section, a series of decreasing-temperature separators is used to separate the gaseous fraction. To recover the hydrogen, this gas fraction is scrubbed, re-concentrated, compressed and recycled. The liquid fractions from the separators are flashed and distilled.

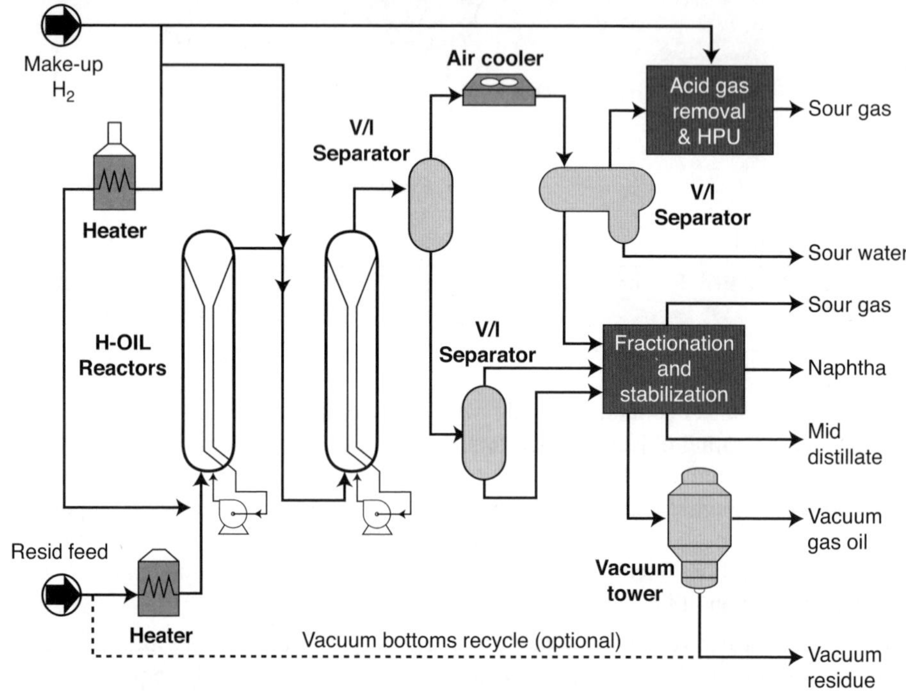

Figure 18.8

Example of a Typical Ebullated Bed Process Flowsheet (H-Oil$_{RC}$ process).

Ebullated bed processes have high selectivity for middle distillates and vacuum gas oil. Other yields include a gasoline and a gas fraction, containing H_2S, NH_3 and H_2O, which are formed through the heteroatom removal reactions. However, the yield pattern is a strong function of the level of residue conversion. At higher conversions, the selectivity towards lighter products, including light gases, increases rapidly. As for the unconverted residue fraction, it is upgraded relative to the vacuum residue fraction in the feedstock and thus has a greater utility and value. Most commercial applications are in the 50-70 vol% residue conversion range with a desulfurization level of approximately 70-85 wt%. Chemical hydrogen consumption is typically 200 Nm^3/m^3 of feed or 1,200 SCF/bbl. Relative to fixed bed processes, this value appears high; however, most of the additional hydrogen is a result of the higher vacuum residue hydrocracking severity. The typical performance of an ebullated bed unit is shown in Table 18.3.

Table 18.3 Typical Performance of an Ebullated Bed Unit.

Residue conversion	vol%	45-85
Hydrodesulfurization	wt%	55-90
CCR conversion	wt%	45-65
Nitrogen removal	wt%	25-35
Metal removal	wt%	65-95
Chem. H_2 consumption	Nm^3/m^3	130-300

To summarize, ebullated bed processes offer several important advantages as compared to fixed bed processes:

- Can process feeds that are very rich in metals and sediments without plugging problems.
- Reactor pressure drops are much lower.
- Since exothermicity is controlled and there is no bed coking and plugging to be feared, higher temperatures may be used, which is favorable for free radical cracking reactions and allows the achievement of much higher conversions.
- Continual catalyst addition allows the achievement of constant activity and hence constant process performance.

Several drawbacks should also be mentioned:

- Reactors are highly back-mixed reactors, which have a lower efficiency than plug flow reactors, although this is partially compensated for by using two or three reactors in series. The catalyst is therefore less effectively used than in a fixed bed reactor, resulting in lower purification performance.
- Reactors are not completely filled with catalyst and the catalyst bed density is 1.3 to 1.5 times less dense than in a fixed bed reactor due to bed expansion.
- Operation at higher temperature not only favors conversion, but also favors condensation reactions, leading to a lower quality residue (viscosity, impurities, etc.).
- Catalyst is in motion, which will cause some attrition, even though its mechanical properties are good. The resulting smaller particles will subsequently be entrained with the fluid flow, increasing catalyst consumption and downstream fouling.

There are currently two ebullated bed residue hydroconversion processes which are industrially available on the market: the H-Oil$_{RC}$ process licensed by Axens, and the LC-Fining process licensed by Chevron Lummus Global (CLG).

18.3.4 Slurry Bed Processes

Slurry bed hydroconversion processes aim to fully convert vacuum resid fractions into lighter fractions in the presence of a dispersed catalyst. To achieve this objective, highly severe conditions in terms of temperature, hydrogen partial pressure and residence time must be used. The dispersed catalyst or catalyst precursor is injected on a continuous basis with the feed, promoting hydrogenation of radicals formed by thermal cracking reactions, and limiting coke formation. The catalyst utilized provides the catalytic activity but also a surface for the deposition of metals and asphaltenes. As the catalyst travels with the feed and passes through the reaction section, it is recovered in the unconverted vacuum residue fraction after the fractionation section. Hence, the slurry bed reactor for conversion of heavy petroleum fractions is in most cases an empty plug-flow type vessel, since the heavy oil/catalyst suspension behaves as a homogenous phase. It is clear that a high solid concentration is required in the hydrocracking reactor to reach near-complete conversion. This can be achieved by combinations of large fresh catalyst make-up rate, high reactor back-mixing and recycling the unconverted residue which contains the catalyst.

The theoretical advantages of slurry bed processes reside in a much better hydrogenation, especially of the heaviest products, thanks to better accessibility of the active sites, resulting in a higher conversion, improved product quality and higher product stability. Moreover, due to the lower catalyst residence time, catalyst deactivation is strongly reduced and relatively high HDS and HDM levels can be achieved, although the HDN level remains low. Overall, hydrorefining (HDS, HDN, HDM and HDAr) is lower in slurry processes than in fixed bed conversion processes, but the quality of the residue at a given conversion is not very different *versus* the quality of the residue from an ebullated bed conversion process. Integrated hydrotreating of the various products therefore seems a necessary component of these highly thermal slurry bed hydroconversion processes. The main drawback, however, lies in the injection of the dispersed catalyst. If high concentrations of catalyst are needed, the operating cost will be high and a catalyst or metal recovery process will need to be installed on the vacuum residue outlet stream.

The main applications for slurry catalyst processes concern the heaviest vacuum residue feeds, which cannot be treated in fixed bed systems and would result in very high catalyst consumption in ebullated bed systems. The main targets for these extra-high conversion processes are applications where there is no market for heavy fuel oils or high sulfur coke.

Currently, all slurry conversion processes are in the development or demonstration stage. Nevertheless two commercialization announcements have been made and these units are now (end of 2009) under early construction with a starting date scheduled in 2012/2013. They are based on the ENI/EST (23,000 bpd) and HDHPLUS®/Intevep-PDVSA 50,000 bpd) processes. Table 18.4 lists the slurry bed processes that have been developed. The most advanced developments seem to be EST (ENI), HDHPLUS® (Intevep-PdVSA), $(HC)^3$ now known as HCAT (Headwaters) and CANMET (Petro-Canada), UOP (SRC Uniflex^{TM}) based on the CANMET technology On the other hand, the process developments for MHR (Idemitsu/Kellogg), M-coke (Exxon), VCC (Veba) and Aurabon (UOP) seem to have stopped. Chevron developed a slurry process called CASH in the 1990s at laboratory scale, recently called VRSH for the industrial process, but limited information is available.

Recently a new process was proposed by Mobis/RIPI (HRH). An example of a slurry process scheme is shown in Figure 18.9.

As mentioned previously, the main drawbacks of slurry bed processes concern the high initial investment needed, high operating cost – mainly due to large catalyst requirements and significant hydrogen consumption – and disposal of the unconverted residue and spent catalyst. To industrialize a slurry conversion process, the need for large amounts of catalyst will need to be counterbalanced. Hence, these processes will need a cheap but active catalyst, which can be recovered fairly easily and, if necessary, regenerated. Alternatively, catalyst consumption could also be decreased by enhancing back-mixing in the reactor to increase catalyst concentration, or by recycling the unconverted residue containing the catalyst to the reactor inlet (EST process). All these actions would provide a more catalytic than thermal conversion of the heaviest product. Finally, a last concern for the scale-up of these slurry bed processes is the ability to ensure long term stable and coke-free operation.

Table 18.4 Slurry bed hydroconversion processes.

Process	Company	Catalyst Precursor(s)	Development Scale	Development Status
Aurabon	UOP	Ni + V	pilot plant	discontinued
SRC Uniflex™ based on the Canmet Technology	Petro-Canada, then UOP (SRC Process)	Fe	5,000 bpd	ongoing (SRC process)
CASH/ VRSH®.	Chevron	dispersed sulfided compound	30 bpd, then 3,500 bpd scheduled	ongoing
Gelcat	HTI	Fe + Mo	2 kg/hr	ongoing
EST	Eni-Snamprogetti	organic (?) Mo species	0.3 bpd & 1,200 bpd	20 kbpd announced/under construction
HCAT™, $(HC)^3$	HTI, Headwaters	oil-soluble Mo compound	1 kg/hr	industrial scale-up tests ongoing
HDH	PDVSA	Fe/clay	150 bpd	see HDHPLUS
HDHPLUS®	PDVSA	metals + additive	0.3 and 10 bpd	50 kbpd announced/under construction
HTC	Toyo – IMP	Fe on carbon support	small pilot plant	discontinued ?
MCoke	Exxon	soluble Mo + *in situ* coke	pilot plant	discontinued
MHR	Idemitsu – Kellogg	supported metal sulfides	10 bpd	discontinued
SOC	Asahi/NMRC	Mo + carbon black	3,500 bpd	discontinued ?
SPH	Kobe Steel Ltd	Fe	3 bpd	discontinued ?
VCC	Veba	Fe + lignite	4,000 bpd	discontinued
HRH	Mobis/RIPI	Nano particles of MoS_2?	pilot plant 2 bpd & 200 bpd	ongoing

Source: Beardon *et al.*, 1997; Benham *et al.*, 1996; Fukuyama *et al.*, 2002; Lott *et al.*, 2000; Marzen *et al.*, 1995; Montanari *et al.*, 2003; Silverman *et al.*, 1995; plus press releases and recent communications to congresses.

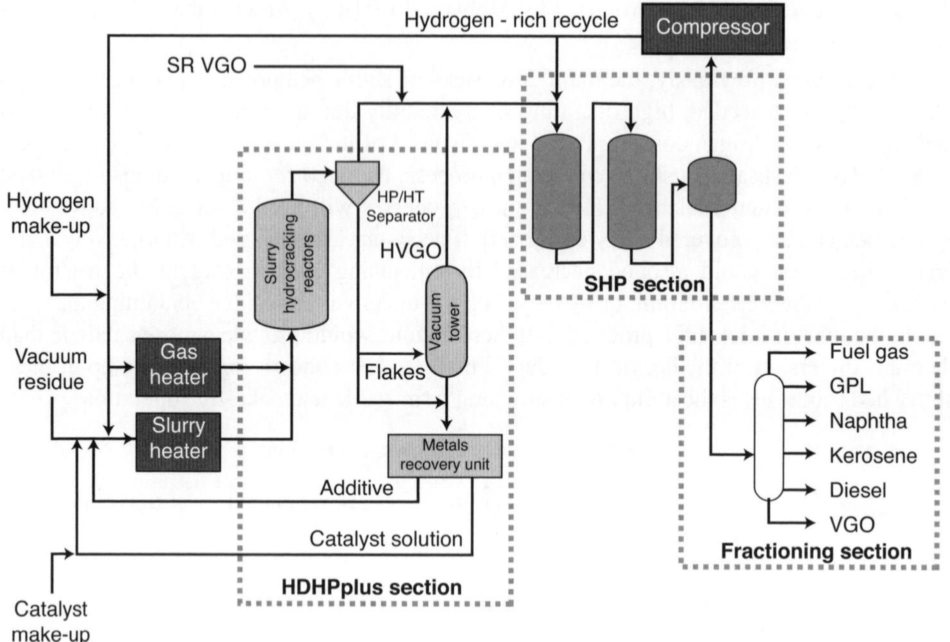

Figure 18.9

Example of a Typical Slurry Bed Process Flowsheet (HDHPLUS® with SHP/ Sequential HydroProcessing).
Source: PDVSA.

18.4 CATALYST CHARACTERISTICS

18.4.1 Catalysts

Most heavy oil catalytic hydroconversion processes involve the use of supported catalysts: the active phase is dispersed on a support (preferably alumina) which has a specific surface and a porosity adapted to heavy oil processing [Ancheyta *et al.*, 2005; Leprince, 2001]. Catalysts are available in various shapes and sizes: cylindrical or multilobe, extrudates and beads (Figure 18.10). or nano-particles (several tens of microns or less) injected as such in the feedstock or formed *in situ* in the reactor from a precursor injected with the feedstock. Most often, these catalysts are self-supported.

18.4.2 Catalyst Active Phase

The function of the catalyst active phase is to: hydrogenate aromatic and heterocyclic structures leading to an increase in the H/C ratio of products (improvement in their quality),

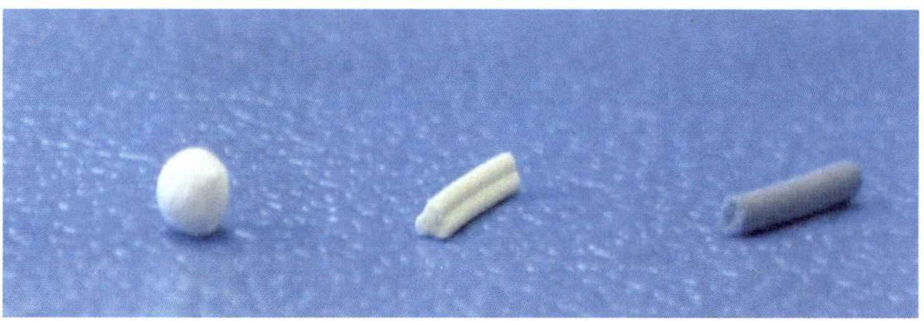

Figure 18.10

Examples of Catalyst Supports.
Left to right: Beads, trilobe extrudates, cylindrical extrudates.

hydrogenate radicals produced by thermal cracking reactions before they can recondense to form coke precursors and sediments, and hydrogenolyze C-N, C-S and N-Metal bonds, thereby allowing N, S and Ni + V contaminants to be eliminated.

The active phases that allow these reactions are composed of group VI transition metal sulfides (most often molybdenum) promoted by group VIII elements (generally nickel and/or cobalt). These phases are also those of distillate hydrotreatment catalysts.

The promoted molybdenum sulfide is shaped as truncated hexagonal and/or triangular layers (2-5 nm in length) organized according to a lamellar structure (with average packing of 1.5 layers), with the promoter atoms "decorating" the layer edges [Topsoe *et al.*, 1996], as shown in Figure 18.11.

Initial active sites are molybdenum atoms at the edge or corner exhibiting coordination insaturation whose activity is elevated by the presence of cobalt and/or nickel atoms (mixed sites). Since they are deficient in electrons, these sites behave like Lewis acid sites capable of adsorbing electron-donating molecules like sulfur and nitrogen containing molecules, as well as aromatic hydrocarbons. The molecules thus adsorbed may undergo hydrogenolysis reactions (*via* E2 elimination mechanisms) [Bataille *et al.*, 2000] and/or hydrogenation reactions *via* the addition of "activated" hydrogen in the form of hydrides ($MoH^{-\delta}$) and protons ($SH^{+\delta}$). These species are the result of the heterolytic dissociation of molecular hydrogen on pairs of Mo and S sulfur anion sites [Kasztelan *et al.*, 1994; Rana *et al.*, 2000].

If the sulfur and nitrogen resulting from the feedstock's sulfur and nitrogen containing molecules are in the form of H_2S and NH_3 in the gas phase, the metals of the feedstock are deposited on molybdenum sulfide crystallites. This metal deposition results in poisoning of the hydrorefining functions of promoted molybdenum sulfide crystallites, as shown in Figure 18.12.

Although this deposition is prohibitive for the activity of residue HDS catalysts, it is an integral part of the process of metal capturing on hydrodemetallation (HDM) catalysts. On the latter, promoted molybdenum sulfide crystallites form anchoring points on which metals from the feedstock will be deposited. After several hundred hours under feedstock flow, the initial active phase is completely covered by metals from the feedstock, thus leading to formation of a phase based on Ni and V sulfides supported on the initial sulfide phase. On

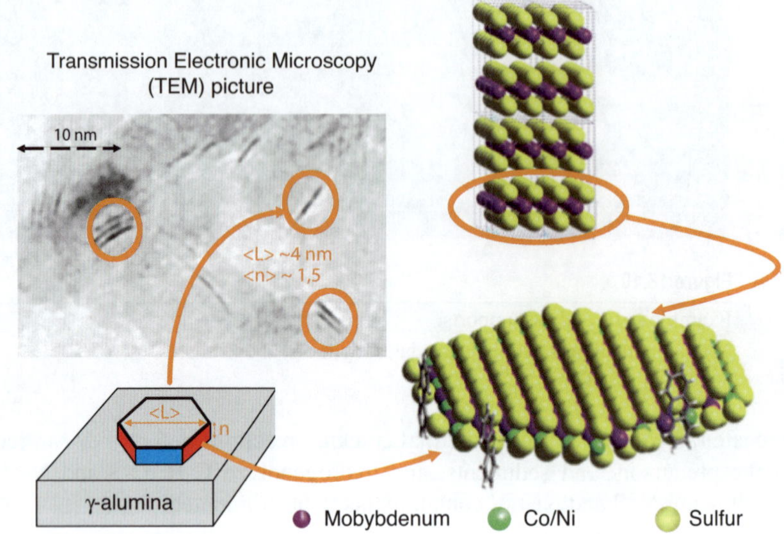

Figure 18.11

Supported MoS_2 Layers Characterized by Transmission Electron Microscopy and Modeled *via ab initio* Molecular Modeling (*IFP Energies nouvelles* project [Raybaud, 2007]).

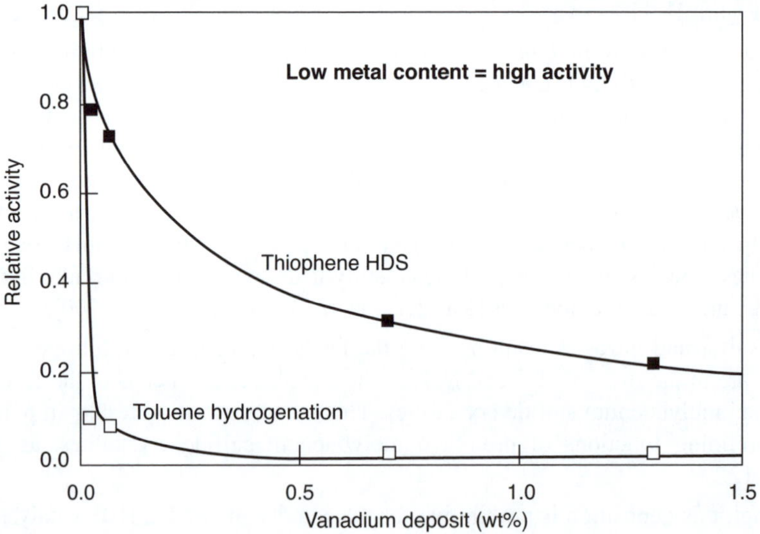

Figure 18.12

Performance of Model Molecules Against Quantity of Vanadium Deposited on an NiMo Catalyst Supported on Alumina During Tests on Atmospheric Residues (*IFP Energies nouvelles* project [Kressmann *et al.*, 1999]).

supported catalysts, this new phase remains sufficiently active in hydrogenolysis and hydrogenation to ensure demetallation up to the maximum metal retention capacity of the support, while limiting excessive coking. The origin of the activity of this phase is subject to debate: activity belonging to a sulfide phase based on vanadium or nickel [Toulhoat *et al.*, 1990] or promoter action of vanadium on the initial molybdenum sulfide [Guillard *et al.*, 1990].

On supported catalysts, dispersion of the initial active phase is a key parameter to maximize metal retention, as is the support's texture and porosity. Metal retention increases with the number of anchoring points on the support. This control of initial active phase dispersion is performed during preparation of the oxide phase precursor of the sulfide phase. In general, this oxide phase is prepared by "dry" impregnation of the support with a solution of metal salts to be deposited, followed by adapted stages of maturation, drying and calcination. Selection of the metal salts to be deposited is an important parameter that will impact both dispersion of the active phase and quality of the promotion that will result after the sulfurization stage [Griboval *et al.*, 1997; Martin *et al.*, 2005; Van de Water *et al.*, 2005].

The sulfiding stage transforms the oxide phase into a sulfide phase, which is the active phase. This processing is generally performed *in situ* in hydrotreatment and hydroconversion facilities. For fixed bed facilities, this stage is performed at temperatures between 350 and 400°C under hydrogen pressure, with a feedstock of gasoil or Vacuum Gas Oil (VGO) plus an organo-soluble compound like dimethyl disulfide (DMDS) which generates hydrogen sulfide (sulfiding agent). In ebullated bed or mobile bed units, only the catalyst inventory is sulfided at start-up of the unit, generally with VGO under conditions comparable to those used to sulfide fixed bed catalysts. On the other hand, the fresh catalyst added as a daily aliquot is not sulfided prior to its contact with the feed. It is injected in the reactor in a mix with the hydrocarbon feedstock [gasoil, VGO or Light Cycle Oil (LCO)] used to carry it between the weigh hoppers and the reactor. Sulfiding occurs at its introduction in the reactor upon contact of the feed and H_2S resulting from hydrogenolysis reactions of sulfided molecules.

Finally, the active phase is adjusted in terms of formulation ("NiMoS", "CoMoS", "NiCoMoS", "MoS_2", etc.) and concentration in accordance with targeted applications: fixed bed, mobile bed, ebullated bed and slurry (transported bed).

18.4.2.1 Fixed Bed Supported Catalysts

The fixed bed catalytic system is a combination of several families of catalysts placed in series in the reactors, with each having specific and complementary roles:

- The first family essentially ensures the conversion of asphaltenes (HDAs: elimination of asphaltenes insoluble in *n*-heptane) and hydrodemetallization (HDM) of the feedstock, i.e. essentially hydrodemetallization of nickel (HDNi) and vanadium (HDV). It is preceded by a guard bed used to eliminate impurities (in the form of solid particles) from the feedstock, such as iron, salts, sediments,...
- A second family is composed of one or several catalysts designed for the deep refining reactions of hydrodesulfurization (HDS), hydrodenitrogenation (HDN) and Conradson Carbon reduction (HDCCR).
- Between these two families, intermediate catalysts can be used to finish the HDM and start the hydrorefining reactions.

These catalysts are in extruded form, generally multilobe or in the form of beads. Their size is adapted to reduce diffusional limitations (small sizes) and to avoid creating a pressure loss that is too large.

A. Guard Materials

Guard materials are used to filter solid particles contained in heavy oils, which will be deposited on the initial layers of the catalyst, resulting in a significant pressure loss or even clogging of the catalytic bed.

These particles have various origins: iron in organic or mineral form, but also sodium chloride, coke and/or products of thermal degradation coming from furnaces, storage, sediments, etc. Their size varies from several microns to several tens of microns.

Guard materials have no catalytic action, and play only a role in filtration: their specific shape is designed to trap these large-sized particles on the external surface, in intergranular spaces or in large macropores. As an example, Figure 18.13 shows various shapes of guard materials offered by catalyst manufacturers.

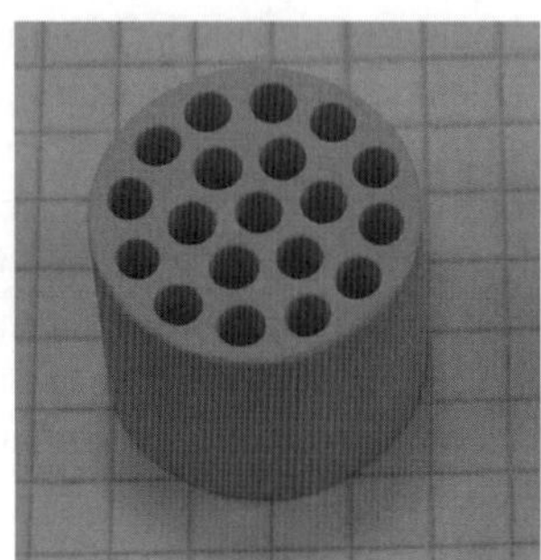

Figure 18.13

Examples of Various Shapes of Guard Materials.
Left to right: Miniliths ($\varnothing_{cyl}$ of 20 to 25 mm), fluted extrudates ($\varnothing_{ext}$ from 10 to 15 mm) and wagon wheels ($\varnothing_{ext}$ from 10 to 25 mm)

B. HDM Catalysts

Asphaltenes, which concentrate most metals of the feedstock, have a harmful action on hydrorefining catalysts. This action is manifested by direct inhibition of active sites *via* selective adsorption, excessive coke deposition, and metal deposition which gradually destroys the activity either by direct inhibition of active sites or over time by obstruction of catalyst pores. This is why the main function of HDM catalysts is the conversion of asphaltenes and the hydrodemetallization of the feedstock. They must also have sufficient metal retention capacity to ensure convenient cycle durations. To a lesser degree, they participate in HDS and HDCCR reactions [Kressmann *et al.*, 1998].

A key characteristic of HDM catalysts is the porous distribution of their support (generally alumina) which must include:

- High porous volume to capture a large quantity of Ni and V sulfides.
- Mesoporosity to develop the active surface.
- Macroporosity to facilitate access of asphaltenes to the active surface.

The mesoporosity must be centered on a population of pores that maximizes the active surface accessible to asphaltenes. As shown in Figure 18.14, this active surface resulting in maximum HDAs and HDV activity is obtained for pore diameters of 100 to 200 angstroms.

Figure 18.15 shows the two types of pore distribution encountered for HDM catalysts: bimodal distribution and multimodal distribution [Kressmann *et al.*, 2004]. Bimodal distribution is characterized by the presence of two separate groups of pores, while multimodal distribution is characterized by a continuum of pores between macropores and mesopores. Catalysts of bimodal porosity are generally more active (greater active surface) than those of multimodal porosity, but have less metal retention capacity despite comparable pore volumes. The metal retention capacity of bimodal catalysts does not exceed 60 to 80% by weight Ni + V as compared to the mass of the fresh catalyst, *versus* retention rates greater than 100% for multimodal catalysts. These differences in retention are mainly associated with differences in texture.

Multimodal distribution is linked to a "chestnut bur" porous structure shown in Figure 18.16. This structure [Toulhoat *et al.*, 1985] is composed of a plurality of juxtaposed agglomerates, each formed from a plurality of acicular platelets, the platelets of each agglomerate being oriented radially in relation to one another and relative to the center of

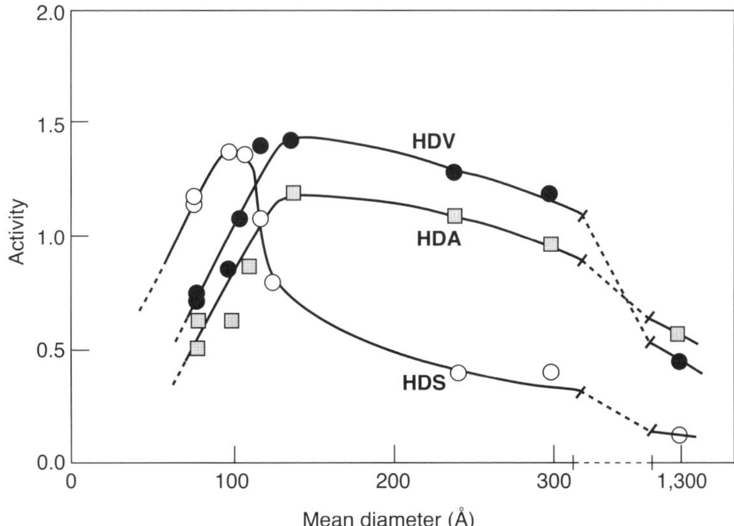

Figure 18.14

Effect of Average Pore Diameter on Catalytic Activity [Plumail, 1983].

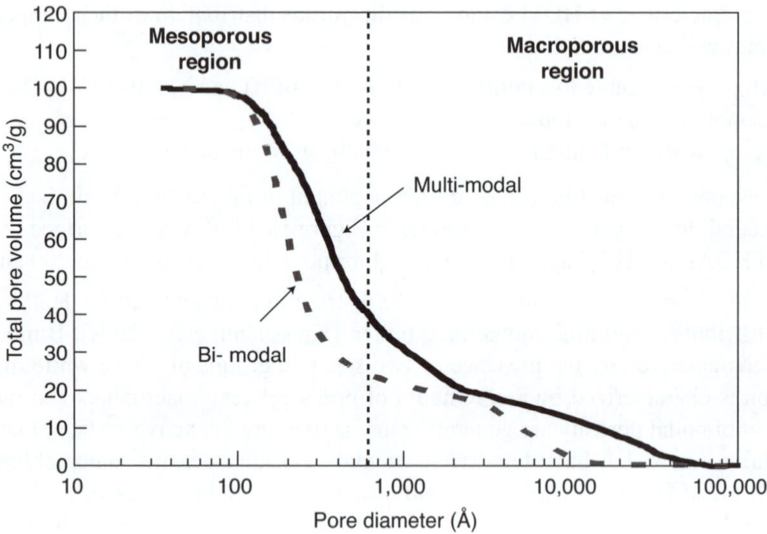

Figure 18.15

Pore Distribution (by Hg porosity) of HDM Catalysts.

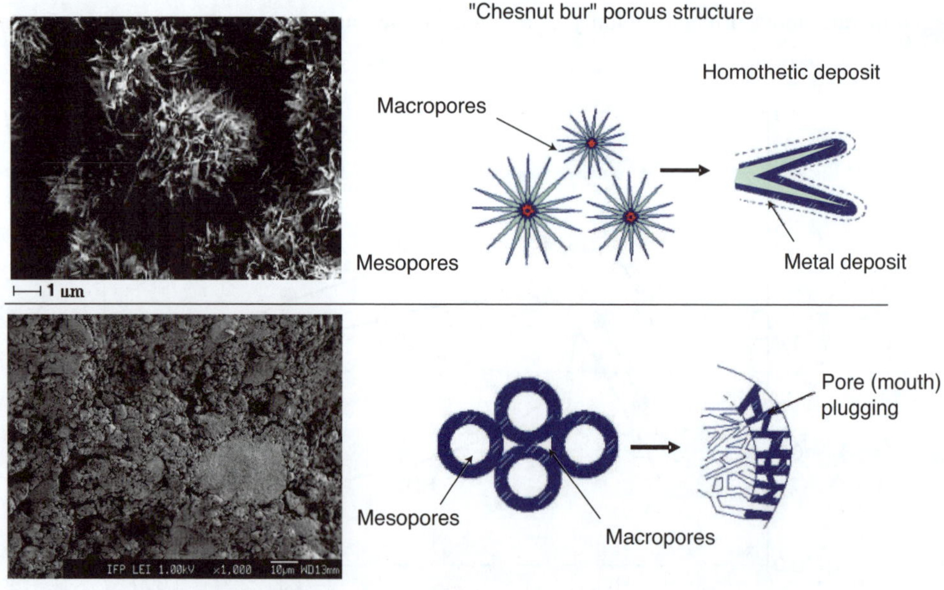

Figure 18.16

Various Textures of HDM Catalysts (Characterized by Electronic Scanning Microscopy) and Consequences for Clogging –
Taken and adapted from [Toulhoat *et al.*, 1985].

the agglomerate. Most pores are composed of free spaces located between the radiating acicular platelets. Since these pores are "in corners", they are of continually variable diameter from 100 to 1,000 Å. The network of interconnected macropores results from the space left free between the juxtaposed agglomerates. This texture allows for homothetic deposition of metals, thereby avoiding the pore clogging phenomena that occur on catalysts of conventional texture. The supports of these catalysts are in fact formed by the agglomeration of mesoporous alumina particles which seal along their periphery. This sealing makes the internal mesoporosity inaccessible, thereby explaining the lower metal retention capacity of catalysts of conventional texture.

These two types of porous structures are obtained *via* different fabrication methods: the chestnut bur texture is obtained with alumina agglomerates prepared by rapid dehydration of alumina hydrargillite flashed and autoclaved, [US 4,552,650 Patent] while bimodal supports are generally prepared by extrusion from an alumina gel of boehmite or pseudoboehmite type, or by co-mixing an alumina feedstock with a binder which may be an alumina gel [US 5,089,463 Patent].

Finally, metal deposition on the HDM catalyst is also accompanied by coking, together leading to inactivation of the catalyst. In the case of fixed beds, the inactivation is compensated for by a gradual increase in temperature. During the catalytic cycle, although the metals are deposited in gradual fashion on the catalyst, the coke is deposited very quickly *via* adsorption of the most polar compounds to the alumina surface, and then reaches a plateau linked to the thermal level [Leprince, 2001]. Inactivation by coke is predominant for high-temperature operations. It is minimized by using a support whose surface is as little acidic as possible (adsorbing few basic compounds of the feedstock, including resins and asphaltenes) and by using a hydrogenating initial active phase based on nickel and molybdenum.

C. HDS Catalysts

HDS catalysts must promote deep hydrorefining reactions: HDS of course, but also HDN and HDCCR. They are placed downstream of HDM catalysts and therefore receive feedstock that is mostly demetallized and de-asphalted. As such, the pore distribution of these catalysts is very different from that of HDM catalysts. It is of mesoporous monomodal type (Figure 18.17), providing a raised surface that is accessible to lighter fractions of the feedstock, but less so to residual heavy fractions. As shown in Figure 18.14, optimal HDS activity, for cylindrical extrudates of 1.2 mm in diameter, is achieved with a mesoporosity centered at 90-100 angstroms. Below this value, the surface is less accessible to light fractions (diffusional limitations) and beyond this value, it is no longer protected from poisoning by the deposition of metals contained in heavy fractions [Kressmann, 1999]. The pore diameter value resulting in optimal activity also depends on the size and shape of the catalyst grains used.

The active phase is generally based on the "CoMoS" phase to promote HDS reactions, or "NiMoS" which is stronger for hydrogenation, HDN and HDCCR reactions. The support is generally an alumina-type support which can be doped by the presence of acidic doping agents to promote hydrogenation reactions. In some cases, more acidic supports, e.g. silica-alumina based, are used downstream of HDS catalysts to convert the feedstock VGO into gasoils.

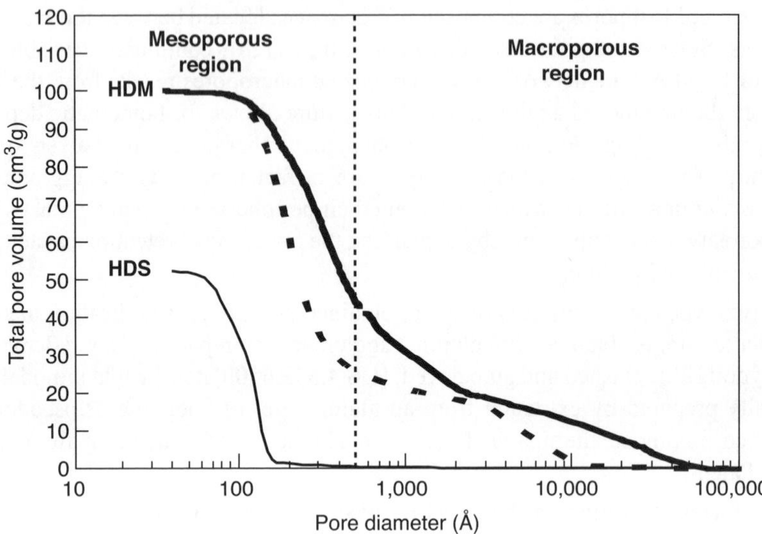

Figure 18.17

Pore Distribution of HDS Catalysts.

HDS catalysts are often used at a lower temperature than HDM catalysts (typically 380°C) and therefore in conditions that are thermodynamically more favorable for hydrogenation reactions.

Finally, so-called intermediate catalysts, located between HDM and HDS catalysts, are used to finish the conversion of asphaltenes and the elimination of metals, and to start hydrorefining reactions. They are generally characterized by monomodal porosity with large mesopores, often comparable to the mesoporosity of bimodal catalysts.

18.4.2.2 Ebullated Bed Supported Catalysts

In contrast to fixed bed implementation, ebullated beds use little or no catalyst chaining. Often, there is a single catalyst which acts as a catalyst of HDM, of retention, of hydrogenation and refining (HDS, HDCCR). However, if the number of reactors exceeds two, there can be several catalysts as in fixed beds. Depending on the target or the feedstock being converted, HDM and retention functions can be emphasized at the expense of refining functions.

The characteristics of supported catalysts used in ebullated beds resemble those of HDM catalysts used in fixed beds, with several adaptations due to the more severe operating conditions of ebullated beds:

- High implementation temperatures (400°-450°C).
- Difficult feedstocks (vacuum residues) with high concentrations of metals (100-500// 600 ppm) and asphaltenes.

As in fixed beds, the catalyst must therefore have a large porous volume in order to capture a large quantity of Ni and V sulfides. Moreover, the macroporosity must be developed to facilitate the access of large molecules from the feedstock in order to promote deposition of metals in the grain, as well as the access of polyaromatic radicals formed by thermal cracking to the hydrogenating active phase (Figure 18.18), which is particularly important in the case of high conversions.

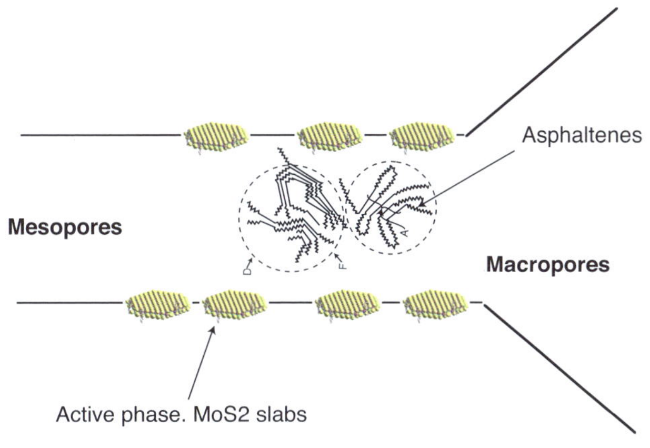

Figure 18.18

Accessibility of the Active Phase.

In the processes of hydroconversion in ebullated beds, the thermal reactions are larger than in fixed beds, and the catalyst must, *via* its hydrogenating function, ensure the stability of the effluents formed. Moreover, note that strong hydrogenating activity can also limit or delay coke deposition on the catalyst, thereby increasing its operating life. We will therefore try to develop porous distribution by creating macropores that allow resins and asphaltenes to quickly progress in the grain mass and mesopores of optimized average diameters so that the molecules can access the maximum possible active surface.

Among the possibilities for porous distributions, the following will be selected:

- Multimodal porosities or "chestnut bur" structures described in the case of fixed beds, which are particularly interesting since they allow for greater accessibility of the active phase and homothetic deposition of metals.
- Bimodal porosities to maximize the active surface in mesopores and protect this same surface – and its refining function – from poisoning by large molecules and metals.

Figure 18.19 shows the distribution of metals according to various types of porosity along the catalyst grains.

In ebullated beds, since pressure loss constraints do not exist, the size of catalysts can be reduced to facilitate their fluidization and to decrease diffusional limitations. On the other hand, the constraint in this type of process involves strong mechanical resistance of the catalysts.

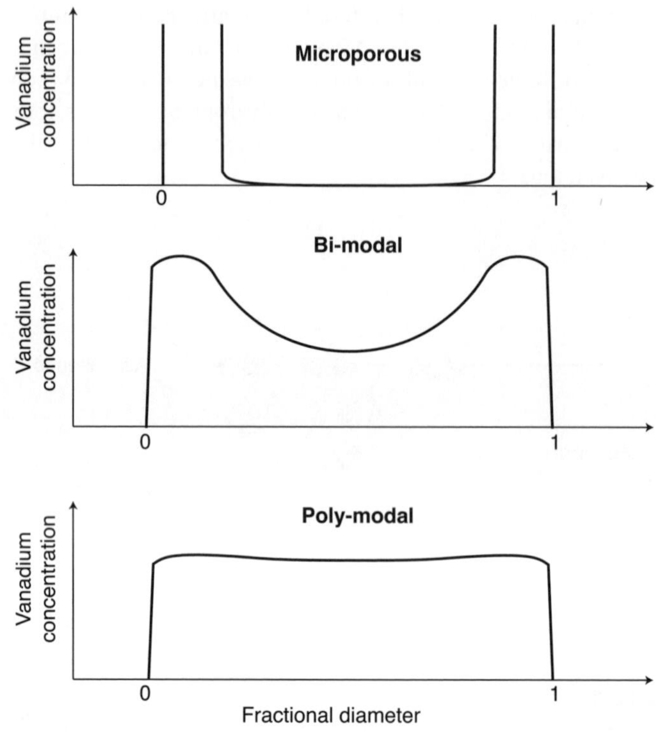

Figure 18.19

Distribution of Vanadium According to Porous Distribution of the Support.

18.4.2.3 Slurry Bed Dispersed Catalysts

Dispersed catalysts are composed of fine particles, injected as such or formed *in situ* in the reactor from a catalytic precursor. The particles are in suspension in the reactor and exit *via* entrainment with the effluents outside the reactor.

This type of catalyst is particularly well-adapted to the ultra deep conversion of heavy oils rich in metals, sediments and asphaltenes which can quickly clog the pores of supported catalysts. They can be implemented at high temperature without risk of agglomeration, thereby allowing for quasi-total conversions. In addition, due to its close mixing with the feedstock, the catalyst can easily play its role of hydrogenation and inhibitor of polycondensation without the accessibility problem encountered for supported catalysts (Figure 18.20).

There are two families of catalysts:

- The first is composed of particles that are inexpensive and have low hydrogenating activity (typically iron-based and generally used in coal liquefaction), with high concentration ranges.

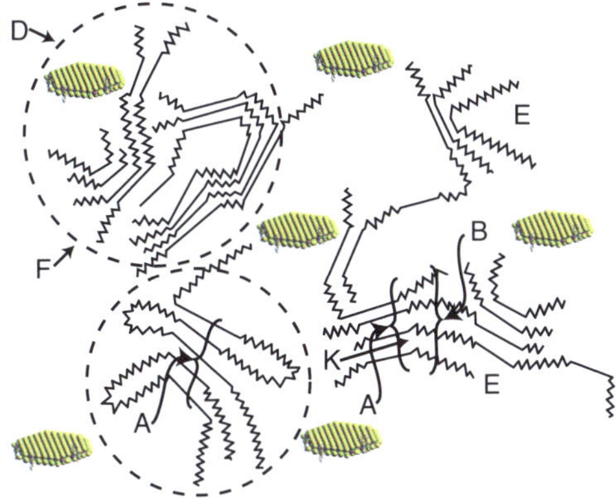

Figure 18.20

Dispersion of the Active Phase in the Feedstock.

– The second family is more complex and more costly. It is often implemented starting from a catalytic precursor that is molybdenum-based, possibly promoted by nickel, at far lower concentrations of 100 to 500 ppm. The precursors may be organo-soluble (e.g. molybdenum naphthenate) or water-soluble (e.g. phosphomolybdic acid or ammonium molybdate) [Panarati *et al.*, 2000; Paez *et al.*, 2000].

By "*in situ*" sulfuriding and decomposition, these precursors are processed into molybdenum sulfide particles, generally associated with a carbon matrix. The addition of a nickel precursor can promote molybdenum sulfide just like on a supported catalyst and increase its hydrogenating activity.

The main problem linked with implementation of these dispersed catalysts is their recovery, particularly when molybdenum is involved: they are found in the unconverted residue and are difficult to separate and/or reuse.

REFERENCES

Ancheyta J, Rana MS, Furimsky E (2005) Hydroprocessing of Heavy Petroleum Feeds: Tutorial, *Catalysis Today*, **109**, pp 3-15.

Bataille F, Lemberton JL, Michaud P, Perot G, Vrinat M, Lemaire M, Schulz E, Breysse M, Kasztelan S (2000) Alkyldibenzothiophenes Hydrodesulfurization-Promoter Effect, Reactivity, and Reaction Mechanism, *Journal of Catalysis*, **191**, pp 409-422.

Beardon R Jr *et al.* (1997) MICROCAT-RC: Technology for Hydroconversion Upgrading of Petroleum Residues, Division of Petroleum Chemistry, *ACS Meeting*, San Francisco, CA (USA), April 13-17, 1997.

Benham NK *et al.* (1996) Canmet Residuum Hydrocracking Advances Through Control of Polar Aromatics, *NPRA Annual Meeting*, San Antonio, TX (USA), March 17-19, 1996, Paper AM-96-58.

Billon A, Bousquet J, Rossarie J (1988) HYVAHL F and T Processes for High Conversion and Deep Refining of Residues, *NPRA Annual Meeting*, San Antonio, TX (USA), March 1988, Paper AM-88-62.

Colyar JJ, Wisdom LI (1997) The H-Oil Process: A Worldwide Leader in Vacuum Residue Processing, *NPRA Annual Meeting*, March 16-18, 1997, San Antonio, TX (USA), Paper AM-97-46.

Fujita K, Abe S, Inoue Y, Plantenga FL, Leliveld B (2002) New Developments in Resid Hydroprocessing, *Petroleum Technology Quarterly*, 2002Q1, 51-58.

Fukuyama H *et al.* (2002) Development of Carbon Catalyst for Heavy Oil Hydrocracking, *NPRA Annual Meeting*, San Antonio, TX (USA), March 17-19, 2002, Paper AM-02-14.

Furimsky E (1998) Selection of Catalysts and Reactors for Hydroprocessing, *Appl. Catal. A*, **171** (2), pp 177-206.

Furimsky E, Massoth FE (1999) Deactivation of Hydroprocessing Catalysts, *Catal. Today*, **52** (4), pp 381-495.

Gauthier Th, Héraud JP, Kressmann S, Verstraete JJ (2007) Impact of Vaporization in a Residue Hydroconversion Process, *Chem. Eng. Sc.*, **62** (18-20), pp 5409-5417.

Griboval A, Blanchard P, Payen Ed, Fournier M, Dubois JL (1997) Alumina Supported HDS Catalysts Prepared by Impregnation with New Heteropolycompounds, *Stud. Surf. Sci. and Catal.*, **106**, pp 181-194.

Guillard C, Lacroix M, Vrinat M, Breysse M, Mocaer B, Grimblot J, Des Courières T, Faure D (1990) *Catalysis Today*, **7**, pp 587-600.

Kasztelan S, Guillaume D (1994) Inhibiting Effect of H_2S on Toluene Hydrogenation over a MOS2 AL_2O3 catalyst, *Ind. Eng. Chem. Res.*, **33**, pp 203-210.

Kressmann S, Boyer C, Colyar JJ, Schweitzer JM, Viguie JC (2000) Improvements of Ebullated Bed Technology for Upgrading Heavy Oils, Oil & Gas Science and Technology – *Rev. IFP*, **55** (4), pp 397-406.

Kressmann S, Guillaume D, Roy M (2004) *14th Annual Symposium, Catalysis in Petroleum Refining & Petrochemicals*, King Fahd University of Petroleum & Minerals (KFUPM), Dharan, Saudi Aramco, December 5th-6th, 2004.

Kressmann S, Harlé V, Guibard I, Tromeur P, Morel F (1999) *218th American Chemical Society National Meeting*, New Orleans, August 22-26, 1999.

Kressmann S, Morel F, Harlé V, Kasztelan S (1998) Recent Developments in Fixed Bed Catalytic Residue Upgrading, *Catal. Today*, **43** (3-4), pp 203-215.

Le Page JF, Chatila SG, Davidson M (1992) Resid and Heavy Oil Processing, Editions Technip, Paris (France).

Leprince P (2001) Conversion Processes, IFP Publications, Editions Technip, Paris (France).

Lott R *et al.* (2000) $(HC)_3$ Process for Partial Upgrading of Bitumen to Meet Pipeline Specifications, *NCUT Symposium*, Edmonton, AB (Canada), September 18-19, 2000.

Martin C, Lamonier C, Fournier M, Mentré O, Harlé V, Guillaume D, Payen Ed (2005) Evidence and Characterization of a New Decamolybdocobaltate Cobalt Salt: An Efficient Precursor for Hydrotreatment Catalyst Preparation, *Chemistry of Materials*, **17**, pp 4438-4448.

Marzen R *et al.* (1995) Heavy Oil Hydroprocessing to Produce Zero Resid, *AIChE Spring Meeting*, Houston, TX (USA), March 19-23, 1995, Paper 45e.

Meyers RA (2004) Handbook of Petroleum Refining Processes, 3rd Ed., McGraw Hill Companies, Inc., New York, NY (USA).

Montanari R *et al.* (2003) Convert Heaviest Crude and Bitumen into Extra-Clean Fuels *via* Eni Slurry Technology, *NPRA Annual Meeting*, San Antonio, TX (USA), March 23-25, 2002, Paper AM-03-17.

Moulijn JA, Makkee M, van Diepen A (2001) Chemical Process Technology, J. Wiley & Sons, Chichester (UK).

Paez R, Luzardo L, Guitian J (2000) *World Petroleum Congress* Proceedings, **3**, pp 78-87, Calgary, Alberta.

Panarati N, Del Bianco A, Del Piero G, Marchionna M (2000) Petroleum Residue Upgrading with Dispersed Catalysts Part 1. Catalysts Activity and Selectivity, *Applied Catalysis A: General*, **204-2**, pp 203-213.

Plumail JC (1983) PhD, Étude de l'influence de la texture poreuse des catalyseurs CoO-MoO-AlO lors de l'hydrotraitement du pétrole brut de Boscan, IFP/ENSPM.

Rana MS, Sámano V, Ancheyta J, Diaz JAI (2007) A Review of Recent Advances on Process Technologies for Upgrading of Heavy Oils and Residua, *Fuel*, **86** (9), pp 1216-1231.

Rana MS, Srinivas BN, Maity SK, Murali Dhar G, Prasada Rao TSR (2000) Origin of Cracking Functionality of Sulfided (Ni) CoMo/SiO2-ZrO2 Catalysts, *Journal of Catalysis*, **195**, pp 31-37.

Raybaud P (2007) Understanding and Predicting Improved Sulfided Catalysts: Insights from First Principles Modeling, *Applied Catalysis A: General,* **322**, pp 76-91.

Ross JL, Harlé V, Kressmann S, Tromeur P (2000) Maintaining On-spec Products with Residue Hydroprocessing, *NPRA Annual Meeting*, San Antonio, TX (USA), March 26-28, 2000, Paper no. AM-00-17.

Ross JL, Harlé V, Kressmann S, Tromeur P (2000) Maintaining On-spec Products with Residue Hydroprocessing, *NPRA Annual Meeting*, San Antonio, TX (USA), March 26-28, 2000, Paper no. AM-00-17.

Scheffer B, van Koten MA, Röbschläger KW, de Boks FW (1998) The Shell Residue Hydroconversion Process: Development and Achievements, *Catal. Today*, **43** (3-4), pp 217-224.

Silverman MA *et al.* (1995) SOC Technology – A Flexible Approach to Residual Oil Upgrading, *AIChE Spring Meeting, Houston*, TX (USA), March 19-23, 1995, Paper 45a.

Speight JG (1991) The Chemistry and Technology of Petroleum, 2nd Ed., Marcel Dekker, Inc., New York, NY (USA).

Speight JG (2000) The Desulfurization of Heavy Oils and Residua, 2nd Ed., Marcel Dekker, Inc., New York, NY (USA).

Topsoe H, Clausen BS, Massoth FE, in Anderson JR, Boudart M (Eds.) (1996) Hydrotreating Catalysis, Science and Technology, Springer-Verlag, Berlin/Heidelberg, **11**.

Toulhoat H *et al.* (1985) *A.C.S. Prepr., Div. Pet. Chem.,* **30**, 1, pp 85.

Toulhoat H, Hudebine D, Raybaud P, Guillaume D, Kressmann S (2005) A New Model for Combined Simulation of Operations and Optimization of Catalysts in Residues Hydroprocessing Units, *Catal. Today*, **109** (1-4), pp 135-153.

Toulhoat H, Szymanski R, Plumail J-C (1990) Interrelations Between Initial Pore Structure, Morphology and Distribution of Accumulated Deposits, and Lifetimes of Hydrodemetallisation Catalysts, *Catal. Today*, **7** (4), pp 531-568.

US 4,552,650 Patent

US 5,089,463 Patent

Van de Water LGA, Bergwerff JA, Leliveld BRG, Weckhuysen BM, de Jong KP (2005) Insights Into the Preparation of Supported Catalysts: A Spatially Resolved Raman and UV-vis Spectroscopic Study Into the Drying Process of CoMo/gamma-Al2O3 catalyst bodies, *J. Phys. Chem. B*, **109**, pp 14513-14522.

Verstraete J, Le Lannic K, Guibard I (2007) Modelling Fixed Bed Residue Hydrotreating Processes, *Chem. Eng. Sci.,* **62** (18-20), pp 5402-5408.

Wisdom LI (1995) Second Generation Ebullated Bed Technology, *AIChE National Spring Meeting*, March 19-23, 1995, Houston, TX (USA), Paper 46d.

Wisdom LI, Colyar JJ (1996) Second Generation Ebullated Bed Technology, *Fuel and Energy Abstracts*, **37** (4), pp 258.

PART 5
Environmental Issues

A.Y. Huc

With regard to the environment, *heavy crude oils* represent a more challenging industrial issue than conventional oils, due to their very nature, composition and subsequent production schemes.

With the increasing environmental concerns the exploitation of *heavy crude oils* is subjected to more and more constraining monitoring and regulations by-laws. Actually this exploitation is controlled by current environmental laws and regulations that apply to conventional oil exploitation, with specific emphasize on land reclamation when strip mining is involved (tar sands).

The potential environmental impact which encompasses the production, transport and upgrading are related to:

- A stronger pressure on the land, due to the drilling of numerous injection and recovery wells with a tight well spacing for the production of heavy and extra heavy oils, and strip mining of tar sands surfaces which disturbs the natural landscape and which needs to be reclaimed (Chapter 19).
- Solids management, resulting from the production of sand during cold production by the CHOPS technology or to a lesser extent associated with steam injection methods, and from the surface mining which produces a huge amount of solid waste and tailings (Chapter 19).

Water management, including consumption (mainly drilling and boiler feedwater for steam generation) and treatment. Even with a strong recycling effort the current consumption of water with the most efficient procedure in Canada ranges from 0.2 barrel of water per barrel of bitumen produced by *in situ* operations up to 2-4 barrels of water per barrel of synthetic crude oil produced from tar sand surface mining.

- A part of the water comes from the oil formation itself, as this water is co-produced with the oil. However, impacts on water resources has to be carefully considered for each project, in the perspective of environmental concerns and competing uses of water. *Heavy crude oil* production can affect both ground and surface water in terms of quantity and quality. Water recovered from steam injection operations and water used to wash bitumen from tar sands contains a variety of contaminants that must be treated before the water can be reused or disposed of. The treatment of production water and of wash-water is required and strictly regulated by law. This operation is more complex with regard to *heavy crude oil* compared to conventional oil, due to the smaller difference in density between *heavy crude oil* and water interfering with the efficiency of fluid-fluid separation procedures (Chapter 19).

- *Heavy crude oils* contain a higher amount of contaminants including sulfur, nitrogen and metals which need to be removed by upgrading and disposed of. In order to monitor and limit pollutant emissions during these refining operation within the stringent regulations framework, adapted techniques, largely extrapolated from the experience of the refining and conversion of conventional crude oils, are used (Chapter 20).
- Large amounts of energy are needed for the extraction and upgrading to synthetic crude. Subsequently substantial amounts of GreenHouse Gas (GHG), specifically CO_2 emissions are obviously a major issue throughout the processes involved in the exploitation of *heavy crude oils*, in particular with regard to extraction (i.e. steam generation) and upgrading. In the perspective of a shortage of global energy supply and impact on climate change, a careful survey of the involved CO_2 and other greenhouse gases emissions is needed (Chapter 21). In this respect *heavy crude oils* exploitation is obviously a niche where CCS (Carbon Capture and underground Sequestration) would be a pertinent technological solution.

The assessment of the CO_2 potentially pumped into the atmosphere, needs to account for a realistic CO_2 mass balance and requires an integrated approach. This approach should include, not only the direct CO_2 emissions associated with the different operations along the production to upgrading workflow, but also all the negative and positive induced effects related to the relevant current and future technologies involved in the manufacturing of the targeted commercial products (Chapter 22).

In the following chapters, references to the Canadian and the Venezuelan heavy oil industries are largely cited since these countries are currently the main players in the field of *heavy crude oils* exploitation.

19 Reservoir and Production

E. Delamaide, J.F. Nauroy, G. Renard, C. Dalmazzone, C. Noïk

Several critical inputs are required in order to produce heavy oils or extract bitumens from oil sands and process them into a product that can be sold and readily shipped to refineries. These are mainly the management of surface reclamation after surface mining, solid byproducts, and natural gas and water consumption. GHG emissions will be covered in chapter 21.

19.1 RECLAMATION AFTER SURFACE MINING

Reclamation is obviously an important and necessary step in the process of mining. Disturbed areas can be quite extensive, e.g. over 7,500 hectares to date for Suncor, and over 15,000 hectares in total for all the oil sands mining operations. The goal of reclamation is "to achieve maintenance-free, self-sustaining ecosystems with capabilities equivalent to or better than pre-disturbance conditions" [Oil Sands Vegetation Reclamation Committee, 1998]. It is thus clear that the requirements go much farther than simply filling up the old mine pits with sand and gravel.

Many different types of land forms need to be reclaimed beyond the pits themselves: the tailings sand dykes and ponds, overburden dumps, etc.

The first step of the process is to reconstruct new soil over the areas to be reclaimed. This is done by mixing overburden, muskeg soil and tailings. The new soils must not only have a "natural" appearance, but also be constructed such that the ecosystem can develop on them. Thus, their chemical composition and organic content must be adequate, drainage needs to be ensured and erosion needs to be controlled.

Erosion can be severe in some places, especially where the slopes are steep. However, it can be controlled by planting shrubs, trees and barley for instance. In this respect, Suncor and Syncrude have planted nearly 6 million trees since the beginning of their operations.

Through these efforts, both wetlands and wildlife natural habitat are developed, which are then suitable to receive their animal inhabitants. Syncrude is focusing on wood bison and now has several of them living on reclaimed land, while Suncor is favoring toads, which apparently adapt very well to reclaimed ponds.

The reclamation process takes many years – so far Suncor has reclaimed only 10% of the land disturbed by its operations – and once completed needs to be monitored carefully to ensure that it has achieved its objectives: returning the mined areas to the animals and the plants.

19.2 MANAGEMENT OF SOLID BYPRODUCTS

The main byproduct of Cold Heavy Oil Production (CHOPS) is obviously sand. In oil sands mining, the most challenging byproducts are the tailings resulting from open pit mining operations, but other byproducts such as sulfur and coke also need to be taken care of. Other production methods such as steam injection can also cause sand production, but the quantities involved are generally much smaller.

19.2.1 Processing of Solids for Cold Production

According to Canadian experience [Dusseault M, 2002], the oil sands produced *via* CHOPS can be:

- Directly incorporated in the roadway foundation layer or in the surface course.
- Spread over confined zones, with or without planned biodegradation.
- Washed to be used in construction or industry.
- Re-injected into saline cavities or aquifers.

Incorporation in the roadway foundation layer, by mixing the sand with gravel and bitumen, only requires transport. On the other hand, using it alone is not recommended since the rather uniform granulometry of sand does not favor a stable driving road.

Sand spread over confined zones or stored in backfill will undergo slow biodegradation which can be accelerated by *ad hoc* bacteria, but there is the possible risk of contaminating neighboring water tables.

Washing sand to obtain a clean material that can be used in construction or industry (glass) requires specialized processing units. Washing with hot water and steam produces dirty, salty water which then must be processed or re-injected in *ad hoc* geological layers.

The sand can also be deposited in cavities created in salt. This solution has been used at several Canadian sites (Alberta and Saskatchewan). This solution is chosen due to the low permeability of salt and the tendency of cavities formed in salt to slowly close again, thereby sealing the deposits. However, creating saline cavities requires a large amount of water (seven times the volume of dissolved salt) which must then be evaporated to recover the salt and other solid constituents that can be used.

The last solution is re-injection of the sand – incorporated in a slurry – into formations *via* hydraulic fracturing. These techniques are the same as those used for Cuttings Re-Injection (CRI) and used at many fields throughout the world. The sand is first screened to remove large particles that could clog the perforations or damage the pumps. The slurry is then prepared with water and 10 to 15% sand. Various additives are added to improve the physical properties of the mix (viscosity, etc.). Finally, the slurry is injected in cycles or in continuous fashion combined with monitoring to adjust operating parameters and ensure injectivity that is reliable over time [Bruno M, 2000].

19.2.2 Tailings Management in Oil Sands Mining

In a Decision dated December 2004, the Alberta Energy and Utilities Board [Alberta, 2004] made a very revealing comment on the tailings situation: "The Board considers that oil sands tailings present unique challenges that no operator has fully resolved, despite ongoing efforts at each operation and with each new project."

As described in Chapter 10, tailings are one of the byproducts of the extraction process in open pit mining. They are composed of water mixed with sand and clay, and a low percentage of bitumen. They are what is left over from the original ore once the bitumen has been extracted.

Tailings are first separated into tailings sand and fine tails. As its name indicates, tailings sand is composed of the coarser material that has been separated from the water and sludge: mostly sand (95%) with roughly 5% of finer material (silt and clay) and a very small percentage of bitumen (0.2%). On the one hand, tailings sand is solid and disposed of to fill the areas from where the oil sands were extracted, in terraced piles that are covered with reclaimed soil and then planted with trees and other vegetation. These piles can be very large, 30-100 m (100-330 ft) high, and can cover several tens of square kilometers; however, they do not present any significant technical challenge.

On the other hand, disposing of the fine tails is much more difficult. Fine tails contain mostly water (85%), and the rest is composed of very fine particles of clay and silt. Different approaches can be considered for disposing of these fine tails.

19.2.2.1 Conventional Methods

According to the composition of a fine tail, it is a sludge that can take dozens of years to settle and is so soft that it cannot be used for reclamation or construction purposes. Fine tails therefore need to be stored in tailings ponds surrounded by dykes, and because of the long time it takes for the sludge – sometimes called Mature Fine Tailings (MFT) – to settle and gain some consistency, the volumes to be stored are large. For every barrel of bitumen produced, there are nearly 1.5 barrels of MFT produced. For example, one of the operating companies in Canada had produced roughly 310×10^6 m^3 (11 billion cu. ft.) up to the end of 2003 – enough to fill a rather large pool measuring 10 km by 10 km (6.2 miles by 6.2 miles), and 3 meters (10 ft) deep. And by the same date they had succeeded in efficiently disposing of just over 50% of that volume.

But volume is not the only issue. The water in the ponds needs to be recycled, but this is a very slow process. After some time, the bitumen that is still trapped in the tailings floats to the surface of the ponds. It is skimmed because it is still a valuable product, but also because it is harmful to birds and other avian creatures. One operator uses scarecrows to frighten birds off its tailing ponds.

Several technologies are being researched to improve the situation: Composite, or Consolidated, Tails (CT), Thickened Tailings (TT), and others.

19.2.2.2 Composite/Consolidated Tails

The CT process consists in mixing sand tailings (70%) and fine tails (30%) and adding gypsum. The gypsum acts as a flocculent on the fines; thus the water is released much faster (one or two years) than with the conventional method and can be recycled. However, the CT mixture is more difficult to reclaim than the sand tailings. It contains high concentrations of ions which result in high electrical conductivity – a factor that is unfavorable for plant growth and re-vegetation. Some people also question the interest of using a good product – sand tails – to make one of lesser quality. Finally, the process is not easy to operate and performance is variable: one company reported in 2004 that it had produced only 41% of the CT it was planning, and only 53% of that was on-spec. Thus, although this technology is currently used in ongoing mining projects, it is not yet considered the ultimate solution.

19.2.2.3 Thickened Tailings

With this technology, fines are thickened by the addition of polyacrylamide polymer and other chemicals, thus accelerating the settling process. The process is also energy-efficient, since the water's heat (which is between 40 and 80ºC at the outlet of the extraction process) can be recovered instead of being lost to the atmosphere when the tailings go directly into tailing ponds. However, pilot tests have shown that the composition of the slurry has a strong influence on the performance of the thickener. Thus it is quite difficult to find a well-adapted product that can cope with variable slurry compositions. A field trial took place in 2001-2002 at the site that successfully produced thickened tailings; it also displayed some of the issues with the process. Therefore, there has been no commercial application of the technology so far.

19.2.2.4 Other Technologies

Other technologies are being tested for reclamation of the fine tails. For example, water-capping – covering the tailing ponds with 3 to 20 m (10 to 66 ft) of natural or processed, non-toxic water – is being tested by one operator in its operations. But the best solution might be to simply reduce the amount of water required in the extraction process. This is what UTS/Petro-Canada are proposing to do with their BITMIN technology (see Chapter 6).

19.2.3 Other Byproducts

19.2.3.1 Sulfur Dioxide

Bitumen contains roughly 3 to 5% sulfur (by weight), which is removed during the upgrading phase in the form of sulfur dioxide (SO_2). For example, in 2004 one operator was producing 67.6 tons of SO_2 per day in its oil sands operations, which corresponds approximately to 250,000 tons per year; SO_2 emission intensity was about 1.85 kg/m^3 of production. Upgraders are now required by the AEUB to reduce sulfur emissions.

Companies are investing significant capital to reduce SO_2 emissions. For example, one operator received approval in 2004 for a CDN $400 million project to reduce sulfur dioxide emission through the use of gas scrubbers. The project should come online in 2009 and should reduce sulfur compound emissions from 250 tons a day to about 100 tons a day.

19.2.3.2 Sulfur

The sulfur market – for the production of fertilizer, road asphalt, sulfuric acid and others – is growing but cannot absorb the quantities currently produced, and unfortunately the increasing production from the oil sands is going to grow faster than the sulfur market. Thus, sulfur needs to be stockpiled. Companies are stockpiling the sulfur in "sulfur cells", large piles of sulfur blocks covered by overburden as well as reclamation material. For example, in its application to the AEUB for its future project, one operator stated that its sulfur stockpile for the first 20 years of operation would cover an area of 80 hectares and reach a height of about 20 m.

There have been some discussions regarding deeper underground storage for sulfur, but there is no clear innovation in the domain. An alternative solution could be to re-inject the SO_2 into formations instead of converting it into sulfur.

19.2.3.3 Coke

Coke is produced as a byproduct of the "coking" upgrading process, and given the volumes of oil currently produced, or those that will start producing from the oil sands in the near future, the quantities generated can be quite significant. For example, one operator for its planned application announced that it would have to stockpile over 3 million tons of coke per year.

Coke is itself a source of energy, and so is used partly to fuel burners or is sold (e.g. in the Petrocedeno (ex Sincor) project in Venezuela), but a significant portion of coke production has to be stockpiled.

At the same time, thermal projects like SAGD need to generate steam and use gas for this purpose; upgraders also require gas. As long as gas prices were low, this was not an issue. But in the past few years, gas prices have risen significantly and thus using the coke byproduct to fire the burners and reduce gas consumption has recently become an attractive alternative. For example, in its initial application for a large project in 2002, one operator had rejected coke gasification because of unfavorable economics, but in 2005 announced its intention to pursue coal gasification in order to reduce its dependency on gas. Another operator has also planned to build a gasifier, and process roughly 20% of its coke production when the gasifier comes on line.

Those projects are not yet operational, and in the meantime coke will continue to be stockpiled.

Managing the byproducts of the oil sands is one of the key issues that need to be dealt with. There is strong pressure on the operators to do as little damage as possible to the environment, and at the same time to keep operating costs as low as possible. The balance is not easy to find and technology will have a significant role to play in that respect.

19.3 NATURAL GAS ISSUE

Natural gas is generally an input to three steps of the production process. It is used as a fuel to generate electricity, which in turn is used to power the mining equipment, produce the steam required for *in situ* production, and produce the hydrogen used in upgrading the bitumen to SCO (Synthetic Crude Oil). Finally, condensates from natural gas are used as a diluent to facilitate pipeline transportation of heavy oil produced.

In the province of Alberta, Canada, where there are many heavy oil projects, as outlined in Chapter 10, the natural gas currently used is from local production. However, with Canadian natural gas production declining (17.4 billion cu-ft/d peak in 2002; forecasts for 2020 around 10 million cu-ft/d), and the needs of the oil sands industry increasing, it will be important to either find alternative sources of natural gas or alternative sources of energy for oil sands operations, and alternative sources of hydrogen for upgrading and refining of heavy oil sands products. A potential new source of natural gas could come from the Arctic region if the MacKenzie valley pipeline under project goes forward.

A proposed alternative to natural gas is synthetic gas (syngas), a gaseous hydrocarbon resulting from the gasification of low-value heavy bitumen residues such as coke and asphaltenes. Gasification consists of incomplete combustion of hydrocarbons at high pressure and temperature in the presence of steam. Due to a lack of oxygen in the reactor, molecular bonds are fully broken without realizing full oxidation of carbon and hydrogen. The process produces a mix of CO_2, H_2, and CO, which is shifted to syngas (CO_2 and H_2) through steam reforming. The hydrogen content of the gas can then be used as a feedstock for upgrading operations, and as a heating fuel for bitumen separation processes in mining operations or for steam production in SAGD.

If executed properly, development of this technology has the potential to provide an alternative to natural gas while utilizing low-value oil bitumen residues. Gasification processes are already being seriously examined by the electric power generation industry due to their role in innovative coal technologies, and developments in this arena are likely to benefit oil sands production. However, gasification is not yet deployed on an industrial scale, and the difficulties inherent in the design of gasification boilers raise major reliability issues. For example, the Long Lake Project [Long Lake website] – the first integrated upgrader/gasifier plant currently jumpstarted by OPTI/Nexen – will rely on three redundant boilers, using only two of them at nominal capacity.

Nuclear power has been identified as a non-fossil fuel based, low carbon energy source that could reduce the oil industry's reliance on limited natural gas resources and decrease its exposure to volatile natural gas prices. Nuclear energy could be effectively considered for electricity and direct generation of steam for heavy oils and/or oil sands production. Its impact on CO_2 emission reduction could also be very large.

In 2003, CERI of Canada [Dunbar RD *et al.*, 2003] looked at the relative economics of the nuclear and gas-fired steam generation options. The nuclear and gas-fired facilities were both configured to produce 78,000 m^3/d (491,000 CWE b/d) of 80% quality steam from which would result a bitumen production rate of 23,200 m^3/d (146,000 b/d) assuming a CSOR of 2.5 m^3/m^3. Given the limitations of transporting steam over long distances, adequate bitumen reserves to support a project of this scale would need to be located within reasonable proximity of the central steam generation site. Steam supply costs per ton were then calculated to be roughly the same, CDN $8.6 for the nuclear facility and CDN $8.7 for the gas-fired facility (no Kyoto compliance costs were assumed).

A sensitivity analysis conducted to identify key variables and determine their influence on steam supply cost showed that steam supply cost from a nuclear facility is very sensitive to capital cost (a 25% increase in capital cost would increase the steam supply cost to CDN $10.3/t – 20% increase) and that steam supply cost from a gas-fired facility is very sensitive to natural gas prices (a 25% increase in the cost of natural gas from CDN $4.25/GJ to CDN $5.21/GJ would increase the steam supply cost to CDN $11/t – 26% increase) and possible Kyoto compliance cost (a CDN $15 per ton of CO_2 emitted would increase the steam supply cost to CDN $10.3/t – 18% increase).

However, in considering the use of nuclear power in Canada, account must be taken of the fact that the province of Alberta currently lacks any nuclear regulatory authority. Moreover, in this country today as in many others, nuclear technology is not considered to meet society's expectations on reliability, flexibility, safety and environmental concerns. Therefore, nuclear energy is not yet considered to be a viable fuel switching option in heavy oil producing countries.

19.4 WATER ISSUE

Everywhere, water is another critical input whose supply needs to be assured, especially for mining and *in situ* production operations.

Water consumption is a major concern for mining operations. The extraction process consumes significant quantities of water. In 2005, Syncrude used 20.6 volumes of water per volume of bitumen produced [Syncrude, 2005]. Most of this water (88%) was recycled; still, 28.2 million m^3 (177.4 million barrels) of water had to be pumped from rivers – about 2.3 volumes of water per volume of production. In its application for Phase I of the Jackpine mine, Shell suggested that it would require 2.8 volumes of fresh water per volume of production. In a comprehensive report released in May 2006 [Pembina Institute website], the Pembina Institute estimated that it requires between 2.2 and 4.4 volumes of water to produce one volume of synthetic crude oil by mining, and that the oil sands mining operators are licensed to use twice as much water as the City of Calgary (1,000,000 people) uses in a year. The operators are striving to reduce water consumption and increase water recycling, but there is no "magic solution" in sight.

Established oil sands mining projects are already licensed to divert 395.7 million m^3 of water a year from the Athabasca River [Brooymans H, 2006], and this is expected to grow to 529 million m^3 a year, given already planned projects [Canadian, 2006]. Although the river has an average flow of 20 billion m^3 per year, the flow rate is highly seasonal, with an average flow in winter months of less than 6.5 billion m^3 per year equivalent on average. Therefore, planned projects at average operating levels would be withdrawing approximately 8%

of average flow; in times of low flow, the diversions would represent an even greater fraction of the flow. Under the Athabasca River Water Management Framework, a new scheme defines limits on withdrawal allowances, including special limitations in periods of low flow which will constrain operations of these planned projects [Alberta, 2007].

In situ production using steamflood or SAGD is also based on extensive use of water. In Canada, regulations impose the treatment and recycling of produced water, though high recycling rates (90-95%) lead to the withdrawal of only 0.2 barrels of water from freshwater aquifers per barrel of bitumen produced. Demand for fresh water linked to *in situ* production processes is however predicted to rise from 5 to 16 billion m^3 per year from 2006 to 2015 if all announced *in situ* projects reach nominal capacity on schedule.

In Canada, environmental concern may become an even more important constraint than competing uses of water [Lacombe RH, 2007]. Downstream from the oil sands area, the Athabasca River feeds into Lake Athabasca through the Peace-Athabasca Delta, south of Wood Buffalo National Park. The region was designated in 1983 a UNESCO World Heritage Site for the biological diversity of the delta and the fact that it is the largest inland delta in the world. Water flow and quality are hence under close scrutiny from governmental and non-governmental groups. Developing additional sources of water supply or alternatives to the current water-intensive processes remains an important short-run challenge to the development of the industry.

Continued expansion of the industry requires development of supply for gas and water, or improvement in the technology to reduce the need for them, since the availability of both could become questionable very soon.

19.5 TREATMENT OF HEAVY OIL PRODUCTION WATER

In recent years, the processing and elimination of production water resulting from oil and gas exploitation have become a significant environmental and economic challenge. Production water represents the largest source of waste (pollutants) from this industry. Traditionally, many options are available for treatment and elimination or recycling, but must be optimized for each specific case, from both a technical and financial standpoint.

With the evolution of environmental regulations, water treatments have become essential and required. Treatments are more complex with heavy oils than with conventional oils [Lin D, 1999]: in fact, the oil/water separation is made more difficult by the small difference in density between the two fluids, high viscosity of the oil phase and presence of many surface-active compounds (e.g. asphaltenes) which promote the formation of stable emulsions. In addition, improving recovery *via* the injection of hot fluids (water or steam) requires the use of large quantities of water, the majority of which will be found in the fluids produced. Thermal processes favor the presence of silica since steam dissolves silica from the reservoir, as well as the formation of fairly stable emulsions. After separation of the crude oil, the production water must be treated in order to be reused in boilers generating hot fluids or eliminated (discharge at the surface or in the ground) [Heins W and Peterson D, 2005]. In terms of quantity, less water is recovered in primary production (cold production) than in thermal production, but must be treated nonetheless.

The main solutions for elimination are generally injection into the subsurface when this is possible, surface evaporation (thermal or solar), ground spreading or wetlands-based treatment, and possibly freeze-thaw treatments or reverse osmosis followed by surface elimination [Simmons BF, 1996].

The possible water treatment schemes are varied and mainly depend on composition of the water, specifications required for the discharge, re-injection, or supply of steam generators in the case of recycling, and whether the "zero liquid discharge" option [Pédenaud P *et al.*, 2005] is chosen.

Since the most severe specifications are those required for water that will supply steam generators, this section will mainly discuss treatments of production water to be used for recycling.

19.5.1 Production Water Quality

Knowing the ionic composition of the production water is important, since information on water hardness and salinity influences the choice of a treatment system [Pédenaud P *et al.*, 2005]. In general, production water is mainly composed of sodium chloride-based brine, with high concentrations of silica, and small amounts of calcium and magnesium. Therefore, it is important to know the concentration of both dissolved and undissolved silica.

High alkalinity is frequently observed, and associated with the presence of carbonates. The total quantity of dissolved solids is roughly 4,000 mg/l, but may be significantly higher depending on the type of geological formation.

Finally, production water contains organic compounds that are dissolved and emulsified (oil) at various concentrations, depending on the efficiency of the oil-water separation processes used [Heins W and Peterson D, 2005]. The concentration of soluble organic compounds varies, but is usually between 100 and 700 mg/l.

19.5.2 Traditional Treatment Associated with Bitumen Produced by Steam Injection

If water is to be reused to generate steam, we must consider the quality of the water required to feed boilers.

Two types of boilers are used for steam production:

- Direct steam generators or "Once Through Steam Generators" (OTSG), which vaporize 80% of the feed water.
- Conventional boilers that vaporize 100% of the feed water.

Table 19.1 lists the various specifications for feed water quality according to boiler type. It is clear that specifications required for 100%-steam generating boilers are much stricter than those for OTSG boilers, particularly with regard to silica.

Table 19.1 Feed Water Specifications According to Boiler Type [Pédenaud P *et al.*, 2005].

Specifications	OTSG Boiler	Conventional 100%-Steam Boiler
Total hardness (mg/l $CaCO_3$)	< 1 (0.5 recommended)	< 0.5
Barium (mg/l)	< 0.1	–
Copper (mg/l)	–	< 0.005
Iron (mg/l)	< 0.25	< 0.02
Free chlorine (mg/l)	< 0.1	–
Oxygen (mg/l)	< 0.02	< 0.02
pH	7.0-9.5	9.0-9.5
Silica	< 100 mg/l	< 20 µg/l
Total dissolved solids (mg/l)	< 12,000 (600 recommended)	–
Hydrocarbons (mg/l)	< 0.5	< 0.3
Conductivity (µS/l)	–	< 0.3

Treatment of production water must comprise several steps, including oil removal and elimination of suspended solids, hardness, silica, salts and oxygen.

Figures 19.1 and 19.2 show basic diagrams of the water treatment chain in the case of OTSG boilers and 100%-steam boilers, respectively.

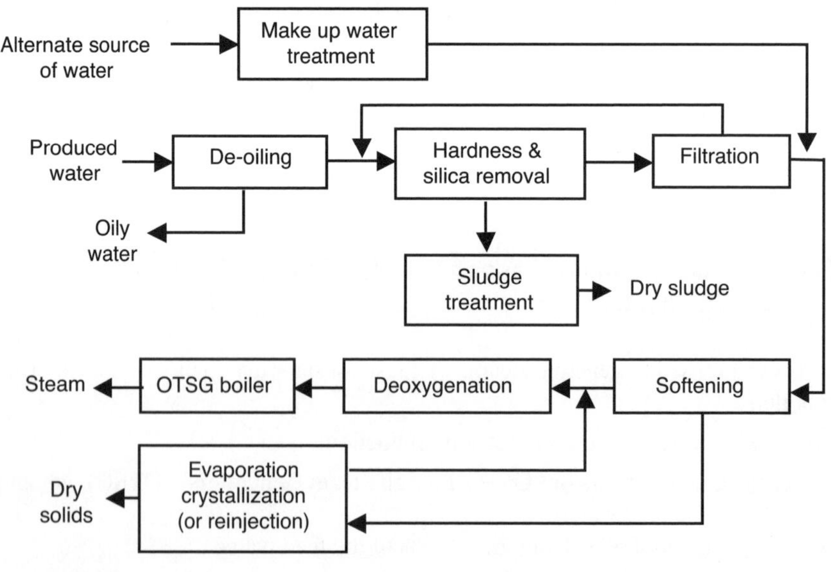

Figure 19.1

Basic Diagram: Treatment of Production Water for Reuse in OTSG Boilers [Pédenaud P *et al.*, 2005].

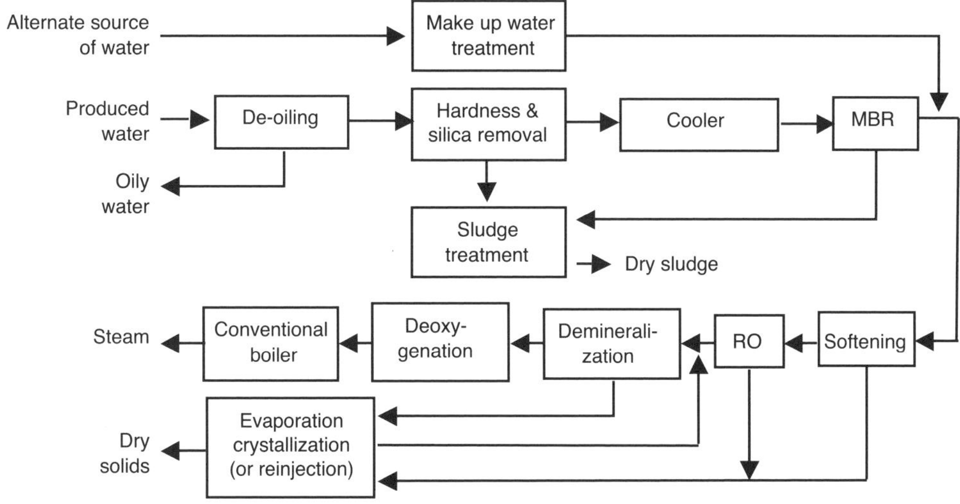

Figure 19.2

Basic Diagram: Treatment of Production Water for Reuse in Conventional (100% Steam) Boilers. RO: Reverse osmosis, MBR: Membrane bioreactor [Pédenaud P *et al.*, 2005].

19.5.3 Oil Removal (De-oiling)

Traditional oil removal processes are generally implemented in three treatment steps. Primary treatment takes place in an API or equivalent skimmer tank, where retention time is typically several hours for output water quality on the order of 500 mg/l for the concentration of organic compounds (HC). Secondary treatment is based on a flotation process which requires retention time on the order of 10 minutes. Generally, flocculent additives are used to improve separation yield and recover water that contains between 30 and 40 mg/l of organic compounds at output. Finally, tertiary treatment uses sand filters to decrease the residual concentration of organic compounds below 10 mg/l and also remove suspended solids. Note that none of these treatments can remove dissolved hydrocarbons or finely emulsified oil droplets. On the other hand, they are used for partial treatment of suspended solids.

19.5.4 Elimination of Organic Matter

Traditional processes used to eliminate organic matter are based on activated carbon filtration. However, more effective biological processes can be used, like MBR (Membrane Bio-Reactor), which combines biological removal of organic compounds with membrane filtration.

19.5.5 Elimination of Silica and Hardness

The processes traditionally used are precipitation processes, e.g. WLS (Warm Lime Softening) or HLS (Hot Lime Softening), or DensaDeg (decantation/precipitation). Dissolved silica is removed *via* precipitation of $Mg(OH)_2$ using $MgCl_2$, while colloidal silica is removed *via* $Fe(OH)_3$ using $FeCl_3$ at basic pH. Water is softened by chemical precipitation with lime. It is important to note that these processes create sludge which must be dehydrated before elimination. Further in this chapter we will see that alternative processes exist, like evaporation and freezing, which allow this step to be skipped. It is also possible to use inhibitors which prevent the polymerization of amorphous silica into colloidal silica [Pédenaud P *et al.*, 2005]. Recently, there was a proposal to replace these silica-removal precipitation treatments with a single treatment for divalent and trivalent cations *via* chelating ion exchange resin in order to reach a maximum concentration of 20 ppb [Bridle M, 2005]. Thus, the absence of these ions should allow OTSG boilers to function at high silica concentrations while avoiding the problems of deposition. However, this method has not been validated experimentally.

Note that the main constraint for boiler feed water is in fact linked to water hardness, which must be very low (on the order of 0.5 mg/l $CaCO_3$, see Table 19.1). The traditional water softening processes discussed above do not allow this specification to be met (best outcome is hardness on the order of 30 to 50 mg/l $CaCO_3$). Thus, carboxylic or chelating-type resins must also be used.

19.5.6 Total Demineralization

Treatment of water for reuse in traditional boilers requires total demineralization. This treatment is performed *via* an ion exchange system. Given the relatively high salinity of production water, it can be advantageous to install a reverse osmosis unit upstream of the demineralization chain, in order to reduce the size of the installation as well as the consumption of chemical products (Fig. 19.2). However, this reverse osmosis unit is not necessary if the treatment consists of water softening by evaporation.

19.5.7 Deoxygenation

Since it is nearly impossible to avoid any contamination of water by oxygen throughout the treatment chain, a deoxygenation operation is often necessary at the end of treatment. A deaerator is generally installed just upstream of the boilers.

19.5.8 Zero Liquid Discharge Option

If the zero liquid discharge option is chosen due to environmental reasons or local regulations, then we must also treat the liquid effluents resulting from various processing equipment in addition to the 10-20% of water not converted into steam in the case of OTSG boilers. The preferred technologies for this are evaporation and crystallization [Heins W and Schooley K, 2004].

19.5.9 Alternative or Emerging Processes

As we have just discussed, the traditional scheme for treating water intended for reuse in boilers is particularly complex. To simplify this scheme, two types of alternative processes may be proposed. The first process is based on evaporation. Traditional evaporation processes generally are not very attractive given the high investment costs required. However, the industry is currently offering vertical tube and vapor compression falling-film evaporators for this application [Heins W and Peterson D, 2005]. Vertical tube falling-film evaporators have excellent heat transfer characteristics. Furthermore, this system, in which surfaces are kept permanently wet *via* the water film, significantly reduces the effects of fouling. The vapor compression cycle is what optimizes the system's energy efficiency, since the amount of energy required for this cycle is only 1/20th the energy required for the evaporation of water. Thus, this type of process can be used to directly treat water which has already been de-oiled and significantly reduce the amount of liquid discharge. It has already been installed in Alberta (Canada) for SAGD applications [Heins W and Peterson D, 2005].

Another type of interesting and emerging process is freezing. In fact, the freezing of ice in contaminated water efficiently separates dissolved and dispersed pollutants by progressively concentrating these impurities in the solution during the process of liquid/solid phase change. The feasibility of using natural freezing processes in cold regions of North America has been studied since 1992 [Boysen JE *et al.*, 1996]. The FTE (Freeze-Thaw-Evaporation) process has been developed, tested in the field and commercially deployed in Wyoming [Harju J, 2001; Boysen JE, 1999]. This natural freezing process consists in forming ice from production water that has been pre-treated (filtered and de-oiled) during the cold season, then recovering the water of very high quality from the thawed ice during the warm season. However, it does not provide continuous water treatment, and consequently would be difficult to implement for *heavy crude oil* exploitation, where water must be continuously recycled. Nonetheless, artificial freezing processes can be envisaged [Heist JA, 1980; Lorain O *et al.*, 2001; Gay G, 2003].

The recovery of heavy oils *via* thermal processes requires the use of large amounts of water as compared to traditional crude oils. Production water then becomes a resource that is essential for generating the steam required for exploitation. Traditionally, in the majority of thermal recovery applications, production water is treated *via* multiple steps including oil removal, treatment of silica and water hardness (calcium and magnesium), as well as treatment of dissolved organic compounds. Currently, the most widely used treatment scheme consists of lime treatment *via* the WLS (Warm Lime Softening) process for recycling water in OTSG boilers. From a financial standpoint, it would seem to be more profitable to use conventional boilers producing 100% steam; but generally, the potential savings from this method is compromised by the fact that the associated water treatments are much more complex, since the water quality specifications are very severe. However, in the future, the development of more economical processes based on evaporation and/or freezing should allow for simplification of the treatment chain.

REFERENCES

Alberta Chamber of Resources (2004) Oil Sands Technology Roadmap: Unlocking the Potential, January 30th, www.acr-alberta.com/OSTR_report.pdf

Alberta Environment (2007) Water Management Framework: In-Stream Flow Needs and Water Management System for the Lower Athabasca River.

Boysen JE, Walker KL, Mefford JL, Kirsch JR, Harju JA (1996) Evaluation of the Freeze Thaw/Evaporation Process for the Treatment of Produced Waters. *Gas Research Institute Report* GRI-97/0081, August 1996.

Boysen JE, Harju JA, Shaw B, Fosdick M, Grisanti A, Sorensen JA (1999) The Current Status of Commercial Deployment of the Freeze Thaw Evaporation Treatment of Produced Water. Paper SPE 52700 presented at the *1999 SPE/EPA Exploration & Production Environmental Conference*, March 1-3, 1999, Austin, TX, USA.

Bridle M (2005) Treatment of SAGD-Produced Waters Without Lime Softening. Paper SPE 97686 presented at the *2005 SPE International Thermal Operations and Heavy Oil Symposium*, Nov. 1-3, 2005, Calgary, Alberta, Canada.

Brooymans H (2007) New Plan Gives Oil Sands Its Fill of Water, Even During a Drought, *The Edmonton Journal*, March 2, 2007.

Bruno M, Reed A, Olmstead S (2000) Environmental Management, Cost Management, and Asset Management for High-Volume Oil Field Waste Injection Projects, SPE 59119 presented at the 2000 IADC/SPE Drilling Conference, New Orleans, Feb. 23-25, 2000.

Canadian National Energy Board (2006) Canada's Oil Sands Opportunities and Challenges to 2015: An Update.

Dunbar RD, Sloan TW (2003) Does Nuclear Energy Have a Role in the Development of Canada's Oil Sands. Paper CIM 2003-096, *Canadian International Petroleum Conference 2003*, Calgary, Alberta, June 10-12, 2003.

Dusseault M (2002) CHOPS: Cold Heavy Oil Production With Sand in the Canadian Heavy Oil Industry (Available an www.energy.gov.ab.ca/OilSands/1189.asp).

Gay G, Lorain O, Azouni A, Aurelle Y (2003) Wastewater Treatment by Radial Freezing with Stirring Effects. *Water Research* 37, pp 2520-2524.

Harju J (2001) The FTE Process-Commercial Deployment in the Rockies. *Gas TIPS*, **7**, 3, pp 25-28.

Heins W and Schooley K (2004) Achieving Zero Liquid Discharge in SAGD Heavy Oil Recovery. *Journal of Canadian Petroleum Technology*, **43**, 8, pp 37-42.

Heins W, Peterson D (2005) Use of Evaporation for Heavy Oil Produced Water Treatment. *Journal of Canadian Petroleum Technology*, **44**, 1, pp 26-30.

Heist JA (1980) Freeze Crystallization Applications for Wastewater Recycle and Reuse. *AIChE Symposium Series*, **77**, 209, pp 259-272.

Lacombe RH, Parsons JE (2007) Technologies, Markets and Challenges for Development of the Canadian Oil Sands Industry. Available at: web.mit.edu/ceepr/www/publications/workingpapers/2007-006.pdf

Lin D (1999) Water Treatment Complicates Heavy Oil Production. *Oil and Gas Journal*, Sept. 20, pp 76-78.

Long Lake web site: www.longlake.ca/

Lorain O, Thiebaud P, Badorc E, Aurelle Y (2001) Potential of Freezing in Wastewater Treatment: Soluble Pollutant Applications. *Water Research*, **35**, 2, pp 541-547.

Pédenaud P, Michaud P, Goulay C (2005) Oily-Water Treatment Schemes for Steam Generation in SAGD Heavy-Oil Developments. Paper SPE-97750-MS presented at the *SPE International Thermal Operations and Heavy Oil Symposium*, Calgary, Alberta, 1-3 November, 2005.

Simmons BF (1996) Produced Water Treatment & Surface Discharge: A Brief Comparison of Disposal Alternatives. 6th Annu. Amer. Filtration Soc. *Prod. Water Seminar*, Jan. 18-19, 1996, League City, Texas, USA.

20 | Upgrading

A. Quignard

European refining in general – and French in particular – is taken as a reference since European regulations for it are among the most stringent in the world and result in the strictest constraints regarding pollutant emissions. All that is discussed within the French and European framework regarding conventional crudes can be widely extrapolated to the refining and conversion of heavy and extra-heavy crudes. With regard to the technological solutions selected to limit and monitor pollutant emissions, only local environmental regulations differ.

20.1 LEGISLATION ENACTED IN EUROPE FOR AIR QUALITY: ACTIONS UNDERTAKEN AND RESULTS OBTAINED

Legislation enacted in France and in Europe, as well as in developed countries, imposes increasingly strict specifications with regard to waste pollutant emissions.

In France and in Europe, environmental legislation that impacts refineries covers two areas: emissions from refineries *per se* and fuel quality. The constraints limiting these two areas are relatively linked.

Thus, the near-total elimination of sulfur from fuels, with gasoline and diesel fuel (including on road and off road applications) at less than 10 ppm in Europe, at less than 10 to 30 ppm in the United States and at less than 50 ppm sulfur content in other developed – or even emerging – countries in the near future, implies sulfur emissions that have been reduced to virtually zero for vehicles. Since all sulfur-containing molecules are removed in the refinery and processed into sulfur, SO_2 emissions of the road transport industry are limited to refinery emissions, which are strictly regulated and monitored. Thus, we can say that the problem of sulfur emissions *via* SO_2 is fully controlled and monitored for road transport in developed countries, and will increasingly be for ocean transport: e.g. the specifications of bunker fuels used for ships in the SECA (Sulfur Emission Control Area: Baltic Sea and in North Sea coastal zones) and ECA (Emission Control Area) that could extend to many other zones such as Mediterranean sea, North America and some sensitive zones in Asia with a very significant decrease from 1.5% S (current specification in SECA) down to 1% in 2010 and to 0.1% only in 2015 and from current 4.5% outside SECA zones, down to 3.5% in 2012 and 0.5% in 2020/2025, as well as air transport. The consequence of these strict regulations on emissions is that in developed countries, "local" and "transborder" pollution is well-controlled, but with an increase in the quantity of energy expended and a concomitant increase in the consumption – and therefore production – of hydrogen, which results in an increase in CO_2 emissions.

Thus, to take the example of France (but the trends can be extrapolated to Europe), emissions objectives regarding long-range air pollution have been achieved or even exceeded (other than Gothenburg and Aarhus for HCBs or hexachlorobenzenes), and emissions objectives regarding the *Large Combustion Plant Directive* (LCPD) have all been achieved for sulfur (see Table 20.1 and references at end of chapter) and according to the French Petroleum Industry Association (Union Française des Industries Pétrolières, UFIP), with measures that have already been decided upon, should be achieved for VOCs (Volatile Organic Compounds).

In contrast, the problem of NOx emissions is more difficult to solve, and again according to UFIP, objectives will not be achieved, regardless of the measures taken.

Table 20.1 Emission objectives for France regarding the NEC Directive as of 2005.

Pollutant	Unit	Emission Limit 1990	Dead Line	Emission Limit 2004	Target
SO_2	kt	1,330	2010	485	375
NOx	kt	1,830	2010	1,176	810
COVNM [a]	kt	2,416	2010	1,360	1,050
NH_3	kt	787	2010	753	780

Source: CITEPA SECTEN February 2005.

[a] *Other than biotic sources: forests, grasslands and forest fires (between 1,200 and 1,400 kt)*

As shown in Table 20.2, emissions from French refineries have significantly decreased since 1990.

In comparison to total French emissions, refineries emit highly variable relative quantities of pollutants, with 24% of SO_2 and 29% of nickel, but only 2% of NOx and 1% of NMVOCs (Non-Methane Volatile Organic Compounds), including 0.5% of benzene and very low quantities (less than or significantly less than 3%) of other metals, PAHs, dioxins, furans, particles, CO, etc.

Once again, SO_2 seems well under control since, for the first time in 2003, fuel oil combustion emitted less SO_2 than all solid fuels burned in France (106 kt vs. 120 kt), and when both diesel fuel and gasoline contain less than 10 ppm of sulfur in 2009, corresponding SO_2 emissions will represent 0.8 kt, or less than 0.2% of current emissions.

This decrease in refinery emissions is partially due to the use of lower sulfur crude oils, but above all to the reduced quantities of heavy fuel oil burned by refineries, which is gradually being replaced by natural gas. As for the decreased sulfur concentration of products, this is mainly the result of investments made roughly ten years ago in deep *desulfurization* of diesel and gasoline cuts, with the sulfur being recovered in the form of solid sulfur in Claus units. Over the medium term, installation of several deep conversion units (also used to produce sulfur-free fuels after increasingly severe hydrogenation/hydrocracking using residual fractions rich in sulfur and metals) can also decrease SO_2 emissions, although to the detriment

Table 20.2 Emissions from French Refineries in 2003.

Pollutants	Refineries 2003	Refineries 1990	Petroleum Industry 2003	Utilities/ Energy 2003	France 2003	France 1990	Comments
SO_2 (kt)	118	186	230*	262	490	1,330	* Emissions on products
NMVOC total (kt)	14.5	24	55*	79	1,400	2,416	* Outside marine terminals
VOC/ Benzene (kt)	0.24*	0.15/ 0.19**	0.4/0.6	1.4	47	83 to 92**	* Secten Feb 05-p. 16 ** Estimated CITEPA 96
NOx (kt)	22.5*	25*	≈ 25*	132	1,176	1,830	* Estimated
$PM_{2.5}$ (kt)	< 3	–	< 10	12	267	359	
CO (kt)	2.0	–	< 5	43	5,633	11,000	Mainly related to transport
Ni (t)	52	55	60	≈ 100	175	325	
As (t)	0.3	0.4	< 0.5	1.5	13	25	
HAP (t)	0.1	0.2	–	0.2	35	44	According to CEE-NU, p. 71 Secten
Pb (t)	0.7	–	–	11	170	4,300	
Dioxins, Furans (g/ TEQ)	0.2	–	–	55	247	1,765	

Source: UFIP.

of CO_2 emissions, due to increased consumption of hydrogen. Industrial hydrogen is produced from hydrocarbonated fossil sources by natural gas reforming or gasification of coal or heavy petroleum residues, which results in very large emissions of CO_2.

The decrease in VOCs is the result of major efforts made by manufacturers to respond to increasingly strict regulations regarding the recovery of *light hydrocarbon vapors* (VOC Directive "Stage One" and "Stage Two", 1994*)* at refineries: in gasoline fuels and bases at storage and processing facilities (diffuse VOCs) and fuel loading stations. This decrease in VOC emissions is also the result of adaptation in fuel and gasoline distribution logistics (when loading storage tanks at service stations and when distributing fuel to vehicles).

For NOx, a "refining" order was issued in France, and sets a "bubble" constraint to significantly limit these emissions, e.g. by decreasing the use of heavy fuel oil rich in organic nitrogenous compounds and instead using natural gas, and by installing low-NOx burners, since a large quantity of NOx emissions comes from nitrogen in the air used for combustion. However, note that this has no major effect on overall French NOx emissions for which refining, with 2% of national emissions, remains a small contributor.

With regard to benzene, implementation of a separator at the end of the 1990s and strict specifications for gasoline containing less than 1% benzene, with the resulting increased use of certain facilities (alkylation, isomerization, etc.), light ether bases (MTBE and ETBE) and ethanol, have significantly limited these emissions.

Finally, dust emissions have been drastically reduced by the use of electrostatic precipitators on FCC units, and possibly on smokestacks, to comply with the integrated "SO_2, NOx and Dust" directive of 2005.

All these trends and observations are also valid for Europe, and in general for refineries in developed countries, regardless of the type of crude processed.

20.2 LEGISLATION FOR PETROLEUM PRODUCTS: EUROPEAN AND WORLDWIDE SPECIFICATIONS

European legislation (Table 20.3) for petroleum products is also the strictest in the world, followed closely by American legislation. However, an entrainment effect is observed, with regulations in other regions of the world following European or American regulations, delayed to a greater or lesser extent depending on the country (Table 20.4). Over the last 20 years, European regulations have resulted in significant changes in refining, in both energy and financial terms. This change has meant that energy consumption by refineries themselves has increased from 4% of crude oil consumed on average in 1980 in French

Table 20.3 Evolution of European Specifications for Fuel.

			1993	1995	1996	2000	2005	2009
EN 228 Gasoline 95/85 without lead								
Sulfur	ppm m/m	max	1,000	500		150	50/10	10
Benzene	%v/v	max	5			1		
Aromatics	%v/v	max				42	35	
Olefins	%v/v	max				18		
Oxygen	%m/m	max	2.5 [a]			2.7		
Vapor Presure (summer)	kPa	max	< 80			60 [b]		
E100	%v/v	max	40 (s)/ 43(w)			46		
Final point	°C	max	215			210		
EN 590 Gasoil								
Cetane Index (calc.)		min	46					
Cetane Number (measured)		min	49			51		
Sulfur	ppm m/m	max	2,000		500	350	50/10	10 [c]
Density	kg/m^3	min	820					
		max	860			845		
95% distillated	°C	max	370			360		
Polyaromatics	%m/m	max				11		
Lubricity	µm 60°C	max			460			

Source: CONCAWE 2003.

[a] *Up to 3.7%, depending on the choice of Member States. Specific limits are set for each compound*

[b] *70 kPa is authorized in arctic conditions or intense winter cold*

[c] *Deadline for passage of all product supplies to sulfur content of 10 mg/kg*

Table 20.4 Evolution of Fuel Specifications throughout the World in 2003.

Current Legislation			EU 09	US Fed. 06[a]	Japan 04	China	India	Brazil	RSA	NZ	WEI 05[b]
Gasoline											
Sulfur	ppm m/m	max	10	30	50	1,000	10,000	1,000	1,000	500	400
Benzene	%v/v	max	1	1	1	2.5	5	1.5	–	5	2.5
Aromatics	%v/v	max	35	(4)	–	40	2.5	45	–	26-48	45
Olefins	%v/v	max	18	(4)	–	35				25	
Oxygen	%m/m	max	2.7	2 to 2.7	7[c]	2.7				0.1	
Gasoil											
Cetane Index (calc.)		min	46	40							
Cetane Number (measured)		min	51		50	45	48	42	45	49	
Sulfur	ppm m/m	max	10	15	50	10,000	2,500	2,000	3,000/50	3,000	2,000
Density	kg/m^3	min	820				820	820	800	820	
		max	845				870	870		860	
Polyaromatics	%m/m	max	11								

Source: CONCAWE 2003.

[a] *US Fed: Must meet the requirements of a complex mathematical model*
[b] *World Bank recommendations*
[c] *MTBE: Methyl Tertiary Butyl Ether*

refineries, to a current 6-7%, which may exceed 10% in the future for conversion refineries or full-conversion refineries processing heavy crudes. The corollary is an increase in CO_2 emissions, although this is somewhat moderated by partial replacement of combustible heavy fuel oil with natural gas. In any event, this case illustrates the antagonism between reducing pollutant emissions and greenhouse gas emissions to take into account the IPPC Directive, which aims to decrease overall pollution, including local pollutant emissions and greenhouse gas emissions.

Qualitatively speaking, in Europe and the United States, then throughout the world, environmental requirements have significantly increased and will continue to do so through the end of the decade. This is evidenced by the decrease in sulfur concentration to less than 1% in heavy fuel oils, a significant decrease in sulfur in marine bunker fuels (fuel oil), the decrease of sulfur concentration to 1,000 ppm in home heating oil, sulfur of 10 ppm in gasoline and diesel fuel by 2009, and the decrease to 50 ppm – then probably to 10 ppm – of sulfur in diesel fuel for non-road mobile machinery (construction and agricultural machinery). This fundamental shift will be reflected by an increase in the energy consumed in refineries resulting – as previously stated – in a significant increase in CO_2 emissions, the majority of which should be compensated for by reduced CO_2 emissions from road transportation due to improved engine yields, and the likely increasing use of electrical propulsion coupled with a thermal engine *via* hybrid vehicles (light-duty vehicles and heavy trucks). Of course, these reductions require large investments to drastically decrease sulfur content, particularly for new hydrotreatment facilities. Thus, more than 1 billion Euro has already been committed by the French refining industry in this area for the period 2000/2005 (see estimates in Tables 20.5 and 20.6).

Table 20.5 Cost Estimate of Environmental Measures for French Refining.

Environmental Constraints		**Estimated Cost**
SO_2	From 1990 limit to 1,000 mg/Nm^3 (bubble concept) limit in 2010	140 MM€/year
NOx	From current limit to 350 mg/Nm^3 (bubble concept) limit	10 to 20 MM€/year
NOx	From current limit to 200 mg/Nm^3 (bubble concept) limit	50 to 80 MM€/year
VOC	VOC Directive Stage 1 & stage 2	700 MM€
Specification Limits on Products		**Estimated Cost**
Benzene in Gasoline	< 1%	500 MM€
Aromatics in Gasoline	From 42 down to 3% v/v	≈ 500 MM€
Sulfur in Gasoline	From 150 to 10 ppm	2,000 MM€
Sulfur in Gasoil	From 350 to 10 ppm	1,200 MM€

Source: CONCAWE 2003.

Table 20.6 Examples of cost and environmental issues for a new desulfurization unit.

Selective desulfurization unit on FCC naphtha: – 1,000 Mt/y unit in an existing refinery: 60 MM€ with an associated CO_2 production increase of 50 kt/year (related to energy and hydrogen consumption).
Deep desulfurization unit on Diesel: – 2,000 Mt/y unit in an existing refinery: 150 MM€ with an associated CO_2 production increase of 200 kt/year (related to energy and hydrogen consumption).

Source: UFIP.

20.3 CONCLUSION

The reduction of "local" pollutants is largely underway in France, Europe and the United States, as well as other countries like Canada, Japan, Australia, South Korea, etc. China is also starting to make its emissions standards and specifications much stricter. This reduction of pollutants by refining and *via* specifications for fuels and road vehicle emissions is the result of efforts by all: manufacturers (automobile manufacturers, heating specialists, oil companies, etc.) and the public authorities. In 2010, we will achieve a very low level of local pollution, and regulations will likely prove of little use or effectiveness at reducing this. In contrast, we should focus on Greenhouse Gas Emissions (GGE), especially since very strict regulations on local pollutant emissions have resulted in increased CO_2 emissions, an effect that will continue.

Pollution reduction is an increasingly complex area, particularly if it is addressed – as it must be – within the framework of Sustainable Development. In order to be effective and remain economically acceptable, it will require good communication between the various stakeholders (researchers, manufacturers, public authorities, citizen representatives, etc.)

impacted by problematics that are complicated and increasingly constraining. It is by succeeding in this approach and operating at the international level, with actions that must be undertaken in coordinated fashion, that Europe and the United States – pioneers in this area – can create "spin-offs" in the rest of the world and propose viable, widely-accepted solutions to generalize regulations on environmental pollution and greenhouse gas emissions.

The problems of refining and conversion of heavy and extra-heavy crudes are merely extensions of current refining problems. The technical solutions for reducing pollutants and greenhouse gases are the same and the same regulations must apply. The only difference is one of degree, due to the very nature of these crudes, which contain more impurities (sulfur, nitrogen, metals and asphaltenes) and far fewer light fractions, as well as a low H/C ratio when compared with conventional crudes. This implies recourse to more severe technical solutions relying largely on hydrotreatment, hydroconversion and hydrocracking processes at high or very high hydrogen pressure, with consumption of very large quantities of this element. This will inevitably lead to far higher energy consumption, in turn resulting in increased consumption by refineries themselves and significantly higher CO_2 production, essentially linked to the very high demand of hydrogen production.

In conclusion, the situation can be summarized as follows: control of pollutant emissions is an accomplished fact in Europe and the United States, and with delays to a greater or lesser extent, should be in the rest of the world, which will very likely follow European and/or American regulations. It can be said that on this front, the battle has been or will be won against pollutant emissions, as far as political authorities of the countries in question are convinced and act in consequence. But we have not won the war since the difficult – but necessary – fight to control and reduce greenhouse gases remains. In addition, the increasing difficulty in accessing energy resources will not facilitate the situation.

REFERENCES

International protocols and conventions

Local pollution

Stockholm (1972)

Transborder pollution

Geneva (1979), Helsinki (1985), Sofia (1988), VOC Protocol (1991), Oslo (1994); Aarhus (1998), Gothenburg (1999)

Global pollution

Vienna (1985); Montreal (1987)

Recent European Directives

- Large Combustion Plant Directive (LCPD), first directive in 1988, revised in 2001.
- National Emission Ceilings Directive (NECD), directive 2001, National Ceilings Plan July 2003.
- VOC Directive "Stage One" and "Stage Two", 1994
- IPPC Directive 1996
- European Directive 2005/33/EC dated July 6, 2005 regarding the sulfur content of marine bunker fuel oil

Summary articles on the problematics of emissions of pollutants, greenhouse gases and water in upgraders

Griffiths M., Dyer S., June (2008) Upgrader Alley report: Oil Sands Fever Strikes Edmonton, Sand Oil Fever Series, the Pembina Institute (available in PDF version).

Len Flint PhD, P. Eng (2004) Bitumen & Very Heavy Crude Upgrading Technology – A Review of Long Term R&D Opportunities. March 31, 2004, LENEF Consulting (1994) Limited (available in PDF version).

21 | Greenhouse Gas Emissions

G. Renard, A. Sanière, E. Delamaide

While there are many challenges that the oil industry needs to address, concern for climate change has pushed compliance with the Kyoto Protocol to the forefront. Companies producing heavy oils or bitumen in countries having signed this protocol, as well as others, are faced with the obligation to reduce their absolute GreenHouse Gas (GHG) emissions. For instance, although it is not the only country, this is particularly true for Canada, which has huge reserves of such heavy oils and bitumen, the production of which is rapidly increasing as shown in Chapter 10. In 2003, its oil sands industry accounted for 3.4% of Canada's total CO_2E [1] emissions. Using 2006 projections, this could amount to 8% of Canada's business-as-usual emissions by 2012, or 12% of Kyoto target emissions [Pembina Institute, 2006].

For example, mining operations generate large quantities of greenhouse gases. In fact, Syncrude and Suncor were # 6 and #7 respectively in the Top 10 of greenhouse gas emitters in Canada in 2004 [Canadian Environmental Law Association website]. Syncrude reported that in 2005 its operations released 9.9 million tons of greenhouse gases, which corresponds to 0.8 ton per m^3 of oil produced (0.125 ton per bbl). The companies are trying to reduce those numbers by adding scrubbing equipment whenever possible and through various other alternatives, but again, no major technical step change is expected soon.

As a rule, operators that produce heavy oil or bitumen are looking for new technologies to reduce the environmental impact of their industry. Governments have addressed this problem and are setting new regulations to encourage lower emission levels. It is essential for the oil industry to seek out and encourage development of GHG emission-reduction strategies or technologies, as well as pursuing sequestration for the remaining emissions.

21.1 ANALYSIS OF GHG LIFECYCLE EMISSIONS

To evaluate GHG emissions from the heavy oil industry and develop effective processes and technologies to reduce their amount, it is imperative to list available point sources and look

1. "CO_2E" (CO_2 Equivalent) is the preferred term used in crude oil and natural gas lifecycle work. In addition to CO_2, it includes the other officially recognized primary greenhouse gases: methane (CH_4) and nitrous oxide (N_2O).

at their respective contribution for the overall crude oil lifecycle as the oil is produced, from reservoir through refining and final consumption (end use). Such a contribution has been published in Table 21.1 [McCann T and Magee P, 1999].

Table 21.1 CO_2E Sources from Production to End Use [McCann and Magee, 1999].

	Greenhouse Gases		
Source Sector	**CO_2**	**CH_4**	**N_2O**
Heavy oil fields	Boilers Heater-treaters Flares	Battery and well Unrecovered associated gas Casing vent gas (not recovered) Seepage outside casing	Minor from combustion sources
Conventional	Heater-treaters Boilers Flares	Tanks Fugitives Flaring Uncaptured CO_2 from CO_2 EOR	Minor from combustion sources
Surfaced-mined oil sands	Diesel exhaust Boilers Cogen Fired heaters	Biogenic formation in certain waste ponds Exposed oil sand	Diesel Boilers (coke-fired) Minor from other combustion sources
All Production Operations	Hydrogen production	Fugitive process leaks Tank blanketing	
Transport Oil pipeline Marine Truck	Generally only due to electricity purchase Marine diesels Marine boilers (small today) Diesel	 Very minor Very minor Very minor	 Very minor
Refining	Hydrogen production Fired heaters Boilers Catalytic cracking	Traces from combustion Tank blanketing Tank blanketing Fugitive	Minor Minor Minor Catalytic cracking
Transport fuel use Gasoline Diesel	 Exhaust Exhaust	 Minor Very minor	
By Product fuels Propane Coke Heavy fuel oil	 Combustion Combustion Combustion	 Very minor Very minor Very minor	 Very minor from combustion
Electricity	Coal combustion	Very minor	
Natural gas supply	Dehydrators Field heaters/boilers Compressors in – Field gas compr. – Field refrigeration – Pipeline	Dehydrators Very minor Fugitives from – Field gas – Pipeline – Distribution systems	 Very minor from combustion

Table 21.2 Example of "CO_2E" Emissions for Various Oils [McCann and Magee, 1999].

Crude Element	**Canadian Light**		**Brent North Sea**		**Saudi Light**		**Typical 1995 SCO**		**Typical 2005 SCO**		**Venezuela Heavy (Primary Waterflood)**		**Venezuela Very Heavy, Partly Upgraded**	
	(1)	(2)	(1)	(2)	(1)	(2)	(1)	(2)	(1)	(2)	(1)	(2)	(1)	(2)
Production emissions	0.21	6.1	0.16	4.8	0.25	7.0	0.78	19.7	0.66	17.2	0.22	5.9	0.50	12.4
Transportation emissions	0.06	1.7	0.03	0.8	0.16	4.5	0.05	1.3	0.06	1.6	0.07	1.9	0.05	1.2
Refining emissions	0.19	5.5	0.19	5.7	0.18	5.1	0.17	4.3	0.17	4.4	0.25	6.7	0.16	4
Transport fuel combustion	2.58	75.4	2.58	77	2.58	72.5	2.60	65.7	2.60	67.7	2.69	71.9	2.74	68.1
Byproduct equivalent	0.38	11.1	0.38	11.3	0.38	10.7	0.36	9	0.35	9.1	0.50	13.4	0.57	14.2
Total	**3.42**	**100**	**3.35**	**100**	**3.56**	**100**	**3.96**	**100**	**3.84**	**100**	**3.74**	**100**	**4.02**	**100**

(1) Metric tons of CO2E/cu m of transport fuel used in North America
(2) % of total

Following this point source list of potential CO_2E emissions from an oil production standpoint, McCann and Magee evaluated and compared the emissions for various oils, from light to heavy, and SCO (Synthetic Crude Oil) derived from bitumen or extra-heavy oil (Table 21.2).

This "well-to-wheel" analysis shows that the end use (transport fuel combustion) is largely responsible for greenhouse gas emissions, with heavy crudes overall exhibiting slightly higher emissions as compared to conventional light crudes – roughly 12 and 16% respectively – but with more than double the production-related emissions.

21.2 GHG EMISSIONS DUE TO HEAVY OIL EXTRACTION

The production of heavy oils by thermal means is a significant source of CO_2 emissions. Compared to emissions associated with the extraction of conventional crudes, these are 10 times larger for the production of asphaltic sands *via* SAGD-type (Steam Assisted Gravity Drainage) *in situ* methods, and 5 times larger for exploitation by mining. In fact, for *in situ* methods, the steam needed to heat the oil in place is obtained *via* the combustion of fossil fuels. In Canada, most of the fields currently being exploited *via in situ* methods use natural gas to generate this steam.

CO_2 emissions for the extraction of extra-heavy oil and tar sands using natural gas to generate steam have been evaluated in Table 21.3.

Table 21.3 CO_2 Emissions for Extraction Process Only.

Hydrocarbon and Recovery Type	CO_2 Emissions in kg CO_2/bbl
Tar sand SAGD	65
Tar sand mining	37
Extra heavy oil cold production	10

Source: IFP Energies nouvelles.

Since most of the planned future projects intend to use gas, tension on the gas market will become even stronger. To reduce their gas dependency, some companies in Canada have begun to use other feeds, e.g. in their Firebag plant, Suncor added the ability to burn diesel fuel instead of natural gas to produce steam. Deer Creek planned the Joslyn Creek facilities to include a small steam generator to test the feasibility of using bitumen instead of natural gas as a fuel source. Coal gasification technology is also being developed, but has yet to become financially viable. The Nexen/OPTI Long Lake project is expected to employ its proprietary gasification technology to create synthetic fuel gas and hydrogen from the low-value, heaviest portion of the bitumen barrel. This process will nearly eliminate the need to purchase natural gas. At the same time gasification, which is a technology brick used in the CO_2 capture pathway during pre-combustion, could be exploited in this way to reduce CO_2 emissions from upgrading operations.

Burning coke, bitumen or diesel fuel instead of gas reduces the natural gas dependency of the projects and the production costs, but it greatly increases the CO_2 emissions of the process, since bitumen, coke and diesel fuel have higher carbon emissions factors than natural gas (Table 21.4).

Thus, by attempting to resolve their problem of dependency with regard to gas, operating companies have created another: emitting more CO_2 into the atmosphere. However, solutions exist to capture the CO_2 produced and consequently reduce the emissions.

Table 21.4 CO2 Emissions for SAGD Steam Generation depending on the Energy Source.

Hydrocarbon and Recovery Type	CO_2 Emissions in kg CO_2/bbl
Natural gas	65
Coke	152
Bitumen	122
Diesel fuel	112

Source: IFP Energies nouvelles.

21.3 GHG EMISSIONS DUE TO HEAVY OIL UPGRADING

CO_2 is also emitted during steps in heavy oil processing, during the transformation of a crude of very low H/C ratio into light fuels of high H/C ratio. The models chosen opt either for extraction of part of the carbon in the form of coke *via* a coking facility, or (with or without coker) for hydrogen enrichment *via* deep hydrotreatment processes. In the latter case, the production of hydrogen, generally from fossil fuels (natural gas or residues for issues of cost), is associated with significant production of CO_2. Due to its nature, this technology can also provide an interesting basis for setting up CO_2 capture operations.

In 2004, the Alberta Chamber of Resources produced the Oil Sands Technology Roadmap (OSTRM, 2004) to cover the future technology development needs and opportunities for upgrading in Canada. A follow-up document [Flint L, 2004] reviewed the long-term R&D opportunities for bitumen recovery technology in this country. In this study, two future-oriented cases were presented according to current production practice: a) Mining combined with upgrading, and b) SAGD combined with upgrading (Figure 21.1). In both cases, the assumption was made that the long-term future includes maximum substitution of natural gas with either residues or coal, since production of natural gas in Canada is supposed to sharply decline in the next few years.

The Mining-Upgrader combination without maximum use of residue (or coke) for energy and hydrogen would exhibit calculated CO_2E emissions of about 80 kg per barrel of SCO. Maximum substitution of natural gas by residues, combined with added hydrogen for higher quality SCO (~40° API) will raise this to about 125 kg per barrel.

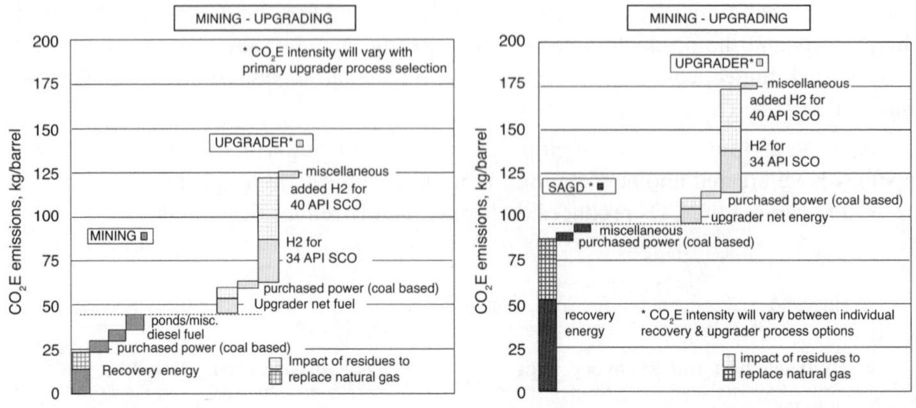

Figure 21.1

Base Case Emissions for Mining-Upgrader and SAGD-Upgrader Combinations. [Flint L, 2004].

In situ (SAGD) recovery energy raises the emissions for this combination to around 175 kg per barrel, of which 70 kg per barrel is due to the anticipated switch from natural gas to residues or coal for all energy and hydrogen generation. The anticipated move to higher quality SCO adds around 20 kg of CO_2E emissions per barrel of SCO, based on residue use.

21.4 REDUCTION OF GHG EMISSIONS THROUGH R&D

All total, the CO2 emitted throughout the chain from extraction of the fossil resource to its processing into fuel for transport will be double for extra-heavy crudes and tar sands produced thermally (*via* steam injection) or by mining as compared to conventional crude oils. Pathways starting from extra-heavy crudes having undergone cold production emit "only" 50% more CO_2 as compared to conventional crude pathways, but provide a lower recovery rate for the resource in place.

CO_2 emissions constitute a real problem for long-term economic and sustainable development of unconventional crude oils. Thus it is crucial to emphasize extraction technologies with low CO_2 emissions and technologies for capturing the emitted CO_2. Although the challenge is difficult to meet, it is not insurmountable. In this regard, taxing the greenhouse gas emissions of crude oil production projects will facilitate the financial profitability of the operation.

An initial approach consists of emphasizing strong overall thermal integration, from production to pre-refining. There is also the possibility of using pathways that do not emit CO_2 (solar, nuclear, etc.) to produce the heat needed for the production of steam for the extraction phase and for the production of hydrogen needed for the pre-refining model. Nuclear energy has the advantage of low associated CO_2 emissions, but creates another problem: radioactive waste and social acceptability. No solution seems truly satisfactory. In the production domain, we must combine the constraints of maximizing recovery and minimizing the environmental impact. Therein lies the importance of research work aimed at avoiding the use of steam processes as much as possible for the recovery of crude oil.

In Venezuela, the production technique currently used – cold production – results in relatively modest CO_2 emissions as compared to those generated by steam injection technologies. In its strategic plan unveiled at the end of 2005, PDVSA discussed its wish to increase the recovery rate and thus plan on the use of steam injection on the new blocks offered. Operators in Venezuela will therefore be confronted with the same problems as their colleagues currently present in Canada: how to produce steam at a lower cost and while minimizing CO_2 emissions?

In his study, Flint outlined that four technology-dependent sources are responsible for 80 to 90% of the CO_2E emissions when producing bitumen from oil sands: 1) SAGD recovery 2) Mining recovery 3) Upgrader process energy and 4) Hydrogen consumption in upgrading.

The most promising area for reduced emissions is *in situ* SAGD recovery, where new technology using solvents is already in advanced stages of piloting (see Chapter 8). This could potentially reduce energy by as much as 50% or more from the current levels. This would remove one-third of emissions for an integrated *in situ*/upgrading project.

The fourth category – reduced net hydrogen consumption – is the second most likely contributor to reduce CO_2E emissions. A combination of selective separations, targeted hydrogen consumption, and recovery or regeneration of hydrogen from upgrader byproducts may all combine to reduce fresh hydrogen addition by a targeted 10%.

Flint thereafter outlined and prioritized R&D areas in terms of GHG emissions reduction potential and project costs *versus* potential payback (Table 21.5).

Table 21.5 Areas for Potential Positive Impact on GHG Emissions [Flint L, 2004].

In situ
Solvent Extraction – Diffusion & Chamber Growth *In situ* Thermal/Solvent Combinations
Mining-Based Extraction & Recovery
Alternative Conditioning via Solvents
Primary Upgrading
Hydro-Retorting Direct Froth Upgrading Upgrading with Oxygen and/or Water Selective Separations
Secondary Upgrading
Membrane Reactors Selective Physical/Chemical Separations Catalysis
Integration Technologies
Novel or Improved Hydrogen Production Advanced Flow Sheet Analysis & Proc. Modeling Light Ends Management Gasification
Enabling Technologies
Bio-Upgrading

21.5 CARBON CAPTURE AND SEQUESTRATION

21.5.1 CO_2 Capture

Carbon Capture and Sequestration (CCS) is another way to reduce CO_2E emissions. However, emitted greenhouse gases are not always easy to capture, and for now, capture is costly and noticeable progress must be made in this domain [Le Thiez P, 2004]. In 2006, separation of CO_2 from flue gas with existing techniques was roughly 70 to 90 US $ per ton of CO_2 avoided. The operation requires gas scrubbing facilities using solvents, especially amines like MEA (monoethanolamine), used to treat natural gas.

Facilities equipped with gas scrubbing capacity have a disadvantage: they are very space-, cost- and energy-intensive. In certain cases, the energy requirement is almost doubled. Moreover, the operating conditions are not very favorable, since they must process large volumes of low-pressure fumes containing diluted concentrations of CO_2. This is why other options are preferred when planning new facilities.

Figure 21.2 shows the three main CO_2 capture options. For an existing facility, unless a complete overhaul is planned, only one solution can be applied directly: installation of a gas scrubbing system that is well adapted to that particular facility. For new facilities, the two other options represented schematically in the figure may be considered.

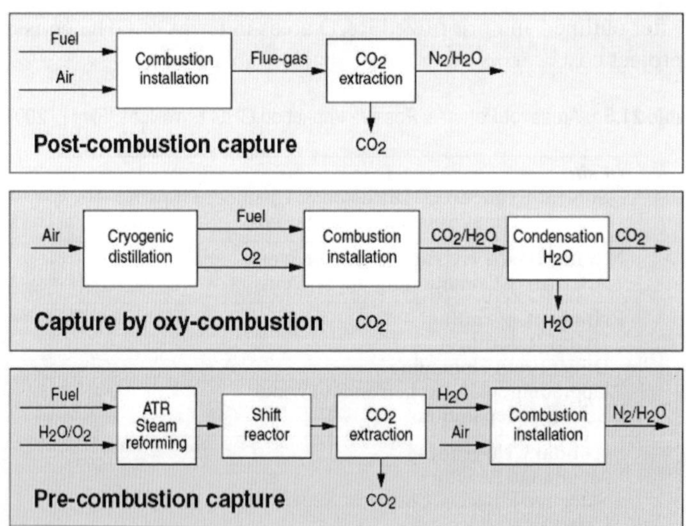

Figure 21.2

Main CO_2 Capture Options [Le Thiez P, 2004].

The fuel can be converted into synthetic gas, a mix of CO and hydrogen, either by steam reforming (natural gas) in the presence of water, or by partial oxidation in the presence of oxygen. The CO in the mix reacts with the water during the shift conversion stage to form CO_2 and hydrogen. Thus, the CO_2 is separated from the hydrogen under favorable conditions (pre-combustion capture) and the hydrogen can be used to produce energy (electricity or heat) without emitting CO_2.

When implementing the oxy-combustion technique, combustion takes place in the presence of pure oxygen, which makes it possible to obtain combustion gases containing concentrated CO_2 that is easy to separate from the steam vapor with which it is mixed.

Table 21.6 [Pembina Institute, 2006] provides a brief description of the aforementioned three systems for capturing CO_2: post-, oxy- and pre-combustion. It outlines the benefits and disadvantages of each technology and provides a range of the cost to remove one ton of CO_2.

Table 21.6 CO_2 Capture Systems [Pembina Institute, 2006].

System	Description	Benefits	Disadvantages	Cost Range (US \$/t CO_2) [a]
Post-combustion	Flue gas is captured after it has been combusted.	System can be fitted on to many of the existing conventional combustion systems in the oils sands. Most mature technology.	Captures up to 20% of available CO_2. Separation is limited to absorption technologies. Other technologies (membranes or cryogenics) are not yet considered commercially viable. High energy input for separation.	44-62
Oxy-fuel	Fuel is combusted in an oxygen-rich environment, creating a high purity CO_2 stream.	CO_2 stream is easily captured. NOx emissions are greatly reduced.	Producing oxygen is energy intensive. High temperature ranges can have adverse effects on materials used.	12-71
Pre-combustion	Fuel is gasified into a 'syngas', after which it undergoes a shift reaction where it is converted to hydrogen and CO_2.	Low incremental energy penalty. CO_2 separation and compression is relatively efficient.	Some questions of suitability of using gasification on lower quality fuels (e.g., coke). Hot gas clean-up, issues related to pure H_2 turbines.	18-44

[a] *Figures may vary in the future due to many reasons*

There are various technologies that can be applied to these three capture systems. There are four categories of carbon capture technologies:

- Absorption – physical and chemical.
- Adsorption – pressure swing adsorption, temperature swing adsorption, electric swing adsorption and vacuum swing adsorption.

- Membrane – gas absorption, gas separation and water gas shift membrane reactor.
- Cryogenics – compression and refrigeration.

The costs provided in Table 21.6 for the different systems account for the use of the various separation technologies. The capture systems technologies most commonly employed today are post-combustion and pre-combustion capture.

An oxy-combustion and CO_2 storage pilot plant project is planned in 2009-2010 by Total in the Lacq area located in the South of France [Aimnard N, 2007]. This technique is being considered by Total for the future steam generation required for its SAGD operations. It is aimed at reducing the cost of CO_2 capture from 75 US $/t CO_2 avoided by post-combustion to 35-45 US $/t by oxy-combustion.

Once the CO_2 is captured, it should be transported and sequestered for long periods, at least throughout the timeframe during which the CO_2 problem is likely to remain critical, which should not exceed one or two centuries. A precautionary measure: consideration is being given to solutions that would sequester CO_2 for thousands of years.

21.5.2 CO_2 Transportation

Once captured, CO_2 can be transported either by pipeline or truck to a storage destination. Since trucking is only financially viable over short-haul distances, transport by pipeline is generally a better option.

A CO_2 pipeline network already exists in some areas, e.g. in the southwest US where CO_2 has been available and used for EOR for many decades. However, there are no pipelines or pipeline network in locations where captured CO_2 from heavy oil production and upgrading could rapidly become a reality: e.g. Canada. In this country, government and industry have been working together to evaluate the concept of establishing an integrated network to transport CO_2 from oil sands facilities in Ft. McMurray and pipeline it to central Alberta for storage [CERI, 2002]. CO_2 can be piped in either a gaseous or a liquid phase. Typically, it is more economical to pipeline it in the dense liquid or supercritical phase. Doing so, however, uses more energy than pipelining it in a gaseous phase.

21.5.3 CO_2 Storage Options

The underground storage which is considered in onshore and offshore situations is feasible in various geological conditions (for a review see the special issue of Science, vol. 325, 2009). These geological repositories include:

- Storage in deep saline aquifers. This solution has the greatest potential in terms of storage capacity (on the order of 10 trillion tons of CO_2). The large geographical range of these aquifers is a favorable factor of proximity between sources of CO_2 and storage sites. For these reasons, storage of CO_2 in deep aquifers is the leading geological option currently being studied, with storage capacities 10 times greater than those of crude oil or gas reservoirs, according to some evaluations. They are composed of

permeable and porous rocks filled with saltwater unfit for drinking. CO_2 can be trapped in the pores of these rocks by flushing out water. Preferably, the CO_2 must be injected beyond 800 m of depth to attain pressure and temperature conditions which determine its transformation to the supercritical state (greater than 31°C and 73 bar). In this state, the CO_2 is dense and occupies a reduced volume. To ensure airtight storage, an impermeable top layer must be present above the rock-reservoir, and composed of clays or salts that prevent the CO_2 from rising to the surface. Over time, some of the CO_2 may dissolve into interstitial fluids and result in geochemical reactions with rock minerals. Although slow, these processes result in trapping of the CO_2 in dissolved form and even in mineral form if conditions are favorable for the formation of carbonated minerals. This increases storage capacity and safety. Nonetheless, we must verify that these geochemical reactions do not have undesirable geomechanical consequences. Initial feedback from operations of injection and storage of CO_2 underground and the analogy with natural CO_2 deposits that exist underground in many countries allow us to confidently envisage the multiplication of pilot storage sites.
- The second option is the storage of CO_2 in unexploited coal seams. Coal adsorbs CO_2 preferentially instead of the methane initially present. This very interesting trapping mechanism also enables the mobilization of trapped methane (Coal Bed Methane), which can be recovered in producing wells, a potentially attractive economic benefit. The key parameter associated with this storage solution is certainly the permeability of this type of formation. In general, it is very low compared to the rock-holding hydrocarbon reservoirs and the kind of aquifers suitable for CO_2 storage. Although difficult to assess, the potential for storing CO_2 in coal seams is on the order of 40 billion tons of CO_2, which is much lower than the potential for storage in aquifers or even in hydrocarbon deposits. This only concerns unexploited formations, since sites which have undergone coal extraction are littered with mine shafts, which constitute ready paths for rapid migration of CO_2 to the surface.
- The last option is the storage of CO_2 in depleted oil and gas fields. This option is particularly interesting. In fact, when CO_2 is injected at the production stage, one can take advantage of injection to effect Enhanced Oil Recovery (EOR). Obviously, any CO_2 escaping with the production fluids must be recovered and returned to the reservoir. The most attractive storage aspect is the natural confinement offered by such structures, which have served as oil or gas traps for several million years. These types of projects are now operating, e.g. Encana's Weyburn project [IEA, 2005].

The use of petroleum reservoirs as CO_2 storage sites has the following advantages:
- Low exploration cost, since the geology is already well-known.
- Proof that the reservoirs have been able to trap liquids and gases for millions of years.
- Equipment for production – and often injection – that is already in place and can be used to transport and inject the CO_2.
- Enhanced recovery of the last reserves of oil and natural gas from deposits.
- Regulations already exist.

However, their extremely unequal worldwide distribution, limited storage capacities as compared to aquifers, and the need to monitor existing wells so that they can not become ready paths for migration of CO_2 to the surface, are all limits for the use of this pathway.

Although providing storage potential that is far more limited in terms of capacity and geographical distribution, this last option allows for additional recovery of oil or natural gas, thereby contributing to the financing of storage operations.

21.5.4 Evaluation of CCS Scenarios in the Context of Heavy Oil

The Pembina Institute [Pembina Institute, 2006] has performed some calculations in order to evaluate different CCS scenarios applied to the Canadian oil sands industry to become carbon neutral by 2020. For that purpose, GHG emissions have been classified into two groups: point and non-point, where non-point sources are geographically dispersed in smaller amounts than point source emissions and therefore technically more difficult to capture. In an oil sands operation, vehicle fleet (e.g. heavy hauler mine trucks) operations and fugitive emissions from tailings ponds and mine faces are examples of potential non-point source emissions.

Assuming a ratio of 55% mining operations and 45% *in situ* operations plus upgrading, typical of the current operations of major Canadian petroleum companies, they evaluated the amount of emissions produced by point and non-point sources (Table 21.7). Two levels of emissions intensity were assumed to reflect the current range of operations. Gasification of coke, a fuel byproduct of bitumen upgrading, was included in the 'Higher Emissions Scenario' for hydrogen production for upgrading.

Table 21.7 Amount of Emissions Produced by Various Point and Non-point Sources of GHGs for Mining and Extraction, *In situ* and Upgrading Oil Sands Processes [Pembina Institute, 2006].

Process	Low Emissions Intensity (kg CO_2E/bbl SCO)		High Emissions Intensity (kg CO_2E/bbl SCO)	
Mining and Extraction				
Point (utility heaters, power)	17	(61%)	24	(62%)
Non-point (fleet, mine face, tailings ponds)	11	(39%)	15	(38%)
Total	28	(100%)	39	(100%)
In situ				
Point (boilers/plant energy)	48	(92%)	55	(92%)
Non-point (fugitive, vehicle fleets, flaring)	4	(8%)	5	(8%)
Total	52	(100%)	60	(100%)
Upgrading				
Point – Hydrogen production	14	(27%)	41	(52%)
Point – Combustion sources (coker, boilers)	37	(72%)	37	(47%)
Non-point (fugitive)	1	(1%)	1	(1%)
Total	52	(100%)	79	(100%)

This table shows that for mining and extraction operations, point source emissions are only 60% of total produced emissions, while they represent more than 90% for *in situ* operations. CCS is therefore supposed to be more efficient at the production level for *in situ* operations.

Emissions amounts were then used to define the costs associated with three carbon-neutral scenarios:

1) CCS Focus: CCS is maximized within oil sands operations, covering emissions from all point sources. Offsets (through CO_2 market trading) are used only where emissions cannot be captured using existing or emerging technologies in the 2020 timeframe.
2) Combined CSS and Offsets: CCS is used on a very limited basis and only captures hydrogen plant operations. Offsets are used for the remaining emissions.
3) Offset Only: Carbon capture is not used. Offsets are purchased for all GHG emissions.

Conclusions of the study indicate that becoming carbon-neutral by 2020 for Canadian oil sand producers appears economically feasible given high oil prices due to subsequent profit margins. Scenario 3 using offset only is the better compromise for the time being. However, it will depend on the future offset prices of CO_2 on the GHG market.

As indicated in Chapter 19, nuclear energy could have an enormous impact on the reduction of CO_2 emissions. However, its acceptance by the population is very low and efforts are needed to change this situation. So, for the time being, nuclear energy cannot be considered as a near-term provider source of energy in countries developing heavy oil assets or oil sands.

Future legislation on total CO_2E emissions to reach the levels required by the Kyoto Protocol, as well as the increasing cost of CO_2, both have the potential for undermining the favorable economics underlying the exploitation of heavy oil assets or oil sands leases. Integrated SAGD/Upgrading, which would be the most heavily impacted technology, could see its supply cost rise by up to US $5 per barrel of SCO if facing a US $30 per ton cost of CO_2 [Lacombe RH, 2007].

However, efforts must be made to reduce GHG emissions. Therefore, the oil industry must take a leadership role to set a target of becoming carbon-neutral as soon as possible. It must support advances in the capture technologies, find quality offsets and help to locally develop a domestic carbon offset trading system. High oil prices are favorable to a continuous decrease in costs for becoming carbon-neutral.

REFERENCES

Aimnard N (2007) Oxy-Combustion and CO2 Storage Pilot Plant Project at Lacq", *2nd International Oxy-Combustion Research Network Workshop*, Windsor, USA, Jan. 2007.

Alberta Chamber of Resources (2004) Oil Sands Technology Roadmap: Unlocking the Potential. Available at: www.acr-alberta.com/ostr/OSTR_report.pdf

CERI (2002) Cost for Capture and Sequestration of CO2 in *Western Canadian Geological Media*, June 2002.

Flint L (2004) Bitumen and Very Heavy Crude Upgrading Technology: a Review of Long Term R&D Opportunities. Available at: www.ptac.org/links/dl/osdfnlreport.pdf

IEA Greenhouse Gas R&D Programme (2005) Weyburn Enhanced Oil Recovery Project.

Lacombe RH and Parsons JE (2007) Technologies, Markets and Challenges for Development of the Canadian Oil Sands Industry. Available at: www.mit.edu/~jparsons/publications/technologies-markets-and-challenges.pdf

Le Thiez P. (2004) The Sequestration of CO2 – IFP Publication – Panorama 2004. www.ifp.fr/IFP/en/files/rechercheindustrie/IFP-Panorama04_10-SequestrationVA.pdf?opendocument&nf=6.

McCann T and Magee P (1999) Crude Oil Greenhouse Gas Life Cycle Analysis Helps Assign Values for CO_2 Emissions Trading", *Oil and Gas Journal*, **97**, Issue 8, Feb 22, 1999.

Pembina Institute (2006) Thinking Like an Owner, Oil Sands Issue Paper No. 3. Available at: http://pubs.pembina.org/reports/Owner_FullRpt_Web.pdf

Pembina Institute (2006) Carbon Neutral by 2020: A Leadership Opportunity in Canada's Oil Sands", Oil Sands Issue Paper No. 2. http://pubs.pembina.org/reports/CarbonNeutral2020_FinalDraft.pdf.

Special Issue on Carbon Capture and Sequestration (2009), Science, **325**, Sept. 25, pp 1644-1659.

22 | CO_2 Mass Balance: an Integrated Approach

J.B. Sigaud

22.1 METHODOLOGICAL APPROACH SPECIFIC TO SYSTEMS OF INTERDEPENDENT PRODUCTS

The energy technologies form a system of interdependent products with possible substitutions between energies of different origin (particularly between biomass and fossil energies), as well as possible conversions from one form of energy to another (e.g. gas to electricity).

It follows that any change concerning a single component of the system (primary energy resources, consumption of energy in its various forms, conversion processes, etc.) may affect all the others. Assessing the impact of such a change on overall CO_2 emissions must therefore take into account the effects caused by this change on the whole energy system.

In addition, since fossil energies (particularly petroleum derivatives) and biomass can also be used as raw materials for the manufacture of chemical products, lubricants, bitumen, etc., such a change may also impact non-energy uses of these resources.

In the absence of a rigorous approach and relevant assumptions, the results obtained and the conclusions drawn may turn out to be completely erroneous, and this is particularly true for estimating the impact on CO_2 emissions.

The much publicized example regarding the use of LPG as a car fuel instead of gasoline provides a good illustration of the need for this type of global approach.

This operation is generally credited with having a beneficial effect on CO_2 emissions, not only on the basis of a balance considering the vehicle itself (which is unanimously recognized today as insufficient), but also the basis of a more appropriate "well-to-wheel" analysis.

However, although the latter approach is considered satisfactory by most experts, it is nonetheless incorrect since it does not take into account the fact that substituting LPG for gasoline as a car fuel has repercussions on other uses associated with the substituted gasoline.

The diagram of Figure 22.1 shows the resources and main uses of LPG, as well as the flow of products affected by this substitution.

LPG is in fact a byproduct of petroleum refining and natural gas processing, when the gas comes from a deposit containing both methane and other light hydrocarbons. Consequently, the

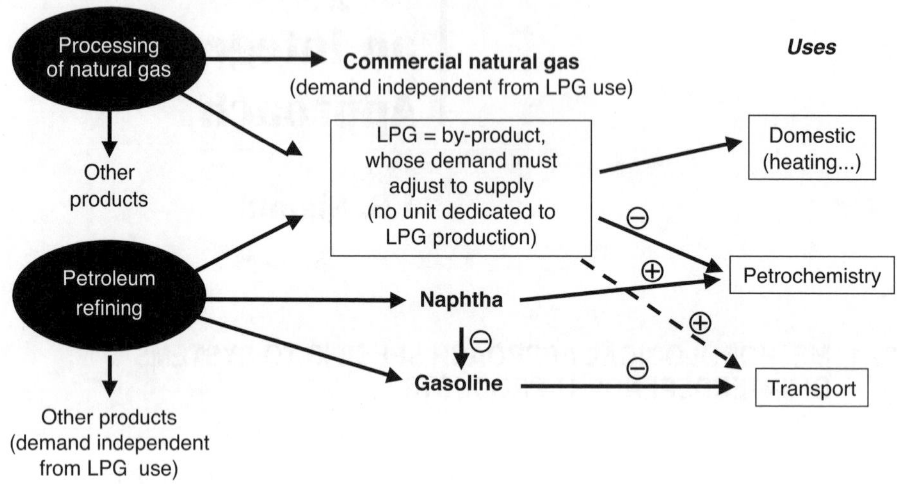

Figure 22.1

LPG Fuel: a "False Good Idea" Ignoring an Integrated Approach of the Various Processes.

LPG production level is not controlled but rather imposed (being a byproduct). In other words, the quantity consumed – all uses combined – must adjust to the supply, rather than the reverse.

As a result, using more LPG as automotive fuel has no impact on the quantities produced and thus implies consuming less for other uses such as petrochemicals or domestic uses.

Since these substitutions between products do not themselves improve the energy efficiency of the processes concerned (vehicle, boiler or steam cracker), this results – all services being equal – in no energy savings and thus no reduction in overall CO_2 emissions.

This example illustrates the most important rules to follow in order to correctly assess the impact of a project on CO_2 emissions, in particular:

a) Carefully define the relevant scope of consolidation. In theory, this includes the entire planet and all activities taking place on it. In practice, the scope must be reduced to a smaller domain corresponding to business sectors that may be significantly impacted by the operation in question. The difficulty lies in not omitting anything important, while limiting the scope to what is absolutely necessary.

b) Clearly identify – with regard to the operation in question – the invariant components of the system (in the example of substituting LPG for gas, these are LPG availability and the energy efficiency of processes impacted by this substitution). This requires a relevant definition of the reference situation (i.e. if the project is not undertaken), as well as the situation resulting from its completion.

In fact, the impact of any project on CO_2 emissions must not be estimated in the absolute, but rather as compared with what the level of emissions will be if the project is not undertaken (reference case). Depending on the nature of the invariants, this reference case may correspond to the mere absence of the project considered (status quo) or implementation of another project.

22.2 EXAMPLE: PETROCEDENO (EX SINCOR) PROJECT

This is an example of the exploitation of heavy oil from the Orinoco Belt in Venezuela. The technology is based on the principle of 'cold' production (i.e. no heating of the deposit) and upgrading of heavy oil into high-quality synthetic crude on the production site.

The relatively high temperature in the reservoir allows the oil to be extracted without heating, and thus without spending energy for this purpose. These favorable conditions, combined with the development of well-adapted drilling and production techniques, limit the energy cost and CO_2 emissions at the crude oil production stage to about the same level as for a conventional crude oil.

In practice, CO_2 emissions are slightly lower in the case of a conventional crude if there is no associated gas flaring, and higher in case of flaring.

However, the need to pre-refine heavy oils locally in order to process them into marketable synthetic crude leads to additional energy consumption, and therefore additional CO_2 emissions as compared with the case of conventional crude (see Figure 22.2).

Main steps of the "heavy crude oils" processes workflow

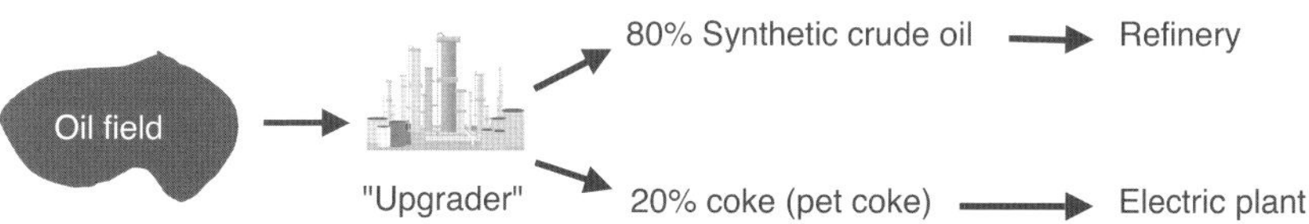

Extra-heavy oils have a higher carbon content than conventional oils.
They cannot be transported as such (too viscous). This implies:

- Pre-refining on the production site
- To get rid of the excess carbon in the form of coke
- Use of the resulting coke instead of coal in electric or cement plants

Figure 22.2

Extra-Heavy Oils: an Additional Step of Processing When Compared to Conventional Crude Oils.

Petrocedeno (ex Sincor) produces a crude oil called Zuata (named after the region of origin) and processes it in a pre-refining facility (see Figure 22.3) located on the seashore 200 km further north, which produces a synthetic crude called Zuata Sweet, resembling a light conventional crude.

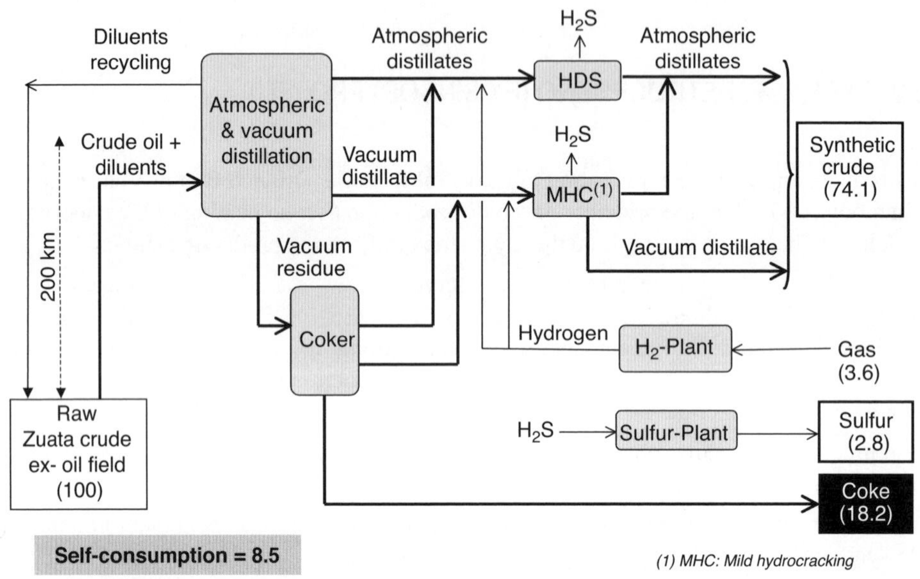

Figure 22.3

Petrocedeno (ex Sincor) Pre-Refining Workflow (Mass Balance for 100 Metric tons of Zuata Crude Oil).

A pipeline is used to transport the oil between the field and the pre-refining facility. Since the crude oil is too viscous to be transported as it is, it is blended with naphtha which acts as a diluent. The naphtha is recovered at the pre-refining site, from distillation of the 'crude oil + naphtha' blend, and returned via a second pipeline to the production field where it is once again mixed with the raw crude oil.

While the energy consumption per ton of crude oil for pre-refining corresponds to that of a simple hydroskimming refinery, the CO_2 emissions correspond to those of a complex refinery: in addition to the CO_2 resulting from energy generation, the emissions related to the production of hydrogen for hydrodesulfurization and hydrocracking facilities must also be taken into account.

In return, the hydrogen enrichment of the synthetic crude which results from this hydrotreatment, the low sulfur content of this crude and the fact that it no longer contains residue result in energy consumption and CO_2 emissions that are reduced at the final refining stage, the CO_2 emissions being reduced to a larger extent than the actual energy consumption due to lower hydrogen needs (and thus less CO_2 coming from hydrogen production).

Table 22.1 compares energy consumption and CO_2 emissions from the oil field to the refinery outlet, at equivalent production of "valuable products", for Zuata crude and Arabian Light crude, assuming in both cases that all the residue is converted into lighter products.

Quite logically, this comparison shows a large handicap for the Zuata crude.

Moreover, pre-refining of Zuata results in significant production of petroleum coke, used as fuel in cement plants or to generate electricity in power plants.

Table 22.1 Energy Consumption and CO_2 Emission Associated with Production and Refining of Zuata and Arabian Light Crudes.

	Energy Consumption (tep/100 metric tons of products)		**CO_2 Emission (tons/100 tons of products)**	
	Zuata	**Arabian Light**	**Zuata**	**Arabian Light**
Production	1.7	1.3	5	4
Pre-refining	6.2	–	27	–
Ultimate refining	5.0	7.5	16	30
Total	13.0	9.0	48	34

Assuming that both crude oils are converted into "white products", coke and sulfur (Figures standardized to the production of "white products": diesel oil and lighter products).

Note: The ratio between the CO_2 emission (associated with the consumption of fuel and gas) and the effective consumption of energy varies according to the type of operation. This ratio increases with the "hydrogenation" effect, since the CO_2 associated with the production of H_2 corresponds to unused energy (C-C or C-H bound energy transforms in H-H bound energy).

Ultimately, a refining plant producing only hydrogen would consume for its own energy needs around 25% of the internal energy of the processed oil, but would reject 100% of the oil carbon as CO_2.

The CO_2 emissions resulting from burning this coke represent roughly double those produced by the Zuata crude pre-refining and refining operations. A CO_2 balance including these emissions therefore appears extremely penalizing for the "heavy oil" technology.

However, it would be incorrect to stop there the analysis, since, as far as CO_2 emissions are concerned, only the global balance at planetary level matters.

As a matter of fact, the production of an additional quantity of petroleum coke has, in itself, no impact on the demand for electricity (nor for cement). It is as though this coke was substituted for a quantity of coal corresponding to the production of the same quantity of electricity.

Consequently, the impact on CO_2 emissions resulting from the production of coke must not be assessed in the absolute, but rather in comparison with CO_2 emissions that would have resulted from the production and consumption of the quantity of coal needed to produce the same quantity of electricity.

For a given power plant technology, the quantity of CO_2 emitted per kWh produced is higher for coal than for petroleum coke, due to the lower H/C ratio and calorific value of coal.

In addition – to make the comparison fully relevant – in the case of coal, the CO_2 balance must include the CO_2 emitted by the power plant, as well as the CO_2 emitted at the production stage (since petroleum coke is comparable with coal that has already been extracted).

Globally, this means that substituting coke for coal in the production of electricity results in some reduction (5% to 10%) in CO_2 emissions per kWh produced. The amount of the reduction depends on the quality of the coke versus the coal, as well as the coal production conditions.

In summary, to assess the impact of heavy oil exploitation on CO_2 emissions, the emissions generated from production of the oil through to combustion of the products must be compared with those that would have been generated from conventional crude and coal, in proportions resulting in the same quantity of petroleum products and electricity (comparison with all services being equal).

Table 22.2 summarizes the figures for both cases, assuming that CO_2 emissions assigned to the "conventional crude" technology correspond to the case of a mix of high sulfur coal and Arab light crude oil with full conversion of the residue of the latter into coke and "white" products (excluding bitumen and heavy fuel oils), in order to remain consistent with the production of a synthetic crude without residue as performed by Petrocedeno (ex Sincor).

Table 22.2 Integrated CO_2 Emission, Comparison Between Zuata and Arabian Light Crudes + Coal "Mix".

	Zuata Crude	**Arabian Light + coal HTS**
Produced Coke (t/100t)	25	6
Coal HTS (t/100t)	–	20 [a]
CO_2 Emission (t/100t):		
– Coal and coke combustion	85	90 (20 + 70)
– Coal Mining	–	5
– Production and refining of crude oil	48	34
Sub-Total	**133**	**129**
Part associated with the petroleum products [b]	38	34
Emission associated with the combustion of products	314	314
Total **CO_2 Emission** "well to wheel"	**352**	**348**

Figures standardized to 100 metric tons of "white products" for a similar production of "white products" and electricity

[a] *Quantity required for the production of the same quantity of electricity that in the Zuata alternative*

[b] *After deducing the emission related to the production of electricity, as calculated in the reference case (Arabian Light + Coal)*

The quantity of CO_2 to be allocated to the manufacture of products is obtained after deducting emissions related to the production of electricity, assuming that this production is, in all cases, equal to the value calculated in the "conventional crude" option used as a reference (since electricity demand is by nature an invariant of the system).

These results show that heavy oil exploitation of the Orinoco basin using the cold production technique can be considered as equivalent to the production of conventional crude oil for roughly 80% and coal for roughly 20% of the total, with no significant difference regarding overall CO_2 emissions.

In both cases, overall CO_2 emissions (including those resulting from the combustion of products) amounts to roughly 3.5 tons of CO_2 per ton of product made available for consumption, compared with a figure of about 4.5 for the production of automotive fuels from gas via the GTL technology using the Fischer-Tropsch process.

Note that if there is no financially attractive outlet for coke, it can be stored near the production site, as in the Fort McMurray region in Canada, which leads to a slightly less favorable result in terms of impact on overall CO_2 emissions (due to the loss of the advantage obtained by replacing a certain quantity of coal by coke for the production of electricity).

22.3 IMPACT OF HEAVY OIL EXPLOITATION ON CO_2 EMISSIONS ACCORDING TO PROJECT TYPE

22.3.1 Methodological Approach

To assess the impact of heavy oil exploitation projects on overall CO_2 emissions, it seems sufficient to limit the scope of consolidation to all energy pathways, while ignoring non-energy uses of products.

In fact, like the Petrocedeno (ex Sincor) project, these projects can be considered as alternatives to the production of conventional crude oil with - depending on the case - associated production of coke and/or consumption of natural gas in order to adjust the carbon/hydrogen ratio to the value corresponding to that of a synthetic crude oil similar to a conventional crude.

All heavy oil exploitation projects can therefore ultimately be analyzed as producing conventional oil (corresponding to the production of synthetic crude) and coal (in a quantity equivalent to the production of coke), with an additional consumption of energy, generally in the form of gas, for the pre-refining and, if necessary, the production of crude oil.

Note that in case of gas shortage, other energy sources can be used for this instead: some of the oil itself, preferably the residual fraction, or even nuclear energy, which has the advantage of not generating CO_2 emissions.

In order to assess the various pathways on a comparable basis regarding their impact on CO_2 emissions, one must always take into account the emissions generated by the entire chain: production of primary energy, processing into marketable products (transportation included) and their consumption, then calculate the value of these emissions with respect to the energy content of the marketed products, i.e. the quantity of products corrected for their calorific value.

22.3.2 Favorable Cases With Regard to CO_2 Emissions

This involves projects using the cold production technique in the case of Venezuela [like the Petrocedeno (ex Sincor) project], or the mining technique in the case of bituminous sand deposits in Canada, although for the latter this includes a significant yet moderate consumption of energy for extraction of the sands and separation between sand and oil.

In addition to consuming the least energy, these techniques also have the advantage of being industrially practiced and proven.

However, at best only 10% of the oil in place can be recovered with these techniques. In the case of cold production in Venezuela, this is due to the low recovery rate (roughly 8%), and in the case of mining techniques for which the recovery rate is, on the contrary, very high, this is due to the fact that only roughly 10% of the accumulations of bituminous sands in Athabasca are suitable for this type of exploitation.

Considering the size of the resources in place, increasing this recovery rate represents a major challenge, which justifies the use of new techniques at development or experimental stage.

22.3.3 Exploitation Techniques Aimed at Improving the Oil Recovery Rate

These techniques are designed to reduce the viscosity of the oil in place by injecting suitable diluents or by increasing the temperature within the deposit (hot production).

The most advanced techniques are based on hot production requiring a very substantial energy input which may seriously downgrade the CO_2 balance of the corresponding projects.

This energy input may be achieved in two ways:

- Partial combustion of the oil *in situ* (still in the development stage): its main advantage is that it would not generate CO_2 emissions.
- Injection of steam produced at the surface: the most efficient technique in terms of recovery rate as well as the most advanced, known as Steam Assisted Gravity Drainage (SAGD), is currently under experimentation in Canada.

With the SAGD technology, if the required energy is produced from natural gas, the resulting CO_2 emissions amount to roughly 1 ton of CO_2 per ton of crude oil, or roughly 1.4 tons of CO_2 per ton of marketable white products.

This leads to "well-to-wheel" CO_2 emissions of about 4.9 tons per ton of white products, or an increase of roughly 40% as compared with conventional crude processing or with a Petrocedeno (ex Sincor) type project.

Although very high, this emission factor remains comparable with that of the GTL (Gas-To-Liquid) technology, which corresponds roughly to 4.5 tons of CO_2 per ton of marketable products. The key point for commercial development of the SAGD technology concerns the availability of natural gas and its price.

In case of gas shortage, steam can be produced by using some of the crude oil, or even vacuum residue or coke resulting from oil pre-refining, but this would nearly double the quantity of CO_2 emitted per ton of steam produced.

Whatever the case, this represents a significant penalty not only from the environmental viewpoint, but also financially for the proportion of the cost that will in the future be associated with CO_2 emissions.

Two types of solution can be considered to avoid CO_2 emissions associated with this steam production: CO_2 capture and sequestration or the use of nuclear energy.

The first solution is still in the experimental stage and in order to be financially justified, will require a sufficiently high CO_2 market price. The current cost of avoiding CO_2 emissions as reflected by market price, indicates that there are other less expensive ways of reducing CO_2 emissions.

While the nuclear option raise other concerns, including the management of radioactive waste, it provides the opportunity to co-generate electricity without emitting CO_2.

22.4 CONCLUSION

Heavy oil exploitation is now technically and financially possible with no significant impact on overall CO_2 emissions as compared with the production of conventional crude oil, but the corresponding technologies only allow a small amount of the oil in place to be recovered.

The major challenge that remains to be met is therefore the development of technologies capable of achieving high recovery rates under conditions which are ecologically satisfactory and financially attractive.

Among the technologies aimed at improving the recovery rate of the oil in place, the SAGD technology with production of steam from natural gas is at an advanced stage of development, but its use at a large scale will require sufficient availability of gas at an acceptable price.

In addition, without capture and sequestration of the CO_2 emitted by the steam generation facility, this technology results in "well-to-wheel" CO_2 emissions roughly 40% higher than those of existing heavy oil upgrading technologies.

Due to commitments to control greenhouse gas emissions, exploitation of the heavy oil reserves in Athabasca will in the future undoubtedly rely on technologies aimed at minimizing both gas consumption and CO_2 emissions, such as *in situ* combustion, CO_2 capture and sequestration and use of nuclear energy.

PART 6

Ongoing Technological Challenges

N. Alazard-Toux

There are significant *heavy crude oil*, tar sands and extra-heavy crude oil resources in the planet's subsurface. Conventional crude oil reserves are currently estimated at 1,000 to 1,200 Gb, representing 40 years of production at the current consumption rate. Renewing these reserves *via* new discoveries is becoming increasingly difficult. Thus, for several years now, there has been increasing interest in identified resources of *heavy crude oil*, extra-heavy crude oil and tar sands. They represent volumes in place in the planet's subsurface that are currently evaluated at nearly 5,000 Gb, including 4,000 Gb for tar sands and extra-heavy crudes (see Chapter 2.2.1.1).

This interest has been heightened further by anticipating a long term increase of the crude oil price on international markets. This trend is exemplified by the substantial jump of the price of oil at the beginning of the XXI century before the economic depression. The development of extra-heavy crudes and tar sands has become a priority for countries holding large resources of them, and major international petroleum companies – after long years of neglect by many – want to invest in their development.

The engineering cost of the heaviest and most viscous crudes – extra-heavy crude oils of Orinoco or tar sands of Canada (projects that are existing or being studied) – varies from 10 to 20 US $/b (2005 costs), a level well above the engineering cost of crude oil produced offshore, which is generally less than 10-12 US $/b.

Furthermore, extra-heavy crude and tar sands exploitation projects are generally very capital-intensive. Investment costs are estimated at 17,000 to 30,000 US $/b/d for *in situ* projects, or double the costs of deep offshore projects. These costs are roughly 40,000 to 50,000 US $/b/d for mining projects (due to integration of pre-conversion facilities constructed on-site).

Although they are still high, the costs of projects for developing the exploitation of very dense and very viscous crude oils have nonetheless significantly decreased since the end of the 1970s. This is true even though the past two years have witnessed cost increases, associated with the increasing price of services, labor and raw materials affecting the entire gas and oil sector.

The challenge in the coming years will be to maintain the effort already achieved, while increasing the rate of recovery of resources in place: two improvements that should help to increase the economical reserves resulting from these resources. In fact, a one-point increase in the recovery rate for heavy crude, extra-heavy crude and tar sands resources in place can represent one to two years of worldwide oil consumption.

Projects and investments to develop these resources are multiplying. Yet optimal exploitation of these crude oils – particularly the most viscous – still poses many challenges due to their very nature. In fact, although the volumes in place in the subsurface are enormous, the economically and technically exploitable reserves are currently much smaller.

First, since heavy oils are denser and more viscous than conventional crude oils, their extraction from the subsurface and transport are more difficult. This generally results in a lower recovery rate for the oil in place and higher development and exploitation costs.

Second, these hydrocarbons intrinsically have a low H/C (hydrogen to carbon) atomic ratio, a characteristic which partially explains their more difficult flow and poses a problem for processing them into light products. Today, this processing generally involves increased CO_2 emissions and thus raises issues from an environmental standpoint. The production of hydrogen requires energy and today, for questions of cost, most often makes use of fossil fuels, with natural gas steam reforming traditionally being the most widely implemented technique.

With regard to technologies, the challenge will therefore be the development of processes, concepts and schemes to decrease exploitation costs and increase ultimate recoverable reserves, while maintaining acceptable overall energy efficiency for the production-transport-processing system and limiting associated greenhouse gas emissions. This challenge can be more easily met today, considering the favorable economic climate: this involves increasing recovery rates, doing away with dilution for transport, processing these crudes into a maximum number of light products in demand by the market, all while controlling costs and environmental impact, particularly greenhouse gas emissions.

Production Domain

Improvement in the efficiency of production technologies and methods must be pursued. Progress must be made on the nature of injected fluids – to lower the viscosity of the oil in place and better drain the reservoir – and on how to inject them, by opting for tools and techniques favoring real-time monitoring and thus better control of the flushing of crude oil in the reservoir. Work must also be done on more complex well architecture. Finally, the recovery of heavy oils can also be optimized *via* better pumping; various types of pumps are currently being studied.

Far upstream in the pathway, we must in the future invest in true pre-development phases and perform more detailed study of the specificities of these crude oils and the characteristics of reservoirs. Taking into account the information acquired during drilling or exploitation to regularly update the geological model would allow us to optimize future production models and thereby to better define the location of production and injection wells. Today, many conventional crude oil deposits located both onshore and offshore are subjected to detailed study in the pre-development phase. This work is even more necessary for fields of unconventional crude oil.

Paradoxically, *heavy crude oil* operators have invested little in pre-development phases up to now. The main explanation is that for a long time, these fields were developed by small companies who did not have the means to undertake detailed studies of deposits before starting entry into production. Today, since major companies are investing in *heavy crude oils*, a transfer of good practices developed for the production of conventional crude oils will become possible.

Increase recovery rates and reduce gas-dependency

Heavy oils have such a high viscosity in reservoir conditions that their mobility is extremely reduced, or even non-existent for bitumen. For this reason, production *via* so-called "cold" primary recovery, i.e. by natural depressurization of the reservoir, is difficult and recovery rates are low (less than 10%). On the other hand, this method has the advantage of not requiring a specific supply of water, gas or other products.

Most reservoirs are exploited *via* enhanced recovery, and more specifically by steam injection: this is used to heat the crude oil, which then becomes less viscous; with the increased mobility, it can be recovered. The oldest method is Cyclic Steam Stimulation (CSS), in which steam is injected *via* one well for several weeks; the same wells then become producers. This technique provides recovery rates on the order of 20 to 25% maximum, but requires fairly significant amounts of water and gas to supply the steam. Steam Assisted Gravity Drainage (SAGD) is a more recent technology in which two horizontal wells are drilled, one above the other. The steam is injected in the upper well, while the lower well recovers the crude oil, whose viscosity has decreased. This technique offers some of the highest recovery rates for *in situ* processes – on the order of 40 to 60% – but water and gas requirements remain significant [1].

Today, all *in situ* development projects in Canada intend to use this technique.

Recovery *via* mining techniques offers the highest recovery rates and only requires fairly modest consumption of water and gas as compared to steam techniques (Table 1). On the other hand, it is only applicable to 20% of volumes identified (certain deposits in Canada located at less than 100 m of depth).

Table 1 Summary of Commercial Extraction Technologies.

Extraction Technology	Recovery Rate %	Gas Consumption Mscf/b	Water Consumption b/b
Mining	90	0.56	1
Cold production	5-10	–	–
CSS	20-25	1 to 1.2	3 to 4
SAGD	40-60	1	2.5 to 4

Source: IFP Energies nouvelles and Alberta National Energy Board.

1. The main disadvantage of steam techniques is the dead loss of one barrel of water per barrel of crude oil, which remains in the reservoir.

Other recovery methods are currently being studied such as *in situ* combustion using a horizontal producer well (THAI^{TM} process: Toe to Heel Air Injection), solvent injection (VAPEX) or use of a catalyst in combination with *in situ* combustion (CAPRI^{TM} process). These processes have the advantage of not requiring recourse to natural gas since steam is not used. The expected recovery rates for these new processes are greater than 60%, but little information is currently available.

Transport Domain

Today, the preferred option for transporting highly viscous crude oils is their dilution *via* the addition of lighter hydrocarbons (gasoline or condensate). To avoid this, many other solutions have been or are being studied, e.g. heating the pipes, transporting the crude oil as an emulsion, core annular flow transport, or on-site pre-processing.

The **dilution method** is the most common transport technology for extra-heavy oil or bitumen because it is very easy to implement. The disadvantage is that dilution involves transporting a larger volume, and requires larger pipeline capacities. Diluent may represent roughly 35% of heavy oil volume. Possible problems could involve availability of the diluent and its recycling. *In situ* projects in Canada use this transportation method. In Venezuela, the four strategic associations use this same procedure and recycle the diluent.

Nearly all current projects (in progress or upcoming) have opted for this solution, which requires the mobilization of significant volumes of diluents (1/4 diluent for 3/4 crude oil). Consequently, one of the current challenges in this area is to develop solutions to reduce diluent volumes and cost.

Innovative approaches are also being investigated in order to provide an alternative to dilution technology:

On-site conversion of heavy oil is commonly used in mega-projects such as those using mining extraction in Canada. The principle is to perform partial upgrading of the heavy crude to produce a synthetic crude oil of sufficient quality to allow transport by pipeline (i.e. API degree greater than 20), and which can be sold to existing refineries for further processing into finished products.

Another method is **emulsion**, wherein extra-heavy crude or bitumen is suspended in water in the form of droplets stabilized by chemical additives. Emulsion technology is well known through its use as a fuel for electrical power plants, e.g. the Bitor company in Venezuela (Orimulsion). One disadvantage is that its use is controversial due to CO_2 emission levels. Solving this problem involves breaking the emulsion, a process which is not currently available. Moreover, this involves additional investment for treating and cleaning the water used.

Research is also underway for a method referred to as **core annular flow**. The principle is that water acts as a lubricating layer which absorbs the shear stress existing between the walls of the pipe and the viscous oil, reducing the resistance to about 1.5 times the flow resistance of water alone. This drastically reduces the pressure drop induced by the viscous fluid. The main problem with this technology is that the oil tends to adhere to the wall of the pipeline upon contact, leading to restriction and eventual blockage of the flow system. Such problems are exacerbated when the flow has to be stopped for a certain period of time.

A very recent idea for the transport of non-conventional oil is the use of a **friction reducing agent** in combination with dilution for optimization. This method is already used for conventional oil, but must be adapted to non-conventional oil.

Core annular flow is potentially the cheapest method, and also exhibits the least contamination, but the risk of pipeline corrosion is increased by the presence of water. Dilution and on-site upgrading, the most widely used methods, involve high investment. Regarding the emulsion method, the challenging issue is water separation and treatment.

Upgrading Domain

In this domain, we must lean toward models that provide increasingly deep conversion.

Heavy and light crude oil processing will provide the same range of refined products, but with significant differences in proportions and qualities. Heavy oils result in far more vacuum residues than lighter ones. These residues have an API degree between 1 and 5 and very high sulfur and metal content, which does not facilitate their processing. Several processes exist to convert vacuum residues: thermal, catalytic or both.

Thermal conversion methods are mature technologies but in general, the products obtained are of lower quality than those obtained *via* catalytic processes. Examples include **visbreaking and coking processes which rely on thermal cracking of the feedstock.**

Solvent De-Asphalting (SDA) is a proven process which separates vacuum residues into low metal/carbon de-asphalted oil and a heavy pitch containing most of the contaminants, especially metals. SDA is the subject of much interest from refiners because it allows for recovery of a substantial quantity of incremental light feedstock, notably for lubricant oil base production from vacuum residue, which means that refinery yield can be increased. Moreover, the pitch can be gasified to achieve zero fuel oil production.

Most recent work has therefore focused on various types of **hydrotreating** processes. The principle is to lower the carbon-to-hydrogen ratio of products by adding a significant amount of hydrogen, as well as desulfurize and remove nitrogen and heavy metals. These processes usually require specific catalyst combinations and operate under high pressure. Three types of reactor technologies exist, depending on how oil and catalyst are placed in contact with each other: fixed bed, ebullated bed and slurry reactor.

Fixed bed process was the first to be developed, but its application is limited for feeds with high metal content. The catalyst lays in the form of a bed

Ebullated bed reactors were first introduced in the 1960s. In this design, hydrogen and oil enter at the bottom of the reactor, expanding the catalyst bed. Catalyst performance can be kept constant by adding fresh catalyst and removing some of the used catalyst on-line. Recent R&D has resulted in substantial improvement in ebullating processes. However, all these processes require large amounts of hydrogen, which means specific production from natural gas and CO_2 emissions.

Slurry reactor, in which the catalyst is finely divided and dispersed as a suspension in the oil. This type of technology could allow high conversion. No commercial process is currently available.

A recent idea is **combining different processes** to optimize heavy crude conversion. Combining the hydrotreating and solvent de-asphalting processes is being widely studied.

These combinations can be used either in refineries or for on-site partial upgrading. The refiner obtains a Syncrude of good quality which can be used as a feed for standard refineries, and a dirty, heavy asphalt phase which can be recovered as solid or liquid fuel for Integrated Gasification Combined Cycle (IGCC), or simply for combustion to generate steam for upstream applications.

Another route for upgrading heavy oil is the process of **gasification,** which consists of using partial oxidation to convert the feed – liquid or solid – into a synthesis gas in which the major components are H_2 and CO. Gasification is a clean, flexible technology that has already been improved for coke or heavy crude. Gasification is now receiving global interest because of the Integrated Gasification Combined Cycle (IGCC), wherein gasification can process low-value refinery streams and increase energy content, along with lower SO_x and NOx content as compared to any other liquid/solid feed technology. In the future, the large CO_2 emissions of such processes will be of major concern. The capital costs of IGCC have fallen by half over a ten-years period. However, the oxygen production phase is still costly, and much research is being carried out to improve air separation and integration of this component with partial oxidation in a single-step reactor.

In this part, we have decided to emphasize two major aspects of the future of *heavy crude oil* exploitation: "Process Workflow Improvement" and "*In-Situ* Upgrading".

REFERENCES

Balier C, Riefacker A, Plouchart G, Fradet A, Dameron V (2005) Le stockage de CO2. *Les Cahiers du CLIP* 17. www.iddri.org/Publications/Autres/les-cahiers-du-CLIP/

Plouchart G (2001) Évaluation des émissions de CO2 des filières énergétiques conventionnelles et non-conventionnelles de production de carburants à partir de ressources fossiles. IFP Rapport 55949.

Saniere A, Lantz F (2006) A Worldwide Economic Analysis of the Non Conventional Crude Oil Supply Based on a Modelling Approach. IAEE – Annual International Conference of the International Association for Energy Economics, Securing Energy in Insecure Times, 29th, Potsdam, 7-10 June, 2006, Proceedings, CD-Rom.

23 *In Situ* Upgrading of Heavy Oil and Bitumen

J.P. Héraud, A. Kamp, J.F. Argillier

INTRODUCTION

Upgrading, or the improvement of oil quality (or "grade"), is usually done at the surface in so-called upgraders or refineries. However, in principle, upgrading may also be achieved within the reservoir. This is called "*in situ* upgrading". The word "upgrading" in this context refers to temporary production of oil of better quality than the overall quality of the oil in the reservoir. Such upgrading may be achieved by physical processes which separate the oil into different fractions, or by chemical processes which involve a modification of molecular structure. Distillation and asphaltene precipitation are examples of physical processes. Examples of chemical processes are thermal cracking, hydrogenation, and microbial upgrading. A preferred process could be some combination of the various processes mentioned above. In this context, the term "*in situ*" means either near the wellbore in inter-well regions, or in the reservoir, and in any case before surface facilities. Currently, very few deliberate applications of *in situ* upgrading exist. However, some operations exhibit a portion of it. Examples are distillation, aquathermolysis, and thermal cracking during steam-based processes and *in situ* combustion. Unwanted asphaltene precipitation in the reservoir and facilities is another example. Because these processes were not optimized for upgrading, the effect is weak, and sometimes undesirable results are obtained.

23.1 NEED FOR NEW RECOVERY TECHNOLOGIES

There is a growing interest in the development of *in situ* upgrading technologies. There are various reasons for this interest. To summarize, the motivation is based on rising energy consumption and declining conventional reserves. It is becoming increasingly difficult for oil companies to find large projects. Since oil prices are high, the exploitation of heavy oil reserves – such as those in the Venezuelan Orinoco belt or the Athabasca oil sands in Canada – has become increasingly interesting. Existing technologies for the exploitation of these reservoirs are insufficient. In Venezuela for example, only about 7% of OOIP will be recovered by the Strategic Associations using primary recovery, which is the exploitation method that is foreseen in the initial development plans. Possibilities for follow-up production techniques do exist, e.g. SAGD, steamflooding (possibly with horizontal wells), and cyclic steam stimulation. Other techniques – which are not ready for application on an industrial scale – look promising. One example is VAPEX technology. SAGD and VAPEX promise recovery factors of up to 60%. However, they have important drawbacks. For SAGD, large quantities of fresh water are needed for the generation of steam. This water is often not available in sufficient quantities. Large quantities of natural gas are burned to generate the steam. This gas is expensive, available in limited amounts, and its burning generates large quantities of greenhouse gases. Water recycling and energy generation from hydrocarbon products may partially reduce these drawbacks. The main problem of VAPEX is the injection of expensive light hydrocarbon vapors in the reservoir. In unfavorable cases, a considerable fraction of these vapors may not be recovered. In addition to this drawback, current estimates yield production rates that are too low to be economical. Some reservoirs, such as thin bedded reservoirs, may also not be candidates for SAGD. It is clear that these technologies will be insufficient to recover a large fraction of the heavy oil and bitumen reserves in place. Technologies are needed that use less fresh water, less natural gas, generate less greenhouse gases and wastes, and apply to reservoirs not addressed by conventional technologies. Of course, such technologies need to be profitable. A solution may come from *in situ* upgrading technologies, which should allow for viscosity reduction of the oil in the reservoir and improved quality in the fluids produced. Such viscosity reduction would facilitate not only the flow of oil in the reservoir and an increase in the recovery factor, but also the transport of oil on surface to the refineries. Indeed, current pipeline transportation of heavy oil and bitumen requires blending with a diluent. Shortages of this diluent in Canada for the purpose of heavy oil and bitumen blending could occur in the near future. *In situ* upgrading would be a possibility to decrease or eliminate the use of diluent for transportation or lifting purposes. Needless to say, *in situ* upgrading technologies will have to be combined with greenhouse gas storage.

23.2 STATE OF THE ART

The idea of *in situ* upgrading is not new, and in past decades many attempts have been made to implement it. However, few projects have been carried out, and often the ideas have been dismissed and qualified as unrealistic. In a more general sense, few breakthrough technologies have been proposed in terms of recovery methods or exploitation strategies for recovery of heavy oils and bitumen. The most important would probably be the development of inclined and horizontal wells, multilaterals wells, and possibly SAGD and VAPEX technologies. With the likely exception of multilateral well technology, the timeframe between invention and commercial application has probably been more than 20 years for most breakthrough technologies. Companies realize that if they want to have new and viable technologies in the medium and long term, their development has to start now. This may be one of the reasons that renewed interest has taken place in the last few years.

23.2.1 Physical Separation

A. Cyclic Solvent Injection

The MIS process (which stands for the Spanish "Mejoramiento *in situ*", or *in situ* upgrading) is a cyclic solvent injection process for *in situ* precipitation of asphaltenes. It has some similarities with the traditional de-asphalting process used in refineries. The principle of this technology is to separate a valuable crude oil and an asphaltene fraction by liquid-liquid extraction with a light paraffinic hydrocarbon solvent (see Chapter 15). Generally, the solvent used is a mixture of propane cut and butane cut.

The first development of this technology applied to *heavy crude oil* recovery was made by Intevep, the research and technology branch of the Venezuelan state-owned oil company, in the early 21st century, and based on a cyclic solvent injection recovery method (see Chapter 10). Thus, the heavy oil recovery is performed in four different steps:

- Solvent injection in the reservoir
- Soaking heavy oil with a solvent
- Upgraded heavy oil production
- Upgraded crude oil and solvent separation

The cyclic solvent injection process diagram is shown in Figure 23.1.

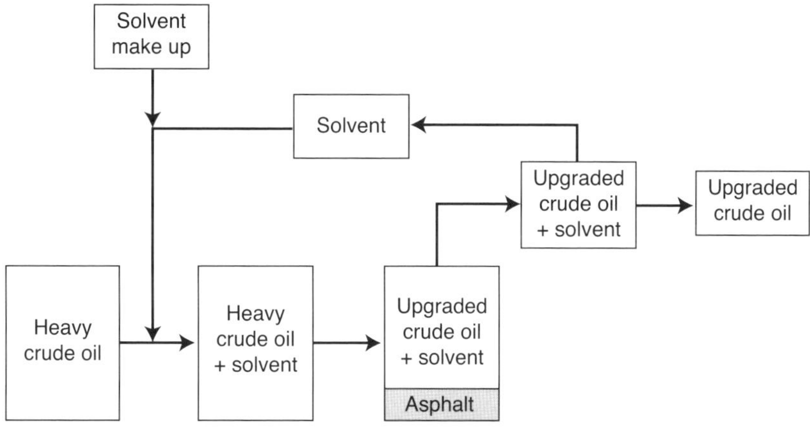

Figure 23.1

Cyclic Solvent Injection Process Scheme.
Source: After INTEVEP, SA.

In the first step, the solvent is introduced into the reservoir *via* a vertical well at reservoir temperature and pressure – typically 60-130°C and 25-100 bar – which are similar to de-asphalting process operating conditions. Since the solvent is cut with light hydrocarbons, it must be liquefied at the surface before its injection. The investment and operating costs of this liquefaction train are critical points of process development.

The amount of solvent injected into the reservoir can be adjusted to obtain the required upgraded *heavy crude oil* quality. As shown in Figure 23.2 and Figure 23.3, the solvent-to-crude oil ratio has a major impact on the quality of the upgraded crude.

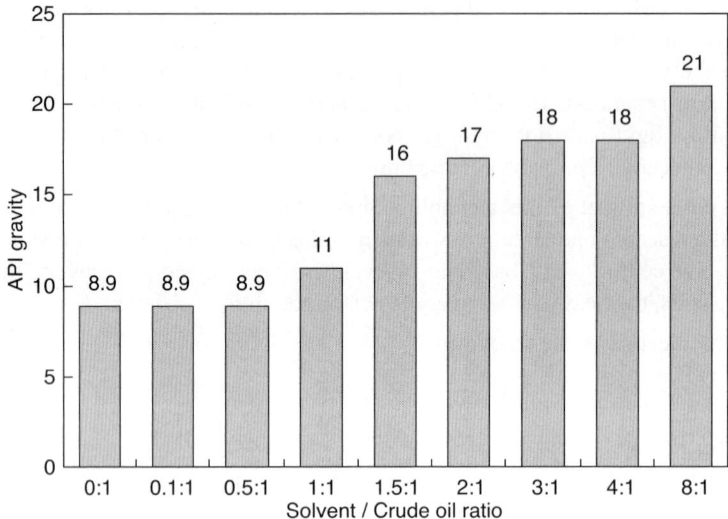

Figure 23.2

Evolution of API Gravity as a Function of Solvent-to-Crude Oil Ratio.
Source: After INTEVEP, SA.

Regarding the API gravity of the upgraded crude oil, a solvent-to-crude oil ratio of 2:1 is required to comply with a transport specification of 17°API. Beyond this ratio, additional quantities of solvent have a minor impact on the API gravity of the upgraded crude oil. Since the quantity of solvent is large, a solvent recycle loop is necessary to recover the solvent from the upgraded crude oil, thereby avoiding solvent availability and purchasing operating costs.

According to Figure 23.3, a solvent-to-crude oil ratio of 2:1 provides compliance with the second transport specification: viscosity.

The second step of cyclic solvent injection involves *heavy crude oil* soaking. This soaking occurs after the well closure and lasts five to ten days. During this period, a fraction of *heavy crude oil* – referred to as asphaltenes – is insoluble in the light hydrocarbon cut. This fraction precipitates in the reservoir. The soluble fraction of the *heavy crude oil* consists of upgraded heavy oil with higher quality than the initial *heavy crude oil*. One risk of *in situ* de-asphalting occurs if asphaltene deposits block the production well or the pores of the producing reservoir, requiring a new well to be drilled. Also, the carbon in these deposits cannot be converted and used to produce utilities like hydrogen or steam *via* gasification.

The third step of this recovery technology is upgraded crude oil production. Since the same well is used for oil production and solvent injection, drilling investment is reduced as compared to SAGD recovery technology, which uses two parallel wells (see Chapter 8).

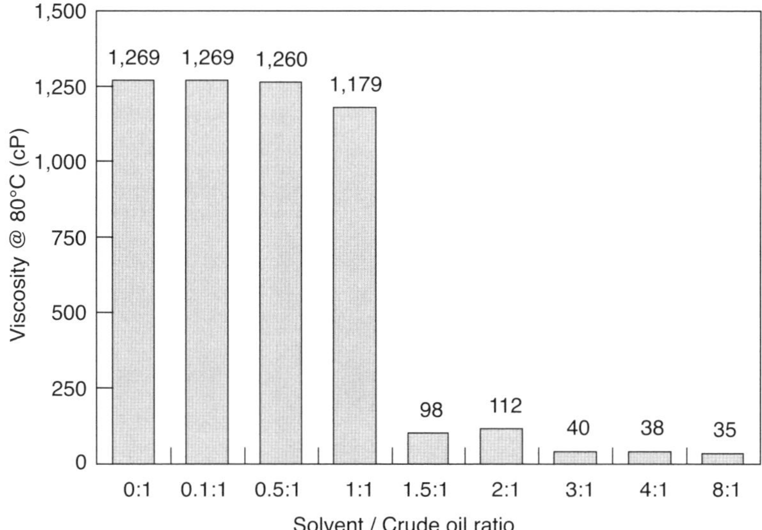

Figure 23.3

Evolution of Viscosity as a Function of Solvent-to-Crude Oil Ratio.
Source: After INTEVEP, SA.

The aim of the last step is separation of the solvent and the upgraded crude oil. Since the solvent is a gas cut – propane or butane – the recovery system is relatively simple and can be done in a flash drum. This step of the process is important to reduce gas make-up+ during solvent injection. Elimination of this light cut from the produced oil can also modify upgraded heavy oil stability. An interesting idea is to apply this technology in patterns, in which different wells are in different recovery phases. A producing well would then provide the solvent needed to inject into a well which is in the injection phase.

B. Vapor Extraction

VAPEX – as in VAPor EXtraction – recovery technology is a processes derived from the SAGD (Steam Assisted Gravity Drainage) process presented in Chapter 8. As in SAGD, two horizontals wells are drilled. However, instead of injecting water vapor in the upper well, a hydrocarbon vapor is injected. This hydrocarbon vapor usually contains a light alkane such as propane or butane. In order to obtain a solvent mixture which is close to saturation pressure at reservoir conditions, methane or carbon dioxide (for example) may be added to it.

The basic scheme of VAPEX technology is shown in Figure 23.4.

The light hydrocarbon solvent injected through the upper horizontal injection well forms a hydrocarbon vapor chamber in the reservoir. This solvent dissolves in the *heavy crude oil.* The diluted crude oil is recovered at the lower production well.

VAPEX technology differs at several points from cyclic solvent injection:

- The solvent is injected in vapor phase. This allows for larger chamber volumes than if the same solvent were in a liquid phase.

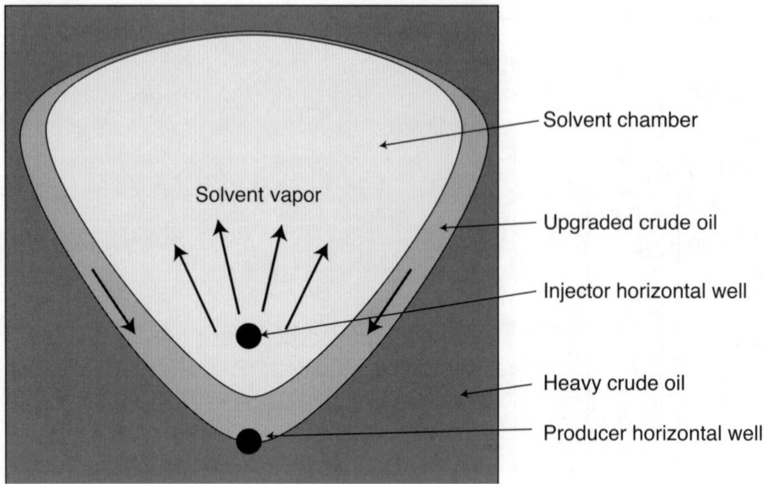

Figure 23.4

Basic Scheme of VAPEX Recovery Technology.

Source: After University of Calgary, Journal of Petroleum Science and Engineering, 21, 1998, pp 43-59.

- In vapor phase, the solvent is lighter than in liquid phase, and there is a larger difference in density with the crude oil. This aids in stabilizing the process by gravity and maintaining continuous recirculation.
- Miscibility is not achieved. However, if the solvent is close to vapor pressure, solubility is relatively high.
- Generally, solvent concentration in the oil is not high enough to obtain significant asphaltene precipitation. The impact of asphaltene flocculation on oil production is a topic of current research.

VAPEX technology is under development at the pilot scale. Various Canadian companies are participating in the Dover VAPEX Project (DOVAP). This cost of this project – led by Petro-Canada – is $30 million and it is funded by public institutes and private companies. The DOVER project began in 2003 and did continue through 2008. The aim of this project is to demonstrate the feasibility and economic advantages of VAPEX technology. Encana is also developing a pilot project based on VAPEX technology at Foster Creek.

One non-condensable gas that can be used in the VAPEX process is CO_2, recognized as the greatest contributor to GreenHouse Gas (GHG) emissions. CO_2 has interesting properties with regard to *heavy crude oil*. In fact, CO_2 exhibits high solubility and is known to provide significant viscosity reduction of heavy oil. The utilization of CO_2 has two advantages: a decrease in operating costs, and an environmental benefit with the reduction of CO_2 emissions and potential CO_2 sequestration.

VAPEX technology may allow recovery of oil from reservoirs in which thermal methods are not profitable, or cost reduction as compared to thermal recovery methods. Currently, recovery rates still seem disappointing. However, future developments may bring solutions. In its current form, under some conditions VAPEX may provide some upgrading of oil.

C. Co-injection of Steam and Solvent

Steam injection processes such as cyclic steam injection, steamflooding and SAGD (see Chapter 8) may be enhanced by co-injecting a solvent. The choice of the solvent is very specific to the various operating pressures. In fact, the water-solvent mixture must be in vapor phase, but the solvent must have high solubility in the oil phase. Consequently, composition of the solvent varies with reservoir properties and over production time.

Moreover, the co-injection of light hydrocarbons with steam impacts the surface facilities. In fact, compared to SAGD or VAPEX technology, in this case two components (water and solvent) must be recovered before processing the *heavy crude oil*. Since the recovery rate is a key point for its economic viability, particular attention must be paid to the recovery facilities. One Petro-Canada study shows that a solvent recovery factor of 70% is necessary to make the process viable.

This technology is currently being tested by Petro-Canada at its McKay River project. Initial results show a production increase of 10% to 50%. Again, such technology may exhibit some *in situ* upgrading, albeit very modest.

23.2.2 Thermal Conversion

A. Hydrovisbreaking

Like cyclic solvent injection, *in situ* hydrovisbreaking recovery technology is based on a traditional refinery process. The aim of the refinery visbreaking process is to decrease the vacuum residuum viscosity and produce a heavy fuel oil by soaking (typical residence time: 30 minutes). Visbreaking operating conditions are a temperature of 410-440°C and a pressure of 2 bar.

When applied *in situ*, the operating conditions for hydrovisbreaking may be closer to reservoir temperature and pressure. However, residence time must then be longer (from a few days to several months).

The company Petroleum Equities is studying the feasibility of hydrovisbreaking technology applied to reservoir conditions at a laboratory scale, and particularly the impact on API gravity and viscosity. Figure 23.5 shows the results obtained *via* a screening test for the API gravity of various crude oils.

Based on results shown in the previous figure, hydrovisbreaking may have a significant impact on API gravity. In fact, an increase of 5 to 10 degrees is observed for the various oils, with an average of 7 degrees. Regarding *heavy crude oil* viscosity, a decrease of up to 99% is noted. Consequently, all upgraded crude oil viscosities fall below the transport specification of 400 cP and API degree of 19.

Petroleum Equities proposed a basic scheme of the process, which is shown in Figure 23.6.

The first step of the technology is injection of steam and gases (mainly hydrogen) into the reservoir *via* a vertical or horizontal well. Hydrogen is used to heat the reservoir to the required temperature, superheat the steam and maintain sufficiently high hydrogen partial pressure. The reaction temperature – roughly 340°C – is achieved *via* combustion of a fraction

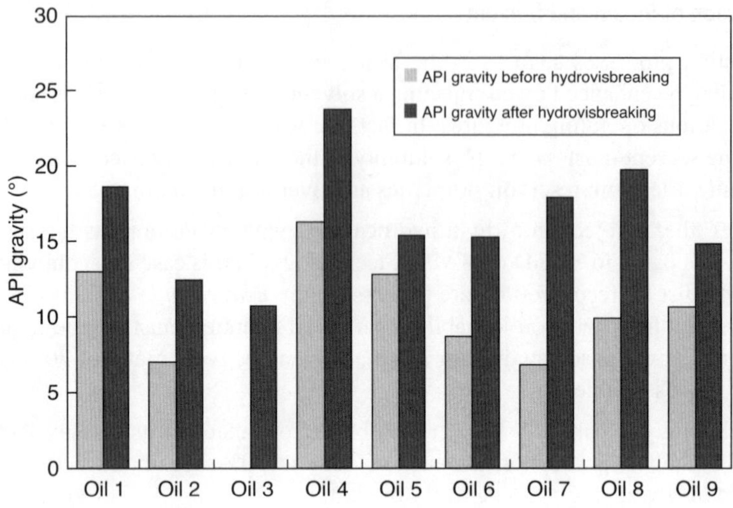

Figure 23.5

Oil API Gravity with or without Hydrovisbreaking.

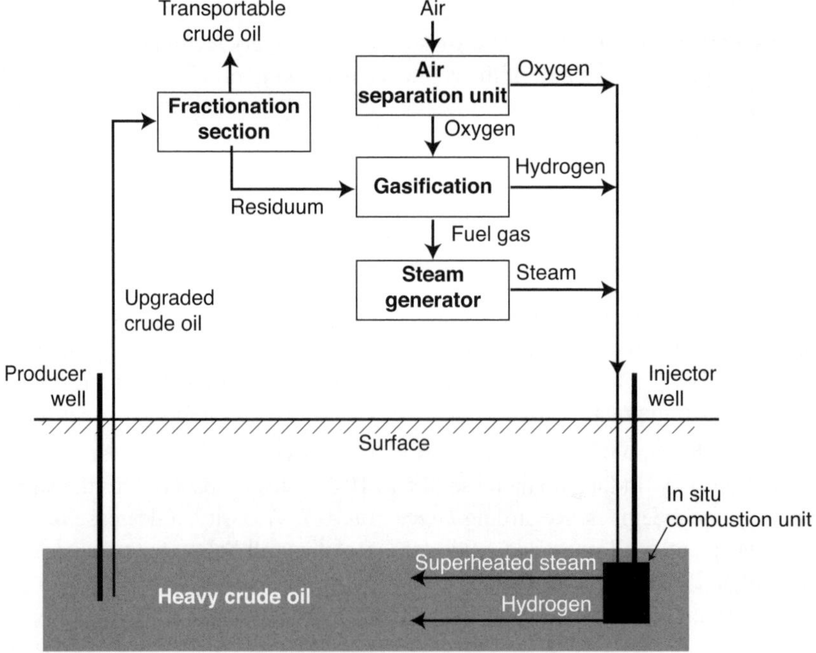

Figure 23.6

Basic Scheme of the Hydrovisbreaking Technology.
Source: After Petroleum Equities.

of the hydrogen in a downhole combustion unit. The oxygen required for combustion is injected along with the steam and hydrogen in the same well.

The second step is *heavy crude oil* conversion. The *heavy crude oil* is upgraded in contact with superheated steam and hydrogen directly in the reservoir. Long residence times are required to achieve increased API gravity and decreased viscosity.

The third step of the hydrovisbreaking recovery process is production of upgraded crude oil. The production well is generally put into contact with the injection well through a horizontal fracture. The distance between the injection well and the production well must be carefully chosen in order to maximize profit.

Afterwards, the upgraded crude oil is fractioned in a topping and vacuum distillation. The aim of this fractionation is to recover the residuum which is sent to a gasification complex. This complex converts the heavy hydrocarbon cut into synthetic gas and produces both hydrogen and steam. Moreover, sulfur, metals and carbon dioxide are extracted and treated separately through specific processes. Due to the integration of gasification, *in situ* hydrovisbreaking produces all required utilities and is independent of natural gas.

The company also created computer simulations to compare the performance of *in situ* hydrovisbreaking with steamflooding technology, and determine the best position of the production well with respect to the injection well. The results of these simulations are shown in Figure 23.7.

According to these results, *in situ* hydrovisbreaking technology provides a significant increase in the heavy oil recovery rate. In fact, after 350 days, 90% of oil in place is recovered for hydrovisbreaking, compared to 50% for steamflooding.

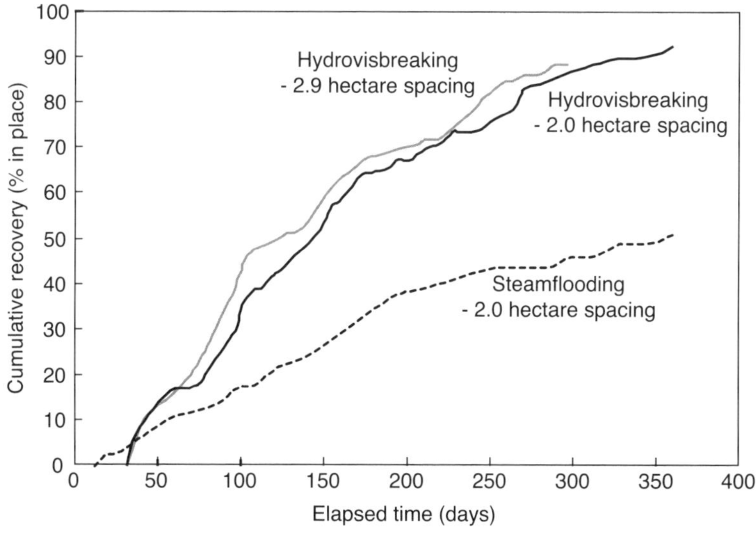

Figure 23.7

Cumulative Recovery as Function of Elapsed Time.
Source: After Petroleum Equities.

The second part of the computer studies concerns the distance between injection and production wells. Simulations were made for 2.0 hectare spacing (100 meters between the two wells), and 2.9 hectare spacing (120 meters between the two wells). For a given elapsed time, the cumulative recovery differs by little.

To conclude, laboratory studies show that the hydrovisbreaking process can be used for *in situ* application, but a pilot study is necessary to demonstrate this technology at field scale. Some pending points must be resolved during the process scale-up, such as safety aspects related to hydrogen injection, maintenance of hydrogen partial pressure in the reservoir, control of downhole combustion and the high reaction temperature in comparison to reservoir temperature.

B. Gasification

The gasification process could be also used for *in situ* upgrading. The aim of gasification technology is the production of a synthetic gas containing hydrogen, carbon monoxide and carbon dioxide from a carbon source like conventional oil, heavy oil or coal.

Subsurface gasification is achieved by injecting steam into the reservoir close to a heating generator. The temperature needed for gasification must be higher than 1,000°C. At this temperature and with an oxidizing agent like steam, carbon is converted through the following reactions:

$$C + H_2O \rightarrow CO + H_2$$
$$CO + H_2O \rightarrow CO_2 + H_2$$
$$CO_2 + C \rightarrow 2CO$$
$$CH_4 + H_2O \rightarrow CO + 3H_2$$
$$CH_4 + CO_2 \rightarrow CO_2 + 2H_2$$

In this case, the produced synthetic gas could be considered a product and directly recovered through a production well. Hydrogen may be extracted from the synthetic gas to be used for surface upgrading, energy generation through combustion, or liquid hydrocarbon generation *via* the Fischer-Tropsch reaction.

The *in situ* gasification process is also considered to be an Enhanced Oil Recovery (EOR) process. The synthetic gas produced in the reservoir creates a pressure increase and facilitates heavy oil recovery. The produced heavy oil is not upgraded and therefore has a composition similar to *in situ* heavy oil.

To conclude, *in situ* gasification seems to be attractive for the production of a synthetic gas, but the feasibility of this technology has yet to be demonstrated at laboratory and pilot scale.

23.2.3 Combustion

A. Toe-to-Heel Air Injection

Toe-to-Heel Air Injection – commonly referred to as THAI™ – was developed in 1993 by Malcolm Greaves from the University of Bath and Alex Turta from the Petroleum Recovery

Institute, now the Alberta Research Council. The company Petrobank purchased the intellectual property rights to this technology and is now developing the process, and particularly its industrial commercialization.

THAI™ recovery technology is a combustion process that benefits from gravity drainage. A vertical well perforated near the top of the reservoir is used to inject air. A horizontal production well located near the bottom of the reservoir allows for production of oil heated by the combustion process. This oil flows toward the production well by gravity. The distance between the injection well and the production well varies from 100 meters to 300 meters.

Before the combustion step, the *heavy crude oil* is preheated *via* steam injection, allowing the oil to become mobile (pre-ignition step). As steam injection associated with traces of oxygen and acid gases is a source of corrosion, the vertical well must be made of specific metals. After preheating, air – from the atmosphere – is compressed in surface facilities and injected into the reservoir. The injected air-to-oil ratio is typically 1,500 Nm^3 per m^3, i.e. 8.4 MSCF per bbl. In contact with the air, *heavy crude oil* ignites, raising the temperature in the reservoir. A combustion front is developed from the injection well with a speed of approximately 20 centimeters per day. The combustion temperature increases with increasing air flow rate.

As shown in Figure 23.8, the region around the combustion front may be divided into three zones:

- Mobile oil zone which contains heated oil subject to some thermal cracking.
- Coke zone resulting from cracking and possibly low temperature oxidation of heavy crude.
- Combustion zone where coke is burned in the presence of air.

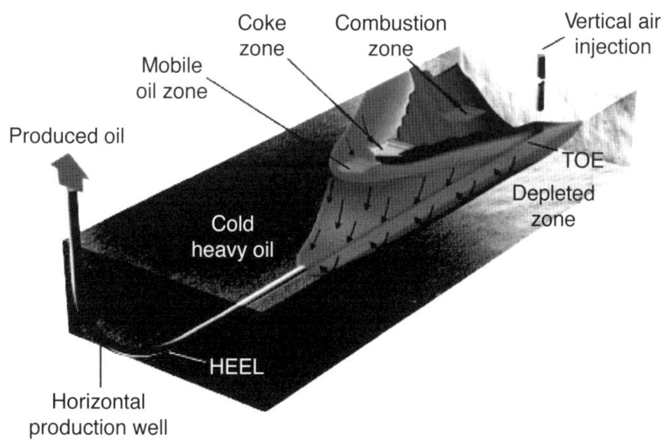

Figure 23.8

THAI™ Process Schematic.
Source: After Petrobank Energy and Resources Ltd.

A significant portion of the mobile oil is recovered from the reservoir, together with hot gases resulting from the combustion process. Surface facilities are required to separate gases from oil.

Typically, the gas has the following composition (by volume): 79% nitrogen, 15% carbon dioxide, 5% methane, 1% carbon monoxide and a low percentage of hydrogen sulfide. A process is needed to eliminate nitrogen and carbon dioxide and allow for utilization of the methane contained in the produced gases. THAI™ technology may be improved by injecting oxygen-enriched air into the reservoir. The use of oxygen reduces the production rate of gases by eliminating useless gases such as nitrogen. However, the *in situ* combustion front and temperature may be more difficult to control at higher oxygen content.

The *heavy crude oil* is partially upgraded in the reservoir with an increase in API gravity from 8 to roughly 16, and 90% removal of nickel and vanadium, as well as 30 to 40% for sulfur.

Reservoirs suited for the application of THAI™ are characterized by the following properties:

- Vertical permeability greater than 0.5 D.
- Oil saturation higher than 50%.
- Heavy oil API gravity greater than 8.
- Heavy oil viscosity lower than 250,000 cP.

In principle, THAI™ technology may be used in deeper reservoirs than those suited for exploitation with SAGD or mining. Development of the THAI™ process will increase *heavy crude oil* recoverable reserves.

The final development of THAI™ recovery technology prior to its commercial application is the current pilot demonstration by Petrobank in its Whitesands lease. The project anticipates production of 1,800 barrels per day of partially upgraded Athabasca bitumen. The *heavy crude oil* is located at roughly 370 meters below the surface. The pre-ignition period – corresponding to steam injection – lasts two to three months, depending on *heavy crude oil* mobility and reservoir properties. Air has been injected since July 2006. During the initial months, the fluids produced were essentially condensed steam coming from the pre-ignition step, reservoir water and bitumen. Then, the quantity of condensed water decreased with an increase in the heavy oil produced. Based on continuous analysis, it was found that reservoir temperature increased to 800°C. An increase in the ratio of carbon dioxide to carbon monoxide indicates a high level of oxygen conversion. The composition of produced gases indicates thermal cracking of heavy oil. These initial results from the pilot experiment suggest that the THAI™ technology is feasible as a commercial recovery process.

To conclude, the THAI™ process seems to be an interesting new recovery technology for *heavy crude oil*: the announced recovery ratio of 70-80% is very high, environmental impact is lower than SAGD or VAPEX and investment is lower as well. The pilot test currently underway has shown promising initial results, but it is still too early to draw conclusions regarding the overall success of this pilot.

23.2.4 Catalytic Conversion

A. Catalytic Injection

Just as several recovery technologies make use of steam or hydrogen injection into the reservoir, the main idea of catalytic conversion is to increase heavy crude quality by adding catalysts directly into the reservoir. Catalysts can be added into the reservoir as a solid component or in a liquid phase.

Solid catalysts may be introduced in the gravel pack, generally used to prevent production of sand and set around a production well (Figure 23.9).

In this case, the gravel pack is used as a catalyst support. Typical catalysts for hydrotreatment – e.g. molybdenum associated with cobalt or nickel metals and supported by alumina or silica-alumina – may be used for this application. Old or regenerated refining process catalysts can be re-used. Laboratory experiments suggest that these catalysts used in reservoir conditions have a lifecycle of the same order as catalysts in refining processes. The main drawback

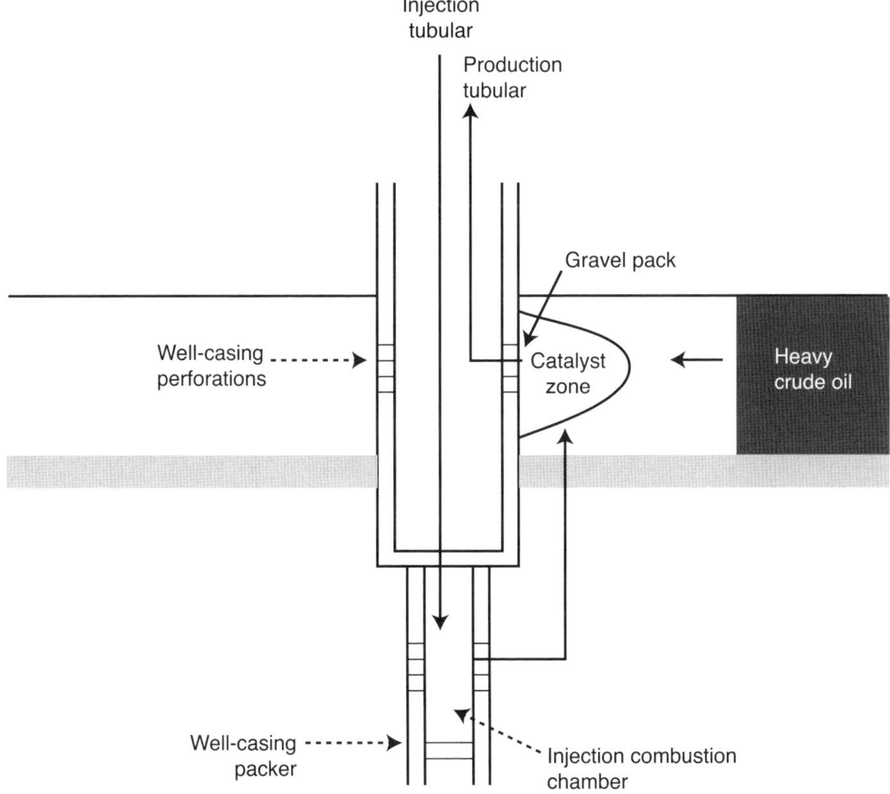

Figure 23.9

Catalytic Gravel Pack Schematic.

Source: After Energy & Fuels, 1996, 10, pp 883-889.

of catalysts in the gravel pack is the possible occurrence of impurity deposits (metals and sulfur) causing an increased pressure drop and thus a decreased production rate. Two methods can be used to unblock a production well: burn the deposits or fracture the gravel pack.

A second injection method consists of introducing the catalyst in an aqueous solution phase. This technique can be profitable if catalysts are homogenously distributed in the heavy oil formation, allowing for upgrading of a large proportion of crude oil. However, the quantity of catalysts injected into the reservoir is higher than that used in the gravel pack system, and contacts between the *heavy crude oil* and catalytic particles are not well controlled. Additionally, the aqueous solution may interact with clays present in the oil formation and result in significantly decreased permeability. The catalytic solution may be injected into the reservoir through steam or water flooding.

Laboratory experiments performed under heterogeneous conditions suggest an increase of 2 degrees in gravity density and sulfur removal of 40-60%, with 20-30% corresponding to thermal activities.

In any event, the main drawback of these catalytic recovery technologies is the energy required to achieve conditions suitable for catalytic reactions. This energy may be provided to the reservoir through steam injection, *in situ* combustion or conductive heating. However, much research is still needed before commercial application may become feasible. Electromagnetic or electrical technologies have been proposed for the gravel pack application system, but they require major research and development programs.

B. CAPRI

CAPRI (Catalysis Petroleum Recovery Institute) recovery technology is a derivative of the THAI™ process discussed previously. The main innovation of the CAPRI process is the incorporation of catalytic modules along the horizontal producer well. The use of a hydrotreatment catalyst (CoMo, etc.) has been proposed for this process. The main objective of the CAPRI process is an improvement in the quality of the produced liquid. In fact, the temperature of the reservoir is close to the refinery hydrotreatment process temperature, and the residence time in or around the horizontal well may be considered sufficient for the reaction. However, since no hydrogen is injected, it must be produced *in situ* by water conversion or other reactions. Catalysts present in the producer well may lead to its clogging *via* sediments originating in the reservoir.

23.2.5 Microbial Conversion

Microbial conversion, also referred to as Microbial Enhanced Oil Recovery (MEOR), is an environmentally friendly way to upgrade *heavy crude oil*. This recovery technology makes use of microorganisms (e.g. bacteria) or enzymes.

Different types of microorganisms can be distinguished depending on the environment in which they live (composition, temperature and pressure):

- Aerobic microorganisms that live in the presence of oxygen.
- Anaerobic microorganisms that live without oxygen.

- Psychotropic microorganisms at a typical active temperature of 15°C.
- Thermophilic microorganisms at a typical active temperature of 30-40°C.
- Hyperthermophylic microorganisms at a temperature higher than 60°C.

Typically, psychotropic microorganisms can be used in Canadian heavy oil reservoirs, whereas thermophilic microorganisms may be used in Venezuelan reservoirs.

Two methods of upgrading *heavy crude oil* by microbial recovery technology are currently being studied: gas and solvent production, and heavy component (e.g. asphaltene) degradation.

Gases and solvents are produced by the action of anaerobic bacteria in two steps. During the first step – the growth phase or acidogenic phase – bacteria produce hydrogen, carbon dioxide and various acetate components. In the second step – the stationary growth phase – bacterial metabolism shifts to solvent production and thus mainly ketone and alcohol components like acetone, ethanol, butanol, etc.

These gases and solvents have benefits for *heavy crude oil* recovery:

- Reduced viscosity of *heavy crude oil via* the dissolution of gases and solvents.
- Increased pressure in the reservoir allowing for recovery of more heavy oil.

Other research studies are being carried out on the degradation of heavy components, particularly asphaltenes which are mainly responsible for high *heavy crude oil* viscosity. In this case, microorganisms have a specific action on the molecule structure. One relatively simple pathway is the oxidation of polyaromatic molecules by bacteria associated with special fungi. Other research studies are being conducted to break sulfur and/or nitrogen atom links in the molecule. These studies need further work because they make use of microorganisms whose genes have been modified *via* genetic transformation.

In conclusion, although microbial oil recovery technology seems attractive from an environmental standpoint, considerable research is needed to make it commercially applicable. Institutes and laboratories from various countries such as Canada, China and Norway believe in this breakthrough recovery process and are working on its development. MEOR may be used in depleted reservoirs in order to increase oil recovery or produce gases for surface technologies.

23.3 FUTURE DIRECTIONS

Technologies may be developed to provide working in-reservoir "refineries" that deliver products with improved quality, at minimal environmental impact, and using energy supplied directly from within the deposit. Ultimately, heavy ends could be discarded *in situ*, water and catalysts could be recycled, and emissions and energy use could be reduced. Some of the technologies that may be developed could also be of interest for the recovery of other resources, such as conventional oil and – potentially – coal deposits, of which Europe has quite a few.

Multidisciplinary, integrated research is needed to develop an *in situ* upgrading technology. If the use of catalysis in the reservoir or the wellbore is a viable option, an adequate catalyst should be designed and optimized, and transport phenomena studied. The stability of fluids produced through *in situ* processing is an important point. Tools and methods for chemical monitoring of heavy oil and bitumen homogeneity and its progressive transformation *in situ* should be developed. Reservoir simulation should be used for tracking compositions and process control.

Over the long term, the design, monitoring, control and construction of *in situ* processing reactors may be mastered to the point where reservoir reactor conditions can be dynamically assessed and controlled. We could even imagine the *in situ* gasification of heavy ends (primarily asphaltenes) or residual materials, as well as the alternative strategy of partial conversion of heavy ends at the surface. We could also imagine re-injecting heavy ends in the reservoir to serve as fuel for a combustion process. The increased degree of *in situ* processing may result in a medium-quality, transportable, and conventionally refinable crude oil from the reservoir. Produced waters should be treated and recycled to the reservoir, with only the concentrated waste stream requiring deep disposal. If ultra-dispersed catalysts are used, surface facilities and processes will have to be developed for the production, recovery, recycling and reactivating of these catalysts. Optimal conditions for *in situ* catalytic upgrading could be achieved through dedicated catalyst design, innovative reactor and process design and construction, and real-time monitoring of *in situ* reactor conditions. In addition to producing high quality oils from the reservoir, strategies may be developed to produce economic quantities of heat and gases from the reservoir.

In 15 years or more, a possible target is a working technology for producing energy and chemicals from *in situ* processing, with an extremely low to zero environmental imprint.

23.4 FUTURE RESEARCH

Heavy oil and bitumen are extremely complex mixtures whose compositions, reactivity, and flow behaviors are dependent on many factors. Basic research must be performed on the physical and chemical phenomena that occur during integrated *in situ* upgrading and recovery of these mixtures. Basic research must be combined with an understanding of chemical engineering, *in situ* process design, complexities of bitumen or heavy oil recovery based on its behavior and composition, and the heterogeneous reservoir environment.

We can anticipate the need for research work in the following areas:

- Fluid characterization and thermodynamics.
- Well-architecture and process design.
- Downhole equipment development.
- Reservoir simulation and process control.
- Nanocatalyst design for *in situ* use.
- Biological or enzyme-based upgrading processes.
- Reservoir and fluid monitoring.
- Transportation and stability of process fluids.

REFERENCES

Butler RM (1995) The Solvent Requirements for Vapex Recovery. *International Heavy Oil Symposium*, Calgary, 19-21 June, 1995.

Das SK (1998) Vapex: An Efficient Process for the Recovery of Heavy Oil and Bitumen. *SPE Journal*, September 1998.

Davidson IDF (2001) Enhanced Oil Recovery by *In Situ* Gasification. Patent EP 1 276 962 B1.

Graue DJ (2001) Upgrading and Recovery of Heavy Crude Oils and Natural Bitumens by *In Situ* Hydrovisbreaking. Patent US 6,328,104 B1.

Greaves M (2002) Toe-to-Heel Air Injection – THAI. Maximising Recovery of Heavy Oil/Tar sand Bitumen and In-Situ Upgrading. *PTAC – ARC*, June 10, 2002.

Greaves M (2001) CAPRI-Downhole Catalytic Process for Upgrading Heavy Oil: Produced Oil Properties and Composition. *Petroleum Society's Canadian International Petroleum Conference*, Calgary, June 12-14, 2001.

Greaves M (2004) THAI-CAPRI Process: Tracing Downhole Upgrading Heavy Oil. *Petroleum Society 5th Canadian International Petroleum Conference*, Calgary, June 8-10, 2004.

Greaves M and Turta A (1997) Oil Field *In Situ* Combustion Process, US Patent N° 5,626,191.

Herron EH (2003) Experimental Verification of *In Situ* Upgrading of Heavy Oil. www.petroleumequities.com/cgi-bin/site.cgi?t=5&p=experimentalverification.html. October 2003.

Herron EH (2003) *In Situ* Hydrovisbreaking. Laboratory Analysis. Commercial Application. *The Statoil Research Summit 2003*. September 2003.

Karmaker K (2003) Applicability of Vapor Extraction Process to Problematic Viscous Oil Reservoirs. *SPE Annual Technical Conference and Exhibition*. Denver, 5-8 October, 2003.

Kirkwood KM (2004) Prospects for Biological Upgrading of Heavy Oils and Asphaltenes. Studies in Surface. *Science and Catalysis*, **151**, pp 113-143.

Liu Y (2002) The Effect of Hydrogen Donor Additive on the Viscosity of Heavy Oil During Steam Stimulation. *Energy and Fuel*, **16**, pp 842-846.

Marzuola C (2003) Lighting a Fire Under Heavy Oil. www.petroleumworld.com/

Mokrys IJ (1993) In-Situ Upgrading of Heavy Oils and Bitumen by Propane De-asphalting: The Vapex Process. Production Operations Symposium. March 21-23 1993.

Ovalles C (2001) Extra-heavy Crude Oil Downhole Upgrading Process Using Hydrogen Donors Under Steam Injection Conditions. *SPE International Operations and Heavy Oil Symposium*. Margarita Island, 12-14 March, 2001.

Ovalles C (2001) Physical and Numerical Simulation of an Extra-Heavy Crude Oil Downhole Upgrading Process Using Hydrogen Donors Under Cyclic Steam Injection Conditions. *SPE Latin American and Caribbean Petroleum Engineering Conference*. Buenos Aires, 25-28 March, 2001.

Ovalles C (2003) Downhole Upgrading of Extra-Heavy Crude Oil Using Hydrogen Donnors and Methane Under Steam Injection Conditions. *Journal of Petroleum Science and Technology*, September 2002.

Rebecca S (1998) Biotechnology for Heavy Oil Recovery. *7th UNITAR International Conference on Heavy Crude and Tar Sands*.

Talbi K (2003) Evaluation of CO2 Based Vapex Process for the Recovery of Bitumen From Tar Sand Reservoirs. *SPE International Improved Oil Recovery Conference*. Kuala Lumpur. 20-21 October, 2003.

Vallejos C (2002) Process for *In Situ* Upgrading of Heavy Hydrocarbon. Patent US 6,405,799 B1.

Walter K (2007) Fire in the Hole. *Science and Technology Review*. April 2007.

Ware CH (1986) Recovery Oil by *In Situ* Hydrogenation. Patent US 4,597,441.

Weissman (1997) Review of Processes for Downhole Catalytic Upgrading of Heavy Crude Oil, *Fuel Processing Technlogy*, **50**, pp 199-213.

Weissman JG (1996) Down-Hole Catalytic Upgrading of Heavy Crude Oil. *Energy and Fuels*, **10**, pp 883-889.

Weissmann (1996) Downhole Heavy Crude Oil Hydroprocessing. *Applied Catalysis*, **140**, pp 1-16.

Xia TX (2002) Upgrading Athabasca Tar Sand Using Toe-to-Heel Air Injection, *Journal of Canadian Petroleum Technology*, **41**, 8, pp 51-57.

Xia TX and Greaves M (2001) Downhole Upgrading Athabasca Tar Sand Bitumen Using THAI – SARA Analysis, Society of Petroleum Engineers, SPE 69693 presented at the *SPE International Thermal Operations and Heavy Oil Symposium*, Portlamar, Venezuela, 12 March, 2001.

Zhao L (2004) Steam Alternating Solvent Process. *SPE International Thermal Operations and Heavy Oil Symposium*, 16-18 March, 2004.

Websites:

www.petrobank.com/webdocs/whitesands/thai_faq.pdf

www.petrobank.com/webdocs/whitesands/whitesands_brochure.pdf

24 | Process Workflows

J.P. Héraud

INTRODUCTION

To transport *heavy crude oils* from production fields to consuming countries, the most widely used method is dilution with light hydrocarbons (condensates, naphtha, etc.). This technique is less costly to decrease the viscosity of heavy crudes and thus facilitate their transport *via* pipeline or seaway. To achieve minimum transport specifications of most pipeline systems, i.e. a viscosity of 400 cP, the viscosity of the mix, referred to as DilBit (Diluted Bitumen), is adjusted from the dilution rate, the respective viscosities and densities of the heavy crude and light hydrocarbons. The dilution rate is generally between 25% and 30% by volume and the dilution naphtha used has a variable density depending on the exploitation countries (roughly 65°API in Canada and 45°API in Venezuela). However, with the high price of naphtha (roughly 400 US $/bbl for a West Texas Intermediate crude at 40 US $/bbl), this dilution method is increasingly costly.

A second method consists of producing a synthetic crude (SCO: Synthetic Crude Oil) at the production field, which is then directly upgradeable at the refinery. The extracted heavy crude is converted *via* an upgrader unit (conversion facility of delayed coking type, ebullated bed, etc.) and hydrotreatment facilities. According to the target market and its constraints, the upgrading site can vary in complexity and the synthetic crude produced will be of greater or lesser quality. For the American market, with its high-conversion refineries (86% of sites are equipped with a Fluid Catalytic Cracking unit), upgrading will be minimal and the SCO produced will comply only with the specifications required for transport. In contrast, since European refineries perform less conversion (56% of sites are equipped with an FCC unit), the synthetic crude produced will be the result of more complex upgrading and of better quality. Although investments and operating costs associated with upgrading are high, the end product has an added value greater than DilBit since it is valued like a conventional crude oil.

An alternative technique that is currently less expensive, based on the method of dilution and *in situ* upgrading, consists of mixing synthetic crude with heavy crude in greater proportions (dilution rate roughly 50% by volume). The transported mix is then called SynBit (Synthetic Bitumen).

According to Figure 24.1, this SynBit production is strongly increasing, from 70 kBPSD, Barrels Per Stream Day, in 2005 to roughly 1,300 kBPSD in 2015, while the production of *heavy crude oils* by the method of dilution with a light hydrocarbon (DilBit) remains constant over the next ten years at roughly 600 kBPSD. During this same period, a moderate increase in the production of synthetic crude is observed from 500 kBPSD in 2005 to 800 kBPSD in 2015.

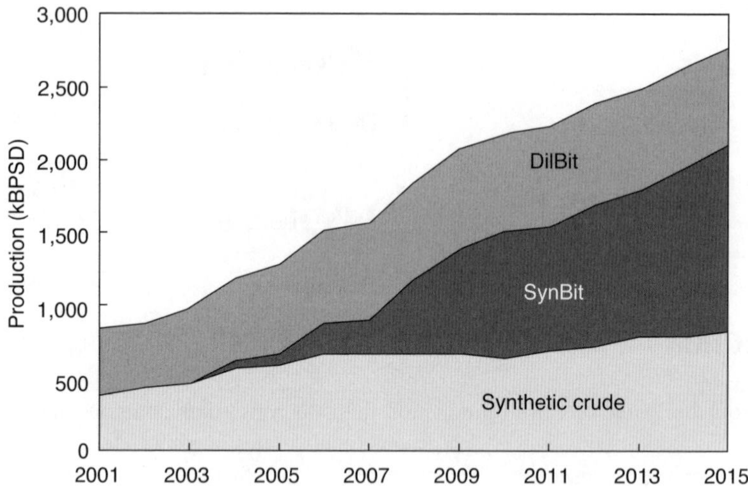

Figure 24.1

Distribution of the Production of DilBit, SynBit and Synthetic Crude in Canada.

Source: Canadian Association of Petroleum Producers.

Over the longer term, the synthetic crude produced will thus be used for direct upgrading (+ 60 % in 10 years) and mainly as a diluent (multiplication of demand by a factor of 18 in 10 years). This strong growth will be accompanied by an increase in the number of upgrader units over the next ten years in both Venezuela and Canada (Table 24.1).

According to Figure 24.2, the various refineries can thus, depending on their level of integration, be supplied with *heavy crude oil* either directly (SCO) or indirectly (DilBit and SynBit).

Table 24.1 Canadian Projects for New Upgraders.

Company	Project	Capacity	Start Date
Suncor Energy	Upgrader 2 expansion	100,000 b/d	2008
Suncor Energy	Voyager expansion	150,000 b/d	2010-2012
BA Energy	Heartland upgarder	77,500 b/d	2008
North West Upgrading	North West upgrader	50,000 b/d	2010
Petro-Canada	Strugeon upgrader	170,000 b/d	2011
Synenco Energy	Northern Lights upgrader	100,000 b/d	2010
Husky Energy	Upgrader expansion	70,000 b/d	2010
Peace River Oil	Bluesky upgrader	25,000 b/d	2010

Source: Oilsands Review.

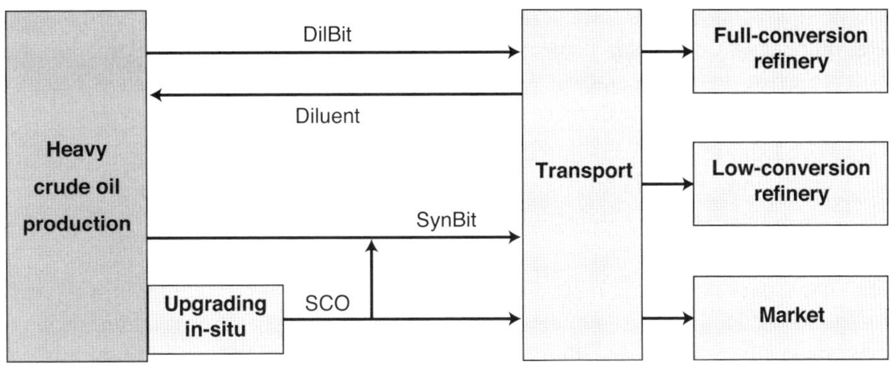

Figure 24.2

Problematics of Supplying Heavy Crude to Refineries.
Source: Bottom of the Barrel Technology Seminar, 2005

Since consumption of both synthetic crude in the form of SynBit and conventional crude are increasing, the various upgrading models used to produce synthetic crudes of low quality (basic upgrading) or high quality (complex upgrading) will be discussed. To decrease the operating costs associated with new extraction methods (SAGD, CSS, etc.) and the increasingly complex processing of *heavy crude oils*, new upgrading models which integrate upstream and downstream will be developed. Over the longer term, ultra-complex upgrading which allow for the *heavy crude oil* producing countries to produce commercial products on specification like gas, kerosene or diesel fuel will appear.

24.1 BASIC UPGRADING

Production facilities for lower quality synthetic crude are mainly located in the Orinoco Belt of Venezuela. In fact, Venezuela's geography does not allow for export of heavy crudes by dilution (DilBit or SynBit) *via* pipeline to consuming countries and as consequence the use of dilution as the technique for transport. Since the seaway is the only means of exporting *heavy crude oils*, basic upgrading is the preferred model in Venezuela. This reduces the viscosity of *heavy crude oils* to produce synthetic crudes, obviously of lower quality, but which have the specifications required for transport. The use of this type of model in Venezuela is also explained by the involvement of American companies at the beginning of two projects. In fact, the Petroanzoategui and Petromonagas projects, initially developed by ConocoPhillips and ExxonMobil, mainly send their Venezuelan production to American refineries in Texas and Louisiana. Since these refineries are heavily converting, the production of synthetic crude of better quality is not necessary.

By producing synthetic crude that only has the specifications required for transport, and thus not requiring significant conversion, the investments and operating costs associated with basic upgrading are reduced as much as possible. Investment for the Petroanzoategui and Petromonagas Venezuelan projects is roughly $2.5 billion including extraction, transport of the crude to the upgrading site (located 200 km from the extraction basin) and upgrading of minimal complexity.

Figure 24.3 shows the production diagram of the Petromonagas project, which is typical of basic upgrading. This diagram includes three steps which are indispensable for sufficiently reducing the viscosity of heavy crude: separation, conversion and hydrotreatment of the naphtha cut.

The first step of separation recovers a large quantity of diluents, required for the extraction of heavy crude. This step includes at minimum an atmospheric distillation, which may be accompanied by vacuum distillation as well (Petroanzoategui project). In both cases, one part of the atmospheric residue is stored then mixed directly in the synthetic crude, while the other part of the atmospheric residue is sent to a conversion facility. This distribution of the stream (roughly 50/50) is used to adjust the end properties of the synthetic crude, limit the capacity of the conversion facility to a minimum and consequently reduce its operating costs.

This conversion facility is considered to be the core of upgrading since it processes part of the extracted heavy crude into products that can directly constitute the final synthetic crude. The investment and operating costs associated with this facility represent a significant portion of the project's final cost. Thermal processes with carbon rejection are the processes widely used in this type of upgrading. The conversion can also be performed *via* other carbon rejection processes like de-asphalting or visbreaking. Catalytic conversion processes are not recommended for this type of upgrading since their operating cost is high and thus their profitability is insufficient. For the Petromonagas project, presented in Figure 24.3, the conversion process used is delayed coking in accordance with technology developed by Foster Wheeler (Sydec®). By operating the facility at low pressure and low recycle rate, liquid production is maximized and coke formation is reduced. The developments brought by Foster Wheeler to this process allow for production of a cut containing little Conradson Carbon or metals. Typical yields of various cuts resulting from processing a Venezuelan vacuum residue in a Sydec® delayed coking facility are shown in the Table 24.2.

Table 24.2 Yields of Various Cuts Resulting From Processing a Vacuum Residue Produced in the Orinoco Belt of Venezuela in a Sydec® Coking Facility.

	Distillation Interval	**Yields**
Gas		5.36 vol%
C_3-C_4		7.04 vol%
Naphtha	C5-177°C	14.07 vol%
LCGO (Light Coker Gas Oil)	177°C-343°C	28.38 vol%
HCGO (Heavy Coker Gas Oil)	343°C-571°C	28.48 vol%
Coke		32.44 wt%

At low pressure 1.03 bar and with a very low recycling rate (1.05), the typical properties of HCGO (Heavy Coker Gas Oil) produced from the previous feedstock in the Sydec® delayed coking facility are listed in Table 24.3.

Table 24.3 Properties of Heavy Coker Gas Oil Resulting From Processing of a Venezuelan Vacuum Residue.

		HCGO
°API		16.6
Sulfur content	(wt%)	4.65
Conradson Carbon	(wt%)	0.31
Metals (Ni + V)	(wt ppm)	0.5

Considering the data presented below, the distillate cut resulting from delayed coking has a good quality, despite a sulfur content that is still relatively high. In the synthetic crude production facility of Petromonagas, this cut is not hydrotreated, in contrast to the naphtha cut.

Hydrotreatment of the naphtha cut is used to desulfurize and stabilize it by saturating the diolefins and the gum precursors. The hydrogen needed by this facility can be supplied by a natural gas steam reforming facility integrated at the site, or purchased from an independent company. For the two Venezuelan projects, the hydrotreatment facility is supplied with hydrogen from outside.

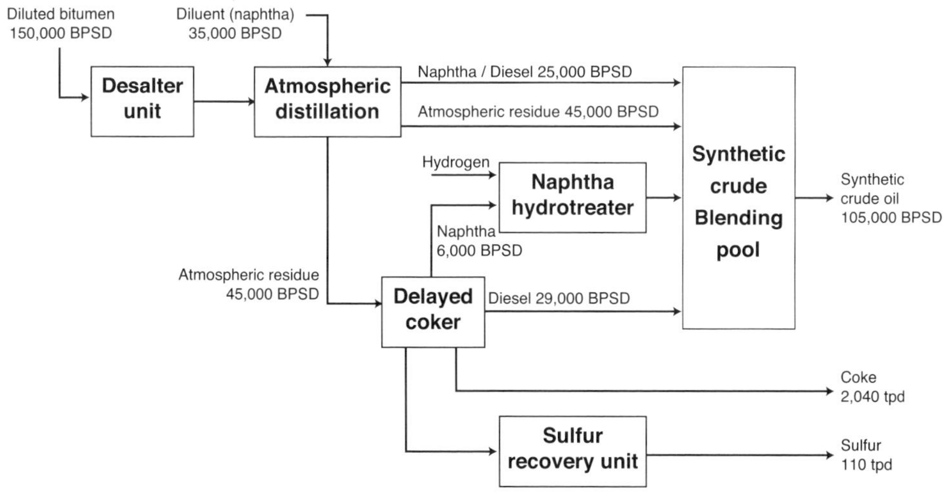

Figure 24.3

Production Diagram for Petro Monagas Project in Venezuela.
Source: Oil and Gas Journal.

This upgrading model is not very complex and, by mixing products obtained during the various steps, allows for production of synthetic crude of sufficient quality for transport. The properties (density and sulfur content) of the synthetic crudes thus obtained may vary slightly depending on the model adopted. In Table 24.4, the characteristics of the Zuata heavy crude extracted in the Orinoco Belt synthetic crudes produced by the Petromonagas and Petroanzoategui projects are listed. Since these two projects use a similar upgrading model, the properties obtained for these two synthetic crudes are also very similar. Considering the sulfur content and density which are still relatively high, these crudes are mainly intended for American refineries, which are highly converting.

Table 24.4 Properties of Two Synthetic Crude Oils Resulting From Basic Upgrading (Petromonagas and Petroanzoategui) as well as the Corresponding Heavy Crude (Zuata).

		Zuata	Petromonagas	Petroanzoategui
°API		8.5	16.5	19-25
Sulfur	(wt%)	4.1	3.3	2.3

Despite the technological improvements made by Foster Wheeler, the quantity of coke formed by the delayed coking process is still high: roughly 2,000 tons per day for the Petromonagas project, or 17 kg per barrel of heavy crude processed.

The producer must therefore find an outlet to commercialize its coke and prevent accumulations of it at the upgrading site. Currently, the United States is a large source of consumption since the coke is used as fuel for power stations. Long-term supply contracts are thus signed between petroleum companies and American companies to sell the coke produced.

The production of a synthetic crude oil of low quality, but which has the specifications required for transport, limits the investments and operating costs. The conversion scheme of the Petromonagas and Petroanzoategui projects are representative of the process workflows possible for this type of upgrading: separation, conversion and hydrotreatment of the naphtha cut.

This type of low-complexity upgrading is found when the target market for the produced synthetic crude is converting, like the American market. For output to European refineries, the synthetic crude must be of better quality, which means that producers must use a more complex upgrading model and make higher investments.

24.2 COMPLEX UPGRADING

The complex upgrading facilities producing a synthetic crude of high quality, i.e. low density and low sulfur content, are mainly located in Canada, in the region of Fort Mc Murray and Edmonton. This location might seem contradictory since these crude oils are mainly exported to the United States. Nevertheless, since transport by pipeline is preferred, the *heavy crude oils* are exported directly in the form of synthetic crude, and increasingly *via* dilution in the form of SynBit, a method which consumes a significant amount of synthetic crude (roughly 50% by volume). In the future, many other complex upgrading projects should be developed due to stagnation of the market for transport heavy crude as DilBit. In fact, the quantity of diluents available can't increase, resulting in tensions in the diluents market and a strong increase in prices.

In Venezuela, two complex upgrading projects are currently being exploited: the Petropiar project (70% PDVSA and 30% Chevron) produces 180,000 b/d and started in 2004, and the Petrocedeno project (60% PDVSA, 30.3% Total and 9.7% Statoil) produces 180,000 b/d and started in 2002. These projects require very high investments and involve major strategic decisions. Investments for the Petrocedeno project, for both upstream and downstream activities, are roughly $4.5 billion, i.e. twice the cost of basic upgrading. Nonetheless, these projects remain profitable since the synthetic crude oils have outlets comparable to conventional crude oils and offer a good return. It is estimated that the Zuata Sweet synthetic crude, resulting from the Petrocedeno project, will have a price slightly higher than Brent (average price of Brent for the year 2009: 61.5US $/bbl).

For these complex upgrading projects, the objective of the producer is not to minimize investment, but to produce a synthetic crude of good quality that is easily commercialized on the world market. The possible processes and workflows allowing the producer to provide the synthetic crude to desired specifications are therefore much more varied. Nonetheless, the three previous step are found: one for separation, one for conversion of the heavy fraction and one for hydrotreatment.

The separation step, similar to basic upgrading, is generally an atmospheric distillation followed by a vacumm distillation. Its purpose is to eliminate the light fraction (naphtha and diesel) contained in the heavy crude and send the entire heavy fraction to the conversion step.

Considering the number of processes available, selection of the technology for the conversion facility is much more complex. Although thermal processes with carbon rejection are still widely used, physical processes with carbon rejection or catalytic processes can also be used.

Like basic upgrading processes, Suncor in Canada and Petrocedeno in Venezuela (Figure 24. 4) have chosen delayed coking for conversion technology. This process has been mastered for many years and is used to convert the residue at lesser cost. Nonetheless, this process has two disadvantages: the coke production and the volume of SCO produced.

The quantity of coke is especially large since the capacity of these projects is higher than a basic upgrading project in order to reduce investments as much as possible (economies of scale). The capacity of the coker unit for the Petrocedeno complex is 89,000 b/d vs. 45,000 b/d for the Petromonagas project. In Canada, coke finds no outlet due to transport difficulties, in contrast to Venezuela where it can be directly exported by seaway. Thus for now, Canadian coke is stored on-site in asphaltic sand mines, but this solution does not appear to be final. Over the longer term, this coke could be used for other processes like gasification.

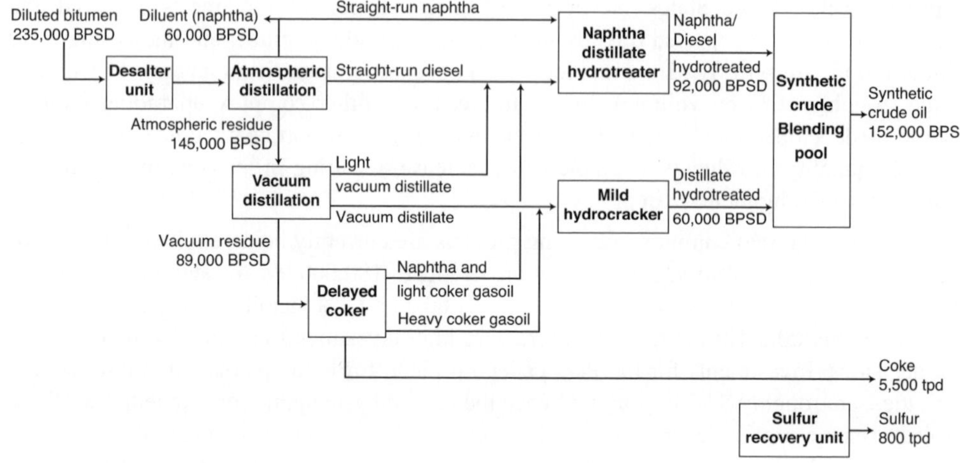

Figure 24.4

Production Diagram for Petrocedeno.

Source: After Upgrading Heavy Crude Oils and Residues – Phase 7, May 2003.

The second negative aspect of using a coker unit as a conversion facility is the negative final liquid yield. Like all carbon rejection processes, whether thermal or physical, the quantity of liquid produced is lower than the feed because a part of the heavy fraction is rejected in the form of coke. Producers who have chosen this conversion process have therefore found a compromise between low operating costs and a final quantity of synthetic crude for resale lower than the quantity could be obtaining with a catalytic conversion process.

In contrast to the Petrocedeno and Suncor projects, the Canadian companies Syncrude and Husky use a catalytic conversion process to convert their heavy crudes (Figure 24.5). The LC Fining® process, developed by ABB Lummus, was selected by Syncrude to increase the capacity of its upgrading site. Husky opted for H-Oil® developed by *IFP Energies nouvelles*. Both of these catalytic conversion facilities operate with ebullated beds and a catalyst based on nickel and molybdenum supported on alumina. To maintain the unit's performances, fresh catalyst is continuously added in the reactor. These catalytic processes consume large amounts of hydrogen, between 2 and 5 wt% of the unit's feedstock flow, depending on the type of feedstock to be converted. However, these ebullated bed conversion units are particularly adapted to processing feedstock with high metal content – a catalyst poison – like *heavy crude oils* extracted in Canada or Venezuela, which have nickel and vanadium content of more than 300 wt ppm. On these units, inactivation of the hydroconversion catalyst is compensated by continuously renewing the catalyst. The hydrotreatment units located downstream and operating as fixed beds are thus protected from these poisons.

These hydrotreatment units are required to stabilize the products and adjust their final properties. In contrast to basic upgrading – which only requires one hydrotreatment of the naphtha cut resulting from the conversion unit – obtaining a synthetic crude of high quality, i.e. one with low sulfur content and density of roughly 30°API, requires several hydrotreatments of naphtha and diesel cuts resulting from atmospheric distillation and the conversion unit, as well

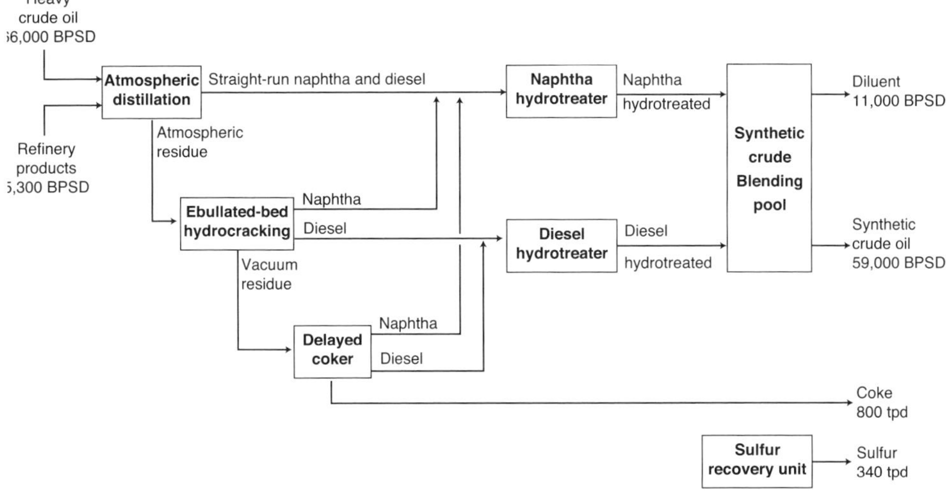

Figure 24.5

Production Diagram for the Husky Lloydminster Site.
Source: After Husky Energy Inc, Lloydminster Upgrader.

as mild hydrocracking of the distillate cut. The large number of units used and operating conditions needed to eliminate metals, sulfur, nitrogen and diolefins and to hydrogenate aromatics are far more severe than those required in basic upgrading. It results in a very significant amount of hydrogen consumption.

The several hydrotreatment units allow to produce a final synthetic crude with a low sulfur content but produce in the same time large quantities of solid or liquid sulfur. This quantity is larger if a catalytic conversion unit is used. In fact, in the case of carbon rejection conversion, a large amount of the sulfur is contained in the coke and/or asphalt, which are stored or sold. This sulfur is a byproduct of the upgrading site and is mainly used for the manufacture of fertilizer (ammonium sulfate nitrate, ammonium sulfate, superphosphate, potassium sulfate). To avoid the storage of its sulfur, Petrocedeno has developed a partnership to resell it.

The problems encountered by basic upgrading regarding the formation of coke are the same for this type of production. In the same way, Petrocedeno has signed contracts to resell its coke, mainly used for power plants and cement plants.

If the synthetic crude final qualities are much higher than those obtained by a basic upgrading, using a catalytic conversion process – which consumes large amount of hydrogen – increases the costs even more. With the increase in the price of natural gas observed in 2005 (roughly 7 US $/MMBtu in Canada), hydrogen produced by natural gas steam reforming is increasingly expensive. To reduce these operating costs, producers may in the future look toward processes like gasification, which will allow them to produce hydrogen while using, in the case of carbon rejection conversion units, the coke or asphalt produced, or for catalytic processes, some of the unconverted residue.

The quality of the synthetic crude produced by this complex process workflow is equivalent to a conventional crude produced in West Africa, with a very high yield of distillate, kerosene and atmospheric gasoil. This distribution of the various cuts is unique as compared to other conventional crude oils, resulting in petroleum that is well-adapted for a market requiring middle distillates, i.e. the European market. The Zuata Sweet crude produced by Petrocedeno is regularly compared to Nigerian Forcados crude. The yields of the various petroleum cuts of these two crudes are very similar (Figure 24.6). Only the quantity of residue varies since the various processes implemented in complex upgrading allow for total depletion of the residue.

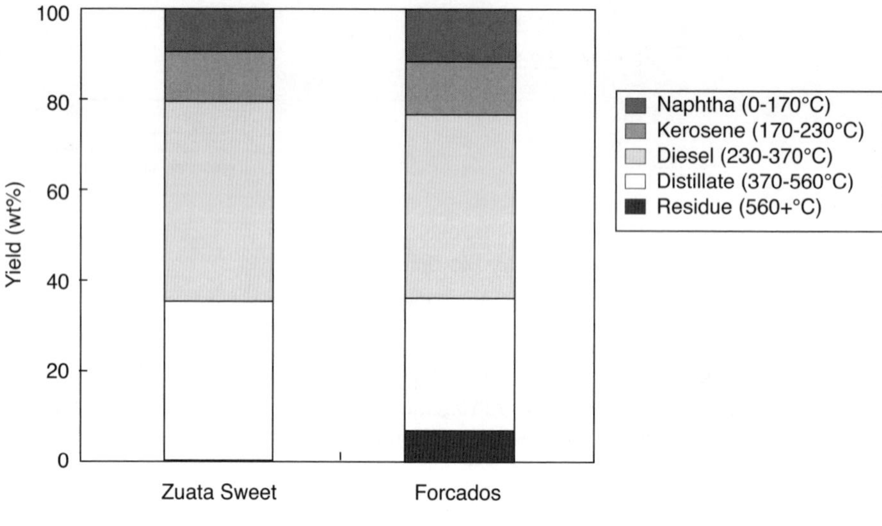

Figure 24.6

Yields of Petroleum Cuts for Zuata Sweet and Forcados Crudes.

These two crude oils of very different origin also demonstrate equivalent overall physico-chemical properties. If we compare the density and sulfur content (Table 24.5), two parameters which are important for defining the quality of the crude, the differences are very low. The various conversion processes can therefore tremendously improve the main qualities of an extra-heavy crude.

Table 24.5 Properties of Zuata Sweet and Forcados Crudes.

		Zuata	Zuata Sweet	Forcados
°API		8.5	31	30
Sulfur	(wt%)	4.1	0.13	0.19
Total nitrogen	(wt ppm)		608	1,182
Vanadium	(wt ppm)		< 1	< 1
Nickel	(wt ppm)		< 1	4.4

Table 24.6 shows the yields and qualities of the various petroleum cuts obtained for the Zuata Sweet crude produced by Petrocedeno. These properties are provided for informational purposes and may vary over time.

Table 24.6 Properties of the Various Petroleum Cuts for Zuata Sweet.

		Zuata Sweet	Forcados
Naphtha			
Yield	(wt%)	9.73	11.86
Density	(–)	0.705	0.764
Sulfur	(wt%)	< 0.01	< 0.01
Total nitrogen	(wt ppm)	25	0.3
RON	(–)	55	68.8
MON	(–)	53	65.5
Kerosene			
Yield	(wt%)	10.82	11.50
Density	(–)	0.817	0.829
Sulfur	(wt%)	0.011	0.031
Total nitrogen	(wt ppm)	44	2.3
Flash point	(°C)	61	62
Freezing point	(°C)	–55	–60
Diesel			
Yield	(wt%)	44.42	40.44
Density	(–)	0.879	0.878
Sulfur	(wt%)	0.08	0.146
Total nitrogen	(wt ppm)	280	117
Viscosity at 50°C	(cSt)	4.6	4.1
Cetane number	(–)	41	42
Distillate			
Yield	(wt%)	35.03	29.10
Density	(–)	0.931	0.941
Sulfur	(wt%)	0.25	0.30
Total nitrogen	(wt ppm)	1,359	2,131
Viscosity at 50°C	(cSt)	53	134
Conradson Carbon	(wt%)	< 0.1	N/A
Vanadium	(wt ppm)	2.1	N/A
Nickel	(wt ppm)	0.8	N.A
Residue			
Yield	(wt%)		6.64
Density	(–)		0.995
Sulfur	(wt%)		0.52
Total nitrogen	(wt ppm)		7,757
Viscosity at 100	(cSt)		1,241
Conradson Carbon	(wt%)		15.5
Vanadium	(wt ppm)		8.2
Nickel	(wt ppm)		65.7

Source: Oil and Gas Journal.

24.3 INTEGRATED UPGRADING

Although complex upgrading produces a synthetic crude whose qualities are comparable to some West African conventional crudes containing a very small residual fraction and a large quantity of middle distillates, the operating costs are also very high. These costs are mainly due to the large consumption of hydrogen, produced by natural gas steam reforming, required for many hydrotreatment units but also for the conversion unit in the case of a catalytic process.

The development of SAGD technology to extract *heavy crude oils* also contributes to the large increase in operating costs. In fact, it is estimated that the quantity of high-pressure steam – produced from a boiler supplied with natural gas – required for the extraction of one barrel of crude varies from two to four barrels, which represents an energy consumption of 1,055 to 1,583 MJ, or 1 to 1.5 MMBtu.

With the strong increase in crude oil and natural gas prices (see Figure 24.7), these operating costs have significantly increased in recent years. The use of natural gas as an energy carrier for the production of hydrogen or steam is therefore becoming increasingly questionable.

Since this special economic climate continues and SAGD technology is more widely used for the recovery of *heavy crude oils*, oil companies want to reduce their operating costs, in particular by avoiding the need to use natural gas. This is why many new projects for upgrading Canadian extra-heavy crude oils include a hydrogen production unit and/or a unit to produce the steam needed for SAGD technology. This production of utilities allows refiners to value the coke which has no outlet in Canada or the residue which has very low added value.

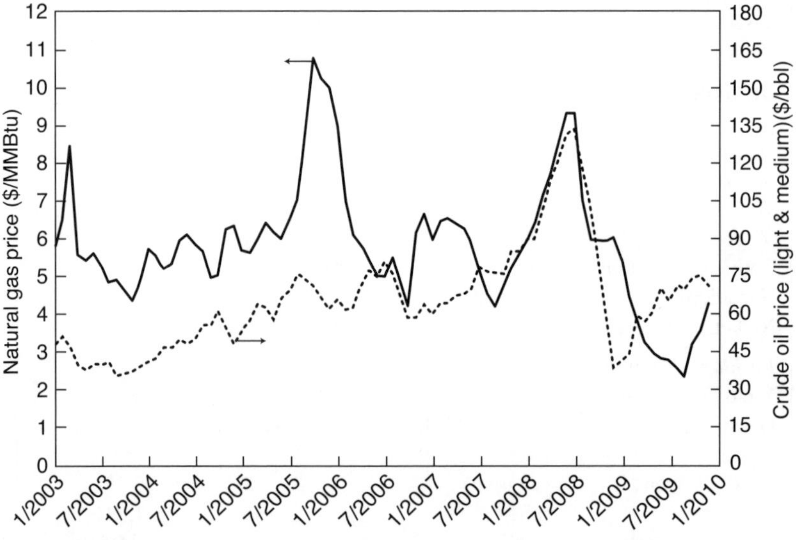

Figure 24.7

Evolution of the Price of Natural Gas and Crude Oil on the Canadian Market.
Source: Alberta Department of Energy.

The gasification process is currently the process most widely cited in various projects (North West Upgrading, Long Lake, etc.). The gasification process uses a low-oxygen environment to burn the carbon contained in the feedstock to initially form a low calorific value gas. This gas is mainly composed of hydrogen and carbon monoxide, which can either be burned in a boiler to produce the steam needed to extract *heavy crude oil via* SAGD technology, or converted into hydrogen *via* the water-gas shift reaction ($CO + H_2O \rightarrow H_2 + CO_2$), then membrane-purified to obtain high-purity hydrogen. Although this process has the advantage of producing both utilities (steam and hydrogen) simultaneously, the investment is very large, in particular due to the cost of the gasifier and the air purification unit. For the gasifier, the materials used must be resistant to very severe operating conditions (temperature of 1,000°C to 1,500°C and pressure of roughly 30 bar) and thus are very costly. The air purification unit – also referred to as an ASU (Air Separation Unit) – is used to separate oxygen from air and use it as comburant. The quantity of gas emitted in the gasifier is therefore reduced and can be more easily upgraded during the rest of the process. By only injecting oxygen in the reactor, the operating conditions of various gasification reactions are well controlled. With the development of this process in other application domains (biomass upgrading, cogeneration, etc.), the cost of the air purification unit will perhaps decrease over the next few years.

Other projects will only generate steam – needed for the extraction of crude oil *via* SAGD technology – from the combustion of coke or residue. Although using coke or residue as fuel is no longer a technological difficulty, processing the flue gas is currently a major economic problem, since the volume of effluent emitted is significant. Absorbent boilers have been developed to recover the sulfur contained in the flue gas, but they use a quantity of absorbents that is too high to be imported to the production site and also produce a large amount of gypsum. Other boilers under development include the regenerative absorption boiler and the oxy-combustion boiler, used to reduce the volume of inert gas, allow the use of conventional methods to process the sulfur dioxide (SO_2) formed by combustion.

To illustrate this topic regarding upgrading without the use of natural gas, two Canadian projects can be mentioned: the NorthWest Upgrading project in which hydrogen is produced by gasification of the residue (Figure 24.8) and the Long Lake project which produces the steam, lean gas and hydrogen required for extraction, conversion and hydrocracking (Figure 24.9).

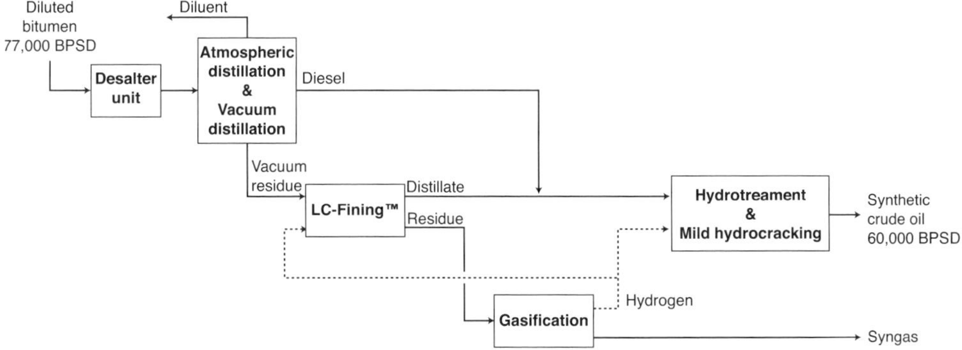

Figure 24.8

Production Diagram for the NorthWest Upgrading Project.
Source: After North West Upgrading Inc.

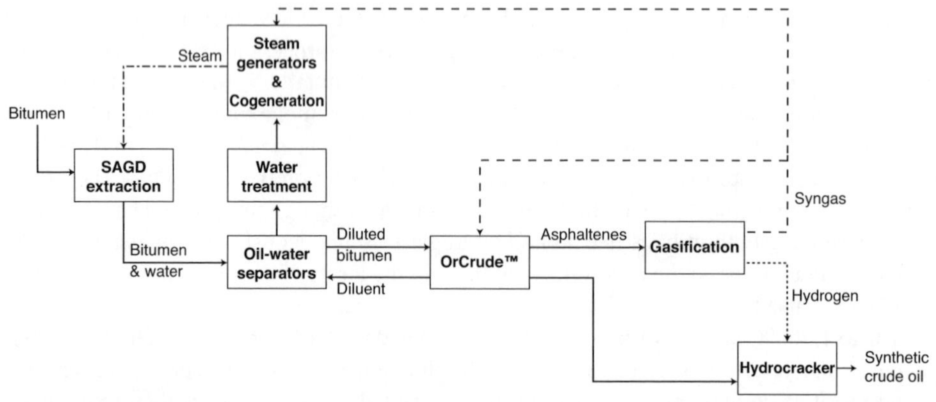

Figure 24.9

Production Diagram for the Long Lake Project.

Source: After Long Lake Project www.longlake.ca/project/bitumen.html

In the NorthWest Upgrading project, the gasifier produces the hydrogen needed for the LC-Fining® process commercialized by Chevron Lummus Global. Two units for the hydrotreatment of naphtha/diesel and distillate cuts allow for the production of a synthetic crude comparable to the crude oils produced in complex upgrading.

The additional lean gas produced is sold to a fertilizer production plant located near the upgrading site.

The Long Lake project uses the gasification process to produce two utilities: lean gas (syngas) and hydrogen. The hydrogen is used by the hydrocracker, which improves the quality of the synthetic crude produced. The syngas is used as a comburant to generate the steam needed for the extraction of heavy crude and also as a fuel for the OrCRUDE™ process. This process is divided into three sections – a distillation, a *deasphalter* and a thermal cracker – and requires a large amount of energy.

With totally or partially integrated project development, the synthetic crude produced is of very high quality, i.e. with low sulfur content and high °API (Long Lake project produces an SCO with °API between 39 and 44). The refiners obtain better value for the investment made in a hydrogen production unit by receiving a higher price of the synthetic crude produced sale.

Like all highly-integrated models, these synthetic crude production models have disadvantages from an exploitation point of view. In fact, all of these processes are highly vulnerable to a full or partial shutdown of a unit, particularly the unit that produces steam and hydrogen. The start-up phase is also a delicate phase since all utilities must be provided immediately. Additional equipment must therefore be planned for this type of model in order to store the feedstock of hydrogen or steam production units, or small additionnal units operating during the start-up phase or in case of problem on the main unit. An alternative solution is to obtain the required utilities from nearby industries.

Thus, models which integrate the extraction of heavy crude and the production of synthetic crude reduce energy costs by decreasing the consumption of natural gas used at the site. In the future, these completely integrated models should significantly develop, particularly in Canada where the local market for natural gas is very tense. In parallel to these workflows which integrate extraction and production, new heavy crude conversion units should be built in the next several years to produce final products which directly have market specifications.

24.4 ULTRA-COMPLEX UPGRADING

In a context of increasingly severe specifications for fuels, we face a progressive decrease of the quality of crude oils processed in refineries (heavier and richer in sulfur) while the demand for petroleum products has regularly increased over the recent years. This trend of the demand should continue in the next few years, with respect to a strong increase in the consumption of refined products by emerging countries such as China or India (Figure 24.10 and Figure 24.11).

Faced with these future constraints, refiners will need to evolve their refining tool and make large investments for the development of new refineries, integration of new hydrotreatment facilities and development of deep conversion facilities.

Refiners located in production countries may then decide to develop a refining tool that can convert the *heavy crude oil* into refined products which directly have the market specifications required by consuming countries. In this context, the American company Chevron, associated with the Spanish oil company Repsol YPF and the Venezuelan state owned company PDVSA, want to build a new heavy crude processing facility in the Orinoco Belt with a capacity of 400,000 b/d. This \$6 billion project implements the new *heavy crude oil* extraction technologies that increase the recovery rate which is currently 6%, but also conversion and secondary processing facilities capable of producing high-quality refined products.

Selection of the ultra-complex upgrading model to be developed will depend on the target final market. Two types of workflows can be presented:

- Gasoline-oriented production for the American market.
- Diesel-oriented production for the European market.

Figure 24.12 shows a model workflow for ultra-complex upgrading. Other workflows can also be planned to achieve this objective.

In Figure 24.12, the heavy crude is separated in an atmospheric distillation followed by vacuum distillation. The vacuum residue is then converted in a deep conversion unit similar to those used in complex upgrading or basic upgrading. The various light fractions output by the deep conversion unit are mixed with straight-run cuts and undergo secondary processing to improve their quality, and in particular reduce their content of impurities.

After mixing, the naphtha cut is then hydrotreated, then sent to a catalytic reforming or isomerization unit. The objective of these two units is to produce a gasoline base of high octane number: roughly 100 for the reformate and 83 to 90 for the isomerate.

The kerosene cut is treated before being sent to the kerosene pool.

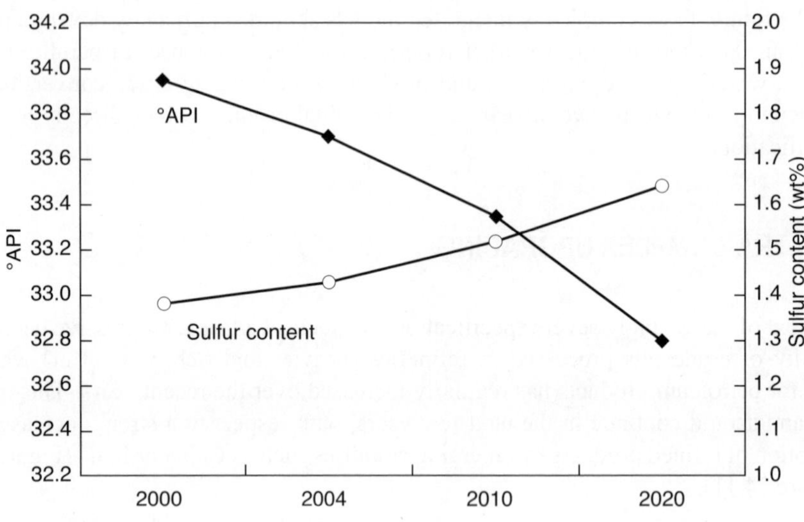

Figure 24.10

Evolution in Sulfur Content and °API of Crude Oils Over the Next Several Years.

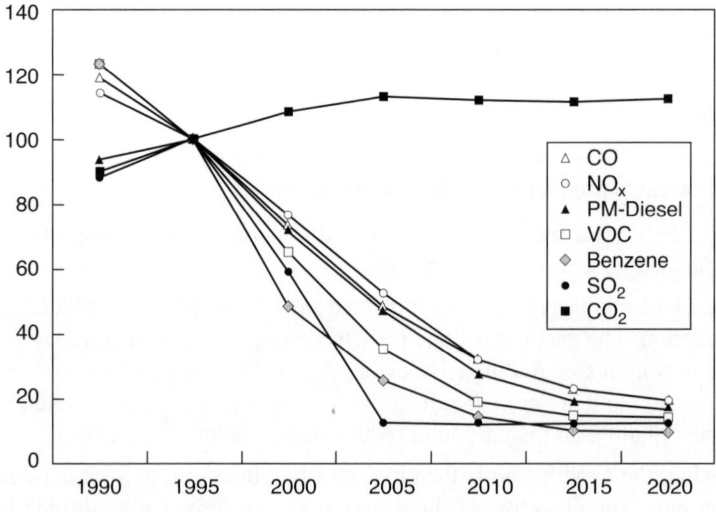

Figure 24.11

Evolution of Pollutant Emissions of Automobile Origin in Europe.

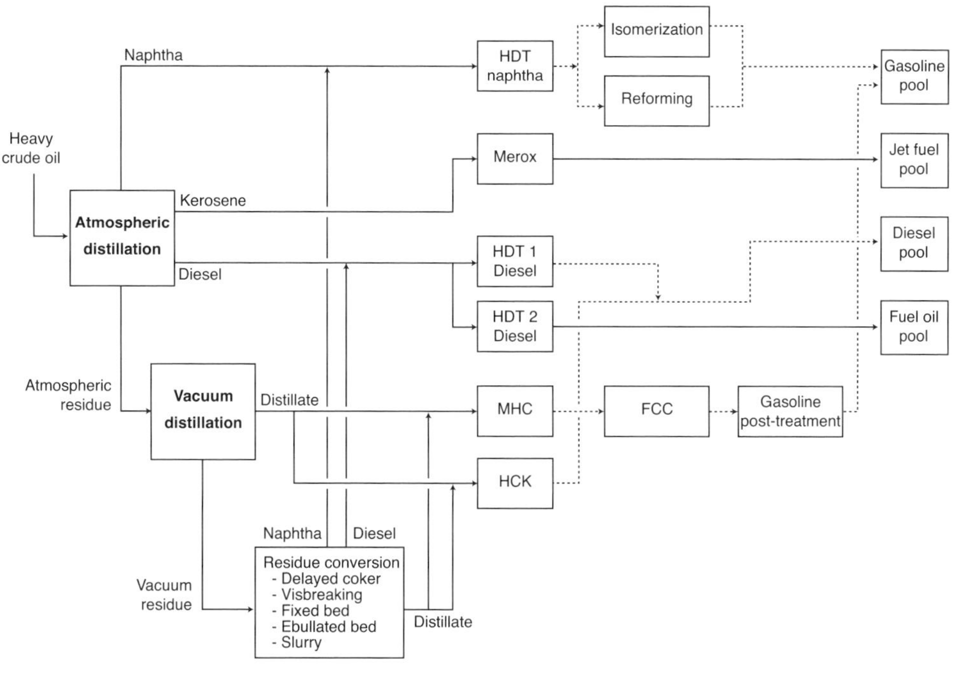

Figure 24.12

Diagram for Ultra-Complex Upgrading.

The diesel cut is hydrotreated at high pressure to produce a diesel base that complies with specifications, or at low pressure to produce an heating oil.

The distillate cut resulting from vacuum distillation and the deep conversion unit can be converted in an FCC (Fluid Catalytic Cracking) unit for a gasoline market or in a hydrocracker for a diesel market.

For production oriented toward a gasoline market, the distillate must first undergo post-treatment in a mild hydrocracking unit before it can be used as FCC feedstock. The gasoline output by the FCC also undergoes post-treatment to produce a gasoline with low sulfur content and high octane number. This gasoline can then be directly incorporated into the upgrader's gasoline pool.

If the target market is diesel, the distillate cut is converted in a high-pressure hydrocracker to produce a high-quality diesel: cetane number of roughly 60, sulfur content less than 10 wt ppm and aromatic content less than 10 wt%. Considering the qualities of the diesel produced, it can be directly incorporated into the upgrader's diesel pool.

Thus, ultra-complex upgrading can produce fuels and fuel oils to the specifications required by the target market, but requires more complex logistics. Since products are exported to the consuming countries mainly by seaway, there is a significant risk of contamination by seawater used for tank cleaning or by the hydrocarbons still remaining in the tanks. Mandatory re-processing of these products in a refinery located in the consuming countries would then be required, thereby reducing the advantages of ultra-complex upgrading.

From a political point of view, the development of ultra-complex upgrading would increase consuming countries' dependence on producing countries. Thus, consuming countries seek to maintain their refining tool, by subsidizing refiners for the development of new refineries or deep conversion units, or by developing alternative energies like bioethanol in Brazil.

24.5 CONCLUSION

With the increase in the price of crude oil and the desire to diversify supply sources, heavy crudes currently represent an abundant resource of petroleum for refiners.

Considering their physico-chemical properties, these crude oils must be partially converted in the production country. Various levels of upgrading according to the target objective could be used:

- Basic upgrading to make the *heavy crude oil* transportable.
- Complex upgrading to produce a synthetic crude of high quality.
- Ultra-complex upgrading to produce products conform to maket specifications.
- Upgrading integrated with *heavy crude oils* production to reduce operating costs related to the use of natural gas.

In any case, there is not only one solution (level of upgrading, selection of conversion and hydrotreatment processes, etc.) and several parameters like evolution of the market or industrial environment must be taken into consideration for the development of an upgrading facility.

The conversion of *heavy crude oils* is performed *via* refining units derived from conventional refining and exploited for many decades now, but used with more severe operating conditions (e.g. hydrogen pressure).

Over the longer term, new conversion technologies have currently being developed and may be commercialized in the next few years (transport bed conversion processes, ultra-rapid thermal processing of the feedstock, hydrothermal conversion with supercritical water, etc.), still using the approach of a unit located *ex situ* near a refining site either near a production site.

Even longer term, research projects undertaken by universities or petroleum companies in the domain of *heavy crude oil* production envisage partial conversion of *heavy crude oils* in the reservoir or *in situ* upgrading.

REFERENCES

Aalund L (1998) Technology, Money Unlocking Vast Orinoco Reserves. *Oil and Gas Journal*.

Butler R (2001) Application SAGD, Related Processes Growing in Canada. *Oil and Gas Journal*, pp 74-78.

Chang T (1998) Upgrading and Refining Essential Parts of Orinico Development. *Oil and Gas Journal.*

Chang T (2001) Sincor to Offer Zuata Sweet Crude in 2002. *Oil and Gas Journal*, pp 52-55.

Fletcher S (2005) N. American Unconventional Oil a Potential Energy Bridge. *Oil and Gas Journal*, pp 22-26.

Fletcher S (2005) Suncor Seeks Gas Supplies for Oil Sand Project. *Oil and Gas Journal*, pp 26-27.

Hawkins D (2004) Canadian Bitumen Stands Poised to Expand to US Markets. *Oil and Gas Journal*, pp 52-56.

Joseph D (1998) Drive to Produce Heavy Crude Prompts Variety of Transportation Methods. *Oil and Gas Journal.*

Moritis G (1998) New Techniques Improve Heavy Oil Production Feasibility. *Oil and Gas Journal.*

Moritis G (2001) Sincor Nears Upgrading, Plateau Production Phase. *Oil and Gas Journal*, pp 47-55.

Nakamura D (2003) Sincor Begins Exporting Petcoke, Sulfur From Jose Crude Upgrader. *Oil and Gas Journal*, pp 58-62.

Oilsands Review. May 2007. pp 8-10.

Soligny JC (2001) Le projet Sincor. *Pétrole et Techniques*, pp 76-79.

Stott J (2002) *In Situ* Projects Gaining Ground in Canadian Oils Sands Development Boom. *Oil and Gas Journal*, pp 24-30.

True W (1998) Orinoco Projects Choose Dilution to Move Production. *Oil and Gas Journal.*

Conclusion

A.Y. Huc

In the perspective of a future decline in the production of conventional oils, the heavy oils, extra-heavy oils and bitumen represent a tempting and probably inescapable alternative source to help supply the world demand for hydrocarbons. If we consider the resource in place (around 5000 Gb), they account for the same amount as their conventional counterpart.

As a consequence of the geological factors controlling their origin, most of the accumulations are favourably located at shallow to moderately shallow depth. The major drawbacks hindering the exploitation of these *heavy crude oils* are their very nature: high to very high viscosity and detrimental chemical composition including low distillate content, low paraffin content, richness in heavy molecular weight compounds, high content in sulphur, nitrogen, metals and often acidic properties. As a result, each step of the exploitation (production, transport and upgrading) is a technical challenge.

Substantial improvements in the technologies have been achieved during the last decades. They are illustrated e.g. by new *in situ* production techniques, original means to facilitate the transport *via* pipelines and upgrading processes. In spite of this pogress, only a small fraction of the resource in place can currently be considered as producible reserves. The main producing countries, Canada and Venezuela, plan to significantly increase their heavy crude projects in the near future. This type of endeavour will have to be backed by extensive and innovative R&D in technical and environmental areas. In order to strictly limit the imprint on the environment, more specifically in the production domain, water and CO_2 management have to be carefully considered and secured.

The huge resource ahead of us and the promising progressing technologies justify an important R&D effort, such as innovative production means, more efficient transport technologies or strategy or better understanding the chemical structure of the heavy end to sharpen our ability to improve the deep conversion of this heavy feedstock. Considerable investments dedicated to projects for the exploitation of *heavy crude oils* are also necessary to support their decisive role in the energy mix needed to accompany the energy transition during this century.

Abbreviations

2D, 3D	Two or three Dimensional in space
5-SPOT, 7-SPOT, 9-SPOT	Patterns involving 5, 7 or 9 wells (see figure 8.8)
ABAN separation	Acid, Basic, Amphoteric, Neutral separation
ACR	Alberta Chamber of Resources
AEUB	Alberta Energy & Utilities Board
AGO	Atmospheric Gas Oil
AMSTM	American Society for Testing and Materials
AOSP	Athabasca Oil Sands Project
AOSTRA	Alberta Oil Sands Technology and Research Authority
APCI-Q-TOF	Atmospheric Pressure Chemical Ionisation / Quadripole / Time Of Flight mass spectrometry
API	American Petroleum Institute
API degree	American Petroleum Institute's inverted scale for denoting the 'lightness' or 'heaviness' of crude oils and other liquid hydrocarbons. °API =(141.5/d)-13.5, where d = relative density at 288.7K
APPI-Q-TOF	Atmospheric Pressure PhotoIonisation / Quadripole / Time Of Flight mass spectrometry
AR	Atmospheric Residue
ARC	Alberta Reseach Council
BB	Bitumen Blend
BBL	Barrel
BIP	Bitumen In Place
BOPD	Barrel of Oil Per Day
BTU	British Thermal Unit
CAF	Core Annular Flow
CAPEX	CAPital EXpenditure
CAPP	Canadian Association of Petroleum Producers
CAPRI	Catalytic variant of the THAI process
CASH	Chevron Activated Slurry Hydroprocessing

CCR	Conradson Carbon Residue
CCS	Carbon Capture and Sequestration
CDN	Canadian dollar
CERI	Canadian Energy Research Institute
CHOPS	Cold Heavy Oil Production with Sand
CLG	Chevron Lummus Global
CLPF	Closed Loop Power Fluid
CNRL	Canadian Natural Resources Limited
CO_2E	CO_2 Equivalent
CONRAD	Canadian Oil Sands Network for Research and Development
COSR	Cumulative Oil Steam Ratio
CPU	Central Processing Unit
CSOR	Cumulative Steam Oil Ratio
CSS	Cycling Steam Simulation
CSTR	Continuous Stirred Tank Reactor
CT	Composite – or Consolidated – Tailings
CTE	Coefficient of Thermal Expansion
CU.M	Cubic Meter
CWE	Cold Water Equivalent
DAO	De-Asphalted Oil
DMDS	DiMethyl DiSulfide
DME	DiMethyl Ether
DOE	Department Of Energy (of United States)
DRA	Drag Reducing Agent
EB	Ebullated Bed
EMRE	ExxonMobil Research and Engineering
EOR	Enhanced Oil Recovery
ESP	Electrical Submersible Pump
ES-SAGD	Expanded Solvent SAGD
EXAFS	Extended X-Ray Fine Structure
FCC	Fluid Catalytic Cracking
FTE	Freeze-Thaw-Evaporation
FTICR MS	Fourier Transform Ion Cyclotron Resonance Mass Spectrometry

FTIR	Fourier Transform InfraRed spectroscopy
GDP	Gross Domestic Product
GHG	GreenHouse Gas
H/C	Hydrogen/Carbon
HCK	HydroCracKing
HDAr	HyDrodeAromatization (aromatic removal by hydrotreatment)
HDAs	HyDrodeAsphatenisation (asphatene removal by hydrotreatment)
HDCCR	HyDrotreatment of Conradson Carbon Residue
HDH	Hydrocracking Distillation Hydrotreating
HDM	HydroDeMetallization (Metal removal by hydrotreatment)
HDN	HyDrodeNitrification (nitrogen removal by hydrotreatment)
HDS	HyDrodeSulfurization (sulfur removal by hydrotreatment)
HDT	HyDroTreatment
HGI	Hardgrove Grindability Index
HLS	Hot Lime Softening
HPLC	High Performance (Pressure) Liquid Chromatography
HTO	High Temperature Oxidation
ID	Inner Diameter
IEA	International Energy Agency
IGCC	Integrated Gasification Combined Cycle
IHS	Inclined Heterolithic Surface
IOL	Imperial Oil Ltd
IOR	Improved Oil Recovery
IPCC	Intergovernmental Panel on Climate Change
IR	InfraRed (spectroscopy)
ISBL	Inside Battery Limits
ISC	*In Situ* Combustion
KBR	Kellogg-Brown-Root
LASER	Liquid Addition to Steam for Enhancing Recovery
LCO	Light Cycle Oil (aromatic gasoil from FCC)
LPG	Liquefied Petroleum Gas
LTBS	Lateral Tie Back System
LTO	Low Temperature Oxidation

LWD	Logging While Drilling
MALDI-TOF	MAtrix Laser Desorption Ionisation / Time Of Flight
MBR	Membrane Bio-Reactor
MEA	Mono Ethanol Amine
MFT	Mature Fine Tailings
MHC	Mild HydroCracking
MWD	Measurements While Drilling
NCG	Non-Condensable Gas
NIR	Near InfraRed (spectroscopy)
NMR	Nuclear Magnetic Resonance
OCR	Onstream Catalyst Replacement
OD	Outer Diameter
OECD	Organisation for Economic Co-operation and Development
OGJ	Oil and Gas Journal
OOIP	Original Oil In Place
OPEX	OPerating EXpenditure
OPF	Open Power Fluid
OSBL	OutSide Battery Limit
OSR	Oil Steam Ratio
OSTRM	Oil Sands Technology RoadMap
OTSG	Once Through Steam Generator
PCM	Pompes & Compresseurs Mécaniques
PCP	Progressive Cavity Pump
PFG- 1H NMR	Pulsed Field Gradient Spin-Echo ^{1}H Nuclear Magnetic Resonance
PONA	Paraffins/Olefins/Naphthenes/Aromatics
POX	Partial OXidation
RCC	Resid Catalytic Cracking
RDS	Residuum DeSulfurization
RES. CONDITIONS	Reservoir conditions
RFCC	Residue Fluid Catalytic Cracking
RICO	Ruthenium Ion Catalysed Oxidation
RON	Research Octane Number
ROSE	Residuum Oil Supercritical Extraction
SAGD	Steam Assisted Gravity Drainage

SANS	Small Angle Neutron Scattering
SAP	Solvent Aided Process
SAR separation	Saturates, Aromatics, Resins separation
SARA fractionation	Saturates, Aromatics, Resins, Asphaltenes fractionation
SAXS	Small Angle X-ray Scattering
SCO	Synthetic Crude Oil
SDA	Solvent DeAsphalting
SEC	Steric Exclusion Chromatography
SGS	Shell Global Solutions
SJVBU	San Joaquin Valley Business Unit
SOR	Steam Oil Ratio
SR	Straight Run (issued from direct distillation of crude oil)
SRC	Saskatchewan Research Council
STOOIP	Stock Tank Original Oil In Place
SWSC	Single Well Steam Circulation
TAN	Total Acid Number (quantity of KOH, in mg, needed to neutralize 1g of oil)
TBP	True Boiling Point
THAI	Toe-to-Heel Air Injection Process
TLC	Thin Layer Chromatography
TT	Thickened Tailings
TVD	True Vertical Depth
UNEP	United Nations Environment Pogramme
UNFCCC	United Nations Framework Convention on Climate Change
UOP	Universal Oil Products
USD	US Dollar
USSR	Union of Soviet Socialist Republics
UTF	Underground Test Facility
VAPEX	VAPour EXtraction Process
VBD	Vibrated Bulk Density
VCM	Volatile Combustible Matter
VGO	Vacuum Gas Oil
VPO	Vapour Pressure Osmometry
VR	Vacuum Residue
WAG	Water-Alternating Gas Process

WASP	Water-Alternating-Steam Process
WBCSD	World Business Council for Sustainable Development
WEC	World Energy Council
WLS	Warm Lime Softening
WMO	World Meteorological Organization
XANES	X-ray Absorption Near Edge Structure (spectroscopy)
XPS	X-ray Photo-electron Spectroscopy

I Glossary

Accommodation space — The space available for sediments to accumulate in a basin. The variation of this parameter through time control the rate of amalgamation of the fluvial channels, and then the vertical connectivity in a fluvial reservoir.

Advanced wells — Wells having several branches in a reservoir drilled from a unique vertical or inclined wellbore.

Alkanes — Saturated hydrocarbons.

API — A measure of liquid gravity common in the oil industry. Water is 10, and a typical light crude is from 35-40. Bitumens are by convention typically from 8-11 API.

Aromatics — Hydrocarbon species which occurs in unusually high concentrations in bitumen and some derived products, and confers poor quality in distillate fuels.

Asphaltenes — The heaviest and most concentrated aromatic hydrocarbon fractions of bitumen.

Back-arc basin — A sedimentary basin located behind of a volcanic arc and characterized by high subsidence and sedimentation rates.

Barrel — Generally accepted measurement of oil. One barrel of oil equals 159 litres One cubic meter contains 6.29 barrels.

Biodegradation — Biological alteration of crude oils mediated by in-reservoir bacterial communities, changing conventional oils into heavy, extra-heavy oils and tar.

Bioturbation — The effect of the animal activity on the sediment. Bioturbation modifies the intial texture of the sediments.

BIP — Bitumen in place.

Bitumen — Naturally occurring viscous mixture of hydrocarbons that contain high levels of sulphur and nitrogen compounds. In its natural state, it is not recoverable at a commercial rate through a well because it is too thick to flow. Bitumen typically makes up about 10% by weight of oil sand, but saturation varies.

Bottom aquifer A layer of 100% water at the bottom of an underground reservoir.

Breakthrough time Time from start of production at which water breaks through in a production well, either coming from an aquifer or from an injection well.

Bubble point pressure Pressure at which the first bubble of gas goes out of a saturated oil.

Carbon dioxide (CO2) see GreenHouse Gases.

Catalyst Used in upgrading processes to assist cracking and other upgrading reactions.

Cetane Number A measure used to describe the combustion characteristics of a diesel fuel. A high Cetane Number indicates a better fuel.

Coke Solid, black hydrocarbon which is left as a residue after the more valuable hydrocarbons have been removed from bitumen by heating the bitumen to high temperatures.

Coking A process commonly used as the first step in bitumen upgrading. The bitumen is cracked by application of high temperatures.

Compaction Reduction of porosity of the porous medium due to inner pressure reduction.

Condensate Mixture of light hydrocarbons recoverable from gas reservoirs. Condensate is also referred to as a natural gas liquid, and is used as a diluent to reduce bitumen viscosity for pipeline transportation.

Conventional crude oil Mixture mainly of pentane and heavier hydrocarbons recoverable at a well from an underground reservoir and liquid at atmospheric pressure and temperature. Unlike bitumen, it flows through a well without stimulation and through a pipeline without processing or dilution. In Canada, conventional crude oil includes light, medium and heavy crude oils, like those produced from the Western Canada Sedimentary Basin. Crude oils containing more than 0.5 percent of sulphur are considered "sour", crudes with less than 0.5 percent are "sweet".

Conventional heavy crude (oil) Oil with a gravity between 22.3 and 10 API, i. e. a specific gravity between 0.920 and 1.000, "*in situ*" level of viscosity is smaller than 10,000 centipoises (cP or mPa.s), it is flowing at reservoir conditions.

Cumulative Oil Steam Ratio (COSR) Ratio of cumulative oil produced to the cumulative steam injected (expressed in Cold Water Equivalent).

Cracking Process of breaking down the larger, heavier and more complex hydrocarbon molecules into simpler, lighter molecules.

Critical rate Maximum oil rate of a well for which there is no water coning in a well produced in a reservoir having a bottom aquifer.

Cyclic steam stimulation Process in which steam is injected in a well, the well shut in for a limited period of time, and finally re-opened for production. Several cycles are generally operated before steam flood replaces cyclic injection.

De-asphalting A family of processes that use light solvents to selectively reject highly aromatic or 'asphaltenic' fractions.

Degree API American Petroleum Institute's inverted scale for denoting the 'lightness' or 'heaviness' of crude oils and other liquid hydrocarbons. °API =(141.5/d)-13.5, where d = relative density at 288.7K.

Depocenter The place where sediment preferentially accumulated in a basin.

Diluent see Condensate.

Dismigration The vertical migration of the hydrocarbons when seals are not efficient.

Ebullated bed process A residue conversion process that employs hydrogen, and keeps the solid catalyst in a semi-fluid state to allow continual addition and removal without halting the process.

Extraction A process, unique to the oil sands industry, which separates the bitumen from the oil sand using hot water, steam and caustic soda.

Extra-heavy oil Oil with a gravity less than 10 °API, i. e. a specific gravity higher than 1.000, "*in situ*" level of viscosity is smaller than 10,000 centipoises (cP or mPa.s), it is flowing at reservoir conditions.

Fines Minute particles of solids such as clay or sand.

Fluid mobility Ability of a fluid to move inside a porous medium.

Fluid viscosity Property of a fluid.

Foamy oil Oil having bubbles of gas trapped in the liquid phase.

Foreland basin A sedimentary basin located ahead of a mountain belt in a compressive setting. The basin is characterized by a high subsidence rate when close to the thrust belt (the "foredeep") and the sedimentary succession progressively reaches the surface hundreds of kilometers away from the mountain belt (the forebulge). Sediments are eroded from the mountain belt and resedimented in the basin as the mountain belt was formed (syn-orogenic sediments). The Orinoco and Canadian heavy oil belts are found in such a setting.

Froth treatment The means to recover bitumen from the water, bitumen and solids froth produced in hot water extraction (in mining-based recovery).

Gas cap A layer of free gas on top of the oil zone in an underground reservoir.

Gas Oil contact Minimum depth at which a 100% liquid - oil plus water - saturation exists in the reservoir.

Gasification A process to partially oxidize any hydrocarbon, typically heavy residues, to a mixture of hydrogen and carbon monoxide. Can be used to produce hydrogen and various energy by-products.

Gravity override Phenomenon of gravity segregation in a reservoir with the steam flowing to the top due to lower density compared to the oil in place.

GreenHouse Gases (GHG) Gases commonly believed to be connected to climate change and global warming. CO_2 is the most common, but include other light hydrocarbons (such as methane) and nitrous oxide.

Heavy crude (oil) In this volume the term *heavy crude oil* refers to *conventional heavy crude (oil)*, *heavy oil*, *extra-heavy oil*, *natural bitumen* or *oil sands* and *tar sands*.

Heavy oil API gravity below 22.3° API, i.e. specific gravity above 0.92.

Heterolithic reservoir A reservoir presenting small-scale lithological heterogeneities, named Inclined Heterolithic Surfaces, making difficult the production of the heavy oil.

Huff & Puff Similar to cyclic steam stimulation.

Hydrocarbons — Organic chemical compounds of hydrogen and carbon atoms that form the basis of all petroleum products. Hydrocarbons may be liquid, solid or gaseous.

Hydrocracking — Refining process for reducing heavy hydrocarbons into lighter fractions, using hydrogen and a catalyst. Can also be used in upgrading of bitumen.

Hydroprocessing — An upgrading process for reducing heavy hydrocarbons into lighter fractions through the catalytic addition of hydrogen.

Inclined heterolithic surface — Facies of low porosity and permeability formed by heterogeneities in lateral accretion surfaces of meandering point bars.

In-situ — In-situ recovery refers to methods to extract the bitumen component of a deposit without removing the rock matrix from its bed. *In situ* deposits are buried too deeply to be mined by open pit techniques.

Light crude (oil) — Oil with a gravity higher than 31.1° API, i. e. a specific gravity below 0.871.

Line drive — Well pattern where injectors and producers are on parallel lines. Direct line drive: injectors and producers are on in front of each other. Staggered line drive: wells are shifted half distance between injectors and producers. See Figure 8.8.

Medium crude (oil) — Oil with a gravity between 22.3° and 31.3° API, i. e. a specific gravity between 0.920 and 0.871.

Multibranch wells — Wells having several branches in a reservoir drilled from a unique vertical or inclined wellbore.

Natural bitumen or oil sands and tar sands — Oil with a gravity less than 10° API, i. e. a specific gravity higher than 1.000, "*in situ*" level of viscosity is higher than 10,000 centipoises (cP or mPa.s), it is not flowing at reservoir conditions.

NSO compounds — Heteroatomic high molecular weight compounds (resins and asphaltenes).

Oil sand mining — Extraction of oil from *ex situ* mines.

Oil sands — Bitumen-soaked sand, located in four geographic regions of Alberta: Athabasca, Wabasca, Cold Lake and Peace River. The Athabasca Deposit is the largest, encompassing more than 42,340 square kilometers.

Pilot plant Small model plant for testing processes under actual production conditions.

Pressure drawdown Difference between the reservoir pressure and bottomhole pressure.

Productivity index Ratio of the instantaneous flow rate of a well to the pressure differential between the reservoir pressure and bottomhole pressure.

Progradation The seaward shift of a shoreline in a deltaic setting.

Pull-apart basin A sedimentary basin bounded by faults showing significant lateral displacement (strike-slip faults). Such basins are characterized by very high rates of subsidence and a high reservoir compartimentalization. The Californian heavy oil reservoirs are found in such a context.

Steam Assisted Gravity Drainage (SAGD) An *in situ* production process using two closely spaced horizontal wells, one for steam injection, the other for production of the bitumen/water emulsion.

Steam drive Similar to steam flood.

Steam flood Process in which steam is injected continuously in wells to displace fluids in place to production wells.

Steam soaking Similar to cyclic steam stimulation.

Subsidence Downward movement of the land caused by compaction of an oil reservoir submitted to production.

Synthetic Crude Oil (SCO) A high quality, light, sweet crude oil, manufactured by upgrading bitumen extracted from oil sands. Mixture mainly of pentane and heavier hydrocarbons derived from crude bitumen through the addition of hydrogen or the removal of carbon.

Thermal recovery Any process by which heat energy is used to reduce the viscosity of bitumen in situ to facilitate recovery.

Transgression The landward shift of a shoreline during a relative sea-level rise.

TV/BIP Corresponds to the ratio of Total Volume (TV) of material to be mined in order to obtain a given volume of bitumen (Bitumen In Place: BIP).

Unconventional crude oil Crude oil which is not classified as conventional. An example would be bitumen and synthetic crude oil.

Upgraded crude oil See Synthetic Crude Oil.

Upgrading	Conversion of bitumen or heavy crude into a lighter, sweeter, high-quality crude oil either through the removal of carbon (coking) or the addition of hydrogen (hydroconversion).
Visbreaking	A process designed to reduce residue viscosity by thermal means, but without appreciable coke formation.
Viscosity	The ability of a liquid to flow. The lower the viscosity, the more easily the liquid will flow.
Water coning/cresting	Phenomenon of water rising in an oil well when a bottom aquifer is present below the oil layer.
Water oil contact	Uppermost depth in the reservoir where a 100% water saturation exists.
Watercut	Fraction of water in the liquid production of a well.
Waterflood, polymer flood, steam flood	Recovery processes involving the injection of water, or polymer or steam.
Wormhole	Zone of a reservoir where there is an increase of porosity due to sand production from a well.

Nomenclature

Symbol	Definition	Unit
°API	Degree API	dimensionless
μ, η	Viscosity	centipoise equivalent to mPa.s
Bb	Volume	Billion barrels (barrel~159 litres)
Bo	Oil volume factor	dimensionless
Ea	Activation energy	KiloJoule / mole
Gb	Volume	Giga barrels (barrel~159 litres)
h	Reservoir thickness	m or ft
Jv, Jh	Productivity Index (vertical or horizontal well)	m^3/d/bar or bopd/psi
k	Permeability	md
Kh	Horizontal permeability	md
Kv	Vertical permeability	md
M	Mobility ratio	dimensionless
m/z	Mass per charge	Dalton
Mb	Volume	Million barrels (barrel~159 litres)
Mcf	Volume	Million cubic feet
Mn, Mw	Molecular weight	g / mole
$\varnothing$	Porosity	dimensionless
Pe	Outer drainage pressure	MPa or bar or psi
Pw	Bottom hole pressure	MPa or bar or psi
Q	Flow rate	m^3/d or bopd
Rev	Reservoir outer radius	m or ft
Rg	Radius of Gyration	Angstrom
Rw	Wellbore radius	m or ft
S	Skin factor	dimensionless
Sgc	Critical gas saturation	dimensionless
So	Oil saturation	dimensionless
Soi	Initial oil saturation	dimensionless

Sor	Residual oil saturation	dimensionless
Sv	Steam saturation	dimensionless
Sw	Water saturation	dimensionless
T	Absolute temperature	Kelvin
T	Temperature	°C or °F
β	Permeability anosotropy ratio	dimensionless
λ	Fluid mobility	mD/mPa.s

Imprimé en France en novembre 2010 par EMD S.A.S.
53110 Lassay-les-Châteaux
N° d'imprimeur : 24126 - Dépôt légal : novembre 2010